CHEMICAL PROCESS INDUSTRIES

VOLUME I

Second Edition

CHEMICAL PROCESS INDUSTRIES

VOLUME I

Second Edition

W.V. MARK

M.Sc., Ph.D., C.Eng.

Edited by

S.C. BHATIA

B.E. (Chem), M.B.A.

CBS

CBS PUBLISHERS & DISTRIBUTORS PVT. LTD.

New Delhi • Bengaluru • Chennai • Kochi • Mumbai • Pune

ISBN : 81-239-1141-6

First Edition: 2001
Second Edition: 2007
Reprint: 2008, 2015

Published by:
Satish Kumar Jain for CBS Publishers & Distributors Pvt. Ltd.,
CBS Plaza, 4819/XI Prahlad Street, 24 Ansari Road,
Daryaganj, New Delhi – 110002, India
delhi@cbspd.com, cbspubs@airtelmail.in • www.cbspd.com
Ph.: 23289259, 23266861, 23266867 • Fax: 011-23243014

Branches:
- ***Bengaluru:*** Seema House, 2975, 17th Cross, K.R. Road,
Bansankari 2nd Stage, Bengaluru - 560070
Ph: +91-80-26771678/79 • Fax: +91-80-26771680
E-mail: cbsbng@gmail.com, bangalore@cbspd.com
- ***Chennai:*** 20, West Park Road, Shenoy Nagar, Chennai - 600030
Ph: +91-44-26260666, 26208620 • Fax: +91-44-42032115
E-mail: chennai@cbspd.com
- ***Kochi:*** 36/14, Kalluvilakam, Lissie Hospital Road, Kochi - 682018
Ph: +91-484-4059061-65 • Fax: +91-484-4059065
E-mail: cochin@cbspd.com
- ***Mumbai:*** 83-C, Dr. E. Moses Road, Worli, Mumbai - 400018
Ph: +91-9833017933, 022-24902340/41 • E-mail: mumbai@cbspd.com
- ***Pune:*** Bhuruk Prestige, Sr. No. 52/12/2+1+3/2,
Narhe, Haveli (Near Katraj-Dehu Road Bypass), Pune - 411041
Ph: +91-20-64704058/59, 32342277 • E-mail: pune@cbspd.com

Printed at :
J.S. Offset Printers, Delhi

Preface to the Second Edition

The new second edition strengthens the text book, which has now been thoroughly revised and edited by incorporating the suggestions made by various professors and students. The salient features of this revised edition are — considerable changes and addition have been made in the entire text to clear the points that were not apparent or needed additional information for better understanding. Some chapter have been completely overhauled and rewritten with the inclusion of additional and latest technology. In fact energy consumption has become synonymous with 'socio-economic development'. A nation which generates and consumes energy many times more than the human or animal muscle–power is considered developed. Keeping this in mind an additional chapter on 'Energy conservation in chemical process industries has been added.

This volume, dealing with inorganic aspects is the first volume of the comprehensive textbook on chemical process industries. This invaluable model textbook is specifically intended for B.E/B.Tech institutions which offer courses in Chemical Engineering in India, including the Indian Institutes of Technology and its primary aim is to provide an exhaustive coverage of principles and concepts which form the basis of chemical process engineering. As engineering education becomes more complex, it is essential for a chemical engineer to be familiar with the realisation of scientific principles in chemical process industries. Even the most sophisticated engineering curriculum must give the student some cultural background of the process industries such as, for example, how sulphuric acid is produced or how petroleum refining and the caustic soda industry have been affected by increasing demands. In addition to academic knowledge, the chemical engineer must have an appreciation of the professional role in the chemical process industries.

The encyclopaedic range of the subject of chemical process industries has been condensed into a handy, easy-to-use reference book, presented in a concise, compact and lucid manner. Stressing the essential facts for the student, it forms the necessary criteria for a systematic study of chemical process industries, the textbook informs, clarifies and explains how chemical processing transforms raw materials into usable and profitable products by chemicals.

Each chapter covers an important aspects of chemical processing industries with an accurate, up-to-date account of each topic Chapters 1 explains the importance of chemical process industry, its origin and development along with systematic analysis of chemical processes. Chapter 2 is devoted to basic laws and processes of chemical technology. Chapter 3 focuses on raw materials, fuel and power where as chapter 4 familiarises the students on methods of water conditioning to suitably prepare water for chemical industry with special emphasis on environmental protection. Chapter 5 describes the importance of catalysts in chemical process industries and chapter 6 is devoted to explosives and propellants which are chemical compounds or their mixtures that rapidly produce large volume of hot gases when properly initiated. Chapter 7 discusses various industrial gases such as carbon dioxide, hydrogen, oxygen, acetylene along with their manufacturing aspects, properties and uses.

Carbon, an extremely useful and versatile element, with its increasingly numerous industrial applications, is the subject of chapter 8 while chapter 9 is devoted to preparation properties and uses of sulphur and sulphuric acid. Chapter 10 focuses on hydrochloric acid and various inorganic chemicals such as alumina, aluminium chloride, aluminium carbonates, hydrofluoric acid, fluorine etc. Chapters 11 and 12 deal with manufacture and uses of products related to nitrogen and phosphorus industries. The student is then familiarised with salt and miscellaneous sodium compounds in chapter 13; the alkali and chlorine industries involved in the manufacture of soda-ash and caustic soda is discussed in Chapter 14.

Chapter 15 explores potassium industries, their raw materials, manufacture and uses, followed by Chapter 16 which deals with barium and its compounds (like barium carbonate, barium chloride, barium nitrate etc.). Chapter 17 focuses on various types of fertilisers their manufacture and uses. Portland cement its manufacturing aspects, properties, and uses along with calcium and magnesium compounds are discussed in chapter 18.

Chapter 19 is devoted to ceramic and refractories followed by chapter 20 which focuses on glass, which has many uses because of its transparency, high resistance to chemical attack etc.

Chapter 21 explores nuclear industries based on radium, uranium and plutonium and emphasise the importance of protection from radioactivity as well as waste disposal. Chapter 22 discusses manufacture of iron and steel. Finally Chapter 23 is devoted to energy conservation in chemical process industries.

The simple, lucid and readable style of text will help the students to comprehend the concepts and information presented, and ensures an easy grasp of key-points which are vividly illustrated by diagrams and tables. A carefully presented index allows the student easy access to each topic covered in the book. Additional information on modern technologies and a special emphasis on environmental issues and energy conservation are the highlights of this well-graded textbook.

While painstacking care has gone into producing a useful and exhaustive textbook, the author would welcome constant constructive criticism and a creative feed back from students and professors, and interaction that will certainly be fruitful.

The text could not have been completed without the help of many colleagues who provided figures, photographs and offered important advice during the course of its preparation.

The Chemical process industries text book first edition printed in 2001 received very well and it is expected that the revised second edition will also be received with same interest.

Appreciation is also extended to Dr. J. L. Mayo who spent many hours preparing high-quality computer graphics when he had much more interesting things to do.

W. V. Mark

Author

Contents at Glance

PREFACE .... III-V
CONTENTS AT GLANCE VI
CONTENTS IN DETAIL .... VII-XXXVIII
CHAPTER 1 Importance of Chemical Process Industry 1-70
CHAPTER 2 Basic Laws and Processes of Chemical Technology 71-86
CHAPTER 3 Raw Materials Fuel and Power for Chemical Process Industry 87-104
CHAPTER 4 Water Conditioning in Chemical Process Industry 105-181
CHAPTER 5 Catalysts and Catalysis 182-212
CHAPTER 6 Explosives and Propellants 213-254
CHAPTER 7 Industrial Gases 255-288
CHAPTER 8 Industrial Carbon 289-312
CHAPTER 9 Sulphur and Sulphuric Acid 313-343
CHAPTER 10 Hydrochloric Acid and Miscellaneous Inorganic Chemicals 344-393
CHAPTER 11 Nitrogen Industries 394-422
CHAPTER 12 Phosphorus Industries 423-449
CHAPTER 13 Salt and Miscellaneous Sodium Compounds 450-468
CHAPTER 14 Alkali and Chlorine Products 469-503
CHAPTER 15 Potassium Industries 504-512
CHAPTER 16 Barium and Its Compounds 513-534
CHAPTER 17 Fertiliser Industries 535-551
CHAPTER 18 Portland Cement, Calcium and Magnesium Compounds 552-574
CHAPTER 19 Ceramic and Refractories 575-607
CHAPTER 20 Glass Industries 608-633
CHAPTER 21 Nuclear Industries 634-670
CHAPTER 22 Iron and Steel 671-695
CHAPTER 23 Energy Conservation in Chemical Process Industries 696-704
REFERENCES 705-708
INDEX 709-715

Contents in Detail

Volume I : Chemical Process Industries

PREFACE .... III-V

CONTENTS AT GLANCE VI

CONTENTS IN DETAIL .... VII-XXXVIII

CHAPTER 1

IMPORTANCE OF CHEMICAL PROCESS INDUSTRY 1-70

Introduction 1

What is Chemical Process Industry ? 1

Origin and Development of Chemical Process Industry 6

Pre-scientific Chemical Industry 6

Scientific Chemical Industry 6

Growth with Restraints 7

Green Challenge to Chemical Industry 9

Indian Chemical Industry Today 10

Mass and Energy Balance 11

Conservation of Mass 11

Conservation of Energy 14

Thermochemistry 15

Chemical Reaction Equilibrium 16

Chemical Kinetics 17

Ideal Gas Laws 18

Phase Equilibrium 20

Unit Operations 21

Classification of Unit Operations 22

Plant Equipment . . . 29
Chemical Reactors . . . 29
Heat Exchangers . . . 32
Mass Transfer Equipment . . . 34
Ancillery Equipment . . . 37
Transportation Equipments . . . 37
Process Diagrams . . . 42
Flow Sheets . . . 42
Instrumentation and Control . . . 52
What is Chemical Engineering ? . . . 54
What Does Chemical Engineer Do ? . . . 54
Research . . . 55
Fundamental Research . . . 55
Exploratory Research . . . 56
Process Research . . . 57
Process Development . . . 57
Process Design and Evaluation . . . 60
Plant Design . . . 61
Production and Supervision . . . 62
Plant Technical Service . . . 64
Product Sales . . . 64
Market Research . . . 64
Product Development . . . 65
Technical Sales and Customers Service . . . 65
Chemical Engineer in the Coming Years . . . 66
General Aspects of Chemical Engineering . . . 69
Communication . . . 69
Human Relations . . . 69
Professional Activities . . . 70
Technical Reading . . . 70

CHAPTER 2

BASIC LAWS AND PROCESSES OF CHEMICAL TECHNOLOGY . . . 71-86
Introduction . . . 71
Homogeneous (Uniform) and Heterogeneous (Non-Uniform) Processes . . . 72

Reversible Processes, Chemical Equilibrium, and Irreversible Processes 74
Rate of Chemical Processes 77
Mixing the Reactants 79
Catalytic Processes 81
Basic Types of Industrial Chemical Processes and Flow Schemes 84

CHAPTER 3

RAW MATERIALS, FUEL AND POWER FOR CHEMICAL PROCESS INDUSTRY 87-104

Introduction 87
Materials for Chemical Processes 88
Basic Principles of Raw Material Benefication 89
Comprehensive Utilisation of Raw Materials 96
Replacement of Edible Raw Materials 97
Water and the Chemical Industry 97
Fuel and Power for the Chemical Industry 100

CHAPTER 4

WATER CONDITIONING IN CHEMICAL PROCESS INDUSTRY 105-181

Introduction 105
Sources of Water Supply 105
- Rain Water 106
- River Water 106
- Spring and Well Water 106
- Sea Water 107
- Surface and Ground Water 107

Characteristic of Water 107
Water for Industrial Purposes 108
- Water for Boilers 108
- Water for Cooling System 109
- Water for Beverages 109
- Water for Distilleries 109
- Water for Paper Mills 109
- Water for Sugar Factories 109
- Water for Dairies and Allied Industries 109
- Water for Laundries 109

Effect of Water on Rocks and Minerals 109
Boiler Scale and Sludges 110
Mechanism of Boiler Scale Formation 110
Effects of Boiler Scales and Sludges 111
Wastage of Fuel 111
Explosion of Boiler 111
Formation of Deposits in Boiler Units 112
Disadvantages of Scale Formation in Boiler 113
Methods of Preventing Deposit Formation 113
Phosphate Treatment of Boiler Water 113
Treatment of Boiler Water with Complexing Agents 114
Alkali Treatment of Boiler Water 115
Preventing Deposit Formation in Turbine Condensers 115
Priming and Foaming 116
Caustic Embrittlement 116
Internal Conditioning of Boiler-Feed Water 117
Factors Promoting Corrosion 118
Corrosion of Boiler 120
Quality of Water 121
Meaning of Pure Water 121
Domestic Use 121
Civic Use 122
Trade or Business Use 122
Commercial or Industrial Use 122
Reasons for the Analysis of Water 123
Impurities in Water 124
Analysis of Water 124
Physical Tests 124
Colour 124
Taste and Odour 124
Temperature 125
Turbidity 125
Chemical Tests 125
Chlorides 126
Dissolved Gases 126

Hardness 127
Hydrogen-Ion Concentration (pH Value) 129
Alkalinity 132
Acidity 133
Total Solids 135
Bacteriological Tests 136
Total Count or Agar Plate Count Test 139
B-Coli Test 139
B-Coli Index 140
Maintenance and Purity of Waters 141
Water-borne Diseases 142
Coagulation of Water 143
Purpose 143
Principle of Coagulation 143
Flocculation 144
Usual Coagulants 145
Filtration of Water 145
Theory of Filtration 145
Types of Filters 146
Disinfection of Water 147
Minor Methods of Disinfection 148
Chlorination 148
Application of Chlorine 149
Water Softening 149
Hard and Soft Water 150
Types of Hardness 151
Temporary Hardness 152
Permanent Hardness 152
Removal of Temporary Hardness 152
Removal of Permanent Hardness 153
Advantages of Hot Lime Soda Process 156
Desalination of Water 163
Miscellaneous Methods of Water Treatment 166
Colour, Odour and Taste Removal 166
Aeration 166

Treatment by Activated Cerbon 167
Use of Copper Sulphate 168
Iron and Manganese Removal 168
Fluoridation 168
Sewage and Sewage Treatment 169
Characteristics of Sewage 169
Physical Characteristics 169
Chemical Characteristics 170
Disposal of Sewage 170
Land Disposal 171
Dilution Technique 172
Oxidation Ponds 172
Sewage Treatment 173
Classification of Sewage Treatment 173
Screening 174
Chemical Precipitation 174
Skimming Tank 176
Imhoff Tanks 176
Clarigester 176
Sludge Treatment and Disposal 177
Concentration 177
Digestion 178
Conditioning 179
Dewatering 179
Oxidation 180
Ultimate Sludge Disposal 181

CHAPTER 5

CATALYSTS AND CATALYSIS 182-212
Introduction 182
Chemical and Physical Aspects 182
Techniques 183
Preparation 184
Organic and Inorganic Chemicals 184
Transition-Metal Complexes—Homogeneous Catalysis 184
Oxidation Catalysis 185
Catalysis Applied to Fuels 185

Zeolites (Molecular Sieves) 186
Dual-Function Catalysts 186
Multimetallic Catalysis 186
Synthetic Fuels 187
Fuel Cells 188
Environmental Control 188
Phase-Transfer Catalysis 188
Mechanism of Phase-Transfer Catalysis 189
Factors Influencing the Usefulness of Phase-Transfer Catalysis. 189
Uses 190
New Ion-Exchange Resin Catalysts 190
Reactors used for IER Catalysed Reactions 191
Distillation Column Reactors (DCR) 191
Ion Exchange Resin Catalysts in Production of Bisphenols 192
Transalkylations 192
Reaction of Ethylene Carbonate with Methanol 192
Higher Olefins to Secondary Alcohols 192
Hydration of Acrolein and *A*-Methyl Styrene (AMS) 193
Acetals and Ketals 193
Dimerisation of Isoamylene and Cross-Dimerisation of Isoamylene with AMS 193
Reactions with Polymeric Substances or Resulting in Polymeric Substances 193
Separation of Close-boiling Substances 193
Alkylation of Aromatic Amines 194
Some Engineering Aspects of the Use of Cationic Ion-exchange Resins as Catalysts 194
Recent Developments of Ion-exchange Resin Catalysts 195
Enzyme Catalysed Selective Oxidation Reactions 198
Enzymatic Oxidation Reactions 203

CHAPTER 6

EXPLOSIVES AND PROPELLANTS 213-254
Introduction 213
General Characteristics 213
Explosive detonation 214
Primary Explosives 215

Mercury Fulminate 215
Lead Azide 216
Silver Azide 216
Lead Styphnate 216
Diazodinitrophenol (DDNP) 216
Tetrazene 217
Secondary Explosives 217
Aliphatic Nitrate Esters 217
Nitroglycerine 217
Nitrocellulose 218
Nitramines 219
RDX and HMX 220
Nitroguanidine (NQ) 221
Tetryl 221
Nitroaromatics 221
Trinitrotoluene 222
Picric Acid and Ammonium Picrate 222
Metegang Methods 223
Safety and Environmental Considerations 225
Uses 225
Propellants 225
Principle of the Rocket 228
Types of Propellants 229
Solid Propellants 229
Manufacture 232
Liquid Propellants 233
Selection 239
Burning Process 239
Burning-rate equations 239
Mechanism of Burning of Nitrocellulose-based Propellants 240
Mechanism of Burning of Composite Propellants 240
Manufacture 241
Nitrocellulose Propellants 241
Solvent Extrusion Batch Process 242
Continuous Solvent Extrusion Process (for Single-bas Propellants) 242

Solventless Extrusion Process 243
Nitrocellulose-based Cast Propellants 244
Polymer-based Cast Propellants 244
Black powder 244
Manufacture 245
Benite 246

CHAPTER 7

INDUSTRIAL GASES 255-288
Introduction 255
Hydrogen (Manufacturing Aspects) 255
From Hydrocarbons and Steam (Steam Reformer Process) 255
From Hydrocarbons by Partial Oxidation 258
From Water Gas and Steam 259
Steam-Iron Process 261
From Water by Electrolysis 261
Thermal Decomposition of Hydrocarbons 261
Catalytic Reforming of Petroleum Stocks 262
From Methanol and Steam 262
From Ammonia by Dissociation 262
High-purity Hydrogen 263
Liquid Hydrogen 263
Uses 264
Carbon Dioxide (Manufacturing Aspects) 264
From Steam and Natural Gas 264
From Fermentation 265
From Sodium Carbonate and Phosphoric Acid 265
Recovery of Carbon Dioxide 266
From Flue or Kiln Gases by Absorption 266
Uses 268
Properties 268
Oxygen (Manufacturing Aspects) 268
From Air by Liquefaction (Modified Linde-Frankl Cycle) 268
Conventional Linde-Frankl Cycle 272
Other Cycles 272

Claude Cycle . . . 273
Uses . . . 274
Properties . . . 274
Acetylene (Manufacturing Aspects) . . . 274
From Calcium Carbide . . . 274
From Paraffin Hydrocarbons by Pyrolysis (Wulff Process) . . . 275
From Natural Gas by Partial Oxidation (Sachsse Process) . . . 277
From Hydrocarbons by the Arc Process . . . 279
Uses . . . 279
Properties . . . 279
Rare-Earth Elements . . . 279
Occurrence . . . 280
Properties . . . 280
Physical . . . 280
Chemical . . . 280
Production and Processing . . . 281
Liquid-liquid Extraction . . . 281
Health and Safety . . . 282
Uses . . . 282
Helium . . . 282
Properties . . . 282
Preparation . . . 282
Uses . . . 283
Neon . . . 283
Properties . . . 283
Preparation . . . 283
Uses . . . 283
Argon . . . 283
Properties . . . 283
Preparation . . . 284
Methods of Purification . . . 284
Uses . . . 284
Krypton . . . 284
Properties . . . 284
Preparation . . . 284

- Uses ... 285
- Xenon ... 285
  - Properties ... 285
  - Preparation ... 285
  - Uses ... 285
- Radon ... 285
  - Properties ... 285
  - Preparation ... 286
  - Uses ... 286
- Sulphur Dioxide ... 286
  - Properties ... 286
  - Preparation ... 286
  - Uses ... 286
- Carbon Monoxide ... 287
  - Properties ... 287
  - Preparation ... 287
  - Uses ... 287
- Nitrous Oxide ... 287
  - Properties ... 287
  - Preparation ... 288
  - Uses ... 288

CHAPTER 8

INDUSTRIAL CARBON ... 289-312

- Introduction ... 289
- Carbon and Artificial Graphite ... 290
  - Structure, Terminology, History and Applications ... 290
  - Activated Carbon ... 292
  - Baked and Graphitised Carbon ... 294
  - Processing of Baked and Graphitised Carbon ... 295
  - Properties of Manufactured Graphite ... 296
  - Other Forms of Carbon and Graphite ... 297
- Carbon Black ... 299
    - Carbon Black Manufacturing Processes ... 302
  - Uses ... 303

Diamond (Natural) 304
Diamond (Synthetic) 306
Catalysed Synthesis 306
Direct Graphite-to-Diamond Process 307
Synthetic Diamond Industry 307
Natural Graphite 308

CHAPTER 9

SULPHUR AND SULPHURIC ACID 313-343
Introduction 313
Sulphur 314
Process of Manufacture 314
From Hydrogen Sulphide by Oxidation 314
Properties 316
Uses 316
Grades 316
Packing 317
Sulphur Dioxide 317
Sulphur Trioxide 318
Properties 320
Physical Properties 320
Chemical Properties 321
Manufacture of Sulphuric Acid 322
Catalysts 322
Contact Process 323
Wet-catalysis Processes 330
Chamber Process. 332
Manufacture of Oleum 333
Equipment 334
Catalysts. 334
Storage and Handling. 336
Specifications and Standards 337
Applications. 339
Indirect Uses. 340
Direct Uses 341

Control of Sulphur Dioxide Pollution due to the Manufacture of Sulphuric Acid . . . 341
Latest Trends Achievements in Emissions of SO_2 per ton of Sulphuric Acid . . . 342

CHAPTER 10

HYDROCHLORIC ACID AND MISCELLANEOUS INORGANIC CHEMICALS . . . 344-393

Introduction . . . 344
Hydrochloric Acid . . . 344
Manufacture . . . 344
From Organic Chlorinations . . . 344
From Chlorine and Hydrogen . . . 346
Anhydrous Hydrochloric Acid . . . 347
Uses . . . 347
Hydrofluoric Acid . . . 348
From Fluorspar and Sulphuric Acid . . . 348
Uses . . . 351
Properties . . . 351
Fluorine . . . 351
Preparation . . . 351
Properties . . . 353
Reactions . . . 353
Uses of Fluorine and its Compounds . . . 353
Alumina (Aluminium Oxide) . . . 354
Properties . . . 354
Preparation . . . 354
Manufacture of Alumina . . . 355
Electrolysis of Alumina . . . 355
Aluminium Sulphate . . . 357
From Bauxite . . . 357
Process . . . 357
Uses . . . 358
Properties . . . 359
Alums . . . 359
Chrome Alum . . . 360

Aluminium Chloride 360
From Aluminium Metal and Chlorine 360
Process 360
Uses 361
Properties 361
Aluminium Carboxylates 361
Aluminium Chloride Hexahydrate 362
Aluminium Bromide and Aluminium Iodide 362
Aluminium Nitrate 362
Aluminium Hydroxides 363
Activated Aluminas 363
Calcined Alumina 365
Copper Sulphate 366
Lithium and Lithium Compounds 367
Sources 367
Production 367
Recovery from Ores and Brines 367
Other Recovery Processes from Silicate Ores 368
Recovery from Brines 368
Lithium Metal 368
Properties 368
Manufacture 368
Inorganic Lithium Compounds 369
Lithium Acetate 369
Lithium Amide 369
Lithium Borates 370
Lithium Carbonate 370
Molybdenum Compounds 371
Halides and Oxyhalides 371
Oxides and Hydroxy Compounds 371
Molybdenum Trioxide 371
Hydroxides 371
Molybdates 371
Normal or Orthomolybdates 372
Heteropolymolybdates 372

Sulphides, Selenides, and Tellurides 372
Cyanides, Thiocyanates, and Carbonyls 372
Organomolybdenum Compounds 372
Uses 373
Strontium and Strontium Compounds 373
Strontium Carbonate 374
Strontium Chlorate 374
Strontium Chloride 374
Boron 374
Boron Compounds 375
Boron Oxides 375
Boric Acid 376
Solutions of Boric Acid and Borates 377
Sodium Borates 377
Alkali-Metal and Ammonium Borates 378
Calcium-Containing Borates 379
Borate Melts and Glasses 380
Other Metal Borates 380
Zinc Borates 380
Boron Phosphate 381
Boric Acid Esters 381
Properties 381
Preparation 382
Uses 382
Boron Halides 382
Rare-Earth Elements 383
Properties 384
Physical 384
Chemical 384
Production and Processing 386
Liquid-liquid Extraction 386
Health and Safety 386
Uses 386
Sodium Chromate 387
From Chromite Ore 387

Hydrogen Peroxide 389
Manufacturing Process 389
By Oxidation of Alkylhydroanthraquinones 391
By Oxidation of Isopropyl Alcohol 392

CHAPTER 11

NITROGEN INDUSTRIES 394-422
Introduction 394
Cyanamide 395
Reactions and Energy Changes 395
Properties 397
Reactions 397
Manufacture 398
Dicyandiamide 398
Reactions 398
Melamine 399
Health and Safety 399
Uses 399
Synthetic Ammonia 399
Reactions and Equilibriums 400
Rate and Catalysis of the Reaction 401
Ammonia 402
Physical Properties 402
Chemical Properties 403
Manufacture 404
Health and Safety 408
Uses 408
Ammonium Compounds 408
Ammonium Chloride 408
Ammonium Nitrate 409
Ammonium Sulphate 410
Health and Safety 411
Uses 411
Urea 411
Manufacture 412

Waste-water Treatment 414
Finishing Processes 414
Uses 415
Nitric Acid 416
Raw Materials 416
Reactions and Energy Changes 416
From Ammonia 417
Process 417
Uses 420
Properties 420
Sodium and Potassium Nitrates 420
Sodium Nitrate 420
Manufacture and Processing 421
Health and Safety Factors 422
Uses 422

CHAPTER 12

PHOSPHORUS INDUSTRIES 423-449
Introduction 449
Phosphorus and Phosphides 424
Properties 424
Manufacture 425
Uses 426
Metallic Phosphides and Alloys 427
Phosphorus Compounds 427
Phosphorus Sulphides 428
Phosphorus Halides 429
Phosphorus Oxides 431
Phosphorus Acids 431
Phosphazenes and Other Phosphorus-Nitrogen Compounds 432
Phosphine and its Derivatives 432
Phosphoric Acids and Phosphates 432
Phosphoric Acid 434
From Phosphate Rock by Ele^tric Furnace Process 434
Wet Process 437

Other Phosphorus-Containing Acids 440
Raw Material 440
Uses 440
Polyphosphoric Acids 441
Polyphosphates 441
Pyrophosphates 442
Environmental Considerations 444
Pharmaceutical and Food Grades 444
Sodium Phosphates 444
Process 445
Uses 447
Properties 448
Monosodium Phosphate 448
Disodium phosphate 448
Trisodium phosphate 448
Tetrasodium pyrophosphate 448
Sodium Acid Pyrophosphate 449
Sodium Metaphosphate 449
Sodium Tripolyphosphate 449

CHAPTER 13

SALT AND MISCELLANEOUS SODIUM COMPOUNDS 450-468
Introduction 450
Sodium Chloride (Common Salt) 451
Manufacture 451
From Saturated Brine by Multiple-Effect Evaporation Process 451
From Saturated Brine by Open Pan (Grainer) Process 453
From Rock Salt by Mining 454
From Sea Water by Solar Evaporation 454
Other Processes 455
Uses 455
Sodium Sulphate (Salt Cake) 455
From Natural Brines 455
From Salt and Sulphuric Acid 456
From Rayon Spin Bath 457

Other Processes 457
Uses 458
Properties 458
Sodium Thiosulphate 458
From Soda Ash and Sulphur Dioxide 458
From Sodium Sulphite and Sulphur 459
From By-Product of Sodium Sulphide 459
From By-Product of Sulphur-Dye Manufacture 459
Uses 460
Properties 460
Sodium Silicate 460
From Sodium Carbonate and Silica (Sand) 460
Uses 462
Properties 462
Sodium Sulphate 463
Properties 463
Preparation 463
Uses 463
Sodium Bisulphate Decahydrate 464
Properties 464
Preparation 464
Uses 464
Sodium Sulphite 464
Properties 464
Preparation 464
Uses 464
Sodium Sulphide 465
Properties 465
Preparation 465
Uses 465
Sodium Hydrosulphide 465
Properties 465
Preparation 465
Uses 465
Sodium Nitrite 465

Properties 465
Uses 466
Sodium Peroxide 466
Properties 466
Preparation 466
Uses 466
Sodium Perborate 466
Properties 466
Preparation 467
Uses 467
Sodium Amide 467
Properties 467
Preparation 467
Uses 467
Sodium Cyanide 467
Properties 467
Preparation 468
Uses 468
Sodium Ferrocyanide 468
Properties 468
Uses 468

CHAPTER 14

ALKALI AND CHLORINE PRODUCTS 469-503
Introduction 469
Soda Ash (Sodium Carbonate) 470
Properties 470
Physical Properties 470
Chemical Properties 471
Health and Safety 472
Manufacture 473
Leblanc Process 473
Modern Processes 473
Natural Soda Ash Processes 474
From Ores 474

Mining ... 474
Purification of Trona ... 475
Production from Brine ... 476
Oil Shale Deposits ... 477
Synthetic Processes ... 477
Solvay Process ... 477
Limestone Burning and Lime Slaking ... 480
Brine Preparation ... 480
Ammonia Absorption ... 481
Precipitation of Bicarbonate (Making) ... 481
Filtration of Bicarbonate ... 482
Calcination of Bicarbonate ... 482
Recovery of Ammonia ... 483
Waste Disposal ... 484
Modified Ammonia-Soda Processes ... 484
Other Synthetic Routes ... 485
Production from Nepheline ... 485
Huls Process ... 485
Akzo Process ... 486
Natural vs Synthetic ... 486
Analyses and Quality Specifications ... 486
Storage and Transport ... 488
Applications ... 489
Sodium Bicarbonate (Baking Soda) ... 490
Manufacture ... 490
From Sodium Carbonate (Soda Ash) ... 490
Uses ... 491
Properties ... 491
Chlorine ... 492
From Salt by Electrolysis ... 492
From Potassium Chloride and Nitric Acid ... 493
From Hydrogen Chloride and Air (Kel-Chlor Process) ... 495
Uses ... 496
Properties ... 496
Sodium Hydroxide (Caustic Soda) ... 497

From Salt by Electrolysis 497
Uses 501
Properties 501
Bleaching Powder 501
Sodium Hypochlorite 502
Sodium Chlorite 503

CHAPTER 15

POTASSIUM INDUSTRIES 504-512
Introduction 504
Potassium Chloride 504
From Sylvinite by Fractional Crystallisation 504
From Sylvinite by Flotation 505
From Searles Lake Brine 507
Various Potassium Salts 508
Potassium Hydroxide 508
Potassium Carbonate 510
Potassium Sulphate 511
Potassium Acid Tartrate 511
Potassium Nitrate 511

CHAPTER 16

BARIUM AND ITS COMPOUNDS 513–534
Introduction 513
Benefication of Barytes 515
Manufacture of Barium Compounds 515
Feed Preparation for Reduction Furnace 517
Reduction Process Operation 518
Barium Compounds 521
Barium Carbonate 521
Sources of Carbon Dioxide 523
Solid Waste Management 523
Barium Chloride 524
Manufacture from Black Ash 524
Manufacture from Solid Waste Generated in $BaCO_3$ Unit 526
By-product Recovery 527

Barium Nitrate 528
Barium Ferrite 530
Blanc Fixe or Precipitated Barium Sulphate 531
From Fresh BaS Solution and Iron Free Good Quality Sodium Sulphate (Preferably from Rayon Process) 531
From BaS Liquor and 70% Sulphuric Acid 531
From $BaCl_2$ Solution and Sulphuric Acid (90–98%) 531
Barium Hydroxide 532
From Barium Chloride by Reaction with Caustic Alkali 533
Barium Stearate 533
Barium Chromate 534

CHAPTER 17

FERTILISER INDUSTRIES **535–551**
Introduction 535
Classification of Fertilisers 536
Phosphorus Fetilisers 537
Production of Superphosphate 539
Manufacture of Phosphoric Acid 543
Manufacture of Concentrated Phosphorus Fertilisers (Triple Superphosphate, etc.) 544
Nitrogen Fertilisers 545
Manufacture of Ammonium Nitrate 545
Manufacture of Carbamide (Urea) 547
Manufacture of Ammonium Sulphate 548
Potassium Fertilisers 549
Prospective Developmnts in Mineral Fertilisers Manufacturing 550

CHAPTER 18

PORTLAND CEMENT, CALCIUM AND MAGNESIUM COMPOUNDS **552–574**
Introduction 552
Clinker Chemistry 552
Phase Equilibria 552
Clinker Formation 553
Phases Formed in Portland Cements 553

Raw Material Proportions 553
Hydration 554
Calcium Silicates 554
Tricalcium Aluminate and Ferrite 555
Other Phases 555
Hydration at Ordinary Temperatures 555
Cement Paste Structure and Concrete Properties 555
Unit Operations, Chemical Conversions, Energy Requirements 556
Manufacturing Processes 557
Manufacturing Procedures, The Wet and Dry Processes 557
Compounds in Cements 559
Setting and Hardening of Cement 560
Special Cements 561
Lime 562
Lime Manufacture 563
Raw Materials 563
Energy Changes, Physical Operations and Chemical Conversions 563
Manufacturing Procedures 565
Gypsum 567
Calcination of Gypsum 567
Hardening of Plaster 567
Special Purpose and Blended Cements 568
Non-Portland Cements 569
Uses 569
Miscellaneous Calcium Compounds 569
Calcium Carbonate 569
Preparation 569
Properties 569
Uses 570
Calcium Sulphide 570
Preparation 570
Properties 570
Uses 570
Calcium Arsenate 570
Halide Salts 570

Miscellaneous Magnesium Compounds 571
Magnesium Carbonate 571
Preparation 571
Properties 571
Uses 571
Magnesium Chloride 571
Preparation 571
Properties 572
Uses 572
Magnesium Oxide (Magnesia) 572
Preparation 572
Properties 572
Uses 572
Magnesium Hydroxide 573
Preparation 573
Properties 573
Uses 573
Magnesium Sulphate 573
Preparation 573
Properties 573
Uses 573
Magnesium Silicate 573
Preparation 573
Properties 574
Uses 574

CHAPTER 19

CERAMIC AND REFRACTORIES 575–607
Introduction 575
Raw Materials 575
Clays 575
Chemical Conversions, Including Basic Ceramic Chemistry 578
Whitewares 580
Manufacture of Porcelain 581
Non-Clay Minerals 583

Silica . . . 583
Feldspar . . . 583
Other Non-Clay Minerals . . . 583
Special Materials . . . 584
Super Refractories . . . 584
Nuclear Ceramic Materials . . . 584
Electronic Ceramic Materials . . . 584
Forming Processes . . . 584
Plastic Forming Processes . . . 584
Extrusion Processes . . . 584
Soft Plastic Forming . . . 584
Dry Pressing . . . 585
Casting Processes . . . 585
Thermal Treatment . . . 586
Methods of Thermal Treatment . . . 586
Kilns . . . 586
Hot-Pressing Processes . . . 586
Physical and Chemical Changes During Thermal Treatment . . . 587
Properties and Applications . . . 587
Composition and Microstructure . . . 587
Chemical Properties of Ceramic Materials . . . 587
Optical Properties . . . 588
Thermal Properties . . . 588
Elastic Properties . . . 589
Engineering with Ceramics . . . 590
Electrical and Magnetic Properties . . . 590
Composites and Cements . . . 590
Uses . . . 590
Ceramic as Electrical Materials . . . 591
Electrical Conduction Phenomena . . . 591
Ionic Conduction in Ceramics . . . 591
Electron Conduction in Ceramics . . . 592
Non-stoichiometric and Solute-controlled Electronic Ceramics . . . 593
Ceramics with High Electronic Conductivity or with Non-linear Behaviour . . . 594
Mixed Conduction in Ceramics . . . 594
Colourants for Ceramics . . . 595
Colourant Systems . . . 596

Black 596
Blue 596
Brown 596
Gray 596
Green 596
Pink 596
Red 597
White 597
Yellow 597
Refractories 597
Physical Forms 98
Raw Materials 98
General Properties 598
Physical Properties 598
Manufacture 600
Processing 600
Speciality Refractories 600
Health and Safety Factors 601
Selection and Uses 601
Silica Refractories 602
Semisilica Refractories 602
Fireclay Refractories 602
High Alumina Refractories 602
Chrome Refractories 602
Magnesite Refractories 602
Dolomite Refractories 602
Spinal Refractories 602
Forsterite Refractories 602
Silicon Carbide Refractories 603
Zirconia Refractories 603
Refractory Coatings 603
Microstructure of Coatings 605
Applications 605
Refractory Fibres 606
Health and Safety Factors 607

CHAPTER 20

GLASS INDUSTRIES **608–633**
Introduction 608
Structure 610
Composition 610
Single-Phase Glasses 611
Glass-Ceramics and Phase-Separated Glasses 612
Properties 613
Manufacture and Processing 616
Glass-Ceramics 618
Glass-Ceramic Process 619
Postforming 619
Nucleation 619
Crystallisation 619
Properties 620
Thermal Properties 620
Optical Properties 620
Chemical Properties 620
Mechanial Properties 620
Electrical Properties 621
Magnetic Properties 621
Commercial Applications 621
Glassy Metals 621
Properties 622
Thermal Behaviour 623
Formation 623
Preparation 623
Applications 624
Annealing 625
Finishing 625
Manufacture of Special Glasses 625
Fused Silica Glass 625
High Silica Glass 626
Optical Glass 626
Borosilicate Glass 627
Lead Glasses 627
Soda Lime Glass 627

Alkali Silicates 627
Coloured Glass 628
Opal Glass 628
Safety Glass 629
Fibre Glass 629
Glass Laminates 629
Glass Wool 630
Flint Glass 630
Bottle Glass 630
Jena Glass 630
Pyrex Glass 631
Rare Earth Glass 631
Photosensitive Glass 631
Photochromic Glass 631
Froast Glass 632
Ground Glass 632
Insulating Glass 632
Vitreosil (99.5% Silica Glass) 632

CHAPTER 21

NUCLEAR INDUSTRIES 634–670
Introduction 634
Uranium and Uranium Compounds 635
Occurrence 635
Resources 636
Properties 636
Isotopes 637
Extraction from Ore 637
Conversion and Purification 638
Manufacture of the Metal 638
Isolation of Specific Isotopes 639
Uranium Compounds 639
Fluorides 639
Chlorides 640
Bromides and Iodides 640
Oxides 640
Nuclear Fission 641
Important Feature of Nuclear Fission 642

Chain Reaction 642
Energy Released During Fission Process 642
Fission Fragments 643
Nuclear Fusion 643
Main Features of Fusion Process 643
Drawbacks of Fusion Process 644
Mass Defect and Binding Energy 644
Binding Energy 645
Binding Energy and Nuclear Stability 646
Plutonium and its Compounds 647
Plutonium 647
Sources 648
Physical Properties 648
Thermal 648
Thermodynamic 648
Electrical and Magnetic 648
Radioactive Self-Heating 649
Spectroscopic 649
Chemical Properties 649
Extractive Metallurgy and Conversion Chemistry 649
Plutonium Compounds 650
Health and Safety Factors 650
Nuclear Reactors 650
Safety Aspects 650
Heat Removal 650
Fuels 651
Special Materials 651
Clear Fuel Reserves 651
Uranium 651
Resources 651
Availability 652
Low Grade Uranium Resources 652
Thorium 652
Resources 652
Demand and supply 652
Water Chemistry of Light-Water Reactors 652
Isotope Separation. 655
Fuel-element Fabrication 655

Chemical Reprocessing . . . 656
Purex Process . . . 656
Fast Breeder Reactors . . . 657
Design . . . 657
Breeder Reactor Types . . . 657
Safety in Nuclear Facilities . . . 658
Nuclear Fuel Cycle . . . 659
Safety Philosophy . . . 659
Safety Features . . . 660
Accidents . . . 661
Other Nuclear Fuel Cycle Facilities . . . 661
Special Engineering for Radiochemical Plants . . . 661
Radiation Resistance . . . 661
Safety Design . . . 662
Occupational Radiation Exposure . . . 662
Environmental Impact . . . 662
Remote Operation and Maintenance . . . 663
Heavy Water for Nuclear Power Industry . . . 663
Properties of Heavy Water . . . 663
Analysis of Heavy Water . . . 664
Production of Heavy Water . . . 665
Uses . . . 667

CHAPTER 22

IRON AND STEEL . . . 671–695
Introduction . . . 671
Impurities in Iron Ore and their Effects . . . 671
Fuel . . . 672
Fluxes . . . 672
Manufacture of Pig Iron . . . 673
Disposal of Pig Iron and Slag . . . 675
Cast Iron . . . 675
Properties of Cast Iron . . . 676
Varieties of Cast Iron . . . 676
Wrought Iron . . . 680
Pudding Process . . . 680
Astron or Byers Process . . . 681
Uses of Wrought Iron . . . 681

Steel 681
Production of Steel 682
Cementation Process 682
Crucible Process 682
Bessemer Process 682
Open Hearth Process 687
Electric Process 690
Oxygen Steel-Making or L.D. Process 690
Duplex Process 691
Effect of Elements on Steels 692
Alloying Elements in Steel 693
Heat Treatment of Steel 694
Annealing 694
Normalising 694
Hardening 694
Tempering 695
Case Hardening 695
Nitriding 695

CHAPTER 23

ENERGY CONSERVATION IN CHEMICAL PROCESS INDUSTRIES **696–704**
Introduction 696
Process Design and Energy Consumption 698
Heavy Water Process 698
Ammonia Production 699
Energy Management Principles 700
Energy Policy 701
Organisation Structures for Implementing the Energy Conservation (Encon) Policy 702
Methodology Adopted for Energy Management 702
Motivation 703
Information System 703
Investment 704

REFERENCES **705-708**
INDEX **709-715**

Chapter 1

Importance of Chemical Process Industry

INTRODUCTION

Products of the chemical process industry are used in all areas of everyday life. The raising of food plants and animals requires chemical fertilisers, insecticides, food supplements, and disinfectants. Many building materials have been chemically processed, for example, metals, concrete, roofing materials, paints, and plastics. Clothing utilises many synthetic fibres and dyes. Transportation depends upon gasoline and other fuels. Written communication uses paper and printing ink; and electronic communication requires many chemically processed insulators and conductors. The nation's health is maintained by drugs and pharmaceuticals, soaps and detergents, insecticides and disinfectants—all products of the chemical process industry. In addition, many chemicals never reach the consumer in their original form but are sold within the industry for further processing or use in the production of other chemicals for consumer use. It is often said that the chemical industry is its own best customer.

The chemical process industry is a sprawling complex of raw-material sources, manufacturing plants and distribution facilities which supplies thousands of chemical products, most of which were unknown 100 years ago. No clear limit can be used to define the chemical process industry and within the industry it is not possible to classify completely the various industries.

WHAT IS THE CHEMICAL PROCESS INDUSTRY ?

Any definition or description of the chemical process industry is bound to be incomplete. Most processes in the chemical industry involve a chemical change. The term "chemical change" should be interpreted to include not only chemical reactions but also physico-chemical changes, such as the separation and purification of the components of a mixture. Purely mechanical changes are usually not considered part of the chemical process, unless they are essential to later chemical changes. As an example, the manufacture of the plastic polythene, using ethylene produced from petroleum or natural gas, involves a chemical process. On the other hand, the moulding and fabrication of the resulting plastic resin into final shapes for consumer products would not be considered part of the chemical process.

Some industries which depend on chemical changes are not usually considered part of the chemical process industry, because of tradition, special processes, or large volume of a particular product, such as paper or steel. Many complex chemical processes in the food industry, for example, cheese making, which involves a fermentation reaction, are not considered to be part of the chemical process industry. On the other hand, the fermentation of sugars to produce beverage and industrial alcohols is often listed as a chemical industry. The huge metallurgical industry is usually distinguished from the chemical process industry because of the special nature of the processes and the vast quantities of the products; but the processing of metals might also be considered to be merely one segment of the chemical industry.

One possible list of the important chemical process industries is given in Table 1.1. Not all industries which utilise chemical processing are included, nor are all parts of the listed industries necessarily part of the chemical process industry. An examination of the list will show the tremendous scope of the industry and will also demonstrate the problems involved in any such listing or classification. A consideration of a few of the products of the industries listed in Table 1.1 will show the wide scope of chemical processing. Only a few typical products are listed.

Table 1.1. Chemical process industries.

Industry	*Typical products*	*End uses*
Inorganic chemicals	Sulphuric acid	Fetilisers, chemicals, petroleum refining, paints, pigments, metal processing, explosives
	Nitric acid	Explosives, fertilisers
	Sodium hydroxide	Chemicals, rayon and film processing, petroleum refining, pulp and paper processing, lye, cleansers, soap, metal processing

(Contd...)

Industry	*Typical products*	*End uses*
Organic chemicals	Acetic anhydride	Rayon, resins, and plastics
	Ethylene glycol	Antifreeze, cellophane, dynamite, synthetic fibres
	Formaldehyde	Plastics
	Methanol	Formaldehyde manufacture, antifreeze, solvent
Petroleum and petrochemicals	Gasoline	Motor fuel
	Kerosene	Jet fuel
	Oils	Lubricating, heating
	Ammonia	Fertiliser, chemicals
	Ethyl alcohol	Acetaldehyde manufacture, solvent, other chemicals
	Alkyl aryl sulphonate	Detergent
	Styrene	Synthetic rubber, plastics
Pulp and paper	Paper	Books, records, newspapers, etc.
	Cardboard	Boxes
	Fibreboard, etc.	Building materials
Pigments and paint	Zinc oxide	
	Titanium dioxide	Pigments for paint, ink, plastic rubber, ceramics, linoleum
	Carbon black	
	Lead chromate	
	Linseed oil	Drying oil
	Phenolic resins	Basic lacquer, varnish, and enamel constituents
	Alkyd resins	
Rubber	Natural rubber (isoprene)	Automobile tyres, moulded and sheeting, footwear, insulation
	Synthetic rubbers (GR-S, neoprene, butyl)	
Plastics	Phenol-formaldehyde	
	Polystyrene	Various uses in all areas of everyday life
	Polymethylmethacrylate	
	Polyvinyl chloride	
	Polyethylene	
	Polyesters	
Synthetic fibres	Rayon	
	Nylon	
	Polyesters	Cloth and clothing
	Acrylics	
Minerals	Glass and ceramics	Windows, containers, bricks, pipe
	Cement	Concrete for construction of buildings, highways, etc.
	Coal	Fuel, cock and its by-products.
Cleansing agents	Soap	

(Contd...)

Industry	*Typical products*	*End uses*
	Synthetic detergents (sodium alkyl) aryl sulphonates)	Household and industrial cleaning
	Wetting agents	
Biochemicals	Pharmaceuticals and drugs	Health and medicinal applications
	Fermentation products:	
	Penicillin	Medicinal use
	Ethyl alcohol	Solvent, beverage
	Food products	Human sustenance
Metals	Steel	
	Copper	Building material, machinery, etc.
	Aluminium	
	Zirconium	
	Uranium	Nuclear fuel

The largest tonnage inorganic chemical is sulphuric acid. Little sulphuric acid reaches the public; nearly all of it is consumed by industry in the manufacture of other products. Large quantities are used by the petroleum and metals industries. Important organic chemicals include alcohols, dyes, and intermediates used to produce other chemicals. Many organic chemicals are produced by several industries listed in other categories. For example, plastics and petroleum are organic materials. Here the division is arbitrary. Ethyl alcohol, an organic chemical, was primarily produced by biochemical fermentation before the Second World War; it is now produced primarily from petroleum or natural gas. Therefore, ethyl alcohol might be listed under organic chemicals, petrochemicals, or biochemicals. Ethyl alcohol production also illustrates two major approaches to organo-chemical manufacture. In fermentation the starting material is a natural product (grain, molasses, etc.); whereas manufacture from petroleum involves synthesis of the molecule from elementary mineral building blocks (ethylene and water).

The primary petroleum products are gasoline, other fuels, lubricants, and petrochemicals. Since the Second World War petrochemicals have assumed a tremendous role in the International economy. They are usually organic chemicals made from components of petroleum or natural gas. Paradoxically, the largest tonnage petrochemical is the inorganic chemical ammonia, produced by reaction of hydrogen from natural gas or petroleum with nitrogen from air. It is used as a fertiliser and as a starting material for other products. Many plastics, such as polyethylene and polypropylene, derive their starting materials from natural gas or petroleum. Primary components of many synthetic detergents are produced in the oil refinery.

The pulp and paper industry uses many unusual processes and is often considered separate from the chemical process industry, even though it employs many chemical engineers. The manufacture of paint involves combination of inorganic pigments or organic dyes for colour with various natural or synthetic organic vehicles, extenders, and dyieng agents. This very old industry still relies heavily on the "art" or "practice" in its production of paints, varnishes, and other surface coatings; although many newer developments are based on a better understanding of the scientific principles.

The division between the rubber and plastics industries is not distinct. The various synthetic rubbers are plastics with properties similar to natural rubber. Natural rubber is made from the sap of a tree, whereas most synthetic rubbers are derived from petroleum. Synthetic fibres are usually polymeric organic compounds. Rayon made from natural cellulose was the first large-volume synthetic textile fibre, and nylon was the first purely synthetic textile fibre.

Important chemical process industries utilising mineral raw materials are glass and ceramics, cement, and coke and its by-products. The important by-products of coke are benzene and toluene compete with the same materials produced in the oil industry. Petroleum, natural gas, and sulphur are also minerals. The trend in the field of cleansing agents has been away from soap produced from natural fats toward synthetic detergents, many of which are produced using raw materials from the oil refinery.

The biochemical process industry includes pharmaceuticals, fermentation products, and food. These industries either use processes involving biological action (such as fermentation to produce penicillin) or produce products that are biologically active (such as penicillin in its medicinal effects). Pharmaceuticals are produced by controlled natural biological processes or by synthetic organic chemistry. Although much of the food industry is not generally considered part of the chemical process industry, those products which have been highly processed, such as sugar and hydrogenated shortening, may be considered to be part of the industry. Fermentation processes produce industrial and beverage alcohols, acetone, and acetic acid by biological action on various sugars. Most of the industrial demand for these materials is now met from petrochemical sources.

As stated earlier, the metals industry is usually considered apart from the chemical process industry, although recovery and purification of many of the newer metals involve much chemical processing. Many chemical engineering principles are applied in the production of all metals. The production of iron and steel involves many high-temperature chemical reactions and phase equilibrium. Aluminium is produced by electro-chemical means. Many new metals, such as uranium, zirconium, and titanium, are produced by processes involving standard chemical engineering operations. Many new metallurgical processes are being developed by chemical companies, rather than by the traditional metals companies. From the above discussion it is evident that any classification of chemical process industry is arbitrary and subject to ambiguities. Perhaps a satisfactory definition of a chemical process industry is "an industry whose principal products are manufactured by processes based upon the chemical and physical principles included in the field of chemical engineering."

ORIGINS AND DEVELOPMENT OF THE CHEMICAL PROCESS INDUSTRY

The chemical industry dates back to prehistoric times when man first attempted to control and modify his environment. Its development can be divided into two periods. The prescientific period and scientific period.

Pre-Scientific Chemical Industry

Probably the oldest chemical process industry is fermentation, although in its earliest times it was more a folk craft than an industry. Vinegar (dilute acetic acid) was the earliest known acid since it was formed on the oxidation of fermented liquors. At this time nitric acid was made from saltpetre (KNO_3) and ferrous sulphate by heating the mixture and condensing the distilled nitric acid. The first industrial application of nitric acid was in the separation of gold from silver. Industrial use of sulphuric acid developed later became important, when it was used to produce chlorine for bleaching cloth. Hydrochloric acid became the cheapest and most widely used mineral acid. Early cleansing agents included many natural materials, such as the alkali found in wood ashes.

Scientific Chemical Industry

Progress and growth of the early chemical industry were slow because there was little understanding of the scientific principles underlying the various processes. The increased understanding of chemical science led to improvements and new developments in chemical processing. The principal chemical industries in the early nineteenth century manufactured alkalies, acids and metals.

INDIAN CHEMICAL INDUSTRY TODAY

In keeping with the nation's endeavour for industrial and economic development, the Indian chemical industry has also over the years waded through phases of trials and turbulence. Starting way back in the mid-thirties with batch process for indigenous production of inorganic chemicals it gradually extended to organic chemicals, fertilisers and agro-chemicals, drugs and pharmaceuticals, paints and varnishes, toiletries and cosmetics to coal and rubber chemicals, fine and speciality chemicals, petroleum refining, plastics, man-made fibres and a wide range of petrochemicals. Side by side, thanks to the government initiatives and support, a well planned network of specialised institutions of learning and research was laid out to cope up with the need for technological transformation of the industry. At the same time, the Indian chemical process industry has been ever open to induction of the latest know-how in process and product developments from all over through technical and financial collaborations. By virtue of the range of its products and also the inputs, the chemical industry is not only closely linked up with other industries and sectors of the economy, but also intimately relates with the well-being and quality of life of the people.

Growth with Restraints

The track record of growth cited above should be viewed against several restraints that apply to the Indian chemical industry. It has to enlarge the product range and improve the quality to match the international standards and meet the rapidly changing demand pattern and preferences of the customers. This calls for continuous upgradation of process technologies which in turn often entail additional investments. While doing so, under Indian conditions it has to put up with high costs of borrowings together with the inadequate, inefficient and yet expensive infrastructure and utilities like power, water, transport, etc., all of which erode the industry's competitiveness vis-a-vis imported goods. While it is denied the level playing field on these counts, the high levels of excise duties and local levies together add to the cost to ultimate consumers, and removal or reduction of customs tariff further add to its woes.

Besides, unlike many other industries and services, the chemical industry has to all along ensure utmost safety of health and environment within the plants and in the area surrounding the chemical complexes. The laws in this regard enforced by the Central and State regulatory authorities are becoming increasingly stringent by the day. Even the existing chemical units have to put up additional safeguard facilities at extra costs which were not envisaged originally when these units were set up. Moreover, the NGOs like the Greenpeace brigades often opposte and resist the setting up of new projects at places which have certain locational advantages. And moving over to any alternate location not only add to the capital costs of the projects but also put a perpetual strain on their operating expenses. As a consequence, the cost of compliance to regulations has been increaisng beyond the reach of the industries, particularly when in our country the production capacities are traditionally low.

The chemical industry is also under obligation to abide by several international treaties like the Chemical Weapons Convention, Basel Convention, Montreal Protocol on *Ozone Depleting Substances*, and conventions on Prior Informed Consents in respect of Dual Purpose Chemicals, Persistent Organic Pollutants, etc. to which the country is bound either as a signatory or as a member of the negotiating committees. While planning its activities the industry has to therefore constantly keep in view the stipulations of these treaties which have long term implications. The industry also has to remain sensitive about the provisions of World Trade Organisation and the Intellectual Property Rights, and such other non-tariff barriers. It has to keep a vigil on the possible dumping of goods that hurt the interest of the industry and promptly notify the same to the designated authorities to initiate anti-dumping and safeguard procedures.

The voluntary disciplines like the Responsible Care and ISO Certification are also becoming preconditions when it comes to international trade and dealings. With rapid strides made in information technology leading to global connectivity, the international trade through the medium of E-commerce is becoming exceedingly competitive as bids can be struck instantly with the most preferred offerer and even the goods on the high seas can be promptly directed for delivery at the deisred destination with the least costs of transportation and time. Hence, a company can easily get marked out for any lapse in compliance of disciplines like ISO or other such obligations. With the passage of time, customers and consumers are becoming ever more demanding. Particularly for the chemical industry they are discriminating in favour of not only products matching high quality standards but also for safer products and cleaner and environmentally benign processes, with conscious commitment about social responsibility as well on the part of manufacturers. Institutional investors and bankers too look for such assurances. As a result, the erstwhile quality control methods have given way to strict quality assurance systems and further to Total Quality Management systems in the chemical process industries. In the changing context, ISO culture is becoming more relevant at all levels of organisational structure. Also consequently, in the chemical process industries, well conceived programmes of continuing education, orientation and training for all categories of work force, followed by rigorous performance assessment, are becoming inevitable. Awareness among the formulators and product users is also the responsibility devolving on the manufacturers.

As technologies become more complex and equipment more sophisticated, the industry can ill-afford any laxity or lapses at the operational levels. For, the consequences could be extremely harmful and expensive. The regulatory authorities are becoming more vigilant and the society more sensitive to having chemical units in the neighbourhood as the Bhopal disaster still remains a dreadful reminder to them. When the stakes could be so high, enforcement of strict discipline in chemical process industry is inescapable.

Green Challenge to Indian Chemical Industry

None can deny that the Indian chemical industry is steadily and increasingly becoming internationaly competitive on account of innovative low cost technologies developed by the industry. While our aim is to ensure that the momentum of growth, particularly of the exports is maintained, there is another threat or challenge, by whatever term you may choose to call it, which the industry has to face squarely in coming years. It has two facets. One side is the challenge to the industry from the developed world through several international conventions already in force or under discussion. The second one is endogenous arising out of the way the Indian chemical industry is structured with a very large segment comprising small and medium scale enterprises which are not yet geared to meet the minimum safety standards of environment and health protection laid down in our environment and labour laws.

The green challenge to the Indian chemical industry is indeed real. To combat it, the industry would need to be vigilant all the time to make sure that it does not blunt the cutting edge of the industry and vibrant exports. People are becoming more vigilant on hazards of pollution. Whether the industry likes it or not, demand for pollution free processes will be the over riding factor in future. No industry, however big or small, can be oblivious to this. Research and development, particularly on totally clean technologies and pollution free alternatives have to be an integral part of industry's business.

THE SYSTEMATIC ANALYSIS OF CHEMICAL PROCESSES

Early chemical processes usually involved a few simple steps. Materials were processed in small batches. Process improvements evolved slowly from experience gained by the operators. As the demand for chemical products increased and the chemical industry grew, processes had to be developed to produce large quantities at the lowest possible cost. Improvements based on experience were no longer sufficient; it became necessary to build new plants for products never before produced. Systematic analysis of chemical processes elucidated many underlying principles which could be used in the synthesis of new processes.

Modern chemical processes are often extremely complex operations involving hundred of pieces of equipment. Without a systematic approach, it would be impossible to analyse an existing process or to design a new process. Therefore, chemical processes are broken down into individual steps that recur in many other processes. The general principles of these steps have been carefully studied and are frequently well understood. If the principles are known, it is possible to design the step to do the best possible job. The typical chemical process is analysed with the following interdependent considerations :

(1) Mass and energy balances;
(2) Thermochemistry;
(3) Unit operations;
(4) Plant equipment;
(5) Ancillery equipment;
(6) Process flow diagrams;
(7) Instrumentation and control; and
(8) Economics.

Mass and Energy Balances

Here some of the fundamental principles that engineers and scientists employ in performing design calculations and predicting the performance of plant equipment are discussed. Topics include the conservation laws for both mass and energy, thermochemistry, chemical reaction equilibrium, chemical kinetics, the ideal gas law, and phase equilibrium.

Conservation of mass

The conservation law for mass can be applied to any process or system. The general form of this law is given in equation (1.1).

mass in – mass out + mass generated = mass accumulated ...(1.1)

Equation (1.1) may be applied to the total mass involved or to a particular species, on either a mole or mass basis. The conservation law for mass can be applied to steady-state or unsteady-state process and to batch or continuous system. To isolate a system for study, we separate it from the surroundings by a boundary or envelope. This boundary may be real (e.g., the walls of a vessel) or imaginary. Mass crossing the boundary and entering the system is part of the mass-in term in equation (1.1) whereas that leaving the system is part of the mass-out term. Equation (1.1) may be written for any compound whose quantity is not changed by chemical reaction, or for any chemical element, regardless of whether it has participated in a chemical reaction. It may be written for one piece of equipment, around several pieces of equipment, or around an entire process. It may be used to calculate an unknown quantity directly, to check the validity of experimental data, or to express one or more of the independent relationships among the unknown quantities in a particular problem.

A steady-state process is one in which there is no change in conditions (temperature, pressure, etc.) or rates of flow with time at any given point in the system. The accumulation term in equation (1.1) is then zero. If there is no chemical reaction, the generation term is also zero. All other processes are classified as unsteady-state. In a batch process, reactants are fed in an unending flow to a piece of equipment or to several pieces in series, and products are continuously removed from one or more points. A continuous process may or may not be steady-state. As indicated previously, equation (1.1) may be applied to the total mass of each stream (referred to as an overall or total material balance) or to the individual components of the streams (referred to as a componential or component material balance). Often the primary task in preparing a material balance is to develop the quantitative relationships among the streams.

The conservation law for mass finds its major application in performing pollution prevention audits. A pollution prevention assessment is a systematic, planned procedure with the objective of identifying ways to reduce or eliminate waste. First the procedure consists of a careful review of a plant's operation and waste streams, and the selection of specific areas to assess. Second, after a specific waste stream or area is established as the focus, a number of options with the potential to minimise waste are developed and screened. Third, the selected options are evaluated for technical and economic feasibility. Finally, the most promising options are selected for implementation. Additional details on the assessment process, providing a direct application of the conservation law for mass, follow.

As discussed above, pollution prevention assessments are procedures for identifying and assessing pollution prevention opportunities within a plant site. This process consists of a careful review of the plant's operations and waste streams and the selection of specific streams and/or operations to assess. It is an extremely useful tool in diagnosing how the facility can reduce or eliminate (hazardous) wastes. A pollution prevention assessment can also focus on how to reduce wastes destined for treatment and/or disposal. This is unlike an environmental audit, which focuses on regulatory compliance and environmental protection. Whereas audits are often thought of in terms of legal compliance, a pollution prevention assessment is commonly referred to technically as a search for pollution prevention opportunities, generally without concern for legal implications.

The assessment process should characterise the selected waste streams and process–an ongoing part of this type of analysis. Some of the data may be collected during the first review of site-specific data; however, much of the information may not be collected until the actual audit inspection. Regardless, as much information as possible should be collected. The information-gathering process can be helpful in establishing specific goals for the programme. Information about the facility's waste stream can come from a variety of sources, such as personnel involved with the process and the waste stream, hazardous waste manifests, biennial reports, environmental audits, emission inventories, engineering studies, waste assays and permits.

Simplified mass balances (application of the conservation law for mass) should be developed for each of the important waste-generating operations to identify sources and gain a better understanding of the origins of each waste stream. As described earlier, a mass balance is essentially a check to make sure that what goes into the process (i.e., the total mass of all raw materials, water, etc.), leaves the process (i.e., total mass of the product wastes, and by-products). The material balance should be made individually for all components that enter and leave the process. When chemical reactions take place in a system, there is an advantage to doing "elemental balances" for specific chemical elements in a system. Material balances can assist in determining concentrations of waste constituents where analytical test data are limited. They are particularly useful when there are points in the production process where it is difficult (because of inaccessibility) or uneconomical to collect analytical data.

These mass balance calculations are particularly useful for quantifying fugitive emissions, such as evaporative losses. Waste stream data and mass balances will enable one to track the flow and characteristics of the waste streams over time. Since in most cases the accumulation equals zero, it can then be assumed that buildup is actually leaving the process through fugitive emissions or other means. This will be useful in identifying trends in waste/ pollutant generation and will also be critical in the task of measuring the performance of the pollution prevention options that are implemented.

The result of this activity is a catalog of waste streams that provides a description of each waste, including quantities, frequency of discharge, composition, cost of management, and other important information. An assessor should also collect information about the facility itself: process design, raw material and production data, operating costs, environmental reports and permits, and policy and organisational information.

The result of the waste stream selection and information-gathering process should be a well-defined "needs" list. The needs list is made up of questions to be resolved and further information to be obtained. It will help the assessor to focus on which data need to be collected during an assessment inspection. One may then convert the list into informal inquiries or perhaps more formal questionnaires to be distributed to appropriate personnel at the facility.

Once the process has been developed and completed, attention can be turned to assessing the various options. Alternatives that do not merit further attention can be weeded out, and the remaining choices ranked in the order of desirability. Ranking can be performed by taking each option in turn and examining how it compares on key criteria such as the following :

(1) Effectiveness for reducing waste
(2) Technical risk
(3) Extent of current use in the facility
(4) Industrial precedent
(5) Capital and operating costs incurred
(6) Effect on the quality of product
(7) Impact on plant operations
(8) Required time for implementation
(9) Other aspects important in a particular situation

Conservation of energy

A presentation of the conservation law for energy would be incomplete without a brief review of some introductory thermodynamic principles. Thermodynamics is defined as the science that deals with the relationships between the various forms of energy.

A system may possess energy as a result of its temperature, velocity, position, molecular structure, surface, and so on. The energies corresponding to these conditions are internal, kinetic, potential, chemical, surface, and so on. Engineering thermodynamics is founded on three basic laws. Energy, like mass and momentum, is conserved. Application of the conservation law for energy gives rise to the first law of thermodynamics. This law, in steady-state form for batch processes, is presented here (potential, kinetic, and other energy effects are not considered).

$$\Delta E = Q - W \quad ...(1.2)$$

For flow process,

$$\Delta H = Q - W_s \quad ...(1.3)$$

Where Q = heat energy transferred across the system boundary

W = work energy transferred across the system boundary

W_s = mechanical work energy transferred across the system boundary

E = internal energy of the system

H = enthalpy of the system

ΔE, ΔH = changes in internal energy and enthalpy during the process

The internal energy and enthalpy in equations (1.2) and (1.3), as well as in the other equations in this discussion, may be on a mass or a mole basis, or they may represent the total internal energy and enthalpy of the entire system. Most industrial facilities operate in a steady-state flow mode. If no significant mechanical or shaft work is added or withdrawn from the system, equation (1.3) reduces to

$$Q = \Delta H \quad ...(1.4)$$

If a unit or system is operated adiabatically, where Q = 0, equation (1.4) then becomes

$$\Delta H = 0 \quad ...(1.5)$$

Although the topics of material and energy balances are covered separately in this section, it should be emphasised that this segregation does not occur in reality; one must work with both energy and material balances simultaneously.

Perhaps the most important thermodynamic function the engineer works with is the enthalpy. The enthalpy is defined by

$$H = E + PV \quad ...(1.6)$$

where P and V are pressure and volume, respectively.

The terms E and H are state or point functions. By fixing a certain number of variables on which the function depends, we automatically fix the numerical value of the function; that is, it is single-valued. For example, fixing the temperature and pressure of a one-component single-phase system immediately specifies the enthalpy and internal energy. The change in enthalpy as the system undergoes a change in state from (T_1, P_1) to (T_2, P_2) is given by

$$\Delta H = H_2 - H_1 \quad ...(1.7)$$

Thermochemistry

Consider now the energy effects associated with a chemical reaction. To introduce this subject, the reader is reminded that engineers and applied scientists are rarely concerned with the magnitude or amount of energy in a system; their primary concern is with changes in the amount of energy. It has been found in measuring energy changes for systems that the enthalpy is the most convenient term to work with. There are many different types of enthalpy effects; these include sensible heat, latent heat, and heat of reaction.

The heat of reaction is defined as the enthalpy change of a system undergoing chemical reaction. If the reactants and products are at the same temperature and in their standard states, the heat of reaction is termed the standard heat of reaction. For engineering purposes, the standard state of a chemical may be taken as the pure chemical at 1 atm pressure.

Chemical reaction equilibrium

With regard to chemical reactions, two important questions are of concern to the engineer: (i) How far will the reaction go? and (ii) How fast will the reaction go?

Chemical thermodynamics provides the answer to the first question; however, it tells nothing about the second. Reactions rates fall within the domain of chemical kinetics and are treated later in this section. Both equilibrium and kinetic effects must be considered in an overall engineering analysis of a chemical reaction.

Chemical reaction equilibrium calculations are structured around another thermodynamic term called the free energy. This so-called free energy (G) is a property that cannot be defined easily without some further basic grounding in thermodynamics. Free energy has the same units as enthalpy and internal energy, and may be on a mole or total mass basis.

Consider the equilibrium reaction

$$aA + bB = cC + dD \quad ...(1.8)$$

Where A, B, C, D = chemical formulas of the reactants and products

a, b, c, d = stoichiometric coefficients

and the equal sign is a reminder that the reacting system is at equilibrium. For this reaction, the change in free energy, ΔG, is given by

$$\Delta G = cG_C + dG_D - aG_A - bG_B \quad ...(1.9)$$

where G_i is the free energy of species i.

The following equation is used to calculate the chemical reaction equilibrium constant K at a temperature T.

$$\Delta G_T = RT\ (\ln K) \quad ...(1.10)$$

The problem that remains is to relate K to understandable physical quantities; for gas phase reactions the term K in equation (1.10) may be approximately represented in terms of the partial pressures of the components involved. This relationship is given by equations (1.11) and (1.12).

$$K = K_P \quad ...(1.11)$$

where K is an equilibrium constant based on partial pressures and defined by

$$K_P = K^c{}_C\ P^d{}_D / P^a{}_A\ P^b{}_B \quad ...(1.12)$$

where P_A = partial pressure of component A, and so forth.

The definition of K_P obviously applies to the reaction of equation (1.8). Assuming that a K value is available or calculable, this equation may be used to determine the partial pressures of the participating components at equilibrium. For liquid phase reactions, K is given approximately by

$$K = K_C \quad ...(1.13)$$

Where $K_C = C^c{}_C C^d{}_D / C^a{}_A C^b{}_B$

C_C = concentration of component C (e.g., gmol/L), and so forth.

Chemical kinetics

Chemical kinetics involves the study of reaction rates and the variables that affect these rates. It is a topic that is critical for the analysis of reacting systems. The objective here is to develop a working understanding of this subject that will permit us to apply chemical kinetics principles in the area of pollution prevention. The topic here is treated from an engineering point of view, that is, in terms of physically measurable quantities.

The rate of a chemical reaction can be described in any of several different ways. The most commonly used definition involves the time rate of change in the amount of one of the components participating in the reaction. This rate is usually based on some arbitrary factor related to the reacting system size or geometry, such as volume, mass, or interfacial area. The definition shown in equation (1.14), which applies to homogeneous reactions, is a convenient one from an engineering point of view.

$$R_A = (1/V)(dn_A/dt) \qquad ...(1.14)$$

Where R_A = reaction rate based on component A

V = volume of reacting system

n_A = number of moles of A at time t

t = time

If the volume term is constant, one may write equation (1.14) as

$$R_A = dC_A/dt \qquad ...(1.15)$$

where C_A is the molar concentration of A at time t.

Based on equation (1.15), the reaction rate is positive if species A is being formed (since C_A increases with time), and negative if A is reacting (since C_A decreases with time). The rate is zero if the system is at chemical equilibrium.

An equation expressing the rate in terms of measurable and/or desirable quantities may now be developed. Based on experimental evidence, the rate of reaction is a function of the concentration of the components present in the reaction mixture, temperature, pressure, and catalyst variables. In equation form,

$$R_A = R_A(C_i, P, T, \text{catalyst variables}) \qquad ...(1.16)$$

Equation (1.16) may be condensed into

$$R_A = (\pm)\, k_A f(C_i) \qquad ...(1.17)$$

where k_A incorporates all the variables in equation (1.16) other than the concentration variable. The (±) notation is included to indicate whether component A is being consumed (–) or produced (+). The k_A may be regarded as a constant of proportionality; it is termed the reaction velocity constant. Although this "constant" is independent of concentration, it is a function of other variables. This term is very definitely influenced by temperature and catalyst activity. For most applications, it is assumed that k is solely a function of temperature. Thus,

$$R_A = (\pm)k_A(T)f(C_i) \qquad ...(1.18)$$

The effect of temperature on k is generally represented by the Arrhenius equation and the values of k are obtained from experimental data.

Ideal gas law

The ideal gas law was derived from experiments in which the effects of pressure and temperature on gaseous volumes were measured over a moderate range of temperatures and pressures. As a general rule, this law works best when the molecules of the gas are far apart, that is, when the pressure is low and the temperature is high. Under these conditions, the gas is said to behave ideally. For engineering calculations the ideal gas law is almost always assumed to be valid, since it generally works well for the temperature and pressure ranges used in most applications.

The two precursors of the ideal gas law are Boyle's law and Charles's law. Boyle found that the volume of a given mass of gas is inversely proportional to the absolute pressure if the temperature is kept constant; that is,

$$P_1V_1 = P_2V_2 \qquad ...(1.19)$$

where V_1 and V_2 are the volumes of gas at absolute pressures P_1 and P_2, respectively. Charles found that the volume of a given mass of gas varies directly with the absolute temperature at constant pressure

$$V_1/T_1 = V_2/T_2 \qquad ...(1.20)$$

where V_1 and V_2 are the volumes of gas at absolute temperatures T_1 and T_2, respectively. Boyle's and Charles's laws may be combined into a single equation in which neither temperature nor pressure need be held constant

$$P_1V_1/T_1 = P_2V_2/T_2 \qquad ...(1.21)$$

This equation indicates that for a given mass of a specific gas, PV/T has a constant value. Since, at the same temperature and pressure, volume and mass must be directly proportional, this statement may be extended to

$$PV/nT = R \qquad ...(1.22)$$

where n is the number of moles and R is the universal gas constant. Equation (1.22) is called the ideal gas laws. Numerically, the value of R depends on the units of P, V, T, and n. Tabulated values of R are provided in Table 1.2.

Table 1.2. Values of R in various units.

R	*Temperature scale*	*Units of V*	*Units of n*	*Units of P*
10.73	°R	ft^3	lbmol	psia
0.7302	°R	ft^3	lbmol	atm
21.85	°R	ft^3	lbmol	in. Hg
555.0	°R	ft^3	lbmol	mm Hg
297.0	°R	ft^3	lbmol	in. H_2O
0.7398	°R	ft^3	lbmol	bar
1545.0	°R	ft^3	lbmol	psfa
24.75	°R	ft^3	lbmol	ft H_2O
1.9872	°R	—	lbmol	—
0.0007805	°R	—	lbmol	—
0.0005819	°R	—	lbmol	—
500.7	°R	—	lbmol	—
1.314	K	ft^3	lbmol	atm
998.9	K	ft^3	gmol	mm Hg
19.32	K	ft^3	lbmol	psia
62.361	K	L	lbmol	mm Hg
0.08205	K	L	gmol	atm
0.08314	K	L	gmol	bar
8314	K	L	gmol	Pa
8.314	K	m^3	gmol	Pa
82.057	K	cm^3	gmol	atm
1.9872	K	—	gmol	—
8.3144	K	—	gmol	—

Phase equilibrium

The term phase, for a pure substance, indicates a state of matter—that is, solid, liquid, or gas. For mixtures, however, a more stringent connotation must be used, since a totally liquid or solid system may contain more than one phase. A phase is characterised by uniformity or homogeneity; the same composition and properties must exist throughout the phase region. At most temperatures and pressures, a pure substance normally exists as a single phase. At certain temperatures and pressures, two or perhaps even three phases can coexist in equilibrium.

The most important equilibrium phase relationship is that between liquid and vapour. Raoult's and Henry's laws theoretically describe liquid-vapour behaviour and, under certain conditions, are applicable in practice. Raoult's law, sometimes useful for mixtures of components of similar structure, states that the partial pressure of any component in the vapour is equal to the product of the vapour pressure of the pure component and the mole fraction of that component in the liquid, that is,

$$P_i = P'_i x_i \qquad ...(1.23)$$

Where P_i = partial pressure of component i in the vapour

P'_i = vapour pressure of pure i

x_i = mole fraction of component i in the liquid

This expression may be applied to all components. If the gas phase is ideal, equation (1.23) becomes

$$y_i = (P'_i/P)x_i \qquad ...(1.24)$$

Where y_i = mole fraction of component i in the vapour

P = total system pressure

Unfortunately, relatively few mixtures follow Raoult's law. A more empirical relationship used for representing data on many systems is Henry's law, valid for low x_i

$$P_i = H_i x_i \qquad ...(1.25)$$

where H_i is Henry's law constant for component i (with units of pressure).

Many equilibrium calculations are accomplished using the phase equilibrium constant K_i. This constant has been referred to in industry as a componential split factor, since it provides the ratio of the mole fractions of a component in two equilibrium phases. The defining equation is

$$K_i = y_i/x_i \qquad ...(1.26)$$

This equilibrium constant is a function of the temperature, the pressure, the components in the system, and the mole fractions of these components. However, as a first approximation, K_i is generally treated as a function of temperature and pressure only. For this condition, K_i may be approximated by

$$K_i = P'_i/P \qquad ...(1.27)$$

Unit Operations

In the chemical and other physical processing industries, including the food and biological processing industries, there are many similarities in the manner in which the entering feed materials are modified or processed into final products. These seemingly different chemical, physical, or biological processes can be broken down into a series of separate and distinct steps called unit operations. These unit operations are common to all types of diverse process industries. For example, the unit operation distillation is used to purify or separate alcohol in the beverage industry and hydrocarbons in the petroleum industry. Drying of grain and other foods is similar to drying of lumber, filtered precipitates, and rayon yarn. The unit operation absorption occurs in absorption of oxygen from air in a fermentation process or in a sewage treatment plant, and in absorption of hydrogen gas in a process for liquid hydrogenation of oil. Evaporation of salt solutions in the chemical industry is similar to evaporation of sugar solutions in the food industry. Settling and sedimentation of suspended solids are similar in the sewage and the mining industries. Flow of liquid hydrocarbons in the petroleum refinery and flow of milk in a dairy plant are carried out in a similar fashion. Thus, the most efficient method of organising the subject matter of unit operations is based on two facts: (i) although the number of individual processes is great, each one may be separated into a series of steps, called operations, each of which in turn appears in process after process; (ii) the individual operations have common techniques and are based on the same scientific principles. The unit operation concept, therefore, is this: By systematically studying these operations themselves—operations that clearly cross industry and process lines—the treatment of all processes is unified and simplified.

Classification of unit operations

Unit operations deal mainly with the transfer and change of energy and the transfer and change of materials, primarily by physical means but also by chemical means. Some, but not all, of the important unit operations that can be combined in various sequences in a process are classified and described below:

(1) *Fluid flow:* Concerns the principles that determine the flow or transportation of any fluid from one point to another.

(2) *Heat transfer:* A unit operation that deals with the principles that govern accumulation and transfer to heat and energy from one place to another.

(3) *Evaporation:* A special case of heat transfer, which deals with the evaporation of a volatile solvent, such as water, from a non-volatile solute, such as salt or any other material in solution.

(4) *Drying:* An operation in which volatile liquids (usually water) are removed from solid materials.

(5) *Distillation:* An operation whereby components of a liquid mixture are separated by boiling because of their differences in vapour pressure.

(6) *Absorption:* A process whereby a component is removed from a gas stream by treatment with a liquid.

(7) *Liquid-liquid extraction:* A process in which a solute in a liquid solution is removed by contact with another liquid solvent that is relatively immiscible with the solution.

(8) *Liquid-solid leaching:* It involves treating a finely divided solid with a liquid that dissolves out and removes a solute contained in the solid.

(9) *Crystallisation:* The removal of a solute such as a salt from a solution by precipitating the solute from the solution.

(10) *Mechanical-physical separations:* Involve separation of solids, liquids, or gases by mechanical means, such as filtration, settling, and size reduction, which are often classified as separate unit operations.

Unit operations are applicable to processes that are essentially physical in nature, as well as to those that are chemical. However, a substantial number of the unit operations are concerned with the problems of changing the composition of solutions and mixtures through methods that do not involve chemical reactions. Most frequently these composition changes are analytical; that is, it is often desirable to separate the original substance into its component parts. Such separations may be entirely mechanical, such as the separation of solid from liquid during filtration or the classification of a granular solid into fractions of different particle size by screening. On the other hand, if the operations involve changes in composition of solutions, they are known as diffusional, or mass-transfer, operations, and it is these with which we are most often concerned. The substance undergoing the change in composition is most frequently itself a solution, but it may be a physical mixture or one that exhibits the properties of both.

Diffusional operations are characterised by a transfer of one substance through another, usually on a molecular scale. For example, when water evaporates from a pool into an air stream flowing over the water surface, molecules of water vapour diffuse through those of the gas at the surface into the main portion of the air stream, where they are carried away. On other occasions, one molecular species may diffuse through another which is itself diffusing in the opposite direction. In every case, the mass transfer is a result of a concentration difference, the diffusing substance moving from a place of high concentration to one of low concentration. The concentration difference is thus a driving force for the mass transfer in the same way that temperature difference is a driving force for the transfer of heat.

It is useful to classify the diffusional operations and to cite examples of each. Separations may be brought about by :

(1) Contact of two immiscible phases, with mass transfer or diffusion through the surface between the phases.

(2) Contact of miscible phases separated by a permeable or semipermeable membrane, with diffusion through the membrane.

(3) Direct contact of miscible phases.

The various methods for each category are discussed below.

Contact of immiscible phases

The three states of aggregation permit six possibilities: gas-gas, gas-liquid, gas-solid, liquid-liquid, liquid-solid, and solid-solid.

Gas-gas. Since all gases are completely soluble in each other, this category of operation cannot generally be practically realised.

Gas-liquid. If all components of the system are present in appreciable amounts in both gas and liquid phases, the operation is known as distillation. In this instance the gas phase is created from the liquid phase by application of heat. For example, if a liquid solution of acetic acid and water is partially vapourised by heating, it is found that the newly created vapour phase and the residual liquid both contain acetic acid and water, but in proportions that are different for the two phases and different from those in the original solution. If the vapour and residual liquid are separated physically from one another and the vapour condensed, two solutions, one richer in acetic acid and the other poorer, are obtained. In this way a certain degree of separation of the original components is accomplished. Conversely, should a vapour mixture of the two substances be partially condensed, the newly formed liquid phase and the residual vapour will differ in composition. In both instances an inter-diffusion of both components between the phases eventually establishes their final composition.

All the components of the solutions involved may not be present in appreciable amounts in both gas and liquid phases, however. If the liquid phase is a pure liquid containing but one component whereas the gas contains two or more, the operations are known as humidification or dehumidification, depending upon the direction of the diffusion. For example, contact of dry air with liquid water results in evaporation of some of the water into the air (humidification of the air). Conversely, contact of very moist air with pure liquid water may result in condensation of part of the moisture in the air (dehumidification). In both cases relatively little air dissolves in the water, and for most practical purposes it is considered that only water vapour diffuses from one phase to the other.

Both phases may be solutions, each containing only one common component that distributes between phases. For example, if a mixture of ammonia and air is brought into contact with liquid water, a large portion of the ammonia but relatively little air will dissolve in the liquid, and in this way the air-ammonia mixture may be separated. The operation is known as gas absorption. On the other hand, if air is brought into contact with an ammonia-water solution, some of the ammonia leaves the liquid and enters the gas phase. This operation is known as desorption, or stripping, a common water pollution treatment method.

To complete the classification, we might consider the case where the gas phase contains but one component and the liquid several, as in evaporation of a saltwater solution by boiling. Here the gas phase contains only water vapour, since the salt is essentially nonvolatile. Such operations do not depend on concentration gradients, but rather on the rate of heat transfer to the boiling solution, and are consequently not considered diffusional. Should the salt solution be separated by diffusion of the water into an air stream, however, the operation then becomes one of desorption, or stripping, and is diffusional.

Gas-solid. Classification of the operations in the gas-solid category according to the number of components that appear in two phases is again convenient.

If a solid solution were to be partially vapourised without the appearance of a liquid phase, the newly formed vapour phase and the residual solid would each contain all the original components but in different proportions. The operation is then fractional sublimation. As in distillation, the final compositions are established by inter-diffusion of the components between the phases. Although such an operation is theoretically possible, it is usually not practical because of the inconvenience of dealing with solid phases in this manner.

However, all components may be present in both phases. If a solid that is moistened with a volatile liquid is exposed to a relatively dry gas, the liquid leaves the solid and diffuses into the gas, an operation generally known as drying, sometimes as desorption. An example is the drying of laundry by exposure to air, and there are many industrial counterparts such as the drying of lumber or the removal of moisture from a wet filter cake by exposure to dry gas. In these cases, the diffusion is, of course, from the solid to the gas phase. If the diffusion takes place in the opposite direction, the operation is known as adsorption.

For example, if a mixture of water vapour and air is brought into contact with activated silica gel, the water vapour diffuses to the solid, which retains it strongly, and the air is thus dried. In other instances a gas mixture may contain several components, each of which is adsorbed on a solid but to different extents (fractional adsorption). For example, if a mixture of propane and propylene gases is brought into contact with a molecular sieve, the two hydrocarbons are both adsorbed, but to different extents, thus leading to a separation of the gas mixture.

In a case in which the gas phase is a pure vapour, such as in the sublimation of a volatile solid from a mixture with one that is non-volatile, the operation is dependent more on the rate of application of heat than on concentration difference and is essentially non-diffusional. The same is true of the condensation of a vapour to the condition of a pure solid, where the rate depends on the rate of heat removal.

Liquid-liquid. Separations involving the contact of two insoluble liquid phases are known as liquid-extraction operations. A simple example is a familiar laboratory procedure: if an acetone-water solution is shaken in a separatory funnel with carbon tetrachloride and the liquids are allowed to settle, a large portion of the acetone will be found in the carbon-tetrachloride-rich phase and will thus have been separated from the water. A small amount of the water will also have been dissolved by the carbon tetrachloride, and a small amount of the latter will have entered the water layer, but these effects are relatively minor. As another possibility, a solution of acetic acid and acetone may be separated by adding it to the insoluble mixture of water and carbon tetrachloride. After shaking and settling, both acetone and acetic acid will be found in both liquid phases, but in different proportions. Such an operation is known as fractional extraction.

Liquid-solid. Fractional solidification of a liquid, where the solid and liquid phases are both solutions of variable composition containing all the components but in different proportions, is theoretically possible but is not ordinarily carried out because of practical difficulties in handling the solid phase and because of the very slow diffusion rates in the solid.

Cases involving distribution of a substance between the solid and liquid phases are common, however. Solution of a component from a solid mixture by a liquid solvent is known as leaching (sometimes called solvent extraction), and as examples we may cite the leaching of gold from its ores by cyanide solutions and the leaching of cottonseed oil from cottonseeds by hexane. The diffusion is, of course, from the solid to the liquid phase. If the diffusion is in the opposite direction, the operation is known as adsorption. Thus, the coloured material that contaminates impure cane sugar solutions may be removed by contacting the liquid solutions with activated carbon, whereupon the coloured substances are retained on the surface of the solid carbon.

When the solid phase is a pure substance and the liquid solution is being separated, the operation is crystallisation, but as ordinarily carried out this is more dependent on heat-transfer rates and only indirectly on concentrations. The reverse operation is dissolution. No known operation is included in the category involving a pure liquid phase.

Solid-solid. Because of the extraordinarily slow rates of diffusion within solid phases, there is no industrial separation operation in this category.

Miscible phases separated by a membrane

In operations involving miscible phases separated by a membrane, the membrane is necessary to prevent intermingling of the phases, but it must be differently permeable to the components of the solutions if diffusional separations are to be possible.

Gas-gas. The operation in the gas-gas category is known as gaseous diffusion, or effusion. If a gas mixture whose components are of different molecular weights is brought into contact with a porous diaphragm, the various components of the gas will diffuse through the pores at different rates. This leads to different compositions on opposite sides of the diaphragm and, consequently, to separation of the gas mixture. In this manner, large-scale separation of the isotopes of uranium, in the form of uranium hexafluoride, is carried out.

Liquid-liquid. The separation of a crystalline substance from a colloid, by contact of their solution with a liquid solvent with an intervening membrane permeable only to the solvent and the dissolved crystalline substance, is known as dialysis. For example, aqueous beet sugar solutions containing undesired colloidal material are freed of the latter by contact with water with an intervening semi-permeable membrane. Sugar and water diffuse through the membrane, but the larger colloidal particles cannot. Fractional dialysis for separating two crystalline substances in solution makes use of the difference in membrane permeability of the substances. If an electromotive force is applied across the membrane to assist in the diffusion of charged particles, the operation is electrodialysis. If a solution is separated from the pure solvent by a membrane that is permeable only to the solvent, the solvent diffuses into the solution—an operation known as osmosis. This is not a separation operation, of course, but theoretically, at least, one could reverse the flow of solvent by superimposing a pressure to oppose the osmotic pressure.

Solid-solid. The operation in the solid-solid category has found little, if any, practical application in the chemical process industry.

Direct contact of miscible phases

Operations involving direct contact of miscible phases, because of the difficulty in maintaining concentration gradients without mixing of the fluid, are not generally considered practical industrially except in unusual circumstances.

Thermal diffusion involves the formation of a concentration difference within a single liquid or gaseous phase by imposition of a temperature gradient upon the fluid, thus making possible a separation of the components of the solution. It was used, for example, in the separation of uranium isotopes in the form of uranium hexafluoride.

If a condensable vapour, such as steam, is allowed to diffuse through a gas mixture, it will preferentially carry one of the components along with it, thus making a separation by the operation known as sweep diffusion. If the two zones within the gas phase where the concentrations are different are separated by a screen containing relatively large-size openings, the operation is called atmolysis.

If a gas mixture is subjected to a very rapid centrifugation, the components will be separated because of the slightly different forces acting on the various molecules, owing to their different masses. The heavier molecules thus tend to accumulate at the periphery of the centrifuge.

The engineer faced with the problem of separating the components of a solution ordinarily must choose between several possible methods. Although the choice is usually limited because of the peculiar physical characteristics of the materials to be handled, the necessity for making a decision nevertheless almost always exists. Until the fundamentals of the various operations are clearly understood, of course, no basis for such a decision is available, but it is good at least to establish the nature of the alternatives at the beginning.

The engineer may sometimes choose between using a diffusional operation of the sort and a purely mechanical separation method. For example, in the separation of a desired mineral from its ore, it may be possible to use either the diffusional operation of leaching with a solvent or the purely mechanical methods of flotation. Vegetable oils may be separated from the seeds in which they occur by extraction or by leaching with a solvent. A vapour may be removed from a mixture with a permanent gas by the mechanical operation of compression or by the diffusional operations of gas absorption or adsorption. Sometimes both mechanical and diffusional operations are used, especially where the former are incomplete, as in processes for recovering vegetable oils wherein extraction is followed by leaching. A more commonplace example is the wringing of water from wet laundry followed by air drying. It is characteristic that at the end of the operation the substance removed by mechanical methods is pure, whereas if removed by diffusion methods it is associated with another substance.

Frequently a choice may be made between a diffusional operation and a chemical reaction to bring about a separation. Water may be removed from an ethanol-water solution either by causing it to react with unslaked lime or by special methods of distillation, for example. Hydrogen sulphide may be separated from other gases either by absorption in a liquid solvent or by chemical reaction with ferric oxide. Chemical methods ordinarily destroy the substance removed, whereas diffusional methods usually permit its eventual recovery in unaltered form without great difficulty.

There are also choices to be made within the diffusional operations. For example, a gaseous mixture of oxygen and nitrogen may be separated by preferential absorption of the oxygen on activated carbon, by absorption, by distillation, or by gaseous effusion. A liquid solution of acetic acid may be separated by distillation, by liquid extraction with a suitable solvent, by absorption with a suitable solvent, or by adsorption with a suitable adsorbent.

The principal basis for choice in most cases is economics: the method that costs least is usually the one to be used. Other factors, including legal, environmental, and so forth, also influence the decision, however. The simplest operation, although it may not be the least costly, is sometimes desired because it will be trouble-free. Sometimes a method will be discarded because imperfect knowledge of design methods or unavailability of data for design will not permit results to be guaranteed. Favourable previous experience with a particular method is often given strong consideration. Cost, however, remains one of the prime factors.

Plant Equipment

The plant equipment provides details on a number of commonly used process units: reactors, heat exchangers, columns of various types (distillation, absorption, adsorption, evaporation, extraction), dryers, and grinders. The purpose of each unit or operation and the many configurations in which the units can be found are also discussed.

Chemical reactors

Reactor definition

The reactor is often the heart of a chemical process. It is the place in the process where raw materials are usually converted into products, and reactor design is therefore a vital step in the overall design of the process.

The treatment of reactors in this section is restricted to a discussion of the appropriate reactor types for a process. The design of an industrial chemical reactor must satisfy requirements in four main areas.

(1) *Chemical Factors:* These involve mainly the kinetics of the reaction. The design must provide sufficient residence time for the desired reaction to proceed to the required degree of conversion.

(2) *Mass Transfer:* The reaction rate of heterogeneous reactions may be controlled by the rates of diffusion of the reacting species, rather than the chemical kinetics.

(3) *Heat Transfer Factors:* These involve the removal, or addition, of the heat of reaction.

(4) *Safety Factors:* These involve the confinement of any hazardous reactants and products, as well as the control of the reaction and the process conditions.

The need to satisfy these interrelated, and often contradictory, factors makes reactor design a complex and difficult task. However, in many instances one of the factors predominates, hence determining the choice of reactor type and the design method.

Reactor types

The characteristics normally used to classify reactor designs are:

(1) *Mode of operation:* batch or continuous;

(2) *Phases present:* homogeneous or heterogeneous; and

(3) *Reactor geometry:* flow pattern and manner of contacting the phases.

The five major classes of reactors are: (i) batch; (ii) stirred tank; (iii) tubular; (iv) packed bed (fixed); and (v) fluidised bed.

In a batch process, all the reagents are added at the beginning of the reaction; the reaction proceeds, the compositions change with time. The reaction is stopped and the product withdrawn when the required conversion has been reached. Batch processes are suitable for small-scale production and for processes that use the same equipment to make a range of different products or grades. Examples include pigments, dyestuffs, pharmaceuticals, and polymers.

In continuous processes, the reactants are fed to the reactor and the products withdrawn continuously; the reactor usually operates under steady-state conditions. Continuous production normally entails lower production costs than batch production, but lacks the flexibility of batch production. Continuous reactors are usually selected for large-scale production. Processes that do not fit the definition of batch or continuous are often referred to as semicontinuous and semi batch. In a semibatch reactor, some of the reactants may be added to, or some of the products withdrawn from, the batch as the reaction proceeds. A semicontinuous process is basically a continuous process that is interrupted periodically—for instance, for the regeneration of catalyst.

Homogeneous reactions are those in which the reactants, products, and any catalysts used form one continuous phase, gaseous or liquid. Homogeneous gas phase reactors are almost always operated continuously, whereas liquid phase reactors may be batch or continuous. Tubular (pipeline) reactors are normally used for homogeneous gas phase reactions. Both tubular and stirred tank reactors are used for homogeneous liquid phase reactions.

In a heterogeneous reaction two or more phases exist and the overriding problem in reactor design is to promote mass transfer between the phases.

The possible combinations of phases are:

(1) *Liquid-liquid:* with immiscible liquid phases.

(2) *Liquid-solid:* with one or more liquid phases in contact with a solid; the solid may be a reactant or catalyst.

(3) *Liquid-solid-gas:* where the solid is normally a catalyst.

(4) *Gas-solid:* where the solid may take part in the reaction or act as a catalyst.

The reactors used for established processes are usually unique and complex designs that have been developed over a period of years to suit the requirements of a particular process. However, it is convenient to classify flow reactors into the broad categories discussed below.

A stirred tank (agitated) reactor consists of a tank fitted with a mechanical agitator and (usually) a cooling jacket or coils. These are operated as batch or continuous reactors. Several reactors may be used in series. The stirred tank reactor can be considered the basic chemical reactor, modelling on a large scale the conventional laboratory flask. Tank sizes range from a few litres to several thousand litres. This equipment is used for homogeneous and heterogeneous liquid-liquid and liquid-gas reactions, and for reactions that involve finely suspended soils and solids that are held in suspension by the agitation. Since the degree of agitation is under the designer's control, stirred tank reactors are particularly suitable for reactions that require good mass transfer or heat transfer.

When operated as a continuous process, the composition in the reactor is constant and the same as the product stream, and, except for very rapid reactions, this limits the conversion that can be obtained in one stage.

Tubular reactors are generally used for gaseous reactions, but are also suitable for some liquid phase reactions. If high heat transfer rates are required, small-diameter tubes are used to increase the ratio of surface area to volume. Several tubes may be arranged in parallel, connected to a manifold, or fitted into a tube sheet in an arrangement similar to a shell and tube heat exchanger. For high temperature reactions, the tubes may be arranged in a furnace.

There are two basic types of packed-bed reactors: those in which the solid is a reactant and those in which the solid is a catalyst. Many examples of the first type can be found in the extractive metallurgical industries. In the chemical process industries, the designer normally employs the second type, catalytic reactors. Industrial packed-bed catalytic reactors range in size from units with small tubes, a few centimetres in diameter, to a large-diameter packed bed. Packed-bed reactors are used for gas and gas-liquid reactions. Heat transfer rates in large-diameter packed-beds are poor, and where high transfer rates are required, fluidised-beds should be considered.

The essential feature of a fluidised-bed reactor is that the solids are held in suspension by the upward flow of the reacting fluid; this promotes high mass and heat transfer rates and good mixing. Heat transfer coefficients in the order of 200 W/m-°C for jackets and internal coils are typically obtained. The solids may be a catalyst, a reactant (in some fluidised combustion processes), or an inert powder added to promote heat transfer.

Although the principal advantage of a fluidised-bed over a fixed-bed is the higher heat transfer rate, fluidised-beds are also useful when large quantities of solids must be transported as part of the reaction processes.

Operational factors that contribute to waste and emissions in chemical reactors include incomplete conversion resulting from inadequate temperature control, by-product formation resulting from inadequate mixing, and catalyst deactivation resulting from poor feed control or purity control. Wastes are often generated in a chemical reactor if the reactor is improperly designed or the catalyst selection is not correct.

Heat exchangers

The transfer of heat to and from process fluids is an essential part of most chemical processes. The most commonly used type of heat transfer equipment is the shell and tube heat exchanger. The chemical process industries use four principal types of heat exchanger.

(1) *Double-pipe exchanger:* the simplest type, used for cooling and heating.

(2) *Shell and tube exchangers:* used for all applications.

(3) *Plate and frame exchangers (plate heat exchangers):* used for heating and cooling.

(4) *Direct or contact exchangers:* used for cooling and quenching.

The word *exchanger* applies to all types of equipment in which heat is exchanged, but is often used specifically to denote equipment in which heat is transferred between two process streams. An exchanger in which a process fluid is heated or cooled by a plant service stream is referred to as a heater or cooler.

One of the simplest and cheapest types of heat exchanger is the concentric pipe arrangement known as the double-pipe heat exchanger. Such equipment can be made up from standard fittings and is useful where only a small heat transfer area is required. Several units can be connected in series to extend their capacity.

The shell and tube exchanger is by far the most commonly used type of heat transfer equipment in the chemical and allied industries. The advantages of this type are :

(1) Large surface area in a small volume
(2) Good mechanical layout—that is, a good shape for pressure operation
(3) Reliance on well-established fabrication techniques
(4) Wide range of construction materials available
(5) Easily cleaned equipment
(6) Well-established design procedures

Essentially, a shell and tube exchanger consists of a bundle of tubes enclosed in a cylindrical shell. The ends of the tubes are fitted into tube sheets, which separate the shell-side and tube-side fluids. Baffles are provided in the shell to direct the fluid flow and increase heat transfer.

In direct contact heat exchange, there is no wall to separate hot and cold streams, and high rates of heat transfer are achieved. Applications include reactor off-gas quenching, vacuum condensers, desuperheating, and humidification. Water-cooling towers are a particular example of direct contact heat exchange. In direct contact cooler condensers, the condensed liquid is frequently used as the coolant.

Use of direct contact heat exchangers should be considered whenever the process stream and the coolant are compatible. The equipment is basically simple and cheap and is suitable for use with heavily fouling fluids and with liquids containing solids; spray chambers, spray columns, and plate and packed columns are used for this purpose.

Heat exchangers contribute to waste generation by the presence of cling (process side) or scale (cleaning side). This can be corrected by designing for lower film temperature and high turbulence.

Mass transfer equipment

Distillation

Distillation is probably the most widely used separation process in the chemical industry. Its applications range from the rectification of alcohol, which has been practised since antiquity, to the fractionation of crude oil. The separation of liquid mixtures by distillation is based on differences in volatility between the components. The greater the relative volatilities, the easier the separation. Vapour flows up a column and liquid flows countercurrently down the column. The vapour and liquid are brought into contact on plates, or packing. Part of the condensate from the condenser is returned to the top of the column to provide liquid flow above the feed point (reflux), and part of the liquid from the base of the column is vapourised in the reboiler and returned to provide the vapour flow.

In the stripping section, which lies below the feed, the more volatile components are stripped from the liquid. Above the feed, in the enrichment or rectifying section, the concentration of the more volatile components is increased. Fig. 1.1(a) shows a column producing two product streams, referred to as tops and bottoms, from a single feed. Columns are occasionally used with more than one feed, and with side streams withdrawn at points up the column (Fig. 1.1(b)). This does not alter the basic operation, but it does complicate the analysis of the process to some extent. If the process requirement is to strip a volatile component from a relatively non-volatile solvent, the rectifying section may be omitted, and the column is then called a stripping column.

In some operations where the top product is required as vapour, the liquid condensed is sufficient only to provide the reflux flow to the column, and the condenser is referred to as a partial condenser. When the liquid is totally condensed, the liquid returned to the column will have the same composition as the top product. In a partial condenser, the reflux will be in equilibrium with the vapour leaving the condenser. Virtually pure top and bottom products can be achieved by using many stages or, sometimes, additional columns.

Adsorption

In the adsorption process, one or more components in a mixture are preferentially removed from the mixture by a solid (referred to as the adsorbent). Adsorption is influenced by the surface area of the adsorbent, the nature of the solvent being adsorbed, the pH of the operating system (liquid application), and the temperature of operation. These are important parameters to be aware of when designing or evaluating an adsorption process.

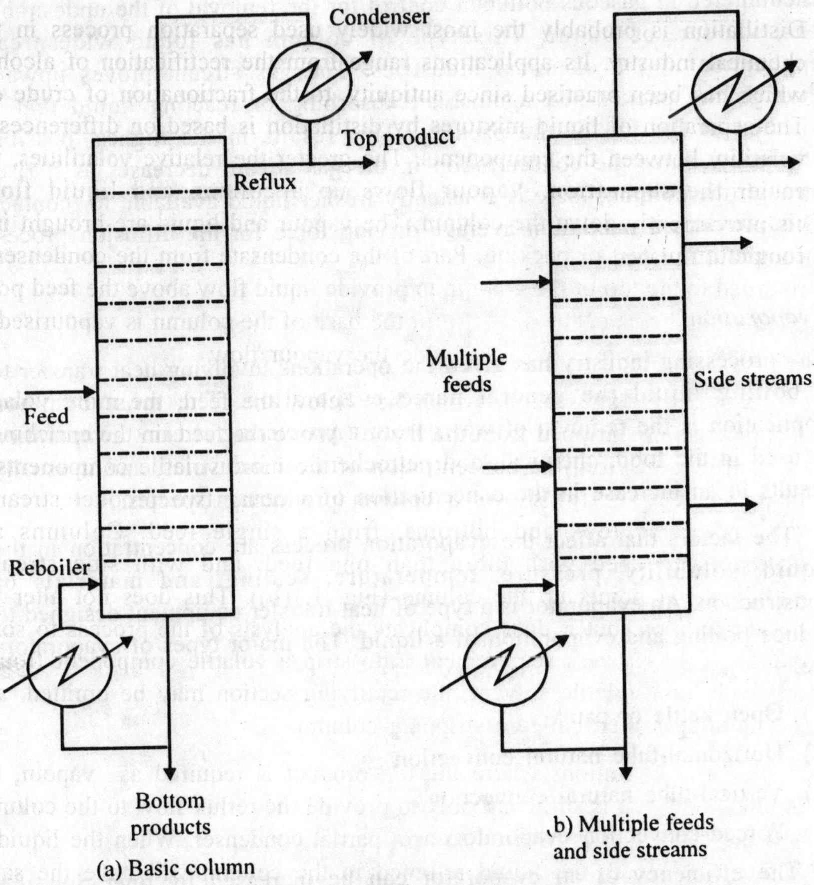

Fig. 1.1. Distillation column.

The adsorption process is normally performed in a column. The column is run as either a packed- or fluidised-bed operation. The adsorbent, after it has reached the end of its useful life, can either be discarded or regenerated. This operation can be applied to either a gas mixture or a liquid mixture.

Absorption

The process of absorption conventionally refers to the intimate contacting of a mixture of gases with a liquid so that part of one or more of the constituents of the gas will dissolve in the liquid. The contact usually takes place in some type of packed column.

Packed columns are used for the continuous contact between liquid and gas. The countercurrent packed column is the most common type of unit encountered in gaseous pollutant control for the removal of the undesirable gas, vapour, or odour. This type of column has found widespread application in the chemical industries. The gas stream moves upward through the packed-bed against an absorbing or reacting liquid that is injected at the top of the packing. This results in the highest possible efficiency. Since the concentration in the gas stream decreases as it rises through the column, there is constantly fresher liquid available for contact. This provides a maximum average driving force for the diffusion process throughout the bed.

Evaporation

The processing industry has given the operations involving heat transfer to a boiling liquid the general name evaporation. The most common application is the removal of water from a processing stream. Evaporation is used in the food, chemical, and petrochemical industries, and it usually results in an increase in the concentration of a certain species.

The factors that affect the evaporation process are concentration in the liquid, solubility, pressure, temperature, scaling, and materials of construction. An evaporator is a type of heat-transfer equipment designed to induce boiling and evaporation of a liquid. The major types of evaporators are:

(1) Open kettle or pan

(2) Horizontal-tube natural convection

(3) Vertical-tube natural convection

(4) Forced-convection evaporators

The efficiency of an evaporator can be increased by operating the equipment in single- or multi effect modes.

Extraction

Extraction (sometimes called leaching) encompasses liquid-liquid as well as liquid-solid systems. Liquid-liquid extraction involves the transfer of solutes from one liquid phase into another liquid solvent; it is normally conducted in a mixer-settler, plate and agitated-tower contracting equipment, or packed or spray towers. Liquid-solid extraction, in which a liquid solvent is passed over a solid phase to remove some solute, is carried out in fixed-bed, moving-bed, or agitated-solid columns.

Drying

Drying generally involves the removal from solids of relatively small amounts of water or organic liquids, whereas evaporation removes larger amounts. Drying removes the liquid as a vapour by warm gas (usually air) currents. Drying can be accomplished on a batch or continuous basis. The drying process is carried out in one of four basic dryer types. The first type is a continuous tunnel dryer. Trays with wet solids are moved through the system, in which warm air is blown over the trays. The second is a rotary dryer, which consists of an inclined hollow cylinder that rotates. The wet solids are fed in one side and hot air is passed countercurrently over the wet solids. The dried solids then pass out the opposite side. A drum dryer is a heat cylinder in which the wet solids are spread across the outside of the hot, rotating drum, are dried on this surface, and are then scraped off. The final type of dryer is a spray dryer. Here a liquid or slurry is sprayed through a nozzle, and the fine droplets are dried by a hot gas. This operation may be run cocurrently, countercurrently, or in some combination of the two modes.

Ancillary Equipment

This section considers devices for transporting gases and liquids to, from, or between units of process equipment. Some of these devices are simply conduits for the moving of material (pipes, ducts, fittings, stacks); others control the flow of material (valves); still others provide the mechanical driving force for the flow (fans, pumps, and compressors). This also covers storage facilities, holding tanks, materials-handling devices and techniques, and utilities (e.g., gas, steam, water) along with air, water, and solid waste control equipment.

Transportation equipment

Pipes, ducts, fittings, pump selection, valves, fans, compressors, and so on, may all contribute to waste generation. Pollution prevention and loss prevention can be implemented by the use of seal-less pumps, bellows-seals valves, and other specified equipment. Selection of proper equipment in the construction and design phase of a transport system is important. Some of the more important transport-type equipment is discussed below.

Pipes

The most common conduits for fluids are pipes and tubing. Both have circular cross-sections, but pipes tend to have larger diameters and thicker walls. Because of their heavier walls, pipes can be threaded, whereas tubing cannot. Process systems usually handle large flow rates that require the larger diameters associated with pipes.

Tubing and pipes are manufactured from many construction materials. The initial selection of the material depends on the corrosivity of the fluid and the flow (system) pressure. If special piping is required to accommodate corrosive liquids or high standards of purity, stainless steel, nickel alloys, or materials of high resistance to heat and mechanical damage are used. Steel pipe can be lined with tin, plastic, rubber, lead, or other coatings for special purposes. If problems of corrosion or contamination are controlling factors, the use of a non-metallic pipe such as glass, porcelain, thermosetting plastic, or hard rubber is often acceptable. There are several techniques used to join pipe sections. For small pipes, threaded connectors are the most common; for larger pipes, flanged fittings or welded connections are normally employed.

Ducts

Pipes and tubing are used as conduits for the transporting of liquids or gases; ducts are used only for gases. Pipes, with their thicker walls, can be used for flows at higher pressures; ducts are always thin walled and are generally used for gas flows close to ambient pressure. Pipes are usually circular in cross-section. Ducts come in many shapes (circular, oval, rectangular, etc.). In general, ducts are larger in cross section than pipes because the fluids are often transported with low densities and high volumetric flow rates. Ducts are most often constructed of field-fabricated galvanised sheet steel, although other materials such as fibrous glass board, factory-fabricated round fibrous glass, spiral sheet metal, and flexible duct materials are becoming increasingly popular. Other duct construction materials include black steel, aluminium, stainless steel, plastic and plastic-coated steel, cement, asbestos, and copper.

In pollution prevention applications, the corrosion resistance of duct materials deserves special consideration. Since material costs generally increase along with corrosion resistance, the selection of material must be determined by the desired life span in the anticipated environment; this environment is a function of the characteristics of the chemical being processed and the operating conditions of the process. For maximum resistance to moisture or corrosive gases, stainless steel and copper are used where their cost can be justified. Aluminium sheet is used where lighter weight and superior resistance to moisture are needed.

Fittings

A fitting is a piece of equipment that has one or more of the following functions:

(1) The joining of two pieces of straight pipe (e.g., couplings and unions).

(2) The changing of pipeline direction (e.g., elbows and Ts).

(3) The changing of pipeline diameter (e.g., reducers and bushings).

(4) The joining of two streams (e.g., Ts and Ys).

A coupling is a short piece of pipe threaded on the inside (some plastics are not) and used to connect straight sections of pipe. A union is also used to connect two straight sections but differs from a coupling in that it can be opened conveniently without disturbing the rest of the pipeline. When a coupling is opened, a considerable amount of piping usually must be dismantled. An elbow is an angle fitting used to change flow direction, usually by 90°C. A T (shaped like the letter *T*) can also be used to change flow direction but is more often used to combine two streams into one. A reducer is a coupling for two pipe sections of different diameter. A bushing is another connector for pipes of different diameter, but, unlike the reducer coupling, it is threaded on both inside and outside. The large pipe screws onto the outside of the bushing, and the smaller pipe screws into the inside. A Y (shaped like the letter *Y*) is similar to the T and is used to combine two streams.

Valves

Valves have one main function in a pipeline: to control the amount of flow. They can also redirect flow. There are many different types, but the most commonly used are the gate valve and the globe valve. The gate valve contains a disk that slides at right angles to the flow direction. This type of valve is used primarily for on-off control of a liquid flow. Because small lateral adjustments of the disk cause extreme changes in the flow cross-sectional area, this type of valve is not suitable for adjusting flow rates.

Unlike the gate valve, the globe valve is designed for flow control. Liquid passes through a globe valve via a somewhat circuitous route. In one form, the seal is a horizontal ring into which a plug with a slightly beveled edge is inserted when the stem is closed. Good control of flow is achieved with this type of valve, but pressure loss is greater than with a gate valve.

Some other types of valve are the check valve, which permits flow in one direction only; the butterfly valve, which operates in damperlike fashion by rotating a flat plate to either a parallel or a perpendicular position relative to the flow; the plug valve, in which a rotating tapered plug provides on-off service; the needle valve, a variation of the globe valve, which gives improved flow control; and the diaphragm valve, a valve specially designed to handle fluids such as very viscous liquids, slurries, or corrosive liquids that might clog the moving parts of other valves.

Fans

The terms fans and blowers are often used interchangeably, and no distinction is made between them in the discussion that follows. Fans are used for low pressure (drop) operation, generally below 2.0 psi. Blowers are generally employed when generating pressure heads in the range of 2.0 to 14.7 psi. Operations at higher pressures require compressors.

Fans are usually classified as either the centrifugal or the axial-flow type. In centrifugal fans, the gas is introduced into the centre of the revolving wheel (the eye) and is discharged at angles to the rotating blades. In axial-flow fans, the gas moves directly (forward) through the axis of rotation of the fan blades. Both types are used in industry.

Pumps

Pumps may be classified as reciprocating, rotary, or centrifugal. The first two are referred to as positive-displacement pumps because, unlike the centrifugal type, when they are operated the liquid or semiliquid flow is broken into small portions as it passes through these pumps.

Reciprocating pumps operate by the direct action of a piston on the liquid contained in a cylinder. As the liquid is compressed by the piston, the higher pressure forces it through discharge valves to the pump outlet. As the piston retracts, the next batch of low-pressure liquid is drawn into the cylinder and the cycle is repeated.

The rotary pump combines rotation of the liquid with positive displacement. The rotating elements mesh with elements of the stationary casing in much the same way that two gears mesh. As the rotation elements come together, a pocket is created that first enlarges, drawing in liquid from the inlet or suction line. As rotation continues, the pocket of liquid is trapped, reduced in volume, and then forced into the discharge line at a higher pressure. The flow rate of liquid from a rotary pump is a function of size and speed of rotation and is slightly dependent on the discharge pressure. Unlike reciprocating pumps, rotary pumps deliver nearly constant flow rates. Rotary pumps are used on liquids of almost any viscosity as long as the liquids do not contain abrasive solids.

Centrifugal pumps are widely used in the process industry because of their simplicity of design, low initial cost, low maintenance, and flexibility of application. Centrifugal pumps have been built to move as little as a few gallons per minute against a pressure of several hundred pounds per square inch. In its simplest form, this type of pump consists of an impeller rotating within a casing. Fluid enters the pump near the centre of the rotating impeller and is thrown outward by centrifugal force. The kinetic energy of the fluid increases from the centre of the impeller to the tips of the impeller vanes. This high velocity is converted to a high pressure as the fast-moving fluid leaves the impeller and is driven into slower moving fluid in the volute of the diffuser.

Compressors

Compressors operate in a similar fashion to pumps and have the same classifications: rotary, reciprocating, and centrifugal. An obvious difference between the two operations is the large decrease in volume resulting from the compression of a gaseous stream, as compared with the negligible change in volume caused by the pumping of a liquid stream.

Centrifugal compressors are employed when large volumes of gases are to be handled at low-to-moderate pressure increases (0.5–50 psi). Rotary compressors have smaller capacities and can achieve discharge pressures up to 100 psi. Reciprocating compressors are the most common type used in industry and are capable of compressing small gas flows to as much as 3500 psig. With specially designed compressors, discharge pressures as high as 25,000 psig can be reached, but these devices are capable of handling only very small volumes and do not work well for all gases.

Stacks

Gases are discharged into the ambient atmosphere by means of stacks (referred to as chimneys by some in industry) of several types. Stub or short stacks are usually fabricated of steel and extend a minimum distance up from the discharge of an induced draft fan. These are constructed of steel plate, either unlined or refractory lined, or entirely of refractory and structural brick. Tall stacks, which are constructed of the same materials as short stacks, provide a driving force (draft) of greater pressure difference than that resulting from the shorter stacks. In addition, tall stacks ensure more effective dispersion of the gaseous effluent into the atmosphere. Some chemical and utility applications use metal stacks that are made of a double wall with an air space between the metal sheets. The insulating air pocket created by the double wall prevents condensation on the inside of the stack, thus avoiding corrosion of the metal.

Process Diagrams

To the practicing engineer, particularly the chemical engineer, the process flow sheet is the key instrument for defining, refining, and documenting a chemical process. The process flow diagram is the authorised process blueprint, the framework for specifications used in equipment designation and design; it is the single, authoritative document employed to define, construct, and operate the chemical process. This type of diagram is also used in other processes and industries.

Flow sheets

Beyond equipment symbols and process stream flow lines, there are several essential constituents contributing to a process flow sheet. These include equipment identification numbers and names, temperature and pressure designations, utility designations, volumetric or molar flow rates for each process stream, and a material balance table pertaining to process flow lines. The process flow sheet may show additional information such as energy requirements, major instrumentation, and physical properties of the process streams. When properly assembled and employed, this type of process schematic provides a coherent picture of the overall process. It can point up some deficiencies in the process that may have been overlooked earlier in the study—for example, by-products (undesirable or otherwise) and recycle needs. Basically, however, the flow sheet symbolically and pictorially represents the interrelations among the various flow streams and equipment and permits easy calculations of material and energy balances. A number of symbols are universally employed to represent equipment, equipment parts, valves, piping, and so on. These symbols obviously reduce, and in some instances replace, detailed written descriptions of the process.

A flow sheet usually changes over time with respect to degree of sophistication and details. A crude flow sheet may initially consist of a simple, freehand block diagram offering information about the equipment only; a later version may include line drawings with pertinent process data, such as overall and componential flow rates, utility and energy requirements, and instrumentation. During the later stages of a design project, the flow sheet consists of a highly detailed piping and instrumentation diagram (P&ID).

In a sense, flow sheets are the international language of the engineer, particularly the chemical engineer. As already discussed, one can conceptually view a (chemical) plant as consisting of a series of interrelated building blocks that are defined as units or unit operations. The plant essentially ties together the various pieces of equipment that make-up the process. Flow schematics follow the successive steps of a process by indicating where the pieces of equipment are located and where the material streams enter and leave each unit. There are basically four types of schematic diagrams in general use.

(1) Block diagram.

(2) Graphic flow diagram.

(3) Process flow diagram.

(4) Process piping and instrumentation flow diagram.

Block diagrams

The block diagram is the simplest but least descriptive of the schematic diagrams. As the name implies, it consists of neat rectangular blocks which usually represent a single unit operation in a plant or an entire section of the plant. These blocks are connected by arrows indicating the flow sequence. The block diagram is extremely useful in the early stages of a process change and is particularly valuable in presenting the results of economic or operating studies since the significant data can be placed within the blocks. Four different types of operations are discussed below.

Increases in inventory represent accumulation; decreases represent depletion. An operation such as this is defined as a *pure inventory operation*. The flow diagram for this operation is shown in its most generalised form in Fig. 1.2(a). Note that if no chemical change takes place, a starting material can be physically separated into more than one product. This is by definition a *separation operation*. The generalised flow diagram is shown in Fig. 1.2(b). It would apply equally well to the recovery of the two components in an apparatus or the partial separation of an entering feed water into exiting steam and exiting hot water. Two or more material streams that are combined or assembled without chemical reaction is, by definition, an *assembly operation*. A generalised flow diagram is presented in Fig. 1.2(c). With more material streams involved, it would apply equally well to an automobile assembly line or the creation of an emulsion by the intimate mixing of oil, water, and an emulsifying agent. These three types of operation have one important common characteristic, the absence of chemical change. This sets them apart from the fourth type, the chemical operation. In Fig. 1.2(d) two reactants, A and B, are shown entering a chemical reaction producing products C and D. The flow diagram will vary with the number of reactants and products involved.

Graphic flow diagrams

The graphic flow diagram is used most frequently in advertising, company financial reports, and technical reports in which certain features of the flow diagram require extra emphasis. It should present the desired information clearly and in an eye-catching fashion that is both novel and informative.

Process flow diagrams

The process flow diagram, or PFD, is a pictorial description of the process. It gives the basic processing scheme, the basic control concept, and the process information from which equipment can be specified and designed. As described earlier, it provides the basis for the development of the P&I diagram (P&ID) and also serves as a guide for the plant operator. The process flow diagram usually includes:

(1) Material balance data (may be on separate sheets).

(2) Flow scheme equipment and interconnecting streams.

(3) Basic control instrumentation.

(4) Temperature and pressure at various points.

(5) Any other important parameters unique to each process.

Data on spare and parallel equipment are often omitted. Valving is also generally omitted. A valve is shown only where its specification can aid in understanding intermittent or alternate flows. Instrumentation is indicated to show the location of variables being controlled and the location of the actuating device, usually a control valve. To help the reader better understand the process flow sheet, a list of commonly used symbols for valves is presented in Table 1.3.

Process piping and instrumentation flow diagrams

The P&ID, which provides the basis for detailed design, offers a precise description of piping, instrumentation, and equipment. This key drawing defines the plant system, describes equipment, and shows all instrumentation,

Table 1.3. Valve symbols.

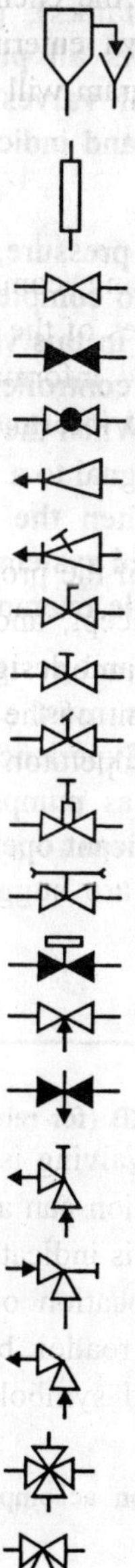

Symbol	Description
	Atmospheric exhaust head
	Silencer
	Gate valve
	Globe valve
	Lubrotite valve
	Check valve
	Stop check
	Plug valve
	Non-lubricated plug valve
	Quick-opening valve
	Self-draining valve
	Chain operated valve
	Reel valve
	Quench valve
	Needle or V-port valve
	Angle non-return valve
	Angle valve
	Angle check valve
	4-way valve
	3-way valve

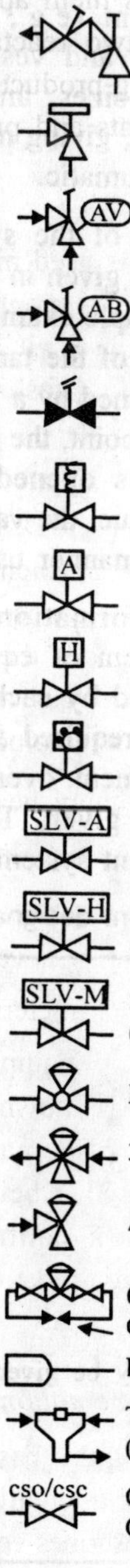

Symbol	Description
	Blow-off valves
	Varec vent valve
	Relief valve
	Vacuum breaker
	Atwood & Morrill straight thru relief valve on exhaust steam (& VE)
	Electric motor operated valve
	Air motor operated valve
	Hydraulically operated valve
	Solenoid valve
	Slide valve (air operated)
	Slide valve (hydraucally operated)
	Slide valve (manually operated)
	Butterfly valve
	3-way control valve
	Angle type control valve
	Control valve assembly Gate va. or globe va.
	Foot valve
	Tempering valve (Taco Type "A")
	CSO = car seal open CSC = car seal closed

instrumentation, piping, and valving. It is used to train personnel and aids in trouble-shooting during start-up and operation. The P&ID assigns item numbers to all equipment (e.g., towers, reactors, and tanks), gives dimensions of equipment and vessel elevations, and shows all piping, including line numbers, sizes, and specifications, and all valves. All instrumentation is covered, giving numbers, function, types, and indicating whether electronic or pneumatic.

A general knowledge of the symbols for flow, level, pressure, and temperature controllers, as given in Fig. 1.3(a–d), is needed to comprehend flow diagrams like the simple example presented in Fig. 1.4. In this vessel, with an inlet feed on top of the tank equipped with a flow controller, the level in the tank is maintained by a level-controlling device. When the level rises above the high level point, the level controller sends a signal to a valve actuator and the valve is opened, to drop the level. When the level approaches a specified value, the valve is closed. Complicated systems can be analysed in the same manner used in this basically simple example.

Certain essential information is often provided adjacent to the representation of each item of equipment. These notations must supply concisely the data required by each usher of the diagram. Experience has dictated the information required for common items such as pumps and vessels. For special equipment, overall dimensions and significant operating characteristics are often given. The following notations are suggested (Table 1.4). Any consistent system is satisfactory.

Table 1.4. Typical equipment designations.

Compressors	K, C
Exchangers	E, C (for condensers) and RB (for reboiler)
Heaters	H
Pumps	P or PU
Reactors	R
Storage tanks	ST
Towers	T
Vessels	V
Compressors	
(This information may be given on the flow diagram or on accompanying attachments).	
Service	
Stages	
Suction conditions	

(Contd...)

First stage suction
Second stage suction
Second stage discharge

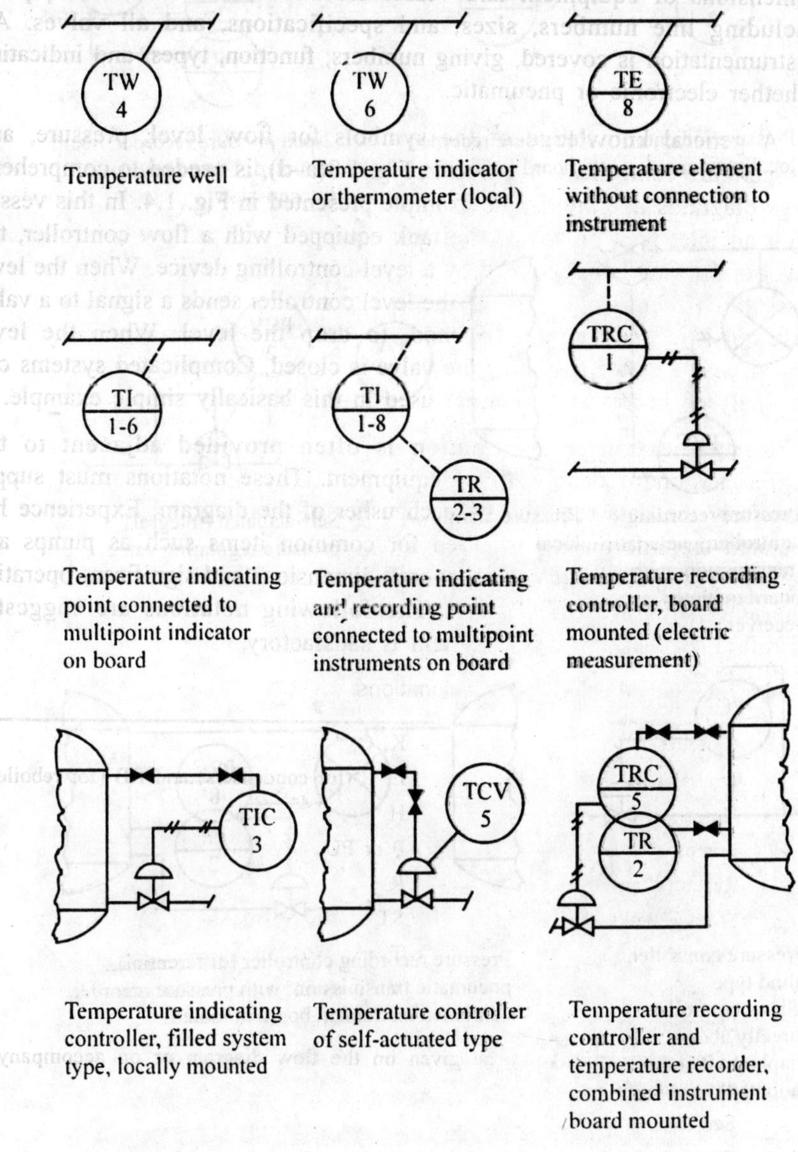

Fig. 1.3(a). Typical instrumentation symbols for temperature.

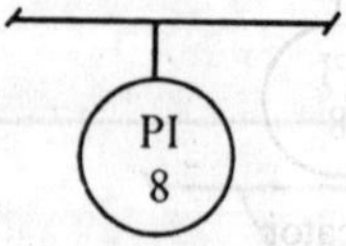

Pressure indicator, locally mounted

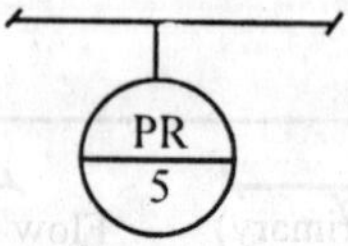

Pressure recorder, board mounted

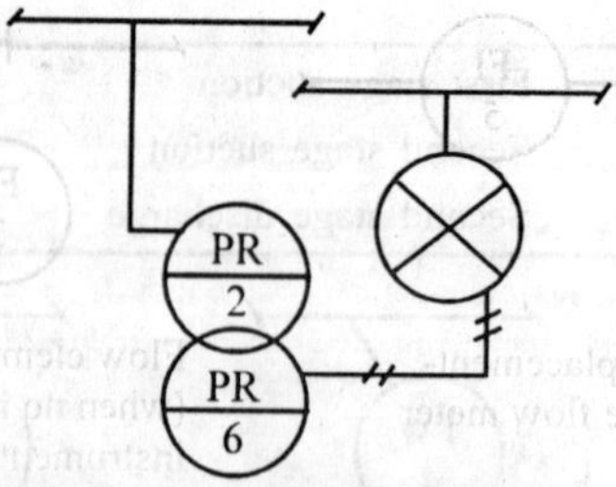

2-pen pressure recorder, board mounted, 1-pen pneumatic transmission

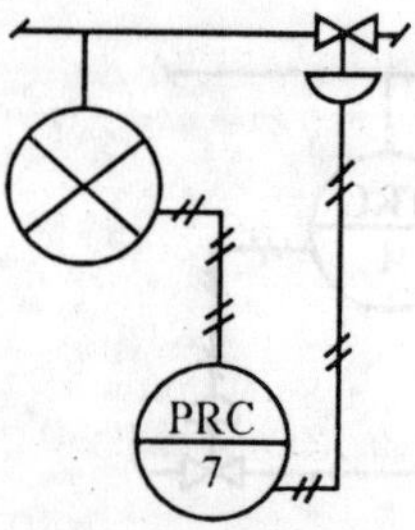

Pressure recording controller, pneumatic transmission, with board mounted receiver

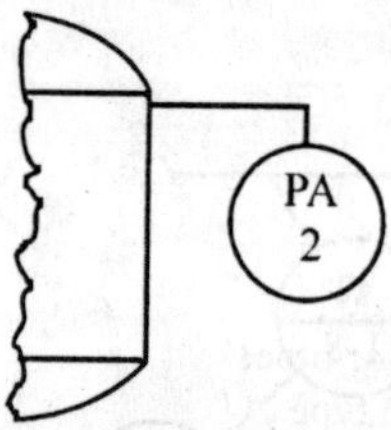

Pressure alarm local

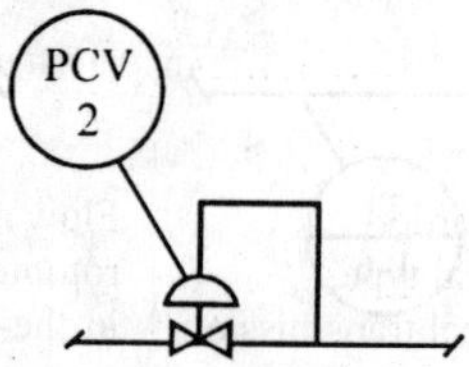

Self-actuated (integral) pressure regulating valve

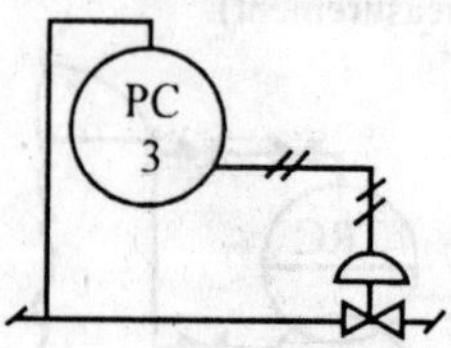

Pressure controller, blind type (Show controller directly above diaphram if so mounted)

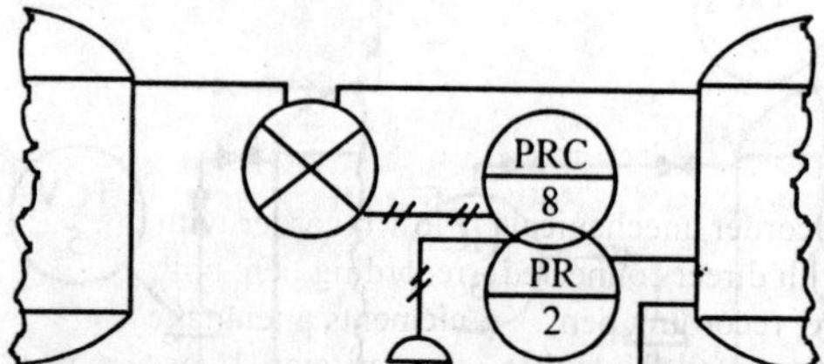

Pressure recording controller (differential), pneumatic transmission; with pressure recorder, combined instrument board mounted

Fig. 1.3(b). Typical instrumentation symbols for pressure.

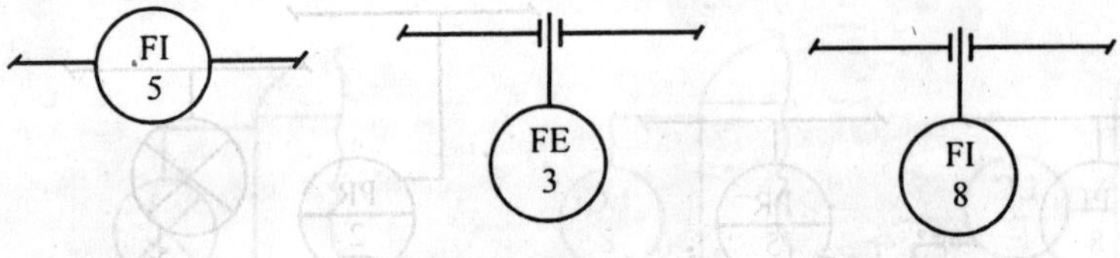

Displacement-type flow meter

Flow element (primary) (when no measuring instrument is provided)

Flow indicator differential type locally mounted

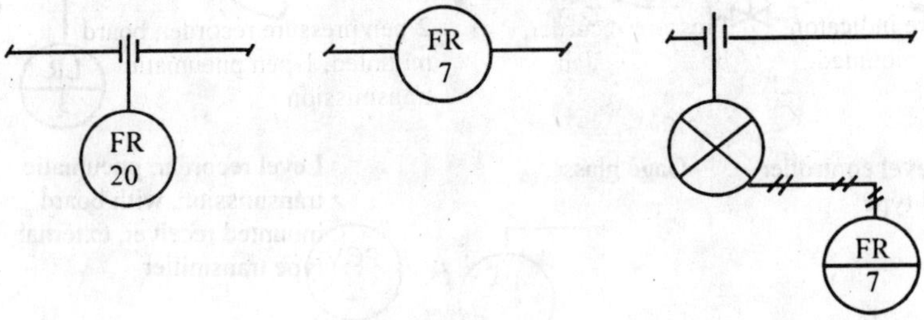

Flow recorder differential type, mechanical transmission, locally mounted

Flow recorder, of rotometer, or other in-the-time type

Flow recorder pneumatic transmission, transmitter local, receiver mounted on board

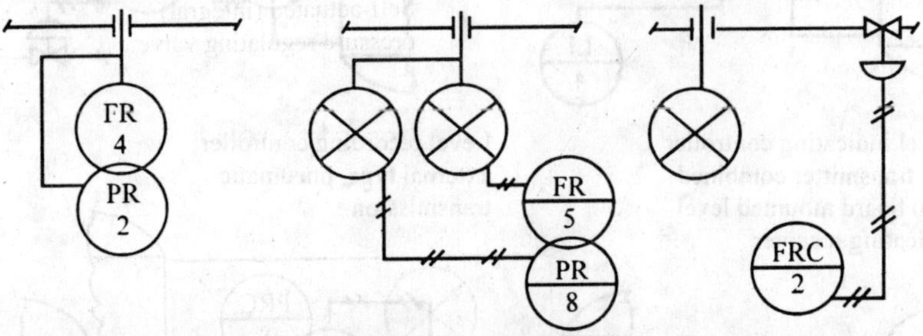

Flow recorder, mechanical type, with direct connected pressure, recording pen, locally mounted (Note that in listing such a combination item in specifications, etc. It would be written as FR-4 and PR-2, thereby treating each element as separate entity)

Flow recorder with pressure recording pen, both elements pneumatic transmission, transmitter local, and receiver board mounted (Receiver should be written as FR-5 and PR-2, and each transmitter identified by its own element)

Flow recording controller, pneumatic transmission with receiver mounted on board and local transmitter

Fig. 1.3(c). Typical instrumentation symbols for flow.

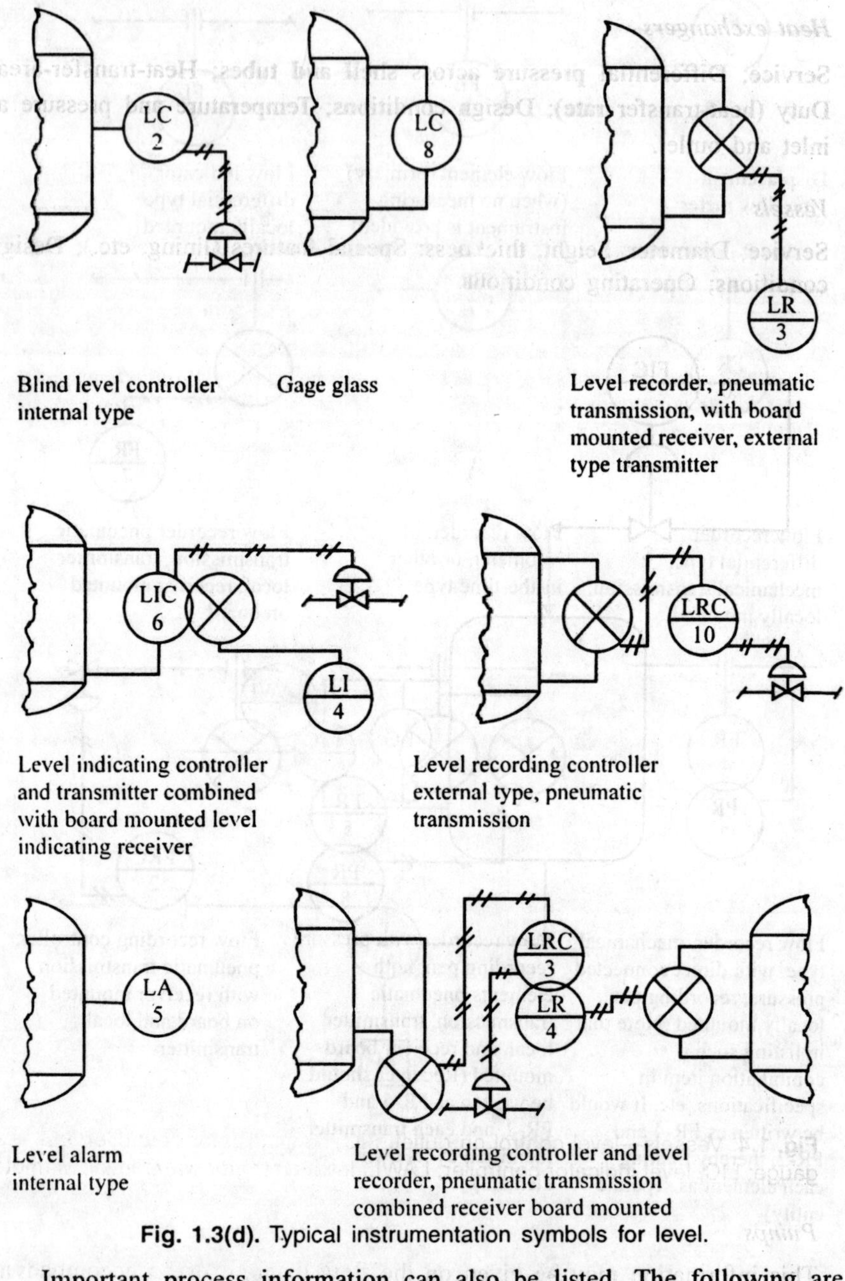

Blind level controller internal type

Gage glass

Level recorder, pneumatic transmission, with board mounted receiver, external type transmitter

Level indicating controller and transmitter combined with board mounted level indicating receiver

Level recording controller external type, pneumatic transmission

Level alarm internal type

Level recording controller and level recorder, pneumatic transmission combined receiver board mounted

Fig. 1.3(d). Typical instrumentation symbols for level.

Important process information can also be listed. The following are suggestions for several common items.

Heat exchangers

Service; Differential pressure across shell and tubes; Heat-transfer-area; Duty (heat-transfer rate); Design conditions; Temperature and pressure at inlet and outlet.

Vessels

Service; Diameter, height, thickness; Special features (lining, etc.); Design conditions; Operating conditions.

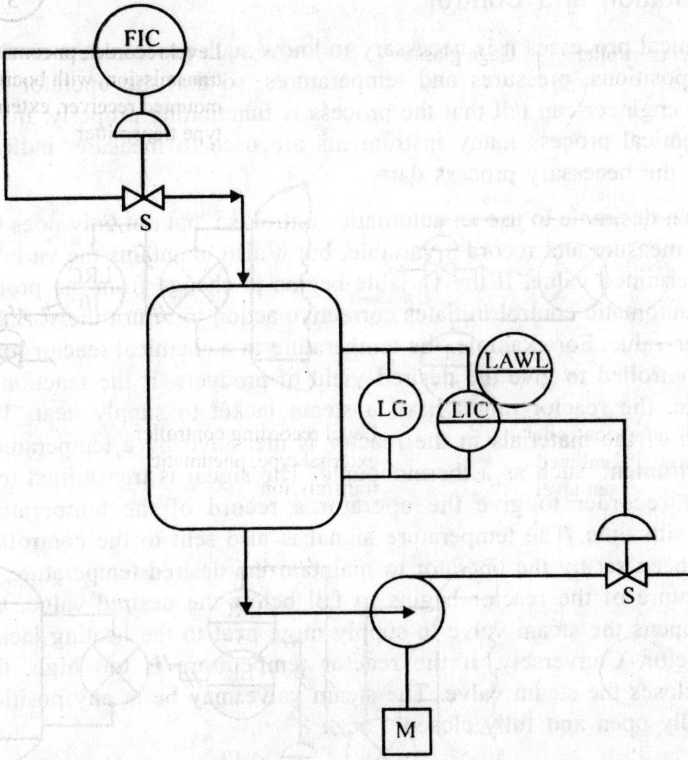

Fig. 1.4. Vessels—level control on outlet. FIC, flow indicator controller; LG, level gauge; LIC, level indicator controller; LAWL, level alarm for water level; M, motor.

Pumps

This information may be given on the flow diagram or on accompanying attachments.

Service
Size and type
Fluid
Pump temperature
Density at pump temperatures
Design GPM (gallons per minute) at pump temperature
Design differential
NPSH (net positive suction head).

Instrumentation and Control

In all chemical processes it is necessary to know such process data as flow rates, compositions, pressures and temperatures, so that the operator and production engineer can tell that the process is functioning properly. In the typical chemical process many instruments are used to measure, indicate and record the necessary process data.

It is often desirable to use an automatic control, so that not only does the instrument measure and record a variable, but it also maintains the variable at a predetermined value. If the variable begins to change from the proper value, the automatic control initiates corrective action to return the variable to its proper value. For example, the temperature in a chemical reactor must often be controlled to give the desired yield of products. If the reaction is endothermic, the reactor might have a steam jacket to supply heat. The temperature of the materials in the reactor is measured by a temperature-sensing instrument, such as a thermocouple. The signal is transmitted to a paper-chart recorder to give the operator a record of the temperature variations with time. The temperature signal is also sent to the controller, which has been set by the operator to maintain the desired temperature. If the temperature of the reactor begins to fall below the desired value, the controller opens the steam valve to supply more heat to the heating jacket on the reactor. Conversely, if the reactor temperature is too high, the controller closes the steam valve. The steam valve may be in any position between fully open and fully closed.

Automatic control may be applied to almost any process variable that can be measured. Control of a variable is often maintained by the measurement of another more easily measured variable. Automatic control reduces the number of human operators. It can give faster and more accurate control than a human operator.

The coupling of automatic controllers to electronic computers makes it possible to run a process with no human control. Such plants are now in operation. The engineer determines the best operating conditions, designs the process, controls and computer to give the optimum results. In operation, the instruments transmit process information to the computer. The computer uses the information to determine whether the process is operating at the best conditions. If it is not, the computer initiates corrective action through the control system.

Although automatic process control has eliminated many low-skill human operators, it is still necessary to have one or more highly skilled operators present to observe that the plant is functioning properly. It is also necessary to have a large highly trained maintenance staff to repair the process and control system when it breaks down. The more complex the control system, the greater the chance for breakdown. Of course, automated chemical plants are more expensive to build and require a much higher degree of knowledge on the part of the design engineer and the operating engineer.

Although the principles of chemical reactors, unit operations and instrumentation and control can be studied separately, in the design of a chemical process all these areas are closely interrelated. It is therefore necessary to consider the interaction of the components of a process to determine the overall behaviour of a process. This is often an extremely difficult and complex job. A consideration of process dynamics must be included in any complete process design.

Economics

No matter how efficiently a process operates to produce a final product of high purity, the process is a failure if the product cannot be sold at a profit. Before a chemical plant is built, a thorough market analysis is made to determine how much of the product can be sold and at what price. Often it is possible to sell more of the product if the price is lower. Present and future competition from other producers must be carefully evaluated. As the plant is designed, many economic analyses are made to determine the least expensive design which will produce the desired quantity of product at a minimum price. If the cost of manufacture is sufficiently below the selling price to give an attractive profit, the plant will be built. If the product is successful and profitable, a competitor may find the market attractive and enter it with perhaps a somewhat better product produced at a lower price by an improved process. It is then necessary for the older producer to improve his process and product, or he will be forced out of the market.

WHAT IS CHEMICAL ENGINEERING?

Chemical engineering is a synthesis of chemistry and engineering. Although the profession grew out of industrial chemistry, today it is based as much on physical principles as on chemical fundamentals. There are many definitions of chemical engineering. Many of them are either too specific or too vague. One popular definition — "a chemical engineer carries out on a large scale reactions developed in the laboratory by the chemist," is true as far as it goes but it does not show the broad scope of problems encountered by the chemical engineer. The unit operations are not obviously a part of this definition. The typical chemist has little knowledge of any of the unit operations in theory and practice. A facetious definition of chemical engineering makes a point, "A chemical engineer is one who talks engineering in the presence of chemists, chemistry in the presence of engineers and politics in the presence of both." A unique characteristic of the chemical engineer is that he can talk to and understand both chemists and other engineers. He must understand the research findings of the chemist and he must be able to use engineering fundamentals to design and operate a new chemical process based on these research findings. It is this versatility that has made the chemical engineer indispensable in the chemical process industry.

The official definition of the American Institute of Chemical Engineers is sufficiently broad to include all aspects of the profession. According to this definition, chemical engineering is "the application of the principles of the physical sciences, together with the principles of economics and human relations, to fields that pertain directly to processes and process equipment in which matter is treated to effect a change in state, energy content, or composition."

Chemical engineering is unique among engineering disciplines. It requires a thorough understanding of chemistry, physics and mathematics, whereas the other fields of engineering usually do not require a thorough understanding of chemistry. For this reason chemical engineering will probably always remain somewhat apart from both chemistry and the other branches of engineering.

WHAT DOES A CHEMICAL ENGINEER DO?

As the chemical industry grew in size and complexity, the work of the chemical engineer required an ever-growing background of education and experience. Today a chemical engineer may work in one of many fields within the profession of chemical engineering. Some of the fields require special training beyond an undergraduate curriculum. The additional training may be formal graduate work or on-the-job training. The major areas of work within chemical engineering are given in the following list :

(1) Research
(2) Process development
(3) Process design and evaluation
(4) Plant design
(5) Construction
(6) Production supervision
(7) Plant technical service
(8) Product sales

The fields are interrelated and often overlap. A chemical engineer may work in a single field with a large industrial organisation, or he may work in several areas simultaneously, especially in a small company. Transfer from field to field is very common, as company needs and the engineer's interests change. In all his work the chemical engineer works closely with specialists in chemistry and other fields of engineering and pure science.

Research

Chemical engineering research may be divided into fundamental, exploratory, and process research.

Fundamental research

Fundamental research is carried out in many areas of chemical engineering. Such research results in the development of new knowledge of the principles of the unit operations, industrial reaction kinetics and chemical process control. In areas subject to experimental analyses, fundamental research is used to develop a new theory and to test it experimentally. For example, in fluid mechanics, turbulent flow is currently the subject of intensive theoretical and experimental study to develop an understanding of the fundamental physical mechanism. The findings of fundamental research are frequently immediately useful, but the research programme is initiated to increase the general knowledge rather than for specific applications. The chemical engineer working in fundamental research must have an excellent background in physics, mathematics, chemistry, as well as the principles of chemical engineering. He often specialises and becomes an expert in one area, such as mass transfer.

Exploratory research

Exploratory research is less frequently the responsibility of the chemical engineer. More typically, it is the work of the chemist. Here, the object is to find a particular reaction which may have commercial possibilities. For example, an oil company may ask a group of research chemists to study the reactions of hydrocarbons found in oil using various catalysts. The object may be to find a particular catalyst, reaction temperature, and pressure which will yield a product of higher octane number than the starting material. The conversion of hydrocarbons to other hydrocarbons of higher octane is called *reforming*. In the commercial process the starting material is naphtha, a mixture of hydrocarbons obtained by the distillation of crude oil. It is a complex mixture and many reactions occur. In exploratory research the chemist may investigate several of the pure compounds which are present in the mixture to determine the reaction of the single component. For example, cyclohexane is a common constituent of naphtha. It has an octane number of 78.6, too low for modern gasolines. It can be reformed under proper conditions to benzene, which has an octane number of 113.6. The reaction is :

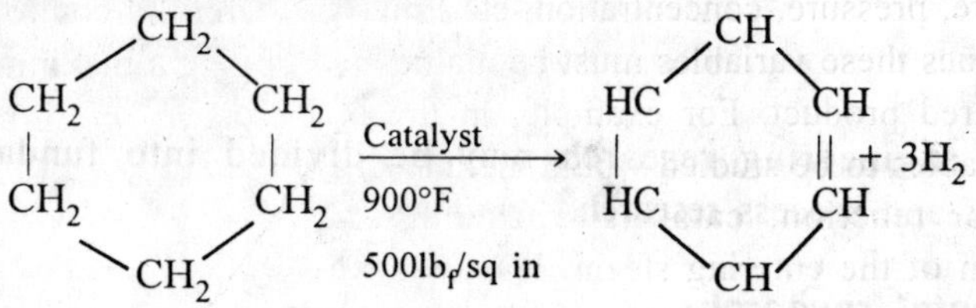

The reaction equilibrium and rate give perhaps a 90% conversion of cyclohexane to benzene at the conditions indicated. Other catalysts and conditions give different conversions. Another reforming reactions is isomerisation. For example, normal heptane (octane number, 0) may be converted to an isomeric heptane with a higher octane number :

$$\underset{\text{Normal heptane}}{CH_3CH_2CH_2CH_2CH_2CH_2CH_3} \xrightarrow[\substack{900°F \\ 500\ lb_f/sq\ in}]{\text{Catalyst}} CH_3\overset{\displaystyle CH_3}{\underset{\displaystyle CH_3}{C}}CH_2CH_2CH_3$$

Isoheptane
(2,2-dimethylpentane)
Octane No. 93

Both the reactant and product contain 7 carbon atoms and 16 hydrogen atoms; they are therefore both isomers of heptane.

The exploratory-research group would try many catalysts and various operating conditions on a small laboratory scale to explore a wide range of possibilities. The research programme may extend over several months or even years. Although many of the attempts prove infeasible, a few results may be commercially promising and these will be passed on to the process-research group for further study.

Process research

Process research takes promising results from exploratory research and intensively studies them on a *bench scale* to determine their commercial feasibility. Process research determines operating conditions for a commercially feasible process, yields data for a preliminary economic evaluation and provides information for design of a pilot plant for further development of the process. Starting with the results of exploratory research, the process-research group will study the process under a wide range of operating conditions in an attempt to discover what variables (temperature, pressure, concentration, etc.) must be carefully controlled and at what values these variables must be maintained to give a maximum yield of the desired product. For example, in the hydrocarbon reforming cited above, variables to be studied would include the temperature, pressure, time allowed for reaction, catalyst type and size, and the flow rate and composition of the entering steam. It may be found that the best yields of higher-octane product are obtained at 900°F and 500 lb_f/sq in. pressure, but that the entering hydrocarbon must be diluted with hydrogen to prevent decomposition to carbon at this elevated temperature. The process-research group would study not only pure hydrocarbons, but also naphtha, the complex mixture of hydrocarbons which would be the feed in a commercial process.

A flow sheet for the bench-scale reactor needed for process research on reforming is shown in Fig. 1.5 The unit is small; the reactor may have a volume of only 100 cm^3 of catalyst. The ultimate pilot plant and commercial installation would have continuously-flow reactors; therefore, the bench scale has continuous feed of hydrocarbons and hydrogen, continuous reaction and continuous withdrawal of products. The hydrocarbon-hydrogen feed is a gas. The gaseous products are condensed to a liquid after the reactor and are separated from the remaining gas, which is mostly hydrogen with some light hydrocarbons.

The process-research group must obtain sufficient quantitative data to make possible a preliminary economic process evaluation. Economic evaluations are made at each step in the development of a process. Because the cost of each step is substantially greater than the previous step, the process must look continually more promising as the work progresses. Exploratory research is *relatively* inexpensive; process research may be more expensive because of more complex equipment requirements and greater operating costs. The next step, process development with a pilot plant, is very expensive, because the equipment is closer to an actual commercial plant. If the results of process research are economically promising, enough data must be obtained to permit design of a pilot plant for process development.

Process Development

After process research has demonstrated the chemical feasibility of a new process and a preliminary economic process and market evaluation has indicated a satisfactory profit, a decision must be made concerning the extent of the process-development programme.

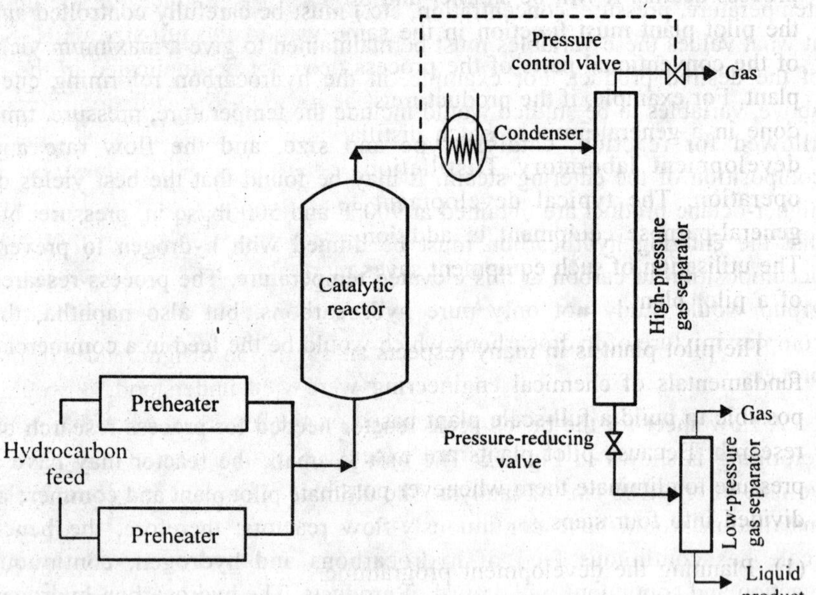

Fig. 1.5. Bench-scale catalytic reactor for process research on hydrocarbon reforming.

It is seldom possible to skip process development and proceed directly from process research to the design and construction of the full-size plant. The large uncertainties regarding process operating conditions and product yield almost always required the construction of a semiworks or pilot plant. The pilot plant is a form of insurance. Although it is expensive to build and operate, it may save much more money by eliminating uncertainties in the construction and operation of the commercial plant. Any difficulties in start-up or operation of the full-size plant results in a great monetary loss in the profit that would have been made on the product.

In addition to supplying process information, the pilot plant may be required to produce the new product for market research. In many cases it is necessary to supply the prospective customer with samples of the new product, so that he may test it and determine whether it satisfies his needs. In this way, a market for the new product is established before the commercial plant starts up. The pilot plant must yield a product which is identical to the expected product of the proposed full-scale plant. Therefore, the pilot plant must duplicate the proposed full-size plant in all important process details. It will be smaller and produce less product, but the units of the pilot plant must function in the same way as the full-size units. Some of the conventional parts of the process need not be duplicated in the pilot plant. For example, if the product must be purified by distillation, it may be done in a general-purpose batch distillation column usually available in a development laboratory. Distillation is a fairly well-understood unit operation. The typical development laboratory has a large variety of general-purpose equipment in addition to its special-purpose pilot plants. The utilisation of such equipment saves money and time in the construction of a pilot plant.

The pilot plant is in many respects an admission of ignorance. If all the fundamentals of chemical engineering were well understood, it would be possible to build a full-scale plant based on the results of extensive process research. Because pilot plants are expensive and time-consuming, there is pressure to eliminate them whenever possible. Process development may be divided into four steps :

(1) Planning the development programme.

(2) Designing and building the pilot plant.

(3) Operation of the pilot plant.

(4) Correlation, presentation and evaluation of the data from the pilot plant.

Process Design and Evaluation

The process-design engineer is concerned with the design of the overall chemical process. Generally, he is not responsible for the detailed design of pieces of equipment which is normally the responsibility of the project engineer in plant design. Because the process-design engineer is interested in producing the desired product at as low a cost as possible, he must examine many alternative process steps to determine an economic optimum. Economic evaluation is an integral part of process design. It may be carried out by the process design engineer or by an economic-analysis group working closely with him. A process design includes the following major items :

(1) Process flow sheet showing all pieces of equipment, instrumentation and control and the operating pressures, temperatures and flow rates.

(2) Mass balances on the overall process and on each unit in the process, showing yields of products and purity of all streams in the process.

(3) Energy balances for all units in the process, including heat-exchanger requirements.

(4) Specification of pump capacities, flow and pressure requirements.

(5) Specification of size and configuration of chemical reactors and storage tanks.

(6) Determination of optimum operating conditions for the mass-transfer operations required for the separation and purification of the raw materials and products.

(7) Estimation of utility requirements, such as steam, water, electricity and fuel.

(8) Economic evaluation with an estimate of capital investment and operating costs.

The process-design engineer utilises all the data of process research and development. He works closely with the development engineer to determine the most economical processing units and the optimum operating conditions. Even with all information at hand, the process-design engineer must use his judgement in filling gaps in data. He never has all the data he needs; he must estimate many quantities, using his previous experience as a basis. The process engineer must be well-grounded in the fundamentals of chemical kinetics and unit operations; he must exercise his imagination and judgement to design a process with often incomplete data. His work involves numerous careful calculations, sometimes using available digital computers for the longer routine calculations.

Plant Design

The plant design group translates results of the process design engineer into complete plans and specifications which will be used by the constructors to build the plant. The plant design must be complete in every detail. It is used to make a firm estimate of the cost of the plant and it may serve as a basis for a contract between the chemical company and the construction firm.

The plant design group may include chemical, mechanical, electrical and civil engineers. The group is supervised by a *project engineer* who is frequently a chemical engineer with a knowledge of the overall process. He must coordinate the activities of the various specialists in his group; he carefully analyses the data supplied to him by the process design engineer and he may make suggestions for modifying the fundamental process which result in substantial savings. The process engineer and the project engineer work closely in analysing such suggestions. The project engineer must concern himself with peripheral problems such as supply of water and other utilities, waste disposal and safcty. IIc may be the only person who has a thorough understanding of and responsibility for the entire design.

Working closely with the project engineer are the *designer* and *draftsman*. The designer is often a specialist in a particular phase of plant design. For example, after a chemical engineer has determined the number, size and spacing of plates in a distillation column, a mechanical designer may specify the physical details of the column, the electrical designer may specify the location and type of instrumentation and control and the structural designer considers the support framework and foundations for the column and auxiliary equipment. The designers make suggestions to the project engineer on specific points where money might be saved. The designers supervise the draftsmen who make the detailed drawings of each unit of the process.

Because of the complexity of a modern chemical plant, some design groups use small-scale models of the plant to study the physical layout of equipment and piping. Such models often bring out weaknesses in layout which would make the plant difficult to operate, maintain, or repair. The project engineer is ultimately responsible for all these details.

The project engineer works closely with the contractor who is building the plant. Many questions arise during construction—materials specified may be unavailable in the time-limit set; further pilot-plant work shows that a change in one unit is essential; the contractor may suggest changes in foundation because of unexpected soil conditions. The project engineer is also present at plant start-up, the completion of his months or years of work.

The young engineer may start out as an assistant in the plant-design group. If he shows interest and aptitude, he becomes a designer. After further experience, he may become a project engineer. Many draftsmen and some designers are not college graduates, but they are highly skilled technicians who make a career in drafting or designing.

Many of the pieces of equipment for the plant, such as pumps, heat exchangers and instrumentation, will be supplied by equipment vendors. An equipment vendor is a company that specialises in the design and construction of a particular type of equipment. The vendor may build a unit to the plant design group's detailed specifications, or he may suggest a standard unit which the company makes. For example, the plant design group specifies the pumping requirements and the vendor suggests a specific pump which will do the job. The vendor employs many engineers in the development, design and sale of his equipment.

Construction

The role of the chemical engineering in the construction field is limited. A general background in engineering is helpful, but it need not be chemical engineering. The construction supervisor is responsible for completing the plant in the shortest time within the allotted budget. He must establish a construction schedule which is realistic and then see to it that the schedule is maintained. He must set up equipment delivery schedules and he must expedite construction in his own shops. The construction supervisor must carefully schedule manpower requirements, keeping in mind the various craft union regulations. He must maintain good labour relations to avoid poor workmanship, slowdowns, or complete work stoppages.

After the plant is completed, the construction engineer tests the equipment to be sure that it meets specifications. He is available for consultation on plant start up. Frequently, a company may buy a complete process which has been developed by another company in the same field or by one of the chemical design and construction firms. These firms employ many chemical engineers in all fields.

Production Supervision

The first job of the production supervisor is to get his new plant running. After days, weeks, or months of intensive work, the plant is finally running smoothly and producing the proper quantity and purity of product. The plant is under automatic control and a few operators occasionally check the instrument panel to be sure everything is functioning properly. The production supervisor checks the daily record. With the plant running smoothly, what more is there for the production supervisor to do? His job has just begun. He tries to improve operation. He is not content with meeting design, specifications; he wants to do better. Because there is always an element of the unknown in any design, the production supervisor knows that there is room for improvement. He adjusts process variables to give optimum conditions; he improves product quality by removing contamination and reducing deterioration; he reduces steam, water, power and materials requirements by careful operation; he reduces labour costs by good labour relations, efficient methods and workable safety practices; he develops efficient maintenance procedures to ensure minimum shutdown time for routine repairs; he sets up a procurement schedule to maintain adequate inventories of raw materials.

After a sustained and long-range effort, the production supervisor and his staff will have the plant running "perfectly". Usually most of the plant components are overdesigned to ensure their doing the job for which they are built and often they can process more. One unit which is designed more closely to the minimum will create a bottleneck when an attempt is made to increase production. The production supervisor and his staff must find and eliminate the bottleneck, often by the addition of equipment. Careful analysis can increase plant capacity substantially and bring greater profits. The production supervisor can see his profits directly in more efficient operation and additional production.

The process-development group continues to work on the pilot plant and may develop an important modification to produce a better product more cheaply. The production supervisor works closely with process development and design in modifying the plant. Process development may come up with a totally new process that economically justifies shutting down the existing process after several years of operation. A competing new product may capture the market and force shutting down the old plant. After years of intensive effort, the production supervisor must be ready to abandon the old plant and move on to a new one.

Because the problems of the production supervisor and his staff of production engineers cover a very wide range, they need a broad background in engineering, rather than specialised training. This broad background is gained by experience. The graduating chemical engineer may start as an assistant production engineer in a small area of the process. As he gains experience, he will move up to production engineer, assistant supervisor, supervisor and perhaps ultimately to plant manager.

Plant Technical Service

The plant technical service group assists the operating engineer in the solution of his start-up and operation problems. This group serves the plant as a whole. When a problem develops in one area, they move in, analyse the problem and suggest a solution to the operator. Their work is identical to the more technical and less routine duties of a production engineer. The boundary of responsibility between the technical service group and the production supervisor varies greatly among companies. Some companies consider technical service to be a part of production and have no over all technical service group. At the other extreme, the production supervisor is a technician with long experience, who turns all his technical problems over to the plant technical service group. He continues to handle routine problems. In between these extremes is a range of division of technical problems between production engineers and technical service engineers. In some cases the technical service group may only make suggestions; the final decision is the production supervisor's. In others the technical service supervisor may initiate changes, especially where the operator is not an engineer.

The technical service group is extremely important in start-up of a new process. Often unusual and complex problems are encountered. Rarely does a large chemical process start without a hitch. The technical service engineer works closely with process development, process design and plant design in starting up a plant. Minor design and construction errors must be corrected. The process may operate but produce a product of inadequate quantity and purity. Start-up may require weeks or even months before the plant is functioning properly. The engineers involved in start-up need a wealth of theoretical and practical knowledge to overcome the difficulties encountered.

Product Sales

The ultimate economic justification of a chemical process is the sale of its products. For the present purposes, product sales may be divided into four closely related areas of interest to chemical engineers—market research, product development, technical sales and customer technical service. If the chemical engineer chooses product sales as his field, he will work closely with business specialists in such fields as economics, marketing and sales.

Market research

The first step in new product sales, begins long before the new process is in operation. It answers the fundamental question, "Will it sell?" The market research group begins its analysis as soon as a promising result is reported by the exploratory research group. If the proposed product is completely new, market research contacts potential users to determine their needs and thereby establish whether a market exists. If management decides to continue the project on the basis of this preliminary market survey, the pilot plant may produce sufficient product to supply samples to potential customers so that they may test the product for their applications. If the proposed product is one already on the market, market research determines how much more of the material could be sold if produced. New uses for the existing product are explored. Market research continually surveys the chemical market to supply management with facts on general trends in new products. It may suggest areas of possible economic return to the exploratory research group.

Product development

Product development finds uses for new products and new uses for established products. It is a form of applied research and may require the solution of complex chemical and engineering problems. It is considered a part of product sales because it is more concerned with the sale of the product than it is with its manufacture. The product development groups assists market research by suggesting and developing new uses for a product; it assists the technical salesman by developing a modified product for the use of a particular customer; it assists the customer service group by suggesting processing methods which the customer might use with the product. Some of the work in developing uses for products is long range and exploratory. Other product development problems need immediate answers. If the problem is a major one, the product development engineer consults with his research department and the customer to work out a solution. For example, a customer may require a very high purity product for his application. The usual product may not be sufficiently pure. The product development engineer will work out means of purifying it either before or after it is sold to the customer. On the other hand, he might suggest a change in the customer's process to eliminate the need for the high purity, thereby saving the customer money and selling the less pure product.

Technical sales and customer technical service

Technical sales and customer technical service are inseparably related. The same engineer may act in both capacities. The technical salesman sells his product on the basis of its proved technical superiority. He must show the customer that it will satisfy the requirement. If the customer tries the product, the technical salesman must retain the market by helping his customer solve any problems that arise in the use of the product. This assistance is customer technical service. Some companies have special groups that assist the salesmen in this area; others expect their salesmen to handle the customer's technical problems. The technical salesman may call on the product development group to answer customers' questions. The product development group is in this sense involved in customer technical service. Often a product is sold because of the service a company is willing to supply to the customer.

Most of the work in product sales involves contact with customers. For this reason, the engineer interested in working in this area must have the interest and personality to work with people. Although a pleasantly impressive personality is an asset in sales, it is the salesman's technical knowledge of his product that keeps the customer satisfied so that he continues to buy.

CHEMICAL ENGINEER IN THE COMING YEARS

Chemical engineering, improving the lifeline, safety, health, energy and environment, is faced with very serious macroeconomic problems to which too little attention is given. The serious macroeconomic problems perceived by an engineer are as follows :

(1) Resources, particularly energy and feedstock for the fertiliser and heavy chemical industry.

(2) Infrastructures for transportation, energy and telecommunications.

(3) Protection of the environment.

(4) Development of agriculture and agro industries.

(5) Transformation of the rural economy, industrialisation and privatisation.

There are a host of other serious problems of less technical content interwoven with the problems listed above. The major politico-economic internal and external problems are as follows :

(1) The Centre versus the States.

(2) Command economy versus liberalisation and privatisation—or the often misunderstood market economy.

(3) Internal (budget) and external balances.

(4) World trade globalisation and India.

(5) Most important and difficult is the overriding problem of Indian competitiveness.

If these problems are not solved progress would at best be erratic. These major technical and economic problems will have to be addressed first. Lack of transport infrastructure and transport fuel blocking interaction with the World Trade Community, lack of electric power may put a nation in uproar. These problems are to be assessed and analysed before they are solved. This is a task for the engineers and economists working together. The collaboration of professional societies like the Institutes is must to help and identify concrete problems and ensure that competent task forces, etc., are set up to deal with them in a way that is realistic and well anchored scientifically. The next task is to identify alternatives for future scenarios and select acceptable ones.

It is urgent to determine the role renewable energy can economically play. Hydropower is the main energy source. Windpower, is not being utilised at all. Solar cells are not much economical until significant technical progress brings investment down. Biomass is also not utilised to its fuller extent. In such a scenario, the options are to use coal, natural gas and oil—or go nuclear. As hydrocarbon resources are limited, either we have to reduce energy consumption or go backwards.

Feedstock for fertilisers is an important problem. Here renewable resources are next to useless. By far the preferred feedstocks are hydrocarbons particularly natural gas where we have to compete with the power industry, which really, for long-term resource conservation reasons, should not be using gas. There is another connected problem, transport fuel, where diesel and gasoline for road vehicles may be partly substituted with electric power and chemicals derived preferably from natural gas. Efforts should be made to use hydrogen converted in fuels cells to electric power driving the vehicle.

For the near future, hydrocarbons are critical because of reduced costs of investments, in industrial plants or vehicles, as compared to coal based facilities. The modest quantity of proven hydrocarbon reserves, for natural gas some 30 trillion cu. ft., may exhaust shortly, so drilling should be increased to search for more gas. Nuclear energy, an ecologically attractive alternative is today of little importance, and obstacles such as public opposition and the high investments required stand in the way.

With this background, the urgent necessity to formulate an updated energy policy is obvious. It must be supplemented by all necessary programmes to implement it. This can only be achieved in collaboration between engineers and economists, on the one hand, and political leaders, on the other. The programmes can focus on the first 10–20 years, for which forecasts for developments in technology are reasonably possible, and they should show the impacts on society including investments. Chemical engineers are well placed to make a major contribution.

Protection of the environment, always and everywhere is in the forefront of debate. Pollution in certain regions is difficult to tolerate, and solid waste and air and water pollution pose big problems. Technically, such problems are not too difficult to solve. Chemical engineers have developed solutions for air and water pollution, but implementation in India, although relatively inexpensive, is not easy to give priority as they are only rarely justified by an economic return. It is, however, very important to note that removal of pollutants with strongly perceived health and welfare hazards is relatively cheap. For instance, sulphur in refineries, obnoxious compounds in industrial exhaust can be controlled with little financial involvement. Giving priority to remedial actions which gives the largest improvement to air purity reduces the investment problem very significantly.

The role of the chemical engineer is evident but the problem of developing laws, standards and trade-offs between perceived or analytical air purity and investments is a problem for government and private institutes, etc., and health hazards pose a major challenge to the medical profession. Dealing with "trade offs" between health risks and the cost of air cleaning is indeed a difficult task for politicians. But there are some serious environment problems—the greenhouse effect, global warming, the CO_2. It is also impossible to determine, whether the many possible effects and particularly warming are good or bad—globally and locally. For instance, El Nino or the Gulf stream and increase in the level of CO_2.

From the changing scenario of the global environment it seems reasonable to assume that a drastic change in the properties of the atmosphere caused by increase in the content of CO_2 and multiplication in the contents of some other greenhouse gases can have a significant impact on the conditions under which man live. What is worse, it appears that an effect could be autocatalytic or self-accelerating. This can be explained by the interaction between the ocean and the atmosphere.

If the oceans are heated up, they will lose part of their CO_2 content to the atmosphere causing further warming which, in turn, will cause further loss of CO_2 from the oceans to the atmosphere. We must assume that greenhouse gases, and notably CO_2 and other gases may cause unacceptable changes. There is no point in stopping the rate of increase in energy use to reduce CO_2 emission. A chemical engineer can help in solving the problem.

Certainly, the chemical engineers and other technological professionals can do a lot to draw attention to the facts discussed above. An understanding between economists and engineers to develop joint-advise to political problems and introduction of innovative technologies have to be worked out by chemical engineers. To name a few, co-production of electric power, chemicals and hydrocarbons—use of DME, i.e. dimethyl ether, as a carrier of energy from, say, the Middle East to India, use of DME in India for generation of electricity and as fuel in diesel engines, use of DME as chemical feedstock, development of long range, high capacity, high voltage DC transmission.

General Aspects of Chemical Engineering

Communications

The successful chemical engineer must be able to express his technical ideas in oral and written communication. Technical brilliance is of limited use if the engineer cannot transmit his ideas to others. Often the major contact the young engineer has with the administrative officer who decides on his promotion is through written reports. All the engineer's reports should be written clearly and concisely with the reader in mind. Writing and speaking are important in all fields of chemical engineering from research to sales.

Human relations

Almost all the engineer's activities require close work with other engineers, scientists and technicians. He must be able to work effectively in a group. He must be able to sell his ideas effectively and tactfully. Many failures in the chemical engineering profession are not due to technical weakness, but can be attributed to failure of the engineer to work effectively in a group. Effective group activity depends primarily on the sensitivity to and respect for the rights of others. The engineer must realise that no matter how lucid his idea is to him, it may not be clear to others and it may not be right.

The increasing complexity of the chemical industry makes group activity more essential than ever before. The research and development engineer frequently discusses his technical problems with his colleagues to get their suggestions; he may supervise a group of technicians in an experimental programmes and he regularly discusses his work with his supervisor. At all levels he must be cognizant of the rights and needs of others. As mentioned earlier, the development and design engineers work closely together and with their respective groups. The production engineer must work with other engineers and with a unionised labour force. The technical service engineer works closely with the operators of a process, carefully explaining suggested process changes. The sales engineer must be particularly sensitive to his customers' needs. The customer is not always right, but it will do no good to tell him point blank that he is wrong.

Professional activities

All engineers should be active in their professional societies.

Technical reading

The chemical engineer should keep up-to-date in his field, not only by attending professional meetings, but also by reading technical journals. There are a number of general publications and many specialised publications in chemical engineering.

Chapter 2

Basic Laws and Processes of Chemical Technology

INTRODUCTION

The processing of raw materials into chemical products should be carried out under economically advantageous (optimum) conditions, i.e. under conditions which result in low consumption of raw materials, an inexpensive and high quality product and a high process rate. Cutting the cost of a chemical product means reducing capital investments, operating expenditures, consumption of raw materials, power and fuel, and labour per unit product. These ends can be accomplished by introducing continuous processes with a maximum degree of mechanisation and automation.

Without knowledge of the laws which govern chemical processes it is impossible to control a modern process and to manufacture chemical products economically. When a new plant is being designed such knowledge aids in selecting a rational flowsheet, i.e. in determining the sequence of production stages, in selecting the design of the equipment to be used and calculating its basic dimensions, in establishing the operating conditions and methods of control, working out instructions for servicing the equipment and supervising the process, etc.

Of the basic laws of chemistry and physics, which serve as the foundation for chemical engineering, the law of conservation of mass, the periodic law and the law of conservation of energy are among the most important.

The law of conservation of mass is used in chemical engineering to calculate the amounts of materials resulting from a process. According to this law, the total mass of the materials entering an apparatus (or a chemical plant) is equal to the mass of the materials leaving it, i.e. output weight is equal to input weight. A material balance for the operation of an apparatus (or plant) is drawn up using this law and taking into account both the main and side reactions. Material balances are necessary for analysing the operation of an apparatus (or chemical plant) and in designing new chemical processes. The balance is usually drawn up for unit weight (kg/t) of the product obtained, or for unit time (hour, day, month) of operation of the apparatus (or plant). After the material balance has been compiled the power balance can be drawn up. The power (in most cases the heat) balance for the operation of an apparatus is based on the law of conservation of energy.

The laws of chemical equilibrium and the theory of chemical process rates, which, is also called chemical kinetics, are of tremendous importance in chemical engineering. Raw materials and reaction products of chemical processes may be in the gaseous, liquid, or solid state. All the chemically or physically interacting substances make up a reacting system.

The physical state of the reacting substances largely determines the group of laws which govern the reactions, as well as the methods, flow diagrams and design of the equipment used for their processing. Therefore, to facilitate the study and discussion of reacting systems and chemical processes, it is convenient to divide them into the two groups of uniform, or homogeneous, and non-uniform, or heterogeneous, processes.

HOMOGENEOUS (UNIFORM) AND HETEROGENEOUS (NON-UNIFORM) PROCESSES

In homogeneous (uniform) systems or processes all the interacting substances are in the same phase: gaseous (G), liquid (L), or solid (S). For instance, the oxidation of NO to NO_2 by the reaction

$$\underset{G}{2NO} + \underset{G}{O_2} \rightleftarrows \underset{G}{2NO_2}$$

is a homogeneous process in the gas phase. If the reacting substances (reactants) are in solution, the process is homogeneous and in the liquid phase.

Heterogeneous (non-uniform) systems consist of two or more phases. Thus, in two-phase systems the reactants may be in the following states: gas-liquid (G-L); gas-solid (G-S); liquid-solid (L-S); liquid-liquid (L-L) (immiscible liquids); solid-solid (S-S). All heterogeneous systems have an interface. For instance, absorption of sulphuric anhydride, SO_3, by water leading to the formation of sulphuric acid, H_2SO_4, proceeds by the reaction (heterogeneous system G-L):

$$\underset{G}{SO_3} + \underset{L}{H_2O} \rightarrow \underset{L}{H_2SO_4};$$

the interaction of oxygen and carbon by the reaction (heterogeneous system G-S):

$$\underset{G}{O_2} + \underset{S}{C} \rightarrow \underset{G}{CO_2};$$

the reaction between sulphuric acid, H_2SO_4, and copper oxide, CuO (heterogeneous system, L-S):

$$\underset{L}{H_2SO_4} + \underset{S}{CuO} \rightarrow \underbrace{CuSO_4 + H_2O}_{L}$$

In all the above reactions there is an interface between the reactants.

In polyphase heterogeneous systems the reactants may be in the gaseous, solid, or liquid states, and several solid, or sometimes liquid, phases may be involved. In the chemical industry two-phase systems (G-L, G-S, L-S) are the most common.

Chemical reactions are, as a rule, accompanied by the release or absorption of heat. Chemical processes in which heat is released are called *exothermic*, those with heat absorbed, *endothermic*. For instance, the synthesis of ammonia, NH_3, from nitrogen, N_2, and hydrogen, H_2, by the reaction $N_2 + 3H_2 \rightleftarrows 2NH_3 + Q$ is a homogeneous exothermic process ($+Q$ means that heat is released), while synthesis of nitrogen oxide, NO, from nitrogen and oxygen by the reaction $N_2 + O_2 \rightleftarrows 2NO - Q$ is a homogeneous endothermic process ($-Q$ shows heat absorption).

An example of a heterogeneous (non-uniform) exothermic process is combustion of carbon: $C + O_2 \rightarrow CO_2 + Q$, while heterogeneous endothermic processes can be illustrated by the decomposition of limestone ($CaCO_3$): $CaCO_3 \rightarrow CaO + CO_2 - Q$.

A chemical process generally consists of the following interconnected elementary processes: (i) feeding the reactants to the reaction zone; (ii) chemical reactions; and (iii) withdrawal of the products from the reaction zone.

For example, let us consider the heterogeneous process of burning pyrite, FeS_2, which proceeds by the overall reaction $4FeS_2 + 11O_2 \rightarrow 8SO_2 + 2Fe_2O_3 + Q$ (air is used as the source of oxygen). For chemical reaction between the pyrite and the oxygen, the oxygen must first be transported to the surface of a pyrite particle and, after the reaction is completed, the sulphur dioxide formed must be removed from the surface of the particle. Oxidation gradually spreads into the particles and the oxygen on its way to the unreacted FeS_2 at the centre must now pass through a layer of ash at the surface and a layer of nitrogen, the inert component of the air surrounding the particles. As the layers of ash and nitrogen grow, the penetration of oxygen into a pyrite particle will slow down. The sulphur dioxide formed must also pass through this obstacle on its way to the bulk of the gas. These rather slow processes of penetration of one substance into another are called diffusion processes. If the ash and the inert gas are not removed from the surface of the pyrite, the roasting process will be retarded.

The overall rate of the process, made up of the three elementary processes, will be governed by the slowest of the processes. If it is diffusion, the process is said to lie in the diffusion region. In such a case measures are taken to speed up diffusion, for instance by stirring.

If the overall rate of the process is determined by the rate of the chemical reaction, the process is said to lie in the kinetic region. Then the chemical engineer tries to increase the reaction rate by increasing the concentration of the reactants, raising the temperature, using catalysts, etc. If the rates of diffusion and the chemical reaction are approximately the same (comparable), steps are taken to speed up both diffusion and the chemical reaction. Chemical processes may be reversible or practically irreversible.

REVERSIBLE PROCESSES, CHEMICAL EQUILIBRIUM, AND IRREVERSIBLE PROCESSES

In reversible processes the products of interaction between the initial substances react with each other to reform the initial substances. If initial substances *A* and *B* react, and the products obtained are *C* and *D*, such a reaction will be expressed by

$$A + B \rightleftarrows C + D$$

The reversibility of this reaction is indicated by the opposing arrows. At the beginning of the reversible process, when the concentration of the initial substances *A* and *B* is high, the rate of the reaction from left to right—the forward reaction—will be high but as the substances *C* and *D* are formed the concentration of *A* and *B* will decrease and the rate of the forward reaction will gradually slow down. At the same time the reaction products

C and *D* will interact to form the initial substances *A* and *B*, the rate of the reverse reaction (from right to left) increases with increase in the concentration of *C* and *D*. Finally a point will be reached where the rates of the forward and reverse reactions become equal, i.e. the number of molecules formed per unit time as a result of interaction between the products *C* and *D* will be equal to the number of reacting molecules of the initial substances *A* and *B*. Such a condition of the reacting system is called chemical equilibrium.

Reversible reactions do not result in complete conversion of the initial substances *A* and *B* into the reaction products *C* and *D*. The product yield for given conditions (temperature, pressure) is maximum at equilibrium and it is called the equilibrium yield. If the conditions of the reversible process (temperature, pressure or concentration of one of the reaction components) are changed, the rates of the forward and reverse reactions will change differently and therefore their rates will be equal at other ratios of the concentrations of the initial substances and the reaction products.

Two-phase and polyphase heterogeneous processes are called reversible if transfer of a substance (or heat) from one phase to the other is possible in both directions.

The effect of temperature and pressure on the chemical equilibrium of homogeneous and heterogeneous processes is determined qualitatively by the Le Chatelier principle: when the equilibrium of a system is disturbed the changes which occur are such as to weaken the disturbance. Thus, when a system in equilibrium is heated the changes will be such that heat is absorbed (the resulting processes will resist a rise in the temperature), while an increase in pressure results in processes which lead to a decrease in the volume of the system. By way of illustration let us consider the effect of temperature and pressure on several reversible exothermic and endothermic reactions.

(1) A reversible exothermic gas phase reaction without a change in volume, i.e. with the volume of the products obtained equal to that of the initial substances. An example is the reaction of CO with water vapour (CO conversion) to produce hydrogen:

$$\underset{\text{vapour}}{CO + H_2O} \rightleftarrows H_2 + CO_2 + Q$$

If the temperature is raised, the equilibrium of this reaction will shift in the direction of the initial substances (i.e. to the left), and the equilibrium yield of hydrogen will decrease. Since the volume of the system does not change in the reaction (two molecules of the initial substances produce two molecules of end products) in this case pressure does not affect the equilibrium (does not shift it): However, the degree, of conversion of one of the Initial products can be increased by using an excess of the other substance (compared with the amount necessary for the reaction). In this particular reaction by using an excess of water vapour the degree of oxidation of CO to CO_2 can be increased and more hydrogen obtained per unit carbon monoxide (CO) spent.

(2) A reversible exothermic gas phase reaction proceeding to the right (forward reaction) with a decrease in volume, for instance synthesis of ammonia, NH_3, from nitrogen and hydrogen:

$$N_2 + 3H_2 \rightleftarrows 2NH_3 + Q$$

In this case four molecules of the reactants produce two molecules of the end product. The equilibrium of this reaction will be affected both by the temperature and the pressure. According to the Le Chatelier principle an increase in temperature will shift the equilibrium to the left, i.e. toward an increase in the concentration of the initial substances; pressure will exert the opposite effect, i.e. its increase will shift the equilibrium to the right and increase the yield of the product, ammonia.

(3) A reversible endothermic gas phase reaction resulting in an increase in gas volume, for instance interaction of methane, CH_4, with water vapour (methane conversion):

$$CH_4 + \underset{\text{vapour}}{H_2O} \rightleftarrows CO + 3H_2 - Q$$

In this reaction two molecules of the reactants produce four molecules of the end products. An increase in temperature will shift the equilibrium of this reaction from left to right, and in pressure from right to left. In practice, high pressure is sometimes used, in such reactions to increase the reaction rate and decrease the size of the equipment.

These examples show that by changing the temperature or the pressure the equilibrium of a reversible reaction can be shifted, and by changing the ratio of the reactant concentrations the degree of conversion of one of them can be increased. Reversible reactions play a very important part in industrial chemistry. If one knows the equilibrium relationships, the equilibrium of a reaction can be shifted in the required, economically advantageous, direction.

In practically irreversible processes at equilibrium there is almost complete conversion of the reactants to the reaction products. Among such reactions are, for instance, the combustion of coal ($C + O_2 \rightarrow CO_2 + Q$), of sulphur ($S + O_2 \rightarrow SO_2 + Q$), etc. It is much easier to select suitable conditions for irreversible reactions than for reversible ones. An increase in the temperature, as a rule, results in an increased reaction rate. The temperature can of course be raised only up to a certain limit depending on the specific nature of the reaction—for instance, the temperature of sintering, slagging, thermal decomposition of the reactants. For reactions in which there is a gas phase an increase in pressure also increases the reaction rate, since an increase in pressure is tantamount to an increase in the concentration of the gaseous reactants.

Knowing how the equilibrium of reversible reactions depends on various factors affecting the process, one can determine the conditions at which the reversible reaction will be shifted toward the formation of the reaction product. This will also make it possible to calculate the maximum yield of the reaction product at given conditions. Process kinetics, i.e. the rate of reaction of the initial substances and the laws governing reaction rates, are no less important than the laws of equilibrium.

RATE OF CHEMICAL PROCESSES

The rate of a chemical process is usually expressed in terms of the amount of the product obtained per unit time. Such conditions for a chemical process should be maintained at which the rate is economically most advantageous. The rate of a chemical process is the resultant of the rates of the forward and reverse reactions. The higher the rate of the forward reaction and the lower the rate of the reverse one, the higher will be the rate of the overall process, i.e. the greater the yield of the product per unit time.

At the beginning of a process, when the concentration of the reactants is high, the reaction rate is considerable, but as the reaction approaches equilibrium it slows down and tends to zero. In industry, it is not profitable to conduct a reaction at a low rate and therefore a reversible reaction is not as a rule brought to a state of equilibrium, i.e. in reversible processes the product yield is in practice always below the equilibrium yield.

The rate of a chemical process depends not only on the concentration of the reactants, but also on a number of other factors, such as the temperature and pressure, the presence of catalysts, the physical and chemical properties of the reactants, the design of the reactor, the rate of flow of the reactants, the efficiency of stirring.

The methods used for increasing the rate of the process depend on the physical state (gaseous, liquid or solid) of the reactants. One of the ways to increase the process rate is to increase the concentration of the reactants, i.e. to enrich solid raw materials and concentrate gaseous or liquid ones. An increase in the concentration of the reactants has a strong effect on the rate of the process. In homogeneous processes the rate of the chemical reaction is directly proportional to the molar concentration of the reactants.

For reversible heterogeneous processes not only the concentration of the reacting component C in the carrier phase at a given moment, but also C_{eq}, the concentration of this component in the same phase at equilibrium must be taken into account. The higher the concentration (C) as compared with the equilibrium concentration (C_{eq}), i.e. the greater the difference $C—C_{eq}$ the higher will the rate of the process be. This difference $C—C_{eq}$ is termed the driving force of the process. At equilibrium the actual concentration and its equilibrium value are equal and, consequently, the driving force of the process $C—C_{eq}$ is zero, i.e. the process has reached equilibrium and its rate will be zero.

In reversible reactions the process rate can be increased by removing the product from the reaction zone and thereby decreasing the rate of the reverse reaction. From a gas mixture the reaction product can be removed by condensation (conversion of the gas to the liquid state) or by sorption with liquids or solids. Sorption of a gas by liquids is called absorption, sorption by solids, adsorption. In many industrial processes removal of the product is achieved by withdrawing the gas mixture from the reaction apparatus, separating the product, and returning the unreacted substances into the reaction zone. The result is a cyclic process. A typical cyclic process is the synthesis of ammonia.

From a liquid mixture the reaction product can be removed by depositing it in the form of crystals (precipitate); the solution remaining after separation of the precipitate is returned to the reaction (for instance, in production of potassium chloride). The reaction product can also be removed from a liquid mixture by evaporating it or by sorption with liquids or solids.

Usually to increase the reaction rate the temperature and pressure are raised, since they greatly affect many chemical processes. Van't Hoff's law can be applied to most reactions. According to this law the reaction rate is increased 2 to 4 times by a 10°C rise in the temperature, while the rate of diffusion-controlled processes increases only 1.1–1.3 times. The large increase in the rate of homogeneous gas reactions with temperature is due to the increased velocity of the gas molecules and the greater number of collisions per unit time. It should, however, be remembered that there is an optimum temperature for many reactions at which the reaction rate ceases to increase and beyond which it begins to drop with a further increase in temperature. This effect of temperature on the rate of some reactions may be caused by various factors one of which is decomposition of the initial substances at high temperatures and a consequent decrease in the concentration of the reactants which greatly affects the rate of the chemical process.

When a gas phase is involved a pressure increase is equivalent to an increase in the concentration of the gas component. Pressure is used to increase the rate of a chemical process primarily in cases where it is impossible or irrational to directly increase the concentration, i.e. to add more of one of the reacting substances, for instance in the synthesis of ammonia or methanol.

Temperature and pressure also strongly affect heterogeneous processes. Thus, an increase in pressure or a decrease in temperature will increase the rate and degree of sorption of gases by liquids and solids, while desorption of a gas (vapour) from a liquid or a solid sorbent is accelerated by raising the temperature or lowering the pressure (down to a deep vacuum).

MIXING THE REACTANTS

The degree of mixing of the reactants strongly influences the rate of chemical reactions. In homogeneous (uniform) processes vigorous stirring results in practically uniform concentrations of the initial substances throughout the entire volume and thus in accelerating their interaction. In heterogeneous (non-uniform) systems stirring increases the interface area of the reacting phases. Active stirring of heterogeneous systems is especially helpful in cases where the process lies in the diffusion region.

There are a wide variety of methods for increasing the interface area of the phases, depending on the physical state of the reactants, and they are usually aimed at increasing the surface area of the heavier phase. Thus, in gas-solid (G-S) and liquid-solid (L-S) reacting systems the solid initial material is crushed or used in the form of porous lumps, in which the internal area of the pores is many times larger than the external surface area of the lumps. Solids are size reduced in crushers and mills of various types.

Solid reactants may be agitated by different methods. The following methods are used for gas-solid (G-S) and liquid-solid (L-S) heterogeneous reacting systems:

(1) A granular solid with a gas or liquid flowing around the particles can be agitated on shelves by mechanical stirrers.

(2) If a finely divided solid is to be mixed in a fluid stream, for a G-S system this can be done by spraying the divided material together with the gas into an empty chamber; an example is the pyrite spray furnace.

For an L-S system the solid is mixed with the liquid by means of an agitator; such a mixer for phosphate meal and sulphuric acid, used in manufacturing superphosphate.

(3) Mixing can be achieved by passing a gas or liquid current through a bed of stationary particles of a divided solid material; equipment operating on this principle is called static bed apparatus; an example is the gas generator for obtaining air producer gas.

(4) A solid granular material can be mixed with a gas or liquid in a suspended or fluidised, bed.

In a gas-liquid (G-L) reacting system the gas-liquid interface area is increased by the following methods :

(1) The liquid is spread over the surface of a packing which fills a tower, or column; the liquid is fed to the tower at the top and, flowing down along the packing, forms a liquid film on its surface. The gas is usually fed to the tower at the bottom, countercurrent to the liquid; the gas-liquid contacting surface depends on the surface area of the packing and a packing is selected which provides a maximum possible surface area. Such an apparatus is called a packed tower or column; Fig. 2.1 shows the operation of a packed tower.

(2) The liquid is sprayed into a hollow tower, or chamber, to which the gas is simultaneously fed (Fig. 2.2); here the gas-liquid interface is equal to the total surface area of the liquid drops suspended in the apparatus volume.

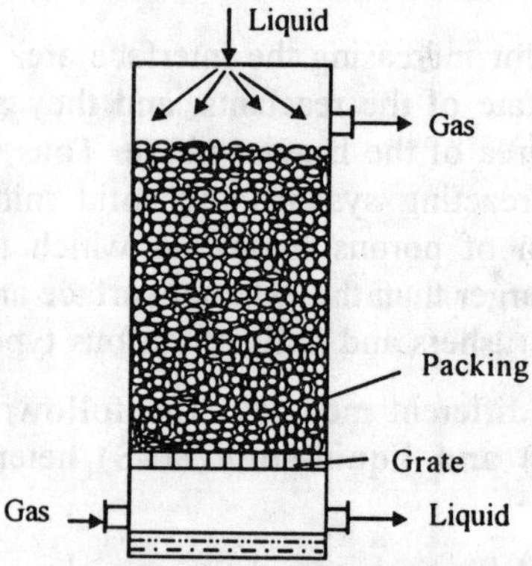

Fig. 2.1. Packed tower.

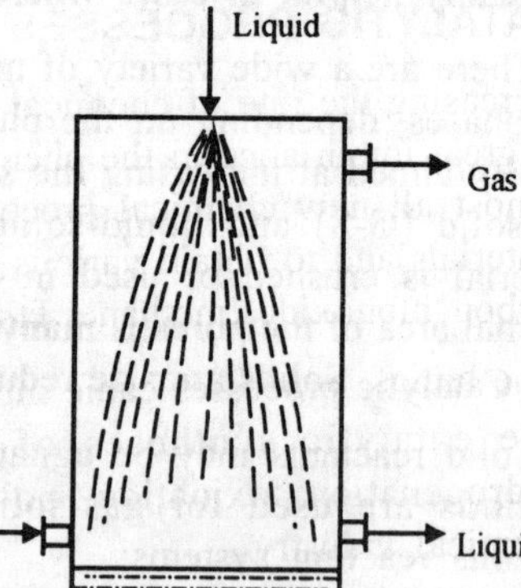

Fig. 2.2. Spray tower.

(3) Gas bubbles are passed up through a liquid layer. To break the gas current into small bubbles sieve trays (Fig. 2.3) or bubble-cap trays are used.

(4) The gas is passed through a liquid layer at such a high speed that the liquid foams, i.e. suspends in the gas current in the form of films, jets and drops; Fig. 2.4 illustrates the operation of such an apparatus.

In a system of two immiscible liquids (L-L) the interface area of the liquids is usually increased by stirring them to form an emulsion in which drops of the heavier liquid are suspended in the lighter one.

In a solid-solid (S-S) system the area of contact of the solids is increased by size reduction and mixing.

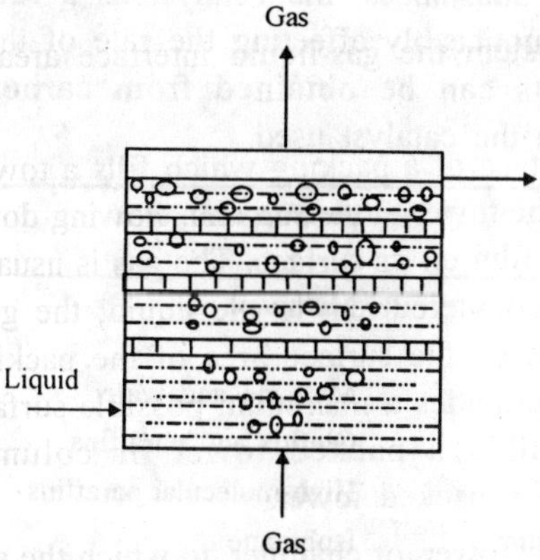

Fig. 2.3. Bubble apparatus (gas is bubbled through liquid layer).

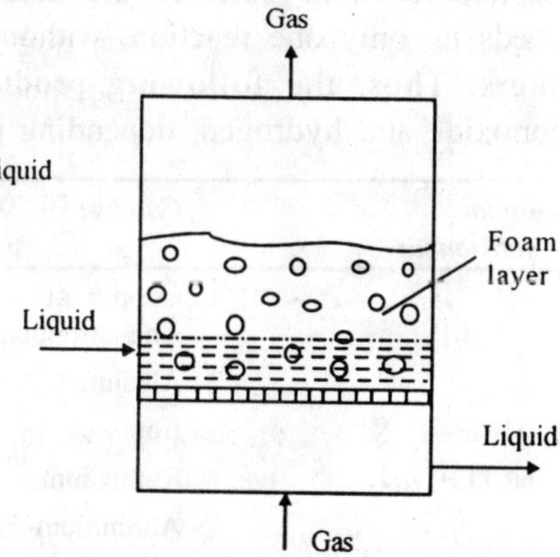

Fig. 2.4. Apparatus with foamed liquid layer.

CATALYTIC PROCESSES

Increasing the rate of chemical reactions by means of so-called catalysts is of great importance in the chemical industry. The utilisation of catalysts in almost all new chemical processes has made it possible to use new raw materials and to obtain various complex products from hydrogen, nitrogen, carbon monoxide, methane, ethylene, etc.

Catalytic processes came into wide use at the beginning of this century. The catalytic syntheses of ammonia and methanol, reactions of hydrogenation, oxidation, polymerisation are used extensively in the chemical industry.

One of the features of a catalyst is that after taking part in a reaction it normally is not changed chemically. One of the examples is the commercial synthesis of ammonia from nitrogen and hydrogen. If we mix nitrogen and hydrogen, no appreciable amount of ammonia can be obtained, even if we raise the pressure to 1000 atm or use high temperatures. This process has become possible solely due to the use as a catalyst of ordinary iron, with small additions of oxides of potassium, aluminium, calcium, silicon, which are called promoters or activators.

An iron catalyst for ammonia synthesis lives a long time and its activity decreases only as a result of mechanical damage, high temperature or impurities in the gas (catalyst poisons).

An important property of catalysts is their selectivity. Of the many possible reactions between the initial substances, the catalyst as a rule speeds up only one reaction, without noticeably affecting the rate of the others. Thus, the following products can be obtained from carbon monoxide and hydrogen, depending on the catalyst used.

Initial substances	*Catalyst*	*Products obtained*
	Copper or zinc-chromium	Methanol
	Alkaline	Methane and water
	Iron and	Olefins and paraffins
mCO + nH$_2$	Ruthenium	High-molecular paraffins
	Aluminium-thorium	Isobutane
	Zinc-thorium	Isopropane

All these products are used in the chemical industry.

We distinguish two basic kinds of catalytic processes: homogeneous and heterogeneous catalysis. Homogeneous catalysis is a catalytic process in which the catalyst and the reactants are in the same physical state: gas-gas, liquid-liquid.

Heterogeneous catalysis is a catalytic process in which the catalyst and the reactants are in different physical states. In industrial heterogeneous catalytic processes the reactants are usually in the gaseous phase and the catalysts are solids. Among such processes are ammonia synthesis, oxidation of SO_2 to SO_3 (the production of sulphuric acid) and oxidation of ammonia (the production of nitric acid).

Some reactions are accelerated in the presence of catalysts thousands and millions of times and proceed at lower temperatures, which results in the saving of power. Catalysts do not change the equilibrium of the reaction, but they speed up approach to equilibrium. The nature of catalytic action is not yet fully understood, and therefore experimental data are of primary importance in selecting a catalyst for a particular reaction.

For a heterogeneous catalytic process in which the reactants are gases and the catalyst is a solid the accelerating effect of the catalyst can be explained in the following simplified way. The molecules of a gaseous substance are in constant chaotic linear motion. When they collide the velocity of the individual molecules changes and they begin to move about at various speeds, from very high to very low ones.

In many cases colliding molecules in a gas mixture lack the necessary kinetic energy to overcome the forces hindering chemical interaction. Therefore the reaction between the gases will proceed very slowly. A rise in temperature will increase the number of molecules of high kinetic energy and thus increase the rate of reaction between them, but this is not always sufficient. Speaking of such gas reactions we say that their activation energy is very high, i.e. for a reaction between such gases to occur high energy of the molecules is necessary to overcome the forces of mutual repulsion between the molecules and to weaken the bonds between the atoms in the molecules. The catalyst reduces the activation energy of the reaction.

Schematically, this can be represented as follows: Two molecules *A* and *B* are reacting in the gas phase to form a new substance *AB* by the reaction $A + B - AB$. The activation energy of the reaction is high, and therefore it proceeds very slowly.

If a catalyst is used, the reaction will proceed in a different way:

$$A + B + K \rightarrow AK + B \rightarrow AB + K$$

where *K* is the catalyst.

It is seen from the above scheme that at first the catalyst reacted with the molecule *A* to form an intermediate compound *AK*, then the second reacting molecule *B* combined with *AK* to form the substance *AB*. After the reaction is completed the catalyst is returned to its original form. With this course of the reaction the activation energy is less and the reaction rate increases.

In heterogeneous catalysis the catalysts, as a rule, have a highly developed surface area. The surface area of an iron catalyst in ammonia synthesis amounts to about 10 m^2/g. There are catalysts with an area of up to 400 m^2/g.

Substances which do not speed up the course of the reaction, but retard (inhibit) it, are called negative catalysts, or inhibitors. For instance, when motor fuels are put away for storage it is advisable to add a small amount of substances which retard oxidation and polymerisation processes.

BASIC TYPES OF INDUSTRIAL CHEMICAL PROCESSES AND FLOW SCHEMES

Industrial chemical processes are classified as concurrent, countercurrent, or crosscurrent, depending on the pattern of flow of the reacting phases.

In concurrent heterogeneous processes (Fig. 2.5) the reacting phases flow in the same direction, in countercurrent processes (Fig. 2.6) they are in opposing flow. In crosscurrent processes (Fig. 2.7) the paths of the reactants intersect at an angle.

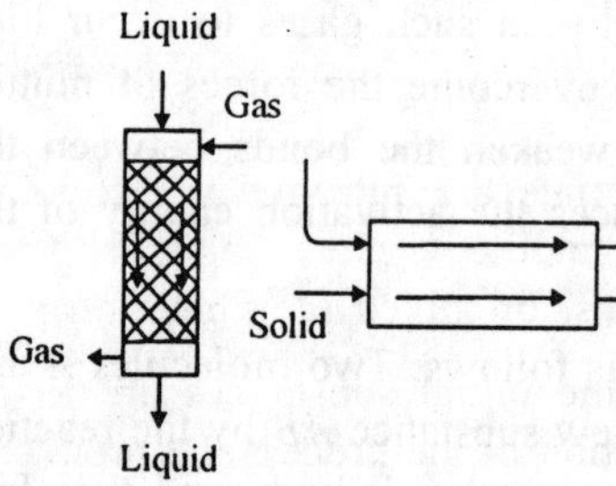

Fig. 2.5. Concurrent flow processes in G-L and G-S systems.

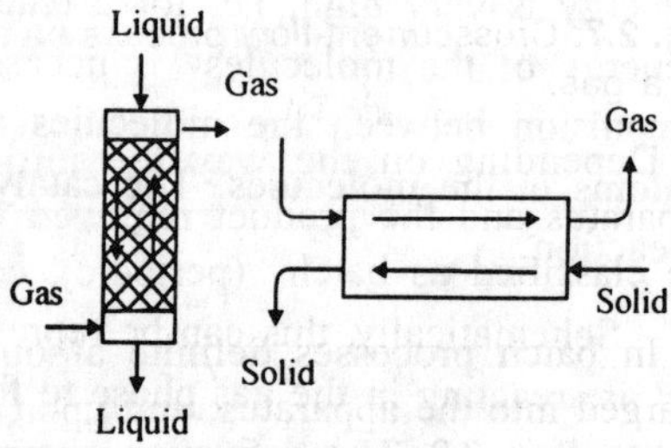

Fig. 2.6. Countercurrent-flow processes in G-L and G-S reacting systems.

Counter-current flow provides better interaction of the reactants than concurrent flow. Therefore in designing industrial chemical processes one should give preference to counter-current interaction of the phases of a heterogeneous system.

Concurrent processes are mainly used where counter-current flow is impracticable owing to some specific feature of the reacting system. Thus, concurrent flow is used for drying ammonium nitrate with heated air and drying flotation pyrite with flue gases. If ammonium nitrate were dried in countercurrent, the hottest air would be in contact with dry and hot ammonium nitrate, and this would result in decomposition of part of the nitrate and increase product losses. When moist ammonium nitrate is introduced into the dryer concurrent with the air the dry ammonium nitrate comes into contact with low-temperature air leaving the dryer, and decomposition losses are reduced. Counter-current drying of moist flotation pyrite would result in ignition of the dry pyrite in contact with the hot (700 to 800°C) flue gases.

Crosscurrent flow usually ensures more complete interaction in heterogeneous systems and in some cases preference is given to a crosscurrent, rather than a countercurrent process.

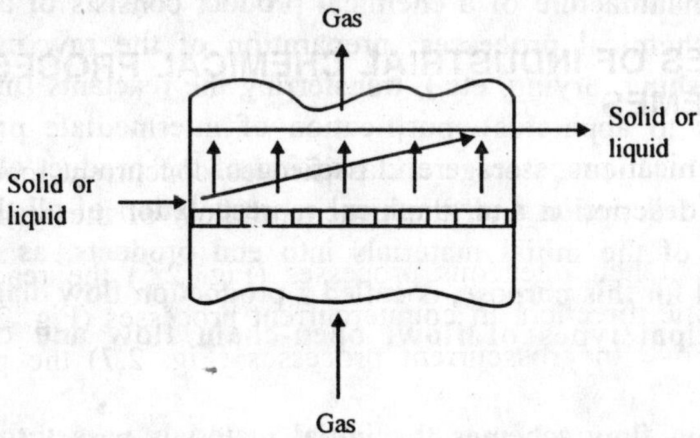

Fig. 2.7. Crosscurrent-flow process on a tray with bed of solid or liquid fluidised by a gas.

Depending on the way the initial materials are introduced into the apparatus and the product removed from it, industrial chemical processes are classified as batch (periodic), continuous, or mixed (combination).

In batch processes definite amounts of the initial substances are first charged into the apparatus, and upon completion of the process the product is withdrawn; after this another batch of the initial materials is introduced, and all the operations are repeated (charging the raw materials—running the process—withdrawing the product). In such processes the equipment is not utilised in the most rational way. The apparatus is idle while the initial materials are being charged into it and while the product is being withdrawn. Besides this, during the halts in the operation of the apparatus much heat or other kinds of energy is lost. In batch processes it is difficult to ensure uniform quality of the products; also mechanisation and automation are more complicated.

In continuous processes introduction of the initial materials, the reactions and withdrawal of the product are all continuous. Continuous processes are usually more economical; they are highly productive, give high yield and uniform quality of the product, are easier to mechanise and automate; the individual stages of a complex process take place continuously and under constant conditions.

In mixed (combination) processes the reacting materials are fed to the process continuously, while the product is withdrawn periodically. In some cases the initial substances (or one of them) are fed periodically, while the products are withdrawn continuously. For instance, coal is charged into a gas producer periodically, while the air-producer gas is removed continuously.

As a rule, manufacture of a chemical product consists of a number of physical and chemical processes: preparation of the raw materials for processing (crushing, drying, etc.), transferring the reactants (moving them from apparatus to apparatus), purification of intermediate products, the main chemical reactions, storage and packing of the product obtained, etc. A step-by-step description and graphical representation of all the stages in the processing of the initial materials into end products, as well as the equipment used for this purpose, is called a production flow diagram. There are two principal types of flow: open-chain flow and cyclic flow (recirculation).

In open-chain flow schemes the initial materials pass through one or more apparatuses in succession. If the degree of processing of the initial materials achieved in one apparatus is insufficient, a succession of similar apparatuses must be used, and the production unit becomes very large.

In cyclic (recirculation) flow schemes, widely used for reversible reaction processes in which the equilibrium is such that a high yield of the product is impossible, the unchanged portion of the reactants is returned to the production cycle after the product has been removed from it.

Cyclic flow schemes are extensively used for the synthesis of ammonia, methanol and various organic compounds. Units using a cyclic scheme are more compact, they consist of a smaller number of apparatuses and ensure a higher degree of conversion of the initial substances into the end products. Capital investments are consequently less.

Chapter 3

Raw Materials Fuel and Power for Chemical Process Industry

INTRODUCTION

The expenditures made by an enterprise in manufacturing and selling a product comprise its full cost; the expenditures directly involved in the manufacturing process alone are known as the plant cost. The principal items of expenditure which together make up the plant cost are :

(1) The raw materials, semi-products and basic materials which directly participate in the chemical reactions.

(2) Fuel and power for technological uses.

(3) Salaries and wages of basic employees.

(4) Sinking-funds set aside to cover wear of basic items of capital investment (buildings, structures, equipment, etc.).

(5) Shop costs—expenditures on maintenance and current repairs of the basic items of capital investment (including salaries and wages of auxiliary- and maintenance-service employees) and also salaries and wages of shop management personnel, health and safety expenditures, etc.

(6) General plant expenditures.

When the cost of the principal product is calculated the value of by-products obtained from the basic raw materials is usually substrated from the full cost. There is great variation in the weight of the items which make up the cost of different chemical products. Usually, however, the first item of the above list is the most important one—on the average, expenditures on materials amount to 60–70 per cent of the cost; the second item (fuel and power) comes to about 10 per cent. Wages and salaries (item 3) average about 4 per cent, and the sinking-fund (item 4) 3 to 4 per cent of the cost.

These figures show that the cost of a chemical product is primarily determined by the cost of the materials used in making it and by the degree of their utilisation; also important are expenditures on fuel and power. The conclusion to be made is that full and comprehensive utilisation of raw materials, i.e. utilisation of all their components, and also cutting expenditures on fuel and power to a minimum are the basic conditions for manufacturing chemical products at lower costs.

MATERIALS FOR CHEMICAL PROCESSES

The materials used in manufacturing chemical products can be divided into the categories of raw materials, semi-products, basic materials and waste products of other processes.

Raw materials are naturally occurring substances which have not been subjected to industrial processing. Semi-products are materials which have undergone some form of preliminary processing at the plant where they are used. A material processed at another plant prior to its use in the chemical process is known as a basic material. Waste products are by-products of other processes.

The requirements imposed on raw materials change with growth and improvement of chemical technology. New raw materials are employed, waste products of various industrial processes are utilised in ever-increasing amounts; the comprehensive utilisation of raw materials and the use of low grade ores and non-metallics after preliminary concentration become more and more important.

The raw materials used in the chemical industry can be divided into the following categories:

(1) raw materials won from the upper layers of the earth:

- (a) naturally occurring fuels, employed either as a chemical material or as a source of power (coal, peat, shale, petroleum, natural gas);
- (b) ores—rock-like materials containing metals which can be economically isolated in the form of a technically pure product (iron ores such as magnetite, red and brown haematites, ores of non-ferrous metals—copper pyrite, zinc-blende, etc.);
- (c) non-metallics—inorganic naturally occurring materials which are utilised otherwise than as a source for producing metals, although many of them contain one or more metal constituents; examples are common salt and other chlorides, sulphates, carbonates, phosphates, alumino-silicates;

(2) vegetable and animal materials—wood, cotton fibre, leather, wool; and
(3) water and air.

Progress in the field of raw materials for the chemical industry means discovering and making use of new kinds of raw materials, concentration of low-grade ores and non-metallics, comprehensive utilisation of raw materials, replacement of valuable edible vegetable materials by mineral ones.

To bring the costs of chemical products down, low-cost materials must be employed. A valuable material is one of low cost. A very striking example of the switch-over of a chemical industry to a cheaper raw material is found in the widespread use of natural gas and casing-head gas for making hydrogen, a component of synthetic ammonia and many synthetic organic products. Until recently the principal material used in producing hydrogen was coke. Calculations show that referred to a standard fuel unit the cost of natural gas is about 12 times less, and that of petroleum about 3.5 times less than the cost of deep mine coal. Moreover, natural gas and petroleum can be transported over long distances using pipelines, whereas railway waggons must be used for coal. This explains why production of petroleum and especially natural gas has been increasing so rapidly in recent years. When natural gas replaces coke as the raw material the cost of synthetic ammonia is cut almost to a half and this results in lower costs of nitrogen fertilisers and the various other products made from ammonia.

According to their physical state raw materials are classified as solid, liquid or gaseous, and according to their chemical composition, as inorganic or organic.

BASIC PRINCIPLES OF RAW MATERIAL BENEFICIATION

One should always strive to employ a raw material with a high concentration of valuable components since this results in compact processing equipment, high throughput, lowered heat and steam consumption, etc. In India raw materials are sometimes shipped over long distances to the place where they are processed. It is evident that transportation costs can be cut by dressing raw materials at the mines. Moreover, the reserves of high grade ores are continuously being depleted. Progress in concentration methods has, however, made it possible to utilise low-grade ores as industrial raw materials. Because of these factors it is common practice to subject raw materials to concentration and to locate the dressing plants at the mines. The product of an ore-dressing process which contains the useful components is known as the concentrate, and the waste part, the useless remainder, as the tails.

A natural material may contain several valuable components. Prior to being used such a material should be separated into fractions each of which is enriched in one of these components; in this way several concentrates are obtained from a single complex raw material. Such preliminary separation results in fuller utilisation of the valuable components. The material to be concentrated may be in the form of a solid (ores of various kinds), gas or liquid solution and the methods employed in each case differ radically.

Solid raw materials are not as a rule pure minerals. They usually consist of a number of minerals, each of which has its own particular physical and chemical properties. These properties—density, hardness, solubility, water repellence, melting point, temperature of sublimation, magnetic susceptibility, electrical conductivity, etc.—can be employed to separate a complex mineral raw material into concentrates and tails.

Prior to its concentration a mineral raw material is crushed to disrupt the bonds between the grains or crystals of the different minerals in it. The following basic techniques are employed in dressing solid minerals: sizing (screening), gravitation, electromagnetic, electrostatic, thermal or chemical separation, flotation.

Separation by sizing (screening) is based on the tendency of the more friable minerals to yield smaller grains or crystals when the material is crushed than the stronger, less friable minerals do. If a material consisting of minerals of different friability is first crushed and then passed through a set of sieves, each having holes of another size, the sieves will retain fractions of the material, each of which is enriched in one of the minerals. Sieves employed for this purpose are called screens. Screening is also used to separate a material of uniform composition (coal or coke, for example) into fractions containing grains or pieces of similar size.

A screen may be in the form of a flat, perforated sheet, slightly inclined so that the larger-size particles will slide down along it, or in the form of an inclined, rotating, perforated cylinder. Flat screens are often supplied with a mechanism for shaking, vibrating or rocking them to expedite passage of the smaller grains through the holes.

Gravity concentrating is based on the difference in the rate with which particles of different density or size will settle down through a stream of a liquid—most often water (hydraulic separation), but sometimes air or an inert gas (dry separation). Fig. 3.1 illustrates the basic principle of wet gravity concentration.

The crushed material (an ore) is mixed with water in an agitated vessel and the resulting pulp (a suspension of solid particles in water) is fed to settling chambers (or basins) I, II, III with cone-shaped bottoms (hoppers). The width of the chambers increases gradually from entrance to exit leading to a drop in the flow rate and expediting the settling of the suspended particles. The heavier (larger-sized) particles settle in chamber I, the medium ones in chamber II and the lighter (fine-grained) ones in chamber III. A classifier of this type may contain two, three or more chambers depending on the number of fractions into which the material is divided. The fractions obtained are discharged through openings a, b and c. Jigs and table concentrators are other types of equipment used for wet gravity separation of materials.

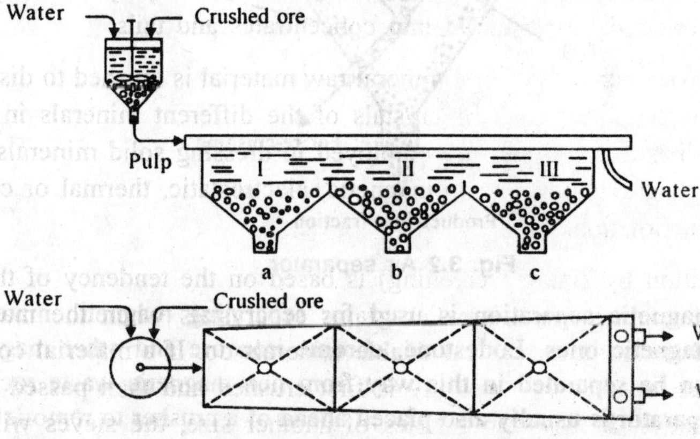

Fig. 3.1 Wet gravity concentration.

For dry gravitational concentration centrifugal air separators are most often used. A diagram of such a separator is given in Fig. 3.2. It consists of cylinder 1 provided with cone bottom 2 (external cylinder and cone). The main cylinder contains a concentrically positioned inner cylinder 4 and cone 3. Disc 5 and fan impeller 6 within the inner cylinder are supported by the separator cover; they are driven by an electric motor located on the cover. When the plate and the impeller rotate air currents (shown by the arrows) develop inside the separator. The crushed material is delivered onto the rotating disc and scattered across the section of the inner cylinder. Small particles are entrained by the air current set up by the impeller and carried into the space between the outer and inner cylinders, where, upon hitting the walls (as the flow rate drops) they slide down them and are removed from outer cone 2 to form the fine fraction. The large particles overcome the force of the air current and drop into inner cone 3, from which they are removed and returned for recrushing.

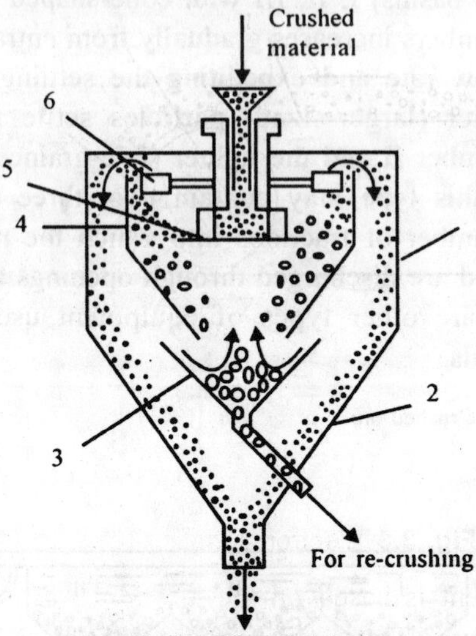

Fig. 3.2 Air separator.

Electromagnetic separation is used for separating magnetic materials from non-magnetic ones. Lodestone, chrome iron ore and other magnetic materials can be separated in this way from non-magnetic waste rock. A magnetic separator is usually also placed ahead of a crusher to remove steel objects that accidentally find their way into the ore during mining and that would damage the crusher.

Fig. 3.3 is a schematic diagram of an electromagnetic separator. The crushed material is fed onto belt conveyor 1, which is provided with drum 2 equipped with electromagnet 3. When a section of the belt passes over the surface of the rotating drum the non-magnetic particles of the material fall from the belt into hopper 4, while magnetic particles, which, so to speak, stick to the belt, continue moving with it untill the belt section passes beyond the magnetic surface of the drum when they drop off it. The magnetic particles fall into hopper 5.

Electrostatic separation is used to separate materials of different electric conductivity. The drum of an electrostatic separator is equipped with a negatively charged electrode connected to an electric current rectifier. Mineral particles of high electric conductivity are charged negatively as the belt moves over the drum surface and as a result are repelled by the belt, move to the top of the layer and are discharged into a hopper placed some distance from the drum, whereas non-conducting particles (dielectrics) fall into a hopper situated closer to the drum.

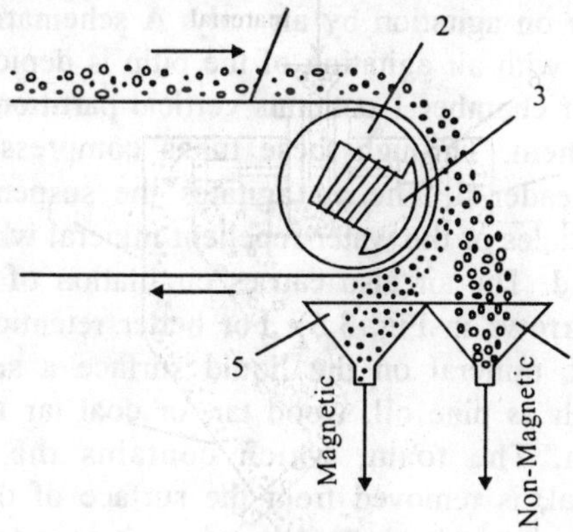

Fig. 3.3 Electromagnetic separator.

Thermal enrichment is based on the difference in the melting point of materials in a mixture. Thus, sulphur, which has a low melting point (119°C) as compared with that of gangue (above 1000°C), can be melted and removed from the ore in the liquid state.

Chemical enrichment methods are extremely varied; they include selective dissolving, decomposition by chemical reagents, roasting, etc. The most important of these methods are discussed dealing with specific industrial processes.

Enrichment by flotation is based on the different wetting properties of the grains of individual minerals. If a small particle of a hydrophobic mineral, A, and that of a hydrophilic one, B, are placed on the surface of water in a vessel, as shown in Fig. 3.4, the hydrophobic particle A will 'press' itself into the liquid and, not being able to overcome the surface tension of the liquid, will remain on the surface. The hydrophilic particle B rapidly becomes covered with a liquid film and slowly moves to the bottom of the vessel by force of gravity. The particles of a hydrophobic mineral can be removed from the surface of the liquid and thereby separated from the particles of a hydrophilic mineral, which settle to the bottom of the vessel.

For flotation to be efficient it is necessary to create conditions ensuring a difference in the affinity for water of the minerals to be separated.

Under industrial conditions a suspension in water of a finely ground rock (of 0.2 mm maximum grain size) is fed into a flotation machine together with certain flotation reagents. Of the two types of machines now in use one is based on mechanical agitation of the pulp with simultaneous supply of air and the other on agitation by air only. A schematic cross-section of a flotation machine with air agitation of the pulp is depicted in Fig. 3.5. A trough-like reservoir chamber 1 contains vertical partitions 2 with a row of tubes 3 between them. Through these tubes compressed air flows into chamber 1 from header 5. The air agitates the suspension, and the air bubbles entrain particles of the water-repellent mineral which float up to the surface of the liquid. The air also causes circulation of the suspension in the chamber (the arrows in Fig. 3.5). For better retention of the particles of the hydrophobic mineral on the liquid surface a small amount of a foaming agent, such as pine oil, wood tar, or coal tar fractions, is added to the suspension. The foam, which contains the particles of the hydrophobic mineral, is removed from the surface of the liquid over the brim of the chamber into trough 4, from where it goes to apparatus for defoaming, thickening and filtering. The mineral particles separated from the foam are dried to obtain the so-called concentrate. The particles of the hydrophilic material settle to the bottom of the chamber and are removed, forming the flotation tails.

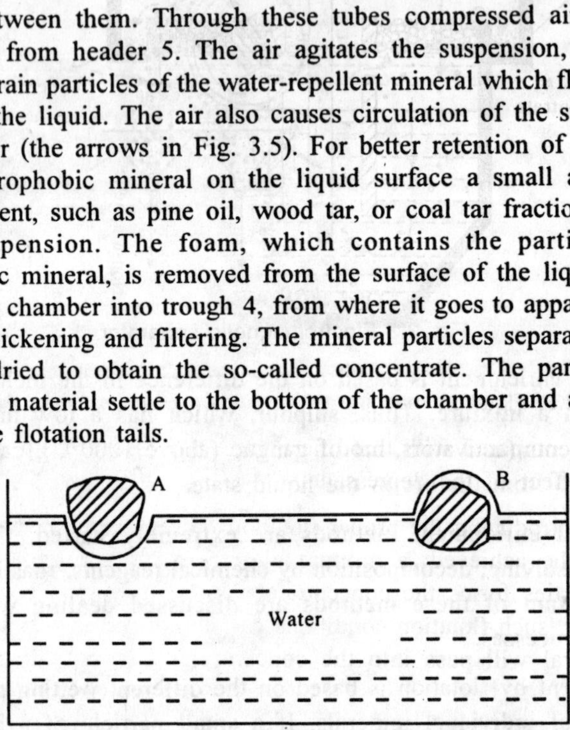

Fig. 3.4. Separation of hydrophobic and hydrophilic minerals.

Comparatively few natural minerals are water-repellent, i.e., naturally flotable. These include native sulphur, graphite, ozocerite, molybdenite, etc. The great majority of minerals are hydrophilic and their separation by flotation requires the addition of special reagents to the suspension to increase water-repellence. These reagents are referred to as collectors. They include oleic acid, naphthenic acids, xanthates, and other organic substances. The collectors coat the surface of the particles of hydrophilic minerals with a hydrophobic film, causing them to go to the top.

In certain cases it is necessary to prevent floating of particular minerals. For this purpose so-called depressants are introduced into the suspension.

Alkalies or salts of alkaline metals (cyanides, silicates, sulphides, etc.) are used for this purpose. They increase the hydrophilic properties of the minerals and as a result impede their floating.

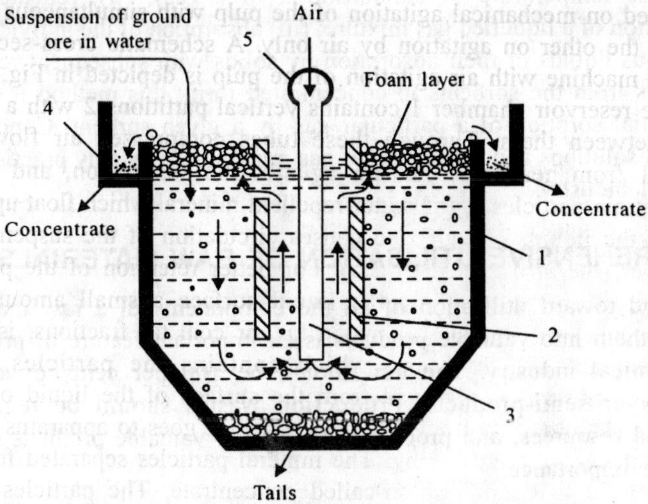

Fig. 3.5. Flotation machine with agitation of the pulp by air.

Other reagents (activators, modifiers, etc.), are also used to enhance or weaken the effect of the depressants.

Thus, by introducing the proper flotation reagents simple, or collective flotation can be achieved: a mixture of minerals which possess similar wetting properties forms the concentrate, while the gangue goes into the tails. However, such flotation conditions can be created in which only one definite mineral will pass into the concentrate. This is called selective flotation.

The wide range of available flotation reagents with their different properties makes it possible to use flotation techniques to enrich or separate a great variety of different ores. The consumption of flotation reagents usually does not exceed 100 g per ton of the rock. Flotation has found extensive application all over the world as a method for concentration and separation of mineral raw materials.

Liquid solutions of various substances are enriched (concentrated) by different methods: by boiling the solvent off, adding a useful component to the solution (saturating), precipitating impurities or transferring them to the gas phase (evaporation of impurities or desorption). Gas mixtures are enriched by separating them, using basically the following methods: (i) fractional condensation by low temperature liquefaction; this separation method is based on a difference in the temperatures of condensation of the components of the gas mixture (cooling is sometimes combined with preliminary compression of the gas mixture, since raising the pressure also raises the temperature of condensation of a gas); (ii), repeated partial evaporation of a liquefied gas mixture; (iii) absorption of the individual gases by various liquids or their adsorption by solids with subsequent removal of the gases from the sorbents in concentrated form; this method is based on preferential sorption of a particular gas by a given sorbent. Concentration of liquid solutions and separation of gas mixtures are widely practised in the chemical industry.

COMPREHENSIVE UTILISATION OF RAW MATERIALS

The trend toward utilisation of all the components of a raw material by making them into valuable products is an important feature of progress of the chemical industry. The aim should be 100 per cent conversion to products or semi-products. Production wastes should be regarded as unutilised resources, and processing them into valuable products as a task of prime importance.

One example of comprehensive utilisation of raw materials is in the processing of apatite-nepheline rock. Formerly apatite, $3Ca_3(PO_4)_2. CaF_2$, was separated by flotation from this rock and then processed into superphosphate, while the nepheline, $(Na_2O, K_2O). Al_2O_3, 2SiO_2$, was discarded, although it contains many valuable ingredients. In recent years a method has been developed for processing nepheline into alumina, Al_2O_3, from which aluminium metal is made; Na_2CO_3, and K_2CO_3, whose consumption is growing year by year; and $CaO.SiO_2$, from which cement is produced.

Another example of comprehensive processing of raw materials is the coking of coal. Coke, coke-oven gas, and coal tar are the products. At one time only the coke was utilised (at iron and steel works), the coke-oven gas was burned as fuel, and the coal tar was considered a waste, a burden for the coke plant. At present the coal tar, which is a complex mixture containing over 300 different organic compounds, is a very valuable raw material. It yields many organic compounds used in the production of intermediate products and dyes, various pharmaceuticals, lubricating oils, etc. From the coke-oven gas we now recover ammonia, hydrogen sulphide, and benzene (aromatic) hydrocarbons, while the remaining part of the gas, which contains principally hydrogen (up to 60%) and methane (up to 30%), is separated into the individual components by cooling to low temperatures, thus serving as a source of hydrogen and other valuable products.

In Indian industry, cooperation between separate establishments to obtain fuller utilisation of raw materials and waste products is very common. For instance, coke plants cooperate with iron and steel works and with nitrogen fertiliser plants.

REPLACEMENT OF EDIBLE RAW MATERIALS

An important task facing a number of branches of the chemical industry is the replacement of edible raw materials by inedible ones. The consumption of such products as ethyl alcohol, casein glue, linseed oil, paints, and soap, formerly made using edible raw materials, has greatly increased. For instance, large amounts of ethyl alcohol are used in the production of artificial rubber, synthetic fibres, plastics, explosives, ethers, etc. In spite of the fact that edible raw materials are being replaced, their consumption is still very high.

The use of edibles as raw materials reduces available food and results in higher costs of industrial products. The production of adequate substitutes for edible raw materials should be forced and a stop made to the use of grain, sugar, vegetable oils and other foodstuffs for industrial purposes.

Edible raw materials can be successfully replaced by mineral ones. Synthetic ethyl alcohol from ethylene is considerably cheaper than alcohol produced from edible raw materials; both capital investments and the product cost are lower. The ethylene for the production of ethyl alcohol can be obtained by processing natural gas, petroleum, coal or shale. A considerable economic effect is attained by substituting synthetic fatty acids for natural vegetable and animal fats.

WATER AND THE CHEMICAL INDUSTRY

Water is one of the raw materials of the chemical industry. In some cases it participates in the chemical reactions directly and enters into the product, for instance, in the production of sulphuric acid ($SO_3 + H_2O \rightarrow H_2SO_4$), or nitric acid ($3NO_2 + H_2O \rightarrow 2HNO_3 + NO$). Water is often used for dissolving solids, liquids or gases for obtaining suspensions or pulps (e.g. in gravitational and flotation concentration). Water is needed for washing solid and gaseous materials to remove impurities, for cooling solids or substances which react with the release of heat. Hot water or water converted to steam is used to heat reacting substances to increase the process rate, or to compensate for heat lost in chemical reactions in which heat is consumed. For cooling reacting mixtures with water or heating them with steam various types of heat exchangers, coolers and heaters are used. Heat exchange between water and reacting mixtures can be either by direct contact of the water with the substance, or through the walls of pipes, vessels, or apparatus.

At large chemical enterprises water is consumed in huge quantities and therefore in many cases spent water, for instance, from coolers, is re-used. This water is called circulating water. It may be cooled in special reservoirs (lagoons); sometimes it is sprayed. Often it is fed to a tower filled with a checker work of wooden grates and battens. Running down the checker work, the water is cooled by a countercurrent of cool air. These are called cooling towers. Water in its natural state usually contains dissolved impurities. We distinguish three kinds of naturally occurring water: atmospheric, surface, and underground. They contain different impurities.

Atmospheric water—rain and snow—contains dissolved gases: oxygen, nitrogen, carbon dioxide, sometimes hydrogen sulphide, and other gases—organic impurities, and dust.

Surface water—river, lake, sea water, in addition to dissolved gases, contains mineral impurities: bicarbonates, sulphates, and chlorides of sodium, potassium, calcium, magnesium, etc. Their content in different water varies widely.

Underground water—spring, well, artesian water, like surface water, contains dissolved mineral salts. The content of organic impurities is lower than in atmospheric and surface water.

Water is considered to be *fresh* if its salt content does not exceed 1 g per litre. The hardness of natural water is due to soluble salts of calcium and magnesium contained in it; such water deposits scale on boiler walls and it does not soap readily. A distinction is made between temporary and permanent hardness of water. Temporary hardness results from the presence of soluble bicarbonates of calcium and magnesium, $Ca(HCO_3)_2$ and $Mg(HCO_3)_2$. When water is boiled these salts decompose: $Ca(HCO_3)_2 \rightarrow CaCO_3 + H_2O + CO_2$. The calcium carbonate, $CaCO_3$, forms a soft precipitate. Permanent hardness is caused by calcium and magnesium chlorides and sulphates present in the water. Water is classified into three groups with respect to the amount of dissolved salts contained: soft water contains up to 3 mg-equivalent/litre of salts, moderately hard (medium) water, 3-6 mg-equivalent/litre, and hard water, over 6 mg-equivalent/litre.

Impurities such as gases, salts, organic compounds and micro-organisms may have an adverse effect on some reactions, cause corrosion of the equipment, form scale, sediments, etc.

Water is conditioned by different operations, depending on the requirements of the chemical process: removal of suspended particles, softening (reducing the content of calcium and magnesium ions), in some cases desalting, neutralisation, degassing, decontamination (removal of organic impurities, bacteria, fungi and other micro-organisms).

Suspended particles are removed by settling or by filtration through a layer of sand and gravel. Impurities (mainly organic substances) contained in the water in a colloid state are usually first precipitated by adding coagulants, for instance aluminium sulphate, after which the sediment is filtered off. Part of the bacteria contained in the water is also removed by such a process.

Softening—reducing the hardness of water (to approximately 0.3 mg equivalent/litre) is achieved by adding lime ($Ca(OH)_2$) or caustic soda ($NaOH$) and soda (Na_2CO_3); calcium and magnesium ions present combine into $CaCO_3$ and $Mg(OH)_2$, which settle and are then filtered off. If the water is to be softened to 0.03 mg equivalent/litre, sodium phosphate, Na_3PO_4, is added; the sediment will consist of $Ca_3(PO_4)_2$ and $Mg_3(PO_4)_2$.

When the water must be desalted more completely use is made of cation- and anion-exchange resins. Cation-resins contain active groups with H^+, Na^+ and NH_4^+ ions; they absorb ions of metals (calcium, magnesium, etc.) in exchange for these cations. Anion-resins contain active groups with OH^-, hydroxyl, ions which are capable of absorbing Cl^-, SO_4^{--} and other ions by exchanging them for the OH^- ions. By passing water first through a filtering layer of a cation exchange resin and then through a layer of an anion-exchange resin both the cations and anions in the water can be removed, i.e. the water can be desalted. When saturation of the resins is approached they are regenerated, i.e. their ion-exchangeability is restored by passing appropriate solutions through them. Thus, a sodium-cation exchange resin is regenerated using a solution of common salt.

Neutralisation is necessary when water is contaminated by acids or alkalies. For instance, in cooling sulphuric acid with water some acid will get into it if there is a leak in the cooler. Acid water cannot be recycled or discharged into the sewage system without neutralisation.

Degassing, i.e. removal of gases such as hydrogen sulphide, sulphur dioxide, carbon dioxide, etc., which may cause damage (corrosion) to the equipment, is effected chiefly by blowing air through the water. The process is referred to as *aeration*. Some chemical enterprises producing pure and chemically pure substances use distilled water.

Depollution of industrial waste-waters is a very important problem. Waste-waters of many chemical enterprises are contaminated by acids, alkalies, salts, and various organic impurities, which may destroy vegetable and animal life. Disposal of such wastes by discharging them to rivers or lakes in the vicinity of the plant is impermissible. Maximum permissible contents of harmful impurities in discharged waste-waters are fixed by local or national laws. Since chemical production is extremely diversified the nature of the pollution in each case also differs widely. This makes it difficult to work out standard methods for water treatment. The methods used are neutralization, sedimentation, chemical conversion of harmful components, etc., depending on the properties of the impurities. Organic impurities are often most effectively removed by *biochemical oxidation*. For instance, bacteria can be used to oxidise phenol, impurities of sulphite pulp mill waste liquors, etc.

How to purify waste-waters of chemical enterprises is a problem that has not been completely solved and it demands continued attention by the staff of chemical plants, research laboratories, and government agencies.

FUEL AND POWER FOR THE CHEMICAL INDUSTRY

Chemical engineering processes are accompanied either by release of heat in exothermic reactions, or by consumption of heat in endothermic reactions. If a reaction proceeds with the release of heat and only the heat of the reaction is utilised for maintaining the process temperature (without the use of heat from the outside) the process is autothermic. For instance, the oxidation of SO_2 to SO_3 proceeds according to the reaction

$$2SO_2 + O_2 \rightleftarrows 2SO_3 + Q$$

The heat (Q) released in this reaction is sufficient to maintain the necessary temperature in the catalytic converters. However, many exothermic reactions do not produce enough heat for autothermicity, and additional heat is necessary to maintain the required process temperature.

Endothermic reactions consume heat (energy), which must be supplied from the outside, the amount of heat necessary is often quite considerable.

However, large amounts of power are consumed at chemical enterprises in transporting raw materials and preparing them for processing, i.e. crushing, grinding and concentrating, and in moving gaseous, liquid and solid materials, compressing gases, etc. The chemical industry is therefore a major power consumer. It uses various kinds of energy :

(1) *Electric power*—for electrolysis of solutions and fused salts, for electro-thermal processes, in which electric energy is converted to heat

(electric furnaces); for electromagnetic and electrostatic processes, etc.; electric power is converted in large amounts to mechanical energy (for carrying out all kinds of physical processes in the chemical industry: crushing, mixing, centrifuging, moving materials, etc.).

(2) *Thermal energy*—for heating, drying, evaporating, melting, and for carrying out endothermic reactions.

(3) chemical energy released in biochemical and exothermic processes.

(4) energy of light—for various photochemical reactions, for instance, the synthesis of HCl from chlorine and hydrogen.

The chemical industry is beginning to employ chemical processes in which radioactive radiation is used for carrying out polymerization, synthesis of phenol, aniline and other products.

Sources of energy are divided into two groups: not renewable and renewable sources. The first group includes deposits of coal, shales, petroleum, natural gas, and nuclear fuels (uranium, thorium); the second group, the energy of rivers, tides, the wind, the sun, heat from the interior of the earth, and vegetable fuels. The energy values of various fuels are listed below :

Coal	8 k Wh/kg
Coke	7.2 "
Lignite	2.5 "
Peat	4.0 "
Coke-oven gas	4.8 kWh/m^3
Natural gas	10.6 "

The share of fossil fuels in the world's power balance is slightly over 80% of total power consumption. Coal ranks first, followed by petroleum, natural gas, and peat. The share of renewable energy sources constitutes about 20%, vegetable fuel (wood) and water power being the most important items.

The power balance of the modern world shows a tendency toward a gradual increase in the share in it of petroleum and natural gas (at the expense of coal) and also toward increased utilisation of water power.

The use of nuclear power is steadily increasing. The conversion of nuclear energy to power no longer presents any unsolvable difficulties. As the power capacity of nuclear plants grows they are becoming competitive to thermal power stations. Of exceptional interest is the problem of utilisation of energy generated by thermonuclear synthesis of helium nuclei from deuterium (heavy hydrogen) nuclei. It is believed that "the utilisation of energy latent in the world's practically limitless resources of heavy hydrogen will provide progressive humanity with a new and inexhaustible source of energy within the next few decades. This source of energy will enable man to control the weather, carry out gigantic reclamation and irrigation projects, synthesise edibles from inorganic raw materials, and realise other bold projects".

Since expenditures on fuel and power usually make up a large part of the cost of a chemical product chemical engineers should strive to increase the efficiency of power utilisation, i.e. to reduce consumption of fuel and power. The power consumption factor is the ratio of the energy theoretically required for obtaining a unit of the product to the amount actually consumed.

To reduce power consumption, chemical plants utilise the heat carried by gases, liquids and solids leaving reaction apparatuses (heat regeneration) and also the heat of compressed gases, thereby reducing losses of energy to the environment.

Hot waste gases are generally used for preliminary heating of materials on their way to reaction apparatus, or for generation of steam necessary for the production process. This is done in heat exchangers, which are divided into recuperators and regenerators, and in waste-heat boilers. In recuperators heat from waste gases is transferred through the pipe walls. Fig. 3.6 shows one such heat exchanger. A heat exchanger (heat recuperator) usually consists of a cylindrical shell 1 containing pipes 3 inserted in the holes of two plates 2. The hot waste gases leaving the reaction apparatus pass through the tubes of the heat exchangers; the cold gases on their way to the reaction apparatus flow through the space between the tubes.

In a regenerator, which represents a chamber filled with a firebrick checker-work, heat exchange is accomplished in a different way. First, the hot waste gases are passed through the chamber, heating the packing, then the flow of hot gases is shut off, cold gases are passed through the chamber where they take up the heat accumulated by the packing. The cooled packing is then heated again by the hot waste gases, etc. Thus, a single regenerator chamber operates as a batch type apparatus, and at least two such chambers are necessary to ensure continuous heat exchange. While the packing of one chamber is being heated, i.e. when hot waste gases are being passed through it, cold gas is passed through the other chamber, where the packing gives up heat to the cold gas. After fixed intervals the gas flow is automatically switched.

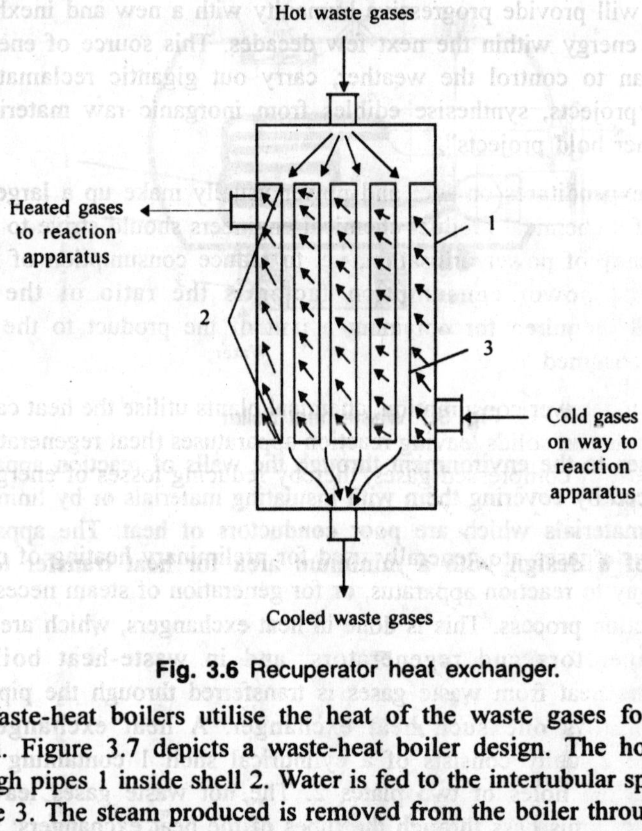

Fig. 3.6 Recuperator heat exchanger.

Waste-heat boilers utilise the heat of the waste gases for producing steam. Figure 3.7 depicts a waste-heat boiler design. The hot gas flows through pipes 1 inside shell 2. Water is fed to the intertubular space through nipple 3. The steam produced is removed from the boiler through moisture separator 5 and valve 4. Waste gases are sometimes used to heat air or water for industrial heating and similar purposes (hot water for showers, hothouses, etc.).

To utilise the available heat of liquids leaving reaction apparatus, heat exchangers of the type are shown in Fig. 3.6, which are usually employed; both gases and liquids can be heated in them.

The available heat of solids is sometimes utilised for heating air used in the production processes. For instance, clinker obtained in cement kilns is cooled by air, and the heated air is used for burning pulverised fuel in the kiln. The energy of exit compressed gases or liquids is used to turn the rotors of gas and water turbines (often mounted on the same shaft with a pump and electric motor) to reduce consumption.

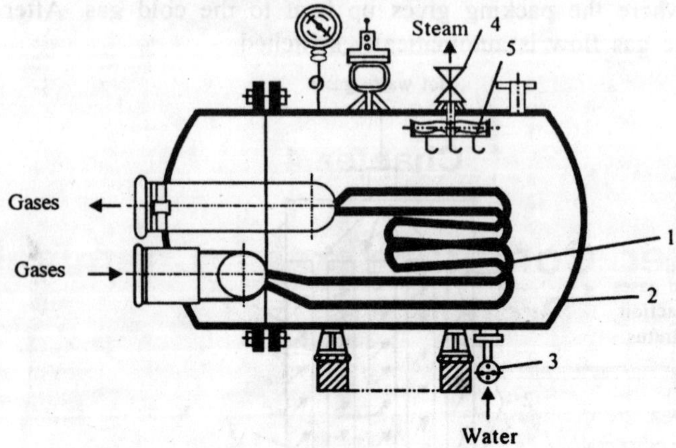

Fig. 3.7 Waste-heat boiler.

Heat losses to the environment through the walls of reaction apparatus can be reduced by covering them with insulating materials or by lining the walls with materials which are poor conductors of heat. The apparatus should be of a design with a minimum area for heat transfer to the environment.

Chapter 4

Water Conditioning in Chemical Process Industry

INTRODUCTION

Water is used for industrial and municipal purposes. The largest water requirement is for municipal use but the standard of purity required for this purpose is quite different from that demanded for industrial and commercial use.

In industry more than half the water is used in chemical plants for the purpose of cooling. A large volume of water is also used for the purpose of dilution and making solutions. While for the first purpose the composition of cooling water is of no importance, in the second, the water used should be reasonably pure.

The availability of water, both in quantity and quality, is one of the prime factors in deciding the growth of towns and cities as well as industries. For chemical industries, the available water must be as near as possible to the factory site and should also be soft. Otherwise the manufacturing cost will increase.

SOURCES OF WATER SUPPLY

In nature water occurs abundantly as: (i) water vapour in the atmosphere; (ii) as liquid water in inland waters and the ocean; (iii) as ice in polar regions; and (iv) as water of hydration in many rocks and minerals in the earth's crust. The original source of all natural water supplies is, however, rain. Rain water, in its passage through the air, dissolves many of the soluble gases present in the atmosphere and in percolating through the various strata of the earth, collects mineral salts and other impurities. These may be either acidic or alkaline. Of the dissolved gases, oxygen and carbon dioxide are the most important. The dissolved minerals include bicarbonates, carbonates, chlorides, sulphates and nitrates of calcium, magnesium, sodium and sometimes iron also. Silica may also be found in places. The water available may also contain suspended matters—both mineral and organic. The quality of water depends entirely upon its source of supply.

Rain Water

Rain water is the purest form of water but to be absolutely pure it should be collected after a heavy shower, because prior to that, the water may contain dissolved gases and also dust. In industrial areas, gases in the atmosphere contaminate rain water, e.g., sulphuric acid. Dissolved carbon dioxide renders water acidic and increases its solvent power. When this acidic water passes through the soil, it reacts with the calcium and magnesium carbonates present in the soil and forms bicarbonates. Pure rain water is soft.

River Water

River and canal waters are inland waters fed by rains. These are less pure than rain water as they contain dissolved chemicals. Their composition is subject to considerable variation depending on the areas over which they are flowing and the soluble chemical salts present in the rocks and soil. Dissolved solids in the river water also vary with the rate of flow. The impurities in river water are generally the chlorides, sulphates, carbonates and bicarbonates of calcium, magnesium and iron. The soluble bicarbonates of calcium and magnesium are formed by the action of dissolved carbon dioxide upon insoluble carbonates. The presence of soluble calcium and magnesium salts in river water makes it hard. Besides this, river and canal waters are liable to be contaminated with sewage and sludges of cities by which they flow and also organic matter, ammonia and nitrites.

Spring and Well Water

This is percolated rain water which, on reaching the ground, passes through the various strata of the soil, thereby dissolving the soluble salts of the soil. This water is extremely clear as a result of natural filtration through sand beds and strata. The chief difference between spring and well water and river water is that the former contains a much higher percentage of dissolved mineral matter although it is free from organic impurities and ammonia, and often has a palatable taste and in most cases, may be used for drinking. Water from springs and wells is extremely hard.

Sea Water

Rivers carry a large volume of water with dissolved impurities into the sea and, as such, the sea contains the maximum amount of dissolved impurities. The total dissolved impurities in sea water are about 3.6 per cent of which 2.6 per cent is common salt; others are magnesium chloride, magnesium sulphate, calcium sulphate, potassium chloride, magnesium bromide and traces of iodides, silica, etc. The quantity of dissolved salts is very much greater in inland seas. Sea water is saline, hence unsuitable for drinking purposes and is hard. Sea water is also faintly alkaline.

Surface and Ground Water

As rain falls on the earth, it flows into streams as surface runoff, passes into the air as vapour and goes down into the ground. It is then recovered as surface water or ground water.

The term surface water indicates natural water in rivers, streams, ponds and lakes, although a part of such deposits of water may have percolated more or less through the soil in the course of its flow along the earth's surface. As a result, the water absorbs a part of the materials with which it comes in contact. Surface water, therefore, indicates the chemical constitution and physical conditions of the area in question.

The water obtained from springs and wells in a given locality may be very different from the local surface water. Ground water also indicates the local geological strata and contains a greater percentage of dissolved salts. Ground water is more nearly constant in composition. Some of the ground water may be hot and may contain dissolved gases. It is sometimes used for medicinal baths. Ground water may be different in taste and chemical character—acid, alkaline, bitter, hepatic, ferruginous, iodinous or siliceous.

CHARACTERISTICS OF WATER

Water freezes at 0°C and boils at 100°C at NTP. This is in sharp contrast to the melting and boiling points of its higher analogues, hydrogen sulphide, hydrogen selenide and hydrogen telluride. Moreover, the latter are bad smelling gases at ordinary temperature and pressure. Water also exhibits many other abnormal properties. Compared with hydrogen sulphide, it has a higher surface tension and a higher dielectric constant. Liquid water shows a maximum density at 4°C, unlike all other substances, whereas ice is less dense than water. The specific heat as well as latent heat of fusion and evaporation of water is all abnormally high. All these facts are explained by attributing to water molecules the tendency to associate, giving rise to polymers of the type $(H_2O)_n$ as a result of hydrogen bondings.

Pure water is virtually a non-conductor of electricity. On heating to a very high temperature water molecules are dissociated into the elements. The formation of water from hydrogen and oxygen is an exothermic reaction:

$$H_2 + 1/2\ O_2 = H_2O\ \text{(Steam)} \qquad \Delta H = 58{,}110\ \text{calories.}$$

Pure water takes part in a variety of chemical reactions. Pure water at room temperature acts on alkaline metals violently and on alkaline earth metals rapidly with evolution of hydrogen. Magnesium and aluminium amalgams are rapidly attacked. Magnesium, iron, zinc and carbon react with water at high temperature. Iron and lead are also attacked by water in the presence of air. Hard water has less reaction on lead than pure and soft water, as the sulphates and bicarbonates present in hard water form a coating of sulphates and carbonates of lead upon the metal which protect it from further action. Hence, soft water should not be supplied for drinking purposes through lead pipes. A solution containing traces of lead is poisonous to the human system.

Water acts as a catalyst for many physical and chemical changes. Perfectly dry hydrogen and oxygen do not react. Although pure water is a non-conductor, water shows a slight dissociation.

$$2H_2O \rightleftharpoons (H_3O)^+ + (OH)^-$$

At 25°C, the ionic product of water is 1×10^{-14}. Water is an excellent electrolytic solvent, particularly for ionic compounds.

WATER FOR INDUSTRIAL PURPOSES

In industries water is used for various purposes. A large amount of water is used in the industries for cooling purposes. Water used for cooling purpose need not be completely free from the impurities. However, when water is used for solutions, carrying out the reactions, dilution etc. should be reasonably pure. Each industry has its own specifications for water. The specifications of water for some industries are given below:

Water for Boilers

The boiler feed water should be free from soluble salts of calcium and magnesium and should not contain nitrates and organic matter.

Water for cooling system

Should be non-corrosive, and non-scale forming.

Water for beverages

It should not be alkaline.

Water for distilleries

As far as possible, water should be free from micro-organisms and should be pure.

Water for paper mills

Should not contain excess of lime and magnesia and should be free from iron salts.

Water for sugar factories

Should not be rich in sulphates, alkaline carbonates and nitrates. Micro-organisms should be minimum.

Water for dairies and allied industries

Water should be colourless, odourless, tasteless and free from pathogenic bacteria

Water for laundries

Should be soft.

EFFECT OF WATER ON ROCKS AND MINERALS

When water flows through the ground rocks or soil, it comes in contact with various mineral constituents and so gets contaminated because of various physical and chemical changes such as dissolution, hydration or action of dissolved CO_2 etc. For example, some minerals such as $CaSO_4$ (anhydrite), Mg_2SiO_4 (olivine) etc. readily undergo hydration when come in contact with water. As a result, products having increased volume are formed and consequently such mineral bearing rocks undergo disintegration.

$$CaSO_4 + 2H_2O \rightarrow CaSO_4.2H_2O \text{ (Gypsum)}$$

The resulting gypsum has a volume about 33% more than the volume of the mineral anhydrite, $CaSO_4$. Similarly, Mg_2SiO_4 undergoes hydration according to the reaction.

$$Mg_2SiO_4 + xH_2O \rightarrow Mg_2SiO_4.xH_2O \text{ (Serpentine)}$$

Mineral constituents of rocks like NaCl (rock salt), gypsum ($CaSO_4.2H_2O$) etc. are readily soluble in water. Dissolved oxygen brings about oxidation as well as hydration reactions. For example:

$$\underset{\text{Magnetite}}{2Fe_3O_4} \xrightarrow[\text{Oxidation}]{1/2\ O_2} \underset{\text{Haematite}}{3Fe_2O_3} \xrightarrow[\text{Hydration}]{2H_2O} \underset{\text{Limonite}}{3Fe_2O_3.2H_2O}$$

Dissolved CO_2 reacts with insoluble carbonates of Ca, Mg and Fe and forms soluble bicarbonates.

BOILER SCALE AND SLUDGES

We know that a large amount of water is used in boilers for steam generation. The boiler feed water should be as soft as possible and also be free from dissolved gases like oxygen, carbondioxide and also it should not be alkaline. Therefore it is necessary to treat the boiler feed water suitably otherwise it causes many problems like :

(1) Formation of boiler scales and sludges.

(2) Boiler corrosion.

The formation of boiler scales and their ill effects are discussed in the following section.

In boilers, water is boiled to generate the steam. The steam is continuously withdrawn from the boiler and fresh water is fed into the boiler. The suspended and dissolved matter present in water are left behind in the boiler. With the increase in concentration of these salts, they are separated in the order of their solubility. If the salts thus separated, form hard and sticky coating on the boiler surface it is called *boiler scale*. On the other hand if the deposited material are in the form of soft mud loosely held on the surface, it is called a *sludge*.

Mechanism of Boiler Scale Formation

When temporary hard water is fed into the boiler, due to boiling the calcium and magnesium bicarbonates present decompose and are converted into respective insoluble carbonates. These carbonates deposit on the boiler surface and form the hard scale.

$$Ca(HCO_3)_2 \longrightarrow CaCO_3 \downarrow + H_2O + CO_2 \uparrow$$

$$Mg(HCO_3)_2 \longrightarrow MgCO_3 \downarrow + H_2O + CO_2 \uparrow$$

When hard water is fed into the boiler, on boiling it evaporates leaving behind the suspended matter and dissolved salts. Consequently, the concentration of these materials goes on increasing, ultimately they are separated and form boiler scale or sludge.

The solubilities of certain substance, like calcium sulphate decrease at the higher temperature. Hence they are separated and form boiler scale.

On the basis of chemical composition, the scales formed in boiler units can be divided into the following four groups :

(1) *Alkali-earth scales:* These scales consist mainly of calcium and magnesium compounds such as $CaCO_3$, $CaSO_4$, $CaSiO_3$, $Ca_3(PO_4)_2$, MgO, $Mg(OH)_2$, $Mg_3(PO_4)_2$ etc. These are further classified as carbonate scale ($CaCO_3$), sulphate scale ($CaSO_4$), phosphate scale [$Ca_3(PO_4)_2$] and other depending upon which of these compounds prevails. If many compounds are present in small amounts, the scale formed is called composite scale.

(2) *Iron oxide scales:* These scales consist of ferric and ferrous compounds such as iron silicates, ferrous phosphate $Fe_3(PO_4)_2$, sodium ferrophosphate ($NaFePO_4$) and iron oxides (Fe_2O_3, Fe_3O_4).

(3) *Copper scales:* Containing considerable amount of copper.

(4) *Silicate scales:* Of various compositions.

The most important property of the scale of any kind is its low thermal conductivity which varies with the nature and porosity of the deposits.

Effects of Boiler Scales and Sludges

Wastage of fuel

The scales formed in the boilers are heat insulators. Due to this, heat supplied to the boiler is not efficiently transferred to water. This results in reduction of the rate of heat transfer. The boiler has to be heated to a higher temperature in order to maintain the supply of the steam and thus more fuel is required.

Explosion of boiler

Due to over heating, the boiler metal becomes weak and cannot withstand the pressure developed in the boiler. Also, due to over heating minute cracks are developed in the boiler scale. Water seepages through these cracks are comes in contact with the over heated boiler plates. This is called *bagging*. Chemical reaction takes place between the hot boiler metal and the water in the cracks with the evolution of hydrogen gas. This causes the erosion of the boiler material. The weakened boiler material will not withstand the pressure developed in the boiler and hence the boiler may burst. Large quantities of sludge formed may block the steam outlets. This is called *plugging*. The plugging may result in the bursting of the boiler.

FORMATION OF DEPOS!TS IN BOILER UNITS

(1) *Formation of alkali-earth deposits:* These deposits are formed in the presence of sufficiently high concentration of scale forming substances or insoluble substances. If boiler water contains cations such as Ca^{2+} and Mg^{2+}, and anions such as OH^-, CO_3^{2-}, SO_4^{2-}, PO_4^{3-}, or SiO_3^{2-}, then the interaction of these ions give rise to a number of scale forming substances such as $CaCO_3$, $CaSO_4$, $CaSiO_3$, $Ca_3(PO_4)_2$, $Mg(OH)_2$, $Mg_3(PO_4)_2$, $3MgO.2SiO_2.H_2O$, which are difficultly soluble compounds.

The rate of formation of alkali earth deposits depends on the concentration of scale forming substances. An increase in the concentration of substances in boiler water is also responsible for the formation of difficultly soluble compounds. The solid phase may precipitate or crystallise from a solution and deposit directly on a heating or cooling surface. This process of deposition of hard scales is known as *primary process of scale formation*. In the secondary *process of scale formation*, suspended particles first precipitate in the bulk of water and subsequently form secondary deposits that strongly adhere to the surface.

(2) *Formation of iron oxide deposits:* The formation of iron oxide deposits are most frequently detected in high duty boiler units, in the water wall tubes of saline sections, on the stretches of tubes exposed to high rates of heat flow, at the spots where the burner flame impinges onto heating surface etc. The important components of iron oxide deposits are iron phosphates, depending upon the conditions of scale formation. The cause of such deposits is the introduction of iron into a boiler unit together with feed water. Iron comes into the feed water due to *corrosion processes* taking place in water condensate circuit of electric power stations and industrial boiler houses.

(3) *Formation of copper scales:* Copper scales, which may contain 20–30% copper, also contain iron oxides and impurities such as calcium and magnesium compounds. Copper scales are usually formed in boiler units operated at different pressures and the main factor responsible for their formation is the introduction of copper compounds into boiler

units with feedwater. In copper scales, copper is present as metal and metal oxides.

Disadvantages of Scale Formation in Boiler

(1) Scale or sludge is a poor conductor of heat. As a result evaporation capacity of boiler decreases. Hence fuel consumed is much more than usual.

(2) Scale or sludge does not allow the water to come in contact with plates and tubes of the boiler. This causes overheating as well as burning out of these plates and tubes.

(3) Tubes and plates are elogged decreasing the efficiency of the boiler.

(4) Tubes may be corroded by some salts such as $MgCl_2$, if deposited as scales.

Methods of Preventing Deposit Formation

The formation of deposits is usually prevented by using the following measures :

(1) Deep softening of make up water.

(2) Preventing contamination of feed water and its components by the products resulting from the corrosion of structural materials.

(3) By eliminating leakage of cooling water in turbine condensers.

(4) By internal boiler water treatment.

(5) Chemical control of boiler water to meet the required standards.

(6) By eliminating high rates of local heat flow through steam generating tubes.

It is not possible to remove completely the scale forming substances from water by softening. The residual hardness of softened water is generally 3–4 μg-equiv/kg and sometimes more. The boiler water treatment makes the scale forming substances harmless, otherwise they penetrate into the boiler unit. The boiler water treatment thus eliminates their deposition on heating surfaces. For boiler water treatment *orthophosphates* or salts of EDTA are introduced into the boiler units.

PHOSPHATE TREATMENT OF BOILER WATER

Cations such as Ca^{2+} and Mg^{2+} entering a boiler unit are capable of producing scale forming compounds with different anions present in the boiler water. These compounds behave differently in boiler water. For example, some of them as $CaSO_4$, $CaSiO_3$ etc. precipitate from the solution and form scales mainly on the surface of steam generating tubes. Other such compounds as $CaCO_3$, $3Ca_3(PO_4)_2$, $Ca(OH)_2$ precipitate mainly as suspension that turn into sludge and is removed from the boiler unit by blow down.

The object of phosphate treatment is to tie up Ca^{2+} ion so that it becomes harmless, (i.e., to create conditions under which scale formers would precipitate as loose sludge). One way to do this is to form a complex containing Ca^{2+} ions. It should be noted that phosphate treatment prevents only the *calcium scale*. Triphosphate magnesium $Mg_3(PO_4)_2$ is capable of yielding *magnesium scale*. *Phosphate treatment* is based on two methods:

(1) *Phosphate Alkaline Method:* in which a certain concentration of PO_4^{3-} ions and free OH^- ions is continuously maintained and this concentration is not bound by PO_4^{3-} ions.

(2) *Purely Phosphate Alkalinity Method:* in which presence of only PO_4^{3-} ions in boiler water is necessary and OH^- ions are bound with PO_4^{3-}. Such conditions are possible only when trisodium phosphate is present in water. Hydrolysis takes place as follows

$$PO_4^{3-} + H_2O \rightleftarrows HPO_4^{2-} + OH^-$$

If such a solution is concentrated by evaporation, the reaction will proceed to the left. OH^- ions will be bound by HPO_4^{2-} ions and PO_4^{3-} ions are formed. This makes the method of purely phosphate alkalinity highly desirable for boiler units operated at 10.0 MPa or higher.

In addition to trisodium phosphate (Na_3PO_4) and hexametaphosphate $(NaPO_3)_6$, tripolyphosphate ($Na_5P_3O_{10}$) has also been used in phosphate treatment.

TREATMENT OF BOILER WATER WITH COMPLEXING AGENTS

The phosphate treatment is not capable of preventing the formation of iron oxide and copper depositions. The conditions of boiler water with complexing agents such as EDTA and disodium salt (trylon B) can secure: (i) scale free and sludge free operation of boiler unit; and (ii) corrosion free operation under certain conditions. The main object of treatment with complexing agents is :

(1) To prevent the deposition of iron oxides in boiler units.

(2) To reduce the carry over of oxides with steam.

(3) To protect the units against steam-water corrosion.

ALKALI TREATMENT OF BOILER WATER

As a result of alkali treatment, the scale forming substances penetrating into the boiler unit precipitate as $CaCO_3$ and $Mg(OH)_2$. In a non-alkaline medium, $CaCO_3$ falls out on heating or cooling surfaces and forms dense scale. With alkaline boiler water, $CaCO_3$ precipitates in the form of a loose sludge not adhering to surfaces. The alkaline treatment of boiler water in boiler units operated at a pressure upto 1.6 MPa is efficient, because stability of CO_3^{2-} ion depends upon temperature, whose increase is followed by decomposition of soda (Na_2CO_3). As a result, the effectiveness of alkaline treatment decreases and steam generated gets contaminated with CO_2. It is to be noted that alkali treatment is useful only when other methods of preventing scale formation prove impracticable.

PREVENTING DEPOSIT FORMATION IN TURBINE CONDENSERS

At electric power stations much water is used as a cooling agent. For the purpose of water economy, closed circuit type cooling water system is used. In this system, cooling water circulates in a closed cycle, i.e., it heats up in the apparatus to be cooled and cools in a supply cooling pond or cooling tower. In such type of water supply systems, various reactions take place in the circulating water. When water is sprayed in a pond or tower, dissolved CO_2 is removed from it. It causes the decomposition of calcium bicarbonate.

$$Ca(HCO_3)_2 \rightarrow CaCO_3 + CO_2 + H_2O$$

This reaction is responsible for the formation of solid carbonate deposits in the tubes of turbine condensers. There is also partial evaporation of water, and thus the concentration of substances in water increases. The formation of carbonate deposits in turbine condensers is prevented by treating makeup water in any of the following ways:

(1) *Acidification:* In this process H_2SO_4 is usually introduced into make-up water. It destroys HCO_3^- ions and forms CO_2.

$$HCO_3^- + H^+ \rightarrow H_2O + CO_2$$

Thus carbonate hardness in water decreases and concentration of free CO_2 increases. All this terminates calcium bicarbonate decomposition.

(2) *Lime Treatment:* The destruction of HCO_3^- ions can also be carried out by treating water with lime, $Ca(OH)_2$, through

$$HCO_3^- + OH^- \rightarrow CO_3^{2-} + H_2O$$

without formation of CO_2. Lime treatment removes all bicarbonate ions from water.

(3) *Phosphate Treatment:* In this process phosphates introduced into water deprive $CaCO_3$ of its scale forming properties.

(4) *Recarbonisation:* Treating water with gases containing CO_2, in particular with the furnace gas from boiler units, is aimed to increase concentration of free CO_2 in water. The process is known as *recarbonisation*, in which CO_2 is blown through water or mixed with water by means of water operated ejector. Thus CO_2 contained in the gas is dissolved in the water. The saturation of water with CO_2 thus prevents the decomposition of bicarbonate.

PRIMING AND FOAMING

The passage of water particles mixed with steam from the boiler is called *priming* which is caused by the presence of large quantity of alkali sulphates and chlorides present in water. Priming is generally associated with *foaming*. When water is associated with suspended fine solid particles, each of these fine particles releases steam bubbles. When these bubbles break, they cause *foam*. The breaking up of bubbles at a certain stage is so quick that the water film enclosing the steam around solid particles, passes out from the boiler along with steam and hence causes *priming*. The dissolved impurities are mainly responsible for priming and foaming and the degree of foaming and priming depends upon the quality and quantity of soluble impurities. As the concentration of water rises in water, the quantity of dissolved impurities increases and hence priming also increases. Priming and foaming can be avoided by: (i) Providing sufficient steam space; (ii) Removing oil or soap from water; (iii) Not allowing accumulation of fine suspended particles in boiler water either by filtration or by efficient softening; and (iv) By blowing off concentrated water for the removal of saline water, at regular intervals. Taking into consideration priming and foaming, water may be graded into three types:

(1) *Non-Foaming Water:* Which can be used in a locomotive boiler for seven days without any foaming.

(2) *Semi-Foaming Water:* Which can be used in a locomotive boiler for two days without foaming.

(3) *Foaming Water:* Which can not be used for two days without blowing off for preventing foaming.

CAUSTIC EMBRITTLEMENT

This is a form of corrosion which may attack boilers. It is caused by a high concentration of sodium hydroxide, which can react with steels stressed beyond their yield point. This particular form of corrosion is characterised by the formation of irregular intergranular cracks at places of high local stress.

If sodium carbonate is present in water, it is hydrolysed to give sodium hydroxide, which reacts with iron to form magnetic oxide.

Under ordinary conditions in unstressed metal a fairly continuous film of oxide is produced, but when the metal is stressed above its yield point, the oxide coating cracks and the chemical attack continues into the metal. The attack is fairly rapid under these conditions due to the large area of metal exposed. It is considered that the presence of a little sodium sulphate or sodium phosphate inhibit embrittlement either by crystallising out and plugging the capillaries and crevices with solid salts before a dangerously high concentration of sodium hydroxide has been produced. Hence, by maintenance of sufficient amount of sodium sulphate or sodium phosphate in the boiler water, embrittlement can be prevented. At low boiler pressures, a small amount of sodium sulphate will be sufficient to prevent the corrosion. The use of an excess of sodium carbonate should be avoided in lime-soda process in boiler feed water treatment. This may also prevent caustic embrittlement.

Internal Conditioning of Boiler-Feed Water

Internal conditioning means correcting the water by adding some chemicals to the water in the boiler itself. It is used for supplementing the external treatment such as lime-soda process, demineralisation, deaeration, etc. As regards hardness, the impurities likely to form boiler scales are converted into sludge-forming impurities or they are complexed (sequestered) so that they remain in solution, but yet they are out of harm's way. The compounds used for the internal conditioning of water for boilers are called boiler compounds. Sodium phosphates, tannin and hydrazine are some of the familiar boiler compounds.

The most important step in the internal conditioning of boiler feed water is phosphate treatment. The three phosphates of sodium dihydrogen sodium phosphate (NaH_2PO_4), disodium hydrogen phosphate (Na_2HPO_4) and trisodium phosphate (Na_3PO_4) are used. Any of these phosphates will react with calcium ions to form a sludge of calcium phosphate.

$$3CaCl_2 + 2Na_3PO_4 \rightarrow Ca_3(PO_4)_2 + 6NaCl$$

Depending on the pH of the water to be treated, one of three kinds of phosphate is chosen. If the water is acidic, trisodium phosphate, which on hydrolysis yields alkali, is added. If the water is neutral, disodium hydrogen phosphate is used. The sludge formed is removed by regular blow-down.

Sludge formation can be avoided if calgon, i.e., sodium hexametaphosphate is used. This compound forms a complex with calcium salts in which the calcium forms part of an anion. Hence it no longer forms scales in the boiler.

$$Ca^{++} + (NaPO_3)_6 \rightarrow Na_2(Ca_2P_6O_{18}) + 4Na^+$$

This complex is water-soluble and hence sludge disposal problem does not arise.

A second boiler compound of frequent use is hydrazine. This chemical combines with dissolved oxygen to form water and N_2. Since this treatment does not increase the dissolved salt concentration of the water, it is always preferred to adding ferrous sulphate or sodium sulphite.

Tannin is another boiler compound used for the complexing of iron. Some coagulants are also used as boiler additives for removal of oil and silica.

Internal conditioning also goes by the name of threshold conditioning because the treatment is done at the gateway to the boiler.

FACTORS PROMOTING CORROSION

The electrochemical corrosion mainly depends upon two factors. These are external factors of corrosion such as nature of the surrounding medium, pH value etc. and internal factors of corrosion such as grades and kind of metal, the presence of stresses, mechanical or heat treatment etc.

Under operating conditions of heat power equipment, the main factor which determines the intensity of corrosion is *oxygen* dissolved in water. Since oxygen is a depolariser, it contributes to the action of other factors of corrosion effectively. Oxygen corrosion of metal first produces ferrous hydroxide.

$$2Fe + O_2 + 2H_2O \rightarrow 2Fe(OH)_2$$

Corrosion of oxygen in neutral or alkaline medium gives rise to the formation of *pits*, i.e., to sharply pronounced local destruction of metal. At pH < 7, oxygen corrosion usually develops more evenly. Since oxygen comes into water from the atmospheric air, care must be taken to preclude exposure of deaerated water or condensate to the atmosphere.

The other factor influencing the development of corrosion is the presence of CO_2 in water. With water, it forms carbonic acid, H_2CO_3 that dissociates to form H^+ ions.

$$CO_2 + H_2O \rightleftarrows H_2CO_3 \rightleftarrows H^+ + HCO_3^-$$

The H^+ ions acidify water and cause corrosion attended with hydrogen depolarisation. CO_2 corrosion is most likely to occur in condensate lines, condensate tanks, air heaters and other similar equipments in which internal surfaces are exposed to condensate. The most important characteristic of CO_2 corrosion is that the above reaction does not proceed to completion. Thus H_2CO_3 is also present in water along with H^+ ions and HCO_3^- ions. As neutralisation of H^+ ions on the cathode continues, the equilibrium gets disturbed and again proceeds from left to right. As a result, new H^+ ions are formed and these H^+ ions maintain the corrosion aggressiveness of the solution at the previous level. This process, which makes up for the decrease in H^+ ions, goes on until H_2CO_3 is completely dissociated. When a strong acid such as HCl is added for water acidification, the process of this kind does not occur. Here complete dissociation of HCl is expected to occur and the decrease in H^+ is not compensated by the process of corrosion developed. As a result, concentration of H^+ decreases gradually and hence corrosion slows down. Hence CO_2 solution is more aggressive than HCl solution at the same pH value.

The HCO_3^- ions are bound by Fe^{2+} ions, which originate in the course of *steel corrosion.* Ferrous bicarbonate also decomposes with the formation of $Fe(OH)_2$ and CO_2:

$$Fe(HCO_3)_2 \rightleftarrows Fe(OH)_2 + 2CO_2$$

CO_2 is now able to react again. This process is however, possible when water contains sufficient amount of O_2 required for oxidising Fe^{2+} to Fe^{3+} and eliminating it from the reaction:

$$2Fe(OH)_2 \rightarrow 1/2\ O_2 + H_2O + 2Fe(OH_3) \downarrow$$

If the corrosion products strongly adhere to the surface of the metal under going corrosion and form a sufficiently dense layer, they may gradually reduce the rate or intensity of corrosion or terminate corrosion altogether.

At electric power stations corrosion occurs in feed lines, steel water economisers, regenerative feed water heaters, heat exchangers and in return condensate lines. The feed line from the deaerator or to the boiler drum is always filled with fresh hot water. When it comes into contact with metal, corrosion takes place if water contains depolarisers. The temperature also increases the rate of corrosion, which also depends upon whether oxygen-containing water is heated in a closed heat exchanger or in a tank open to the atmosphere. In the first case rate of corrosion increases continuously with increase in water temperature. In open system, the rate of corrosion is maximum at 348–358K.

Heaters and evaporators usually suffer corrosion on the side of the heating steam and corrosion is caused by dissolved CO_2 contained in the condensate.

CO_2 finds its way into the steam water circuit of a heat power plant with chemically treated water and due to cooling water leakage in turbine condensers. In both the cases, HCO_3^- and CO_3^{2-} ions are introduced into the system and decomposed in the boiler unit forming CO_2. The most harmful effect of CO_2 corrosion is the contamination of feedwater with iron oxides. This contamination increases with feedwater temperature and time of metal exposure to condensate containing CO_2.

The corrosion of feedwater and steam condensate lines of a power station may be controlled by removing the aggressive gases CO_2 and O_2 from feedwater and condensate. CO_2 corrosion can largely be prevented by carrying out the following measures :

(1) Reducing the amount of free CO_2 in steam by suitable methods of make-up water treatment.
(2) Removing heating steam condensate from regenerative feed water heaters.
(3) Completely venting non-conducting gases from steam space of heaters.
(4) Treating feedwater with ammonium or steam with film forming amines.

CORROSION OF BOILER

Oxygen corrosion occurs in steel water economisers of boiler units. If feedwater is poorly deaerated, steel economisers may fail in 2–3 years after being put in service.

Oxygen corrosion is most commonly encountered in the inlet sections of economisers and iron economisers do not suffer oxygen corrosion. If oxygen concentration in feedwater is suff ently high, oxygen may pass into the boiler unit too. In boiler units corrosion occurs mainly in drums and down comer tubes. An increase in pressure intensifies oxygen corrosion. The composition of water coming into contact with a metal is also important. Small amount of alkali intensifies local corrosion, while chlorides cause corrosion over the entire surface of metal.

Idling boiler units are subject to electrochemical corrosion, which is known as outage or out of service corrosion. The various methods of preventing corrosion when boilers are taken out of service are:

(1) Placing boiler units in dry storage.

(2) Placing boiler units in wet storage.

(3) Coating the internal surfaces of a boiler unit with a film of 10% $NaNO_2$ solution.

(4) Filling the boiler unit with a hydrazine solution.

(5) Connecting the boiler unit to a source of saturated steam at some pressure.

(6) Treating the internal surfaces of a boiler unit with complexing agents.

(7) Preparing for storage by using nitrogen.

Intercrystalline corrosion occurs in riverted and expanded joints of boiler units that are exposed to boiler water. The metal first develops fine cracks which later on grow with time. The cracks run between grains or crystallities of the metal and hence the corrosion is called *crystalline corrosion.* Disintegration caused by intercrystalline corrosion is also called brittle, stainless. This type of corrosion depends mainly upon the following factors:

(1) High tensile stresses in metal.

(2) Leakinesses in riverted or expanded joints.

(3) Aggressive properties of boiler water.

Fractures do not occur at all in the absence of any one of the above three factors. Intercrystalline cracks can also develop under local cooling of small sections of a boiler drum.

QUALITY OF WATER

Meaning of Pure Water

The term *pure water* is a relative term and it has to be interpreted in relation to the use of water. The concept of pure water, potable water or wholesome water with relation to various uses of water is understood as follows:

Domestic use

The water required for domestic consumption should posses a high degree of purity and it should be free from suspended impurities, bacteria, etc. A tolerance of small degree of hardness developed due to certain dissolved salts is however permissible. Thus the drinking water and water used in the food industry and some other industries must meet the highest standard of purity. Following are the requirements of *potable* or *wholesome* water for domestic use:

(1) It should be clear, odourless and colourless.

(2) It should be free from harmful and disease producing bacteria.

(3) It should be free from all objectionable substances.

(4) It should be fresh and cool.

(5) It should be palatable i.e., aesthetically attractive.

(6) It should be tasty.

(7) It should not cause corrosion to the pipes and other fittings.

Civic use

For this purpose, a large quantity of water is required to fulfil various civic purposes such as washing of roads, cleaning of sewers, etc. The nature of use of water is such that any degree of impurity can be tolerated. Hence the water containing large amount of suspended and dissolved impurities may be permitted for this purpose. But the water which is considerably mixed up with sewage and other refuse cannot be tolerated for this purpose.

Trade or business use

The water required for a particular trade will depend upon nature of that trade. For instance, the water required for laundry should not be hard as it will result in more consumption of soap. Similarly the water required for bathing cattles and washing floors in case of stables may contain any type of impurities.

Commercial or industrial use

The water required for this purpose should be chemically pure. The various chemical processes involved in the production make it essential to use chemically pure water. A slight amount of impurity may considerably affect the final results of the product.

The nature of water required for various factories is so variable that some of the factories install their own water supply plant to supply water to their industries. Alternatively, a city with well-developed and highly water-consuming industries should have preferably two systems of water supply: (i) supplying potable water for drinking and domestic use; and

(ii) providing water for industry. Moscow, Paris and a number of other big cities of the world have now adopted this system in practice and in future, it may become more popular as a means to save the drinking water.

Reasons for the Analysis of Water

Following are the reasons or purposes for carrying out the analysis of water :

(1) to ascertain if the supplies maintain the required degree of purity and to find out the extent of any variations which occur

(2) to ascertain the effect of heavy rainfalls or of long-continued drought on river waters

(3) to ascertain the quality of the proposed supply to the new consumers

(4) to decide that the water obtained from some additional source or sources of supply will be pure, wholesome, not too hard and free from the risk of any pollution

(5) to decide the suitability of water for feeding boilers, hot-water pipes, etc.

(6) to examine the effect of pumping on well waters, especially when wells are situated near the sea or an estuary

(7) to find out the organisms responsible for the spreading of the water-borne diseases

(8) to identify the organisms responsible for developing certain effects on water with respect to colour, odour, taste, etc.

(9) to know the characteristics of waters at various depths of deep wells and tube wells

(10) to know the quality of water submitted to the various purification processes

(11) to know the quality of water used or proposed to be used for public swimming baths

(12) to know whether water from a particular supply of water is suitable for specific purpose such as paper making, dyeing, tanning, wool washing, brewing, steam raising, etc.

(13) to study the process of self-purification of streams and rivers

(14) to suggest the best method of purifying, of softening or of preventing action on mains and supply pipes; etc.

Impurities in Water

It is not possible to find pure water in nature. The rain water as it drops down to the surface of earth absorbs dust and gases from the atmosphere. It is further exposed to the organic matter on the surface of earth and by the time, it reaches the source of water supply, it is found to contain various other impurities also.

For the purpose of classification, the impurities present in water may be divided into the following categories: (i) physical impurities; (ii) chemical impurities; and (iii) bacteriological impurities.

Analysis of Water

In order to ascertain the quality of water, it is subjected to the various tests. These tests can be divided into the following categories: (i) physical tests; (ii) chemical tests; and (iii) bacteriological tests.

Physical Tests for Water

Under this category, the tests are carried out to examine water for the following: (i) colour; (ii) taste and odour; (iii) temperature; and (iv) turbidity.

Other physical characteristics for which tests are sometimes carried out are density, electrical conductivity, radioactivity and viscosity.

Colour

The pure water is colourless and following are the sources which contribute colour to the water :

(1) Algae metabolism
(2) End products of degraded organic matter
(3) Discharge of untreated and partially treated waste-water from various industries like food processing, textile industry, tanneries, paper production, etc.
(4) Divalent species containing iron and manganese; etc.

Taste and odour

The water possesses taste and odour due to various causes and they make the water unpleasant for drinking. The test is carried out by inhaling through two tubes of an osmoscope. One tube is kept in a flask containing diluted water and the other one in a flask containing water to be tested. The taste and odour of water may also be tested by threshold number. In this method, the water to be tested is diluted with odour-free water and the mixture at which odour becomes detectable is determined. It indicates threshold number and other intensities of odour are then worked out. The results of test are greatly affected by the sensitiveness of the observer. For public water supply, the threshold number should not be more than three. If an odour of chlorine or iodoform is found, it should always be recorded. The taste is expressed as brackish, saline, salty, etc. If the taste and odour are suspected to be due to the growth of any kind, the cause may be found out by conducting microscopical and biological examinations.

Temperature

The test for temperature of water has no practical meaning in the sense that it is not possible to give any treatment to control the temperature in any water supply project.

The measurement of temperature of water is done with the help of ordinary thermometers. From the study of temperature, the characteristics of water such as density, viscosity, vapour pressure and surface tension can be determined. It also helps in determining the saturation values of solids and gases which can be dissolved in water and also the rates of chemical, biochemical and biological activity.

Turbidity

The colloidal matter present in water interferes with passage of light and thus imparts turbidity to the water. The turbidity in water may also be due to clay and silt particles, discharges of sewage or industrial wastes, presence of large numbers of micro-organisms, etc. and the cloudy appearance developed in water due to turbidity is aesthetically unattractive and it may also be harmful to the consumers. It also disturbs the disinfection process because the solids may partially shield the organisms from the disinfectant.

The measurement of turbidity in the field is done by means of a turbidity rod and it is referred to as the *visual method of turbidity measurement*. For laboratory, the various instruments known as the *turbidimeters* are found out to measure the turbidity of water, the most common being Jackson turbidimeter, Baylis turbidimeter and Nephelometric turbidimeter.

Chemical Tests

Under this category, the tests are carried out to examine water for the following: (i) Chlorides; (ii) Dissolved gases; (iii) Hardness; (iv) Hydrogen-ion concentration (pH value); (v) Alkalinity; (vi) Acidity; (vii) Metals and other chemical substances; (viii) Nitrogen and its compounds; (ix) Total solids.

Chlorides

The chloride contents, especially of sodium chloride or salt, are worked out for a sample of water. The excess presence of sodium chloride in natural water indicates pollution of water due to sewage, minerals, edible oil mill operations, ice cream plant effluents, chemical industries, sea water intrusion in coastal regions, etc. The water has lower contents of salt than sewage due to the fact that salt consumed in food is excreted by body. For potable water, the highest desirable level of chloride content is 250 mg/litre and its maximum permissible level is 600 mg/litre. The measurement of chloride contents is carried out as follows :

(1) 50 c.c. of sample of water is taken by pipette in a porcelain dish.
(2) Two or three drops of potassium chromate solution are added to the sample of water.
(3) The chloride contents are then determined by titrating with standard solution of silver nitrate.

The silver reacts first with all chlorides and silver chloride thus formed then reacts with potassium chromate. The silver chromate appears as reddish precipitate and the amount of silver nitrate required to produce such reddish precipitate determines the amount of chlorides present in water. The chemical reactions are as follows :

$$AgNO_3 + NaCl = AgCl + NaNO_3$$

$$2AgCl + K_2Cr_2O_7 = Ag_2Cr_2O_7 + 2\ KCl$$

The presence of chlorides can corrode and such water cannot be used for boilers because of formation of hydrochloric acid due to presence of magnesium chloride in water. The pH value of sample of water is to be adjusted between 7 and 8 either by adding sulphuric acid or sodium hydroxide solution. Otherwise, the test results are likely to be affected.

Dissolved gases

The water contains various gases from its contact with the atmosphere and ground surfaces. The usual gases are nitrogen, methane, hydrogen sulphide, carbon dioxide and oxygen. The contents of these dissolved gases in a sample of water are suitably worked out.

The nitrogen is not very important. The methane concentration is to be studied for its explosive property. The hydrogen sulphide gives disagreeable odour to the water even if its amount is very small. The carbon dioxide content indicates biological activities, causes corrosion, increases the solubility of many minerals in water and gives taste to the water.

The oxygen in the dissolved state is obtained from atmosphere and pure natural surface water is usually saturated with it. The simple test to determine the amount of dissolved oxygen present in a sample of water is to expose water for 4 hours at a temperature of 27°C with 10% acid solution of potassium permanganate. The quantity of oxygen absorbed can then be calculated. This amount, for potable water, should be about 5 to 10 p.p.m.

Hardness

The hardness of water is divided into the following forms :

(1) Carbonate hardness–metals associated with HCO_3^-.

(2) Non-carbonate hardness–metals associated with SO_4^{--}, Cl^- and NO_3^-.

Now, most of the natural alkalinity in waters is due to HCO_3^- which is produced by the action of ground water on limestone or chalk as shown by the following equation:

$$\underset{\text{(insoluble)}}{CaCO_3} + \underset{\text{(From soil bacteria)}}{H_2O + CO_2} \rightarrow \underset{\text{(Soluble)}}{Ca(HCO_3)_2}$$

Thus we get the following relation :

Total hardness = Carbonate hardness or alkalinity + Non-carbonate hardness

or $$T.H. = C.H. + N.C.H.$$

Following two rules should be remembered in this connection :

(1) When alkalinity < T.H., then C.H. = Alkalinity.

(2) When alkalinity ≥ T.H., then C.H. = T.H.

The excess hardness of water is undesirable because of various reasons such as it causes more consumption of soap, affects the working of dyeing system, provides scales on boilers, causes corrosion and incrustation of pipes, makes food tasteless, etc.

The hardness is expressed as per Clark's scale in terms of degree of hardness. Thus one grain of $CaCO_3$ dissolved in one gallon of water will produce one degree of hardness.

Now, 700 grains = 0.4536 kg = 453600 mg

One grain = 64.8 mg

One gallon (imperial) = 4.546 litres.

Hence one degree of hardness will be equivalent to $\frac{64.8}{4.546} = 14.254$, say 14.3 milligrams of calcium carbonate in a litre of water. The expression p.p.m. is used to mean mg per litre and in that case, one degree of hardness will be equal to 14.3 p.p.m. It is found that each degree of hardness causes wastage of about 0.60 gram of soap.

The hardness is usually measured by the soap solution test. The standard soap solution is added in the sample of water. It is then vigorously shaken for about five minutes and the formation of lather is observed. The difference between the total amount of soap solution and the lather factor indicates the hardness of water.

The water, having hardness of about 5 degrees, is reasonably soft water and a very soft water is tasteless. Hence, for potable water, the hardness should preferably by more than 5 degrees but less than 8 degrees or so. For the purpose of convenience, a tentative scale of hardness may be framed as shown in Table 4.1. The hardness is normally expressed in terms of calcium carbonate. Now the chemical analyses for individual ions are usually given in terms of that ion. It will thus be necessary to convert the analytical results to the common denominator.

Table 4.1. Scale of hardness.

Nature of water	*Hardness in degrees*
Extremely soft	1
Very soft	2
Soft	3
Moderately soft	6
Moderately hard	7
Hard	9
Very hard	11
Excessively hard	15
Too hard for use	17

$$\text{Hardness in mg/l as } CaCO_3 = M^{++} \text{ (mg/l)} \times \frac{\text{equivalent wt. of } CaCO_3}{\text{equivalent wt. of } M^{++}}$$

where M represents any ion or radical.

Now, $\text{Equivalent weight} = \dfrac{\text{Molecular weight}}{X}$

where X = for acids, the number of moles of H^{++} obtainable from 1 mole of acid

X = for bases, the number of moles of H^{++} with which 1 mole of base will react.

Thus, $\text{Equivalent weight of } CaCO_3 = \dfrac{(40 + 12 + 3 \times 16)}{2} = 50$

For solving problems of hardness, the following facts should be noted:

(1) The alkalinity will be caused by positively charged Ca^{++}, Mg^{++} and Sr^{++} ions and negatively charged CO_3^{--}, and HCO_3^- ions only. The readings of other metals should be ignored.

(2) The equivalent weights of Ca^{++}, Mg^{++} and Sr^{++} will be respectively as follows:

Ca = 40/2 = 20;

Mg = 24.4/2 = 12.2; and

Sr = 87.6/2 = 43.8.

(3) For measuring alkalinity, the reading of only CO_3^{--} or HCO_3^- will be required and expressed as percentage, they can be found out from the following relations:

$$\left\{\begin{matrix}\text{Total alkalinity} \\ \text{as } HCO_3 \text{ in mg/l}\end{matrix}\right\} = \left\{\begin{matrix}\text{Bicarbonate alkalinity} \\ \text{in mg/l}\end{matrix}\right\} \times 1.22$$

$$\left\{\begin{matrix}\text{Total alkalinity} \\ \text{as } CO_3 \text{ in mg/l}\end{matrix}\right\} = \left\{\begin{matrix}\text{Carbonate alkalinity} \\ \text{in mg/l}\end{matrix}\right\} \times 0.60$$

$$\left[\begin{matrix}\text{Molecular wt of } HCO_3 = (1 + 12 + 3 \times 16) = 61 \\ \text{Molecular wt of } CO_3 \quad = (12 + 3 \times 16) \quad = 60\end{matrix}\right]$$

Hydrogen-ion concentration (pH value)

The acidity or alkalinity of water is measured in terms of its pH value or H-ion concentration. The pure water (H_2O) consists of positively charged hydrogen or H-ions combined with negatively charged hydroxyl or OH-ions. But the process of dissociation takes place in pure water and hence it contains some uncombined positively charged H-ions and some uncombined negatively charged OH-ions. The water becomes acidic when positively charged H-ions are in excess than negatively charged OH-ions and it becomes alkaline when reverse is the case. For neutral water, the concentrations of H-ions and OH-ions are equal. The water will dissociate into hydrogen and hydroxyl ions as follows:

$$H_2O \rightleftarrows H^+ + OH^- \quad ...(4.1)$$

Now, applying the law of mass action to equation (4.1), we get

$$K = \frac{[H^+][OH^-]}{H_2O} \quad ...(4.2)$$

The brackets indicate concentration of the constituents in moles/litre. A mole is the molecular weight in g. Since the concentration of water in a dilute aqueous system is essentially constant, it can be expressed as follows:

$$K_W = [H^+][OH^-] \quad ...(4.3)$$

where K_W = ionisation constant or ion product of water.

It is found by various experiments that K is approximately equal to 1×10^{-14} at a temperature of 25°C. The equation (4.3) can be used to calculate the hydrogen-ion concentration when the hydroxyl-ion concentration is known and vice versa. Thus a litre of pure water at 25°C contains:

$$10^{-7} \text{ g of H-ion} \times 10^{-7} \text{ g of OH-ion} = 10^{-14} \quad ...(4.4)$$

It is universally acknowledged to express H-ion concentration in terms of logarithm of its reciprocal and thus pH value of water by definition can be expressed as follows:

$$pH = -\log_{10}[H^+] = \frac{1}{\log_{10}[H^+]} \quad ...(4.5)$$

Similarly pOH value of water by definition can be expressed as follows:

$$pOH = -\log_{10}[H^-] = \frac{1}{\log_{10}[OH^-]} \quad ...(4.6)$$

Now as per equation (4.3),

$$[H^+] \times [OH^-] = 10^{-14}$$

$$\text{or} \left(\frac{pH}{-\log_{10}}\right) \times \left(\frac{pOH}{-\log_{10}}\right) = 10^{-14}$$

$$\text{or} \log_{10}(pH + pOH) = 10^{-14}$$

Therefore, pH + pOH = 14 ...(4.7)

With the above relation of pH and pOH, it has become a practice to refer only H-ion concentration in water. The term pH index is used to indicate the reactivity of water with respect to alkalinity or acidity and it is expressed quantitatively by concentration of the hydrogen-ions.

Fig. 4.1. shows pH-scale. A neutral water has pH value of 7. As pH value becomes less, the water becomes acidic and when pH value is zero, it indicates maximum acidity. Similarly the water becomes alkaline as pH values increase and maximum alkalinity is indicated when pH value is equal to 14.

It is desirable to maintain pH value of water very close to 7. The acidic water causes tuberculation and the alkaline water causes incrustation. For potable water, the pH value should be between 6.50 and 8.50 and it is found that if this range is maintained, there is no direct effect on health of human beings. If pH value of water is below 4, it will produce a sour taste and if it is more than 8.50, it will impart a bitter taste. The higher values of pH induces the formation of trihexlomethanes which are responsible for causing cancer in human beings. The lower values of pH starts corrosion in pipes and thereby the toxic metals like Zn, Pb, Cu, etc., are released.

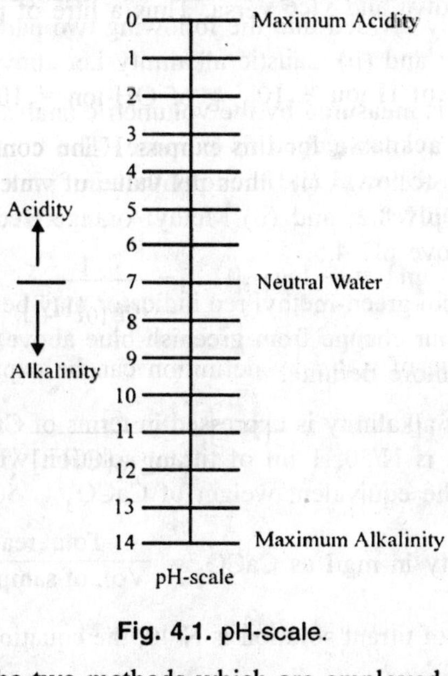

Fig. 4.1. pH-scale.

Following are the two methods which are employed to measure the pH value of water: (i) Electrometric method; and (ii) Colourimetric method.

(1) *Electrometric method:* In this method, the potentiometer is used to measure the electrical pressure exerted by positively charged H-ions. The pH value is then correspondingly expressed.

(2) *Colourimetric method:* In this method, the chemical reagents or indicators are added in water and the colour produced is compared with standard colours of known pH values. A set of sealed tubes containing coloured waters of known pH values is kept in the laboratory for ready reference. This test is simple and hence it is commonly carried out in the public health laboratories. The usual indicators are Benzol yellow, Methyl red, Bromphenol blue, etc. for acidic range and Thymol blue, Phenol red, Tolyl red, etc. for alkaline range.

Alkalinity

The term *alkalinity* with reference to the water and waste-water is defined as the capacity of substances contained in the water to take up hydroxium (H_3O^+) to reach a defined pH value (4.3 to 14). The alkalinity is due to the presence of bicarbonate (HCO_3^-), carbonate (CO_3^{--}) or hydroxide (OH^-).

The determination of alkalinity is very useful in waters and wastes because it provides buffering to resist the changes in pH value. The alkalinity is usually divided into the following two parts: (i) Total alkalinity i.e. above pH 4.5; and (ii) Caustic alkalinity i.e. above pH 8.2.

The alkalinity is measured by the volumetric analysis. The various types of indicators are available for this purpose. The commonly adopted two indicators are as follows: (i) Phenolphthalein: pink above pH 8.2 and colourless below pH 8.2; and (ii) Methyl orange: red below pH 4.5 and yellow-orange above pH 4.5.

The bromocresol green-methyl red indicator may be preferable to methyl orange as the colour change from greenish-blue above pH 4.5 to light pink below pH 4.5 is more definite.

The amount of alkalinity is expressed in terms of $CaCO_3$. If the strength of titrant solution is N/50, 1 ml of titrant solution will be equal to 1 mg $CaCO_3$ because the equivalent weight of $CaCO_3$ is 50.

$$\text{Thus, Alkalinity in mg/l as } CaCO_3 = \frac{\text{Total reading}}{\text{Vol. of sample in ml}} \times 1000.$$

If the strength of titrant solution is N/40, the equation will be as follows:

$$\text{Alkalinity in mg/l as } CaCO_3 = \frac{\text{Total reading} \times (0.025 \times 50 \times 1000)}{\text{Vol. in sample in ml}}$$

$$= \frac{\text{Total reading}}{\text{Vol. of sample in ml}} \times 1250.$$

The neutralisation of OH^- is complete at pH 8.2. The neutralisation of CO_3^{--} is only half complete at pH 8.2 and not fully completed until a pH value of 4.5 is reached. Thus alkalinity on pH scale is represented as follows :

(1) The range of total alkalinity is 4.5 to 14.

(2) The range of bicarbonate i.e. HCO_3^- alkalinity is from 4.5 to 8.2.

(3) The range of carbonate i.e. CO_3^{--} alkalinity is from 8.2 to 10.

(4) The range 0 to 4.5 indicates no alkalinity.

Table 4.2 can be used for finding out the alkalinity of a sample with the process of titration. The letter P indicates phenolphthalein reading and the letter T indicates total alkalinity.

Table 4.2. Determination of alkalinity.

Result of titration	*Hydroxide alkalinity as $CaCO_3$ (OH^-)*	*Carbonate alkalinity as $CaCO_3$ (CO_3^{--})*	*Bicarbonate alkalinity as $CaCO_3$ (HCO_3^-)*
P = 0	0	0	T
P < 0.5 T	0	2P	T – 2P
P = 0.5 T	0	2 (T – P)	0
P > 0.5 T	2P – T	2P	0
P = T	T	0	0

Acidity

The term *acidity* with reference to the water and waste-water is defined as the capacity of substances contained in the water to take up hydroxyl ions (OH^-) to reach a defined pH value (0 to 8.2). The acidity are of the following two types: (i) Carbon dioxide acidity; and (ii) Mineral acidity.

Carbon dioxide acidity: This acidity is due to the presence of CO_2 in ground water and surface water.

Mineral acidity: The mineral acidity is due to the presence of HCl, H_2SO_4, HNO_3 and strong organic acids. The mineral acidity is present in many industrial wastes and the important industries which discharge the acid effluents are electroplating industry, distillery industry, fertiliser industry, mining industry, etc.

The mineral acidity is determined by titrating or neutralising sample with strong base NaOH to pH 4.3. The carbon dioxide acidity in the sample is calculated by neutralising completely by continuing the titration till pH of 8.2. is reached. The amount of acidity is expressed in terms of $CaCO_3$. Then :

$$\text{Total acidity (as } CaCO_3) = \text{Mineral acidity} + CO_2 \text{ acidity.}$$

The determination of acidity of water has got significance because of the following reasons :

(1) It affects the aquatic life.

(2) It affects the biological treatment of sewage.

(3) It corrodes pipes.

(4) It decides the quantity of chemical dosage and line of treatment for industrial wastes.

(5) It interferes in the treatment of water as in case of water softening.

(6) The water containing mineral acidity is unpalatable.

(7) The water having acidity more than 50 mg/l cannot be used for R.C.C. construction.

Metals and other chemical substances

The various tests are made to detect the presence of different metals and other chemical-substances in a sample of water.

Nitrogen and its compounds

The nitrogen is present in water in the following forms: (i) Free ammonia; (ii) Albuminoid ammonia; (iii) Nitrites; and (iv) Nitrates. The amount of free ammonia in potable water should not exceed 0.15 p.p.m. and that of albuminoid ammonia should not exceed 0.3 p.p.m. The term *albuminoid ammonia* is used to represent the quantity of nitrogen present in water before the decomposition of organic matter has started.

The presence of nitrites indicates that the organic matter present in water is not fully oxidised or in other words, it indicates an intermediate oxidation stage. The amount of nitrites in potable water should be nil.

The presence of nitrates indicates that the organic matter present in water is fully oxidised and the water is no longer harmful. For potable water, the highest desirable level of nitrates is 45 mg per litre.

The free ammonia is measured by simply boiling the water. The ammonia gas is then liberated. The albuminoid ammonia is measured by adding strong alkaline solution of potassium permanganate ($KMnO_4$) to the water and then boiling it. The ammonia gas is then liberated. The nitrites and nitrates are converted chemically into ammonia and then measured by comparison with standard colours.

Total solids

The term solid with reference to the environmental engineering is defined as the residue in water left after evaporation and drying in oven at 103°C to 105°C. The total solids consist of dissolved and suspended matter.

Dissolved solids

In natural waters, the dissolved solids mainly consist of inorganic salts like carbonates, bicarbonates, chlorides, sulphates, etc. together with small amounts of organic matter and dissolved gases. For measuring the total dissolved solids, the sample oı water is placed in a clean porcelain dish and it is ignited in a muffle furnace. After partial cooling in the air, it is cooled in a desicator and is weighed.

Then, Total dissolved solids in mg/1 $= \frac{(A-B)}{V} \times 1000$

where A = Final weight of the dish in mg

B = Initial weight of the dish in mg

V = Volume of sample in ml.

It may be noted that mere determination of the dissolved solids does not give a clear picture of the type of pollution. However, many dissolved substances are undesirable in water and they impart displeasing colour, taste and odour. The water with higher content of the dissolved solids has a laxative or sometimes reverse effect on the human body and it takes time for people to adjust with such water.

In a similar way, the high concentration of dissolved solids say 3000 p.p.m. may also produce distress in livestock and may lead to scaling in boilers, corrosion, etc. The estimation of total dissolved solids is useful in determining the suitability of water for drinking purpose as well as for agriculture and industrial processes. The permissible total dissolved solids for drinking water according to BIS to 500 mg/l with tolerable limit as 1500 mg/l.

Suspended solids

In surface water, the suspended solids consist of inorganic matter like silt or organic matter like algae. These materials are generally carried by erosive action of the flowing water over land. The ground water contains negligible quantity of suspended matter because of filtering action of soil strata through mechanical straining action.

The amount of suspended solids in surface waters increases with input of natural and man-made contamination. The term *coefficient of fineness* is sometimes used to indicate the rates of weight of the suspended solids to the turbidity of water.

For measuring suspended solids, the water is filtered through a fine filter and dry material retained on the filter is weighed. The drying is carried out for one hour in an oven at 105°C. The filtered water is evaporated in a muffle furnace at 600°C for 15 to 20 minutes. At 600°C, the organic solids are converted into CO_2 and H_2O and the inorganic salts are remained as ash.

Then, Total suspended solids in mg/l $= \dfrac{(W_2 - W_1)}{V} \times 1000$

Suspended volatile solids in mg/l $= \dfrac{(W_2 - W_3)}{V} \times 1000$

where W_1 = Initial weight in mg

W_2 = Weight of dry material retained on filter in mg

W_3 = Weight after evaporation in mg

V = Volume of sample in ml.

The suspended matter is objectionable in water for the following reasons :

(1) It is aesthetically displeasing
(2) It may include disease causing organisms
(3) It may release obnoxious odours
(4) It provides adsorption sites for chemical and biological agents.

The estimation of total suspended solids is extremely useful in the analysis of polluted waters and for evaluating the efficiency of treatment units.

Bacteriological Tests

The examination of water for the presence of bacteria is very important. The bacteria are very small organisms and it is not possible to detect them by microscopes. Hence they are detected by circumstantial evidences or chemical reactions.

The growth of bacteria takes place by cell division and there are various classifications of bacteria depending upon their shapes, oxygen requirements and effects on mankind. The last classification is important for the water supply engineer from the view point of public health. The bacteria may be harmless to mankind or harmful to mankind. The former category is known as the *non-pathogenic bacteria* and the latter category is known as the *pathogenic bacteria*. It is not possible to isolate pathogenic bacteria with the help of laboratory instruments. Their chances of presence in a sample of water are increased in relation to the amount of non-pathogenic bacteria present in the sample of water.

The combined group of pathogenic and non-pathogenic bacteria is designated by *Bacillus coli* (Bacillus – bacterium and coli = intestine) or *B–coli group*. This group of bacteria is present in intestines of all living warm-blooded animals.

It should be noted that practically all natural waters contain some variety of bacteria and only in rare cases, such as the supply from deep boreholes, the water is found without the presence of any type of bacteria. Even rain before it touches the surface of earth collects dust and bacteria according to the nearness of dense populations, presence of industries, climatic conditions, etc. Subsequently numerous bacteria are added to the water by contact with the soil, vegetation, life and matter of the earth's surface.

The number and species of these bacteria differ to a great extent. Thus many of the bacteria found in water are derived from the air, soil and vegetation. Some of these are able to multiply and continue their existence while the remaining die out in due course of time. Following facts in connection with bacteria in water are to be noted:

(1) A certain number of bacteria cannot be cultivated under laboratory conditions while it is possible to cultivate some varieties of bacteria under special conditions in laboratory.

(2) For surface waters, the seasonal changes have a marked and complex influence on the bacteria.

(3) Many varieties of bacteria produce colonies of various colours such as yellow, brown, red, green and violet.

(4) Only few varieties of bacteria can tolerate any degree of acidity or caustic alkalinity.

(5) Some bacteria continue to live and multiply, others find conditions unsuitable and succumb and there may be many new additions.

(6) The temperature of water has some bearing on bacterial multiplication. Each variety of bacteria has its own optimum temperature for multiplication.

(7) The amount and remoteness of sewage and manurial pollution are the most important sanitary considerations with regard to the bacterial contact of water.

(8) The bacterial purity of water varies according to the source of water i.e., whether from river, stream, lake, shallow well or deep well and these differ greatly among themselves.

(9) The density of a bacterial population in water depends largely on quantity and quality of food available. It may however be noted that the nutritional requirements of bacteria are very simple and as a matter of fact, the organically pure deep well waters can support comparatively large numbers of bacteria and for a certain period even permit rapid multiplication.

(10) The gaseous constituents of water also affect bacterial activity in water to a certain extent. For instance, the oxygen is favourable to some bacteria, but not to others.

(11) The inorganic constituents of the water can also create favourable or unfavourable conditions for the growth of bacteria.

(12) The nature of the organic matter as well as its quantity present in water will determine the degree of bacterial multiplication and also the types or varieties of bacteria that will develop.

(13) The presence of metals in water, especially of copper, silver and aluminium, has usually detrimental effect on bacteria.

(14) The process of sedimentation which takes place during the storage of water is considered to be a potent factor in the bacterial purification of water.

Apart from the main aim of finding out the sewage or manurial pollution, the bacteriology is also of great value for the following purposes:

(1) to ascertain the effect of rainfall or drought on sources of water supply.

(2) to ascertain the efficiency of chlorination of sewage works effluent.

(3) to find out the change in quality of water either at source or after passing through the distribution system of mains and reservoirs.

(4) to study the degree of pollution of river waters.

(5) to study the pollution of sea water near shores by tidal outfalls from sewage works.

(6) to work out the efficiency of purification system of water such as filtration, chlorination, etc.

(7) to work out the pollution by bathers of swimming pool water and to ascertain the efficiency of purification process; etc.

It may be mentioned that the chemical analysis of water affords valuable information of past or remote pollution and as such, it cannot be relied upon the reveal all forms of recent pollution. On the other hand, the bacteriological analysis of water gives less information of the remote history of the water, but it discloses with ease and with great reliability, the immediate or recent causes of water pollution.

It is for this reason that the results of the chemical and bacteriological examinations should always be suitably correlated and an opinion should never be formed or expressed regarding the purity of water on the strength of the results of the chemical analysis only. It is quite likely that chemically pure water may reveal the presence of pathogenic bacteria during its bacteriological analysis and it is for this reason that the bacteriological examination of water has been given considerable significance and importance.

Following are the two standard bacteriological tests for the bacteriological examination of water: (i) Total count or Agar plate count test; and (ii) B-coli test.

Total count or agar plate count test

In this test, the bacteria are cultivated on specially prepared medium of agar for different dilutions of sample of water with sterilised water. The diluted sample is placed in an incubator for 24 hours at 37°C (i.e., blood heat) or for 48 hours at 20°C. These represent the so-called *hot counts* and *cold counts* respectively. The bacterial colonies which are formed, are then counted and the results are computed for 1 c.c. For potable water, the total count should not exceed 100 per c.c.

B-coli test

This test is divided into the following parts: (i) Presumptive test; (ii) Confirmed test; and (iii) Completed test. The presumptive test is based on the ability of coliform group to ferment the lactose broth and producing gas. The confirmed test consists of growing cultures of coliform bacteria on media which suppress the growth of other organisms. The completed test is based on the ability of the culture grown in the confirmed test to again ferment the lactose broth.

(1) *Presumptive test:* Following procedure is adopted in this test:

(a) The definite amounts of diluted samples of water are taken in multiples of ten, such as 0.1 c.c., 1.0 c.c., 10 c.c., etc.

(b) The water is placed in standard fermentation tubes containing lactose broth.

(c) The tube is maintained at a temperature at 37°C for a period of 48 hours.

(d) If gas is seen in the tube after this period is over, indicates presence of *B-coli* group and the result of test is treated as positive. If reverse is the case, it indicates absence of *B-coli* group and the result of test is treated as negative.

(e) A negative result of presumptive test indicates that water is fit for drinking.

(2) *Confirmed test:* This test is carried out in one of the following ways:

(a) A small portion of lactose broth showing positive presumptive test is carefully transferred to another fermentation tube containing brilliant green lactose bile. If gas is seen in the tube after 48 hours, the result is considered positive and the completed test becomes essential.

(b) A small portion of material showing positive presumptive test is marked on the plates containing Endo or eosin-methylene-blue agar. The plates are kept at 37°C for 24 hours. If colonies of bacteria are seen after this period, it indicates positive result and the completed test becomes essential. The colonies are prominent by metallic brightness and dark spots.

(3) *Completed test:* This test is made by introducing or inoculating bacterial colonies into lactose broth fermentation tubes and agar tubes. The incubation is carried out at 37°C for 24 to 48 hours. If gas is seen after this period, it indicates positive result and further detailed tests are carried out to detect the particular type of bacteria present in water. The absence of gas indicates negative result and the water is considered safe for drinking.

The recent development of field measurement of bacteria is that of membrane filter technique. The membrane contains microscopic pores which are capable of retaining bacteria. The water is filtered through the membrane and it is then incubated for a period of 20 hours along with nutrients. The colonies of bacteria, if any, can then be counted by means of microscope.

B-Coli Index

This is an index or number which represents approximately the number of B-coli per c.c. of sample of water under consideration. The presumptive tests are carried out with different dilution ratios of the sample of water with sterilised water. A number of tests is carried out for each proportion and percentage of positive results is recorded. The difference between successive percentages is worked out and it is multiplied by the reciprocal of quantity of solution. The sum of such values indicates B-coli index. For potable water, the B-coli index should be preferably less than 3 and it should not exceed 10 in any event.

In order to arrange correctly a number of analytical results, the laws of probabilities are applied and from the available statistics of test data, the most probable number or M.P.N. is determined by applying the laws of statistics to the results of the test. This number removes the drawback of B-coli index and it indicates the bacterial density which is most likely to be present in water. It should however be remembered that the M.P.N. is not the absolute concentration of organisms that are present, but it only indicates a statistical estimate of that concentration.

For illustration of working out B-coli index, typical test results are recorded in Table 4.3 and B-coli index of the case comes to 5.68.

Table 4.3. B-coli index calculation.

Proportion	*No. of tests*	*Positive results in per cent*	*Difference*	*Reciprocal of column 2*	*Multiplication of columns 5 and 6*
10 c.c.	100	1		0.10	0.03
			0.30		
1 c.c.	100	0.70		1	0.55
			0.55		
0.1 c.c.	100	0.15		10	1.10
			0.11		
0.01 c.c.	100	0.04		100	4.00
			0.04		
0.001 c.c.	100	Nil		1000	–
				Total	5.68

Maintenance and Purity of Waters

Some of the measures which can be taken to maintain the purity of waters are as follows:

(1) The content of dissolved oxygen present in water should be high.

(2) The entry of undesirable elements such as acidic and alkaline discharges, hot discharges from thermal power stations, etc. into source of water supply should be prevented by suitable means.

(3) The presence of algae supplying oxygen to the water is desirable.

(4) The quantity of total dissolved solids, especially the chloride contents, should be very low.

(5) There should be absence of organic wastes exerting oxygen demand on water.

(6) There should be enough reaeration for the purpose of reoxygenation of water.

(7) The temperature of water should be maintained low.

(8) The turbulence of water should be encouraged to make availa le more oxygen from air.

Water-borne diseases

The water is considered to be an important agent for carrying and spreading some of the dangerous diseases. The usual diseases carried by waste are those of intestinal disorders such as cholera, dysentery, typhoid, etc.

The patients suffering from such diseases give out the organisms responsible for such diseases from their intestinal discharges. Such organisms then somehow obtain entrance to the sources of drinking water and the disease is thus further carried forward. The water is also considered partly responsible for other diseases such as hepatitis, tuberculosis, gastro-enteritis, polimyelitis, etc.

The water-borne diseases are caused by pathogenic or harmful bacteria. The bacteria are found everywhere on surface of earth and they are in great excess even in natural unpolluted water bodies. However, the water contaminated with sewage provides an excellent medium for the growth of bacteria.

Out of all bacteria, the pathogenic bacteria form only a very small quantity. Further, the pathogenic bacteria have only a short life in water and they offer less resistance to adverse influences. If the source of pathogenic bacteria is known and if water treatment plant is properly operated, it is possible to detect and to destroy the pathogenic bacteria and thus the danger of water-borne diseases may be avoided.

The most important water-borne bacteria responsible for water-borne diseases are *Salmonella typhi*, *Salmonella paratyphi*, *Shigella dysenterial* and *Vibrio cholerae*. The respective diseases spread by them are typhoid, paratyphoid, dysentery and cholera.

It is significant to note that the water-borne diseases are predominant even in developing countries and it is estimated that about five million babies die annually from water-borne diseases in such countries and that one in six of the world's population suffer every year from water borne diseases.

The surveys conducted by World Health Organisation indicate that about 86% of the rural populations in developing countries are without reasonable access to safe water. It is therefore very clear that the battle with water-borne disease will be finally won only by continued attention to the pollution prevention and by maintaining quality of water upto the required standard. Thus the public health engineer in close co-operation with chemists and biologists will play a great role in preventing the spread of water-borne diseases.

COAGULATION OF WATER

Purpose

The source of water supply for the most of public water supply projects is surface water. This water is turbid and contains many suspended impurities. It also possesses colour which may be due to colloidal matter and dissolved organic material in water. The turbidity is mainly due to the presence of very fine particles of clay, silt and organic matter.

All these impurities are in a finely divided state and it is not possible to detain them in plain sedimentation tanks unless such tanks are designed for longer detention periods. The other alternative to remove such particles is to increase their size so that they become settleable. The purpose of coagulation is thus to make particles of bigger size by adding certain chemicals known as the *coagulants* to the water. The coagulants react with the impurities in water and convert them in settleable size.

The coagulation is to be adopted when turbidity of water exceeds about 40 p.p.m. It should however be remembered that it is not a complete process by itself. It simply assists plain sedimentation and it is to be followed by the process of filtration. Thus the coagulation is merely a process by which impure water is prepared for successful purification by rapid sand filtration.

Principle of Coagulation

The principle of coagulation can be explained from the following two considerations:

(1) *Floc formation:* When coagulants are dissolved in water and thoroughly mixed with it, they produce a thick gelatinous precipitate.

This precipitate is known as the *floc* and this floc has got the property of arresting the suspended impurities in water during its downward travel towards the bottom of tank.

(2) *Electric charges:* The ions of floc are found to possess positive electric charge. Hence they will attract the negatively charged colloidal particles of clay and thus they cause the removal of such particles from water.

Flocculation

The floc produced by the action of coagulants with water is heavy and hence it starts to settle down at the bottom of tank. As it descends, it absorbs and catches more and more suspended impurities present in water. It thus goes slowly on increasing in size. During this process, some amount of bacterial removal also takes place.

The surface of floc is sufficiently wide to arrest colloidal and organic matter present in water. The term *flocculation* is used to denote the process of floc formation and thus the flocculation follows the addition of coagulant and its efficiency depends on the following factors:

(1) *Dosage of coagulant:* The dosage or quantity of coagulant should be carefully determined so as to cause visible floc. The quantity of coagulants should be such that the turbidity of water is brought down to the limit of 10 to 25 p.p.m.

(2) *Feeding:* The feeding of coagulants may be in powder form or in solution form, the latter being more popular.

(3) *Mixing:* The coagulants should be properly mixed with water so as to cause a uniform mass. In the beginning, the mixing may be quick for a period of about 30 to 60 seconds or so.

(4) *pH value:* Depending upon the quality of water and coagulant adopted, suitable pH value should be determined. The pH value should be actually tested in the laboratory at regular intervals. To remove acidity, the lime is added to the water and to remove alkalinity, the sulphuric acid is added to the water.

(5) *Velocity:* The floc should be allowed to more gently after initial quick mixing. The gentle movement of floc results in collision of particles and ultimately, the floc grows in size. The detention period of coagulated sedimentation tanks is about 3 to 4 hours.

The processes of coagulation and flocculation are greatly influenced by the physical characteristics of water, its dissolved constituents and the temperature. The failures in coagulation plant are due to the incorrect dose of the coagulant, inadequate mixing arrangements, improper tank design, etc. Hence the characteristics of water to be submitted to the coagulation plant should be properly studied before deciding the details of the plant.

Usual Coagulants

Following are the usual coagulants which are adopted for coagulation :

(1) Aluminium sulphate
(2) Chlorinated copperas
(3) Ferrous sulphate and lime
(4) Magnesium carbonate
(5) Polyelectrolytes
(6) Sodium aluminate.

FILTRATION OF WATER

The sedimentation tanks remove a large percentage of the suspended solids and organic matter present in raw water. The process of coagulation of water further assists in the removal of impurities present in the water. But even then, the resultant water is not pure and may contain some very fine suspended particles, bacteria, etc.

In order to remove or to reduce the contents of impurities still further, the water is filtered through the beds of fine granular materials like sand. The process of passing the water through the beds of such granular materials is known as the *filtration.*

The filtered water is potable and palatable and it is free from various undesirable impurities like colour, odour, turbidity, pathogenic bacteria, etc.

Theory of Filtration

The process of filtration forms the most important stage in the purification of water. It usually consists in allowing water to pass through a thick layer of sand. It has been noticed from experience that during the process of filtration, the following effects occur on water:

(1) The suspended and colloidal impurities which are present in water in a finely divided state are removed to a great extent.
(2) The chemical characteristics of water are altered.
(3) The number of bacteria present in water is also considerably reduced.

The theory of filtration to explain why such effects take place is based on the following actions :

(1) Mechanical straining

(2) Sedimentation

(3) Biological metabolism

(4) Electrolytic changes.

Types of filters

(1) Slow sand filters

(2) Rapid sand filters (gravity type)

(3) Pressure filters.

Slow sand filters

Purpose: In case of slow sand filtration, the water is allowed to pass slowly through a layer of sand placed above the base material and thus the purification process aims at simultaneously improving the biological, chemical and physical characteristics of water. The slow sand filtration is very well suited for rural areas in developing countries because of its simple operation and maintenance procedures. It thus provides safe drinking water at low recurrent cost.

Rapid sand filters (gravity type)

Purpose: The great *disadvantage* of a slow sand filter is that it requires considerable space for its installation. This requirement makes it uneconomical for places where land values are high.

This difficulty of requiring more space for slow sand filters led the engineers and scientists to find out means to increase the rate of filtration. It was observed that the rate of filtration can be increased in two ways :

(1) By increasing the size of sand so that friction to the water passing through filter media is minimised.

(2) By allowing the water to pass under pressure through the filter media.

The former is achieved in rapid sand filters (gravity type) and it is the most popular method of filtration for public water supply projects. The latter principle is adopted in the working of pressure filters.

Pressure filters

These filters are more or less similar to the rapid sand filters (gravity type). The term pressure filter does not indicate that the water is pumped through the filter under a high pressure loss. But it indicates that a filter is enclosed in space and the water passes under pressure greater than atmospheric pressure. This pressure can be developed by pumping and it may vary from 0.3 to 0.7 N/mm^2.

DISINFECTION OF WATER

When water leaves the filter plant, it is still found to contain some of the impurities. These impurities can be grouped as: (i) bacteria; (ii) dissolved inorganic salts; (iii) colour, odour and taste; and (iv) iron and manganese.

The water should be disinfected before it enters the distribution system. The main purpose of disinfection is to prevent contamination of water during its transit from the treatment plant to the place of its consumption.

The process of disinfection should not be compared with that of sterilisation of water. The destructive effect of disinfection is restricted only to the removal of harmful bacteria. On the other hand, the sterilisation aims at the removal of all sorts of bacteria, whether harmful or harmless to the health.

The materials or substances which are to be used for disinfection are called the disinfectants and the requirements of a good disinfectant are as follows :

(1) Its dose should be such that some residual concentration is obtained to grant protection against contamination in the water during its conveyance and retention.

(2) It should be effective in killing all the harmful pathogenic organisms from the water and make it perfectly safe for use.

(3) It should be harmless, unobjectionable, economical and easily available.

(4) It should be of such a nature that its strength or concentration in the treated water can be quickly determined.

(5) It should not require skilled labour and costly equipment for its application.

(6) It should take only reasonable time in killing the harmful pathogenic organisms at normal temperature.

It has been universally recognised that the chlorine is an ideal material for the disinfection for treating water on a large scale. The disinfection at present therefore is mainly carried out by chlorination. However, there are other minor methods of disinfection.

Minor Methods of Disinfection

Following are the seven minor methods of disinfection: (i) boiling method; (ii) excess lime treatment; (iii) iodine and bromine treatment; (iv) ozone treatment; (v) potassium permanganate; (vi) silver treatment; and (vii) ultra-violet ray treatment. Following factors must be considered in applying the disinfection agents:

(1) concentration and type of chemical agent
(2) contact time
(3) intensity and nature of physical agent
(4) nature of suspending liquid
(5) number of types of organisms
(6) temperature.

Chlorination

In this treatment for disinfection, the chlorine (and its compounds) is used as the disinfecting material. For treatment on large scale, the chlorination is invariable used as treatment for disinfection because of the following facts:

(1) It is easy to apply due to relatively high solubility of about 7000 mg per litre.

(2) It is readily available as gas, liquid or powder.

(3) It is very toxic to most of the micro-organisms and thus metabolic activities are stopped.

(4) It leaves harmless residue in solution, but it provides protection in the distribution system.

(5) It produces desired effects which last for a long time.

(6) The treatment by chlorination is cheap and reliable.

However, chlorine is a poisonous gas which requires careful handling and it may also give rise to the problem of taste and odour in water. Some advantages of chlorination is given as under :

(1) It accomplishes greater bacterial purification in minutes than storage achieves in an equal number of days. It thus avoids the construction of costly storage reservoirs.

(2) It is inexpensive and avoids, wholly or in part, the necessity for raw water storage.

(3) It provides extra security to the water against water-borne diseases.

(4) It serves as a convenient accessory to the process of filtration.

Application of chlorine

Chlorine can be applied in water in one of the following ways :

(1) as bleaching powder

(2) as chloramines and/or

(3) as free chlorine gas.

WATER SOFTENING

The water to be supplied to the public should not be very hard though there is fear of no health hazard, but it is undesirable as it leads to the following economic disadvantages :

(1) It affects the working of dyeing system and leads to the modification of some of the colours.

(2) It causes corrosion and incrustation of pipes and plumbing fixtures.

(3) It causes more consumption of soap in laundry work and hence proves to be uneconomical for washing processes of textile industries.

(4) It increases the fuel costs.

(5) It makes food tasteless, tough or rubbery.

(6) It provides scales on the boilers and other hot-water heating systems.

It is necessary to understand why soap does not form lather with hard water. The ordinary soap consists mainly of sodium or potassium salts of some of the organic acids which are present in fats and fatty oils. Let us suppose that soap is made from sodium and palmitic acid. It will then be the compound sodium palmitate ($C_{15}H_{31}COONa$).

When soap is treated with soft water, it is hydrolysed into caustic soda and palmitic acid. The acid thus liberated unites with a second molecule of the sodium palmitate (soap) and it forms an insoluble substance which, with water, produces the lather. Thus,

Sodium palmitate + $H_2O \leftrightharpoons$ NaOH + Palmitic acid
(soap)

Palmitic acid
+ + Water = Lather
Sodium Palmitate

When soap is treated with hard water, the soluble Ca and Mg salts, present in hard water, react with sodium palmitate forming insoluble Ca and Mg palmitates and they appear as a white precipitate. This process of precipitation and consequent wastage of soap will go on till the whole amount of soluble Ca and Mg salts present in hard water is converted into insoluble Ca and Mg palmitates. The water now becomes soft and will easily form lather with soap.

The hard water is therefore to be softened by some suitable method before it is supplied for consumption. For potable water, it is desirable to bring down the hardness of water to about 5 to 8 degrees. The water having hardness of about 5 degrees is reasonably soft water. But it is found to be tasteless. The water having hardness more than 8 degrees gives undesirable effects as mentioned above. Following are the *advantages* of soft water :

(1) It improves the taste of foods.

(2) It increases the life of textiles which are frequently sent to the laundries.

(3) It leads to overall cleanliness because of the fact that personal washing and domestic cleansing are much more efficient and less laborious with soft water than with hard water.

(4) It makes washing and cleansing easy.

(5) It restricts scale formation and subsequent loss of heat in boilers, hot water pipes, etc. and therefore the economy is achieved in fuel consumption and saving of labour in descaling the affected surfaces.

(6) It results in saving of labour, soap and other detergents.

(7) It undoubtedly proves to be a sound economic proposition to soften public water supplies which are hard in character.

The hard water however possesses the following advantages over the soft water.

(1) It is not desirable to have very soft water for the purpose of drinking because some quantities of calcium salts are required for the growth of the human body, especially in the case of children.

(2) The soft water dissolves lead much more readily than the hard water and hence there is less chance of lead poisoning with hard water than with soft water.

HARD AND SOFT WATER

Water is classed as hard or soft, depending upon its behaviour towards soap solution.

Water that easily forms a lather of films and froths when agitated with a soap solution is known as soft water. Water that reacts with soap solution to form a white scum only without producing a lather easily, is said to be hard water. Hard water does not, therefore, yield a lather easily until a considerable amount of soap has been used up.

The hardness of water is due to the presence of dissolved salts of metals (except the salts of alkali metals such as sodium or potassium), notably those of calcium, magnesium and iron, in water.

Besides the presence of bicarbonates of calcium and magnesium (which cause carbonate or temporary hardness) and sulphates and chlorides of calcium and magnesium (which cause permanent hardness) water may also contain varying amounts of sodium salts, silica, iron, alumina and manganese. Other impurities present in water may be the suspended impurities which cause turbidity, organic matter dissolved gases and colouring matter. The dissolved gases may be carbon dioxide, oxygen, nitrogen and hydrogen sulphide (in sulphurous waters).

As a matter of interest the constituents of sea water will give an idea of the chemical elements present. The major elements present, besides oxygen and nitrogen, are chlorine, sodium, magnesium, sulphur, calcium, potassium; minor elements present are bromine, carbon, strontium, boron and the trace elements are iodine, copper, lithium, silver and gold.

The difficulties involved in removing all these impurities are obvious. While calcium, magnesium and iron are removed by water-conditioning, silica is removed with difficulty by ferric coagulation. Presence of soluble sodium salts do not harm boiler-feed water. Alumina and manganese are seldom removed; if required, manganese is removed by aeration followed by filtration or filtration through manganese-zeolite.

From sea water, potassium and bromine are removed as potassium bromide, sodium and iodine as sodium bromide; sulphur, carbon and chlorine separate as sulphates, carbonates and chlorides of calcium, manganese and iron respectively. Boron, lithium and copper are difficult to separate—these are allowed to remain as such in industry. Picking of gold particles from river water in basins is an age-old process.

The best form of pure water is of course distilled water. Distillation keeps back all types of impurities.

Types of Hardness

Generally there are two types of hardness: (i) temporary hardness; and (ii) permanent hardness.

Temporary hardness

Temporary hardness is due to the presence of soluble bicarbonates of calcium, mangnesium and iron in water. It is also known as carbonate hardness.

Permanent hardness

Permanent hardness is caused by the presence of the chlorides and sulphates of calcium and magnesium in the water as chlorides and sulphates. It is also known as non-carbonate hardness.

Removal of temporary hardness

Temporary hardness can be removed by:

(1) *Boiling:* When the soluble bicarbonates are converted into insoluble carbonates which are precipitated and CO_2 is evolved :

$$Ca\,(HCO_3)_2 = CaCO_3 + H_2O + CO_2$$
$$Mg\,(HCO_3)_2 = MgCO_3 + H_2O + CO_2$$
$$Fe\,(HCO_3)_2 = FeCO_3 + H_2O + CO_2$$
$$(4FeCO_3 + 6H_2O + O_2 = 4Fe(OH)_3 + 4CO_2)$$

While $CaCO_3$ and $MgCO_3$ form white powdery precipitate, iron is precipitated as slimy brown precipitate of $Fe(OH)_3$. The calcium carbonate ($CaCO_3$) and magnesium carbonate ($MgCO_3$) are insoluble in water. They can therefore be easily removed in the sedimentation tanks. The boiling of water on a large scale is impracticable and uneconomical. Hence the addition of lime is preferred to boiling for the removal of temporary hardness.

(2) *Clark's process:* Temporary hard water is softened for industrial purposes by adding the calculated quantity of slaked lime which precipitates the insoluble carbonates.

$$Ca(HCO_3)_2 + Ca(OH)_2 = 2CaCO_3 + 2H_2O$$
$$Mg(HCO_3)_2 + Ca(OH)_2 = MgCO_3 + CaCO_3 + 2H_2O$$

The $MgCO_3$, which is appreciably soluble, then reacts with further quantity of $Ca(OH)_2$ to form insoluble $Mg(OH)_2$.

$$MgCO_3 + Ca(OH)_2 = Mg(OH)_2 + CaCO_3$$

$Mg(OH)_2$ is sparingly soluble and hence removed.

Slaked lime must be added in extremely calculated quantities, as any excess of $Ca(OH)_2$ present in the softened water would again cause hardness. The mixing of lime and water is done in large settling tanks. After the precipitates have settled down, the clear soft water is pumped to reservoirs for distribution.

Removal of permanent hardness

Permanent hardness of water is caused by the presence of the chlorides and sulphates of calcium and magnesium in the water. While temporary hardness can be removed by boiling or adding lime, permanent hardness cannot be removed by those processes since such methods do not precipitate the calcium or magnesium of hard water.

The permanent hardness cannot be removed easily from water. The special treatment, known as the *water softening treatment*, is to be given to the water for the removal of permanent hardness from it. Some processes are adopted for the removal of permanent hardness from water: (i) Lime-soda process; (ii) Zeolite process; and (iii) Demineralisation process. Each of the processes of removing permanent hardness will now be explained.

Lime-soda process

In this process, the lime and sodium carbonate or soda ash are used to remove permanent hardness from water. The hardness is brought down to the level of 3 to 4 degrees (Fig. 4.2). Following are the chemical reactions involved in this process:

$$CO_2 + Ca(OH)_2 = CaCO_3 + H_2O \quad ...(4.8)$$

$$Ca(HCO_3)_2 + Ca(OH)_2 = 2CaCO_3 + 2H_2O \quad ...(4.9)$$

$$Mg(HCO_3)_2 + Ca(OH)_2 = CaCO_3 + MgCO_3 + 2H_2O \quad ...(4.10)$$

$$MgSO_4 + Ca(OH)_2 = Mg(OH)_2 + CaSO_4 \quad ...(4.11)$$

$$CaSO_4 + Na_2CO_3 = CaCO_3 + Na_2SO_4 \quad ...(4.12)$$

$$CaCl_2 + Ca(OH)_2 = Ca(OH)_2 + CaCl_2 \quad ...(4.13)$$

$$MgCl_2 + Ca(OH)_2 = Mg(OH)_2 + CaCl_2 \quad ...(4.14)$$

$$CaCl_2 + Na_2CO_3 = CaCO_3 + 2NaCl \quad ...(4.15)$$

$$MgCl_2 + Na_2CO_3 = MgCO_3 + 2NaCl \quad ...(4.16)$$

The compounds calcium carbonate $CaCO_3$ and magnesium hydroxide $Mg(OH)_2$ are insoluble in water and they can therefore be arrested in the sedimentation tanks. The other compounds formed during the chemical reactions are soluble in water and they do not impart the property of hardness to the water.

Equation 4.8 indicates the chemical reaction between lime and carbon dioxide present in water.

Equations 4.9 and 4.10 indicate the removal of temporary hardness by the action of lime on the bicarbonates of calcium and magnesium.

Equation 4.11 indicates the chemical reaction between lime and magnesium sulphate. The reaction produces calcium sulphate and hence there is no softening of water as such.

Equation 4.12 indicates the chemical reaction between soda ash and calcium sulphate. Thus calcium sulphate already present in water and also formed by equation 4.11 is removed by this chemical reaction.

Equations 4.13 and 4.14 indicate the chemical reactions between lime and chlorides of calcium and magnesium.

Equation 4.15 and 4.16 indicate the chemical reactions between soda ash and chlorides of calcium and magnesium.

Equipments required: Following units of equipment are required for the working of lime-soda process:

(1) *Feeding and mixing apparatus:* The feeding and mixing of lime and soda ash with water should be done with precision. The lime used in this process may be quick lime or hydrated lime. The former is preferable for large plants as it is cheap and less bulky. The required quantities of both these chemicals should be accurately worked out either volumetrically or gravimetrically.

(2) *Settling tank:* This tank is similar to the coagulation sedimentation tank as adopted in case of rapid sand filters. But in this case, the detention period of tank is longer.

(3) *Recarbonation plant:* It is necessary to remove calcium carbonate formed in this process. Otherwise it will precipitate in sand filters and also cause incrustation in pipes. For this reason, the water is allowed to pass through a recarbonation plant after it has passed through the settling tank. In the settling tank, a dose of alum may be given which will produce carbon dioxide. This carbon dioxide reacts with calcium carbonate in the following way and helps to hold it in solution as bicarbonate:

$$CaCO_3 + CO_2 + H_2O = Ca(HCO_3)_2$$

Alternatively, the carbon dioxide gas may be diffused or blown into water in the recarbonation plant.

(4) *Filters:* This unit is required to finish up the treatment. The filters may be of ordinary type rapid sand filter or pressure filter. If the stage of recarbonation does not preceed filtration, it becomes essential to wash the filters frequently and it also results in the frequent replacement of sand bed.

Lime also removes mineral acids, carbon dioxide, dissolved iron and aluminium salts. Lime soda process is employed in two ways: (i) continuous cold lime soda process; and (ii) hot lime soda process.

Continuous cold lime soda process

Hard water is treated with lime and soda ash at room temperature in the treatment tank. The treatment tank is provided with a mechanical agitator. Hard water mixed with the chemicals is agitated thoroughly. Lime and soda ash react with the calcium and magnesium salts present in the hard water. The respective carbonates are precipitated. Coagulating agents like potash alum $Al_2(SO_4)_3.K_2SO_4.24H_2O$ are added. The suspended particles are coagulated and settle to the bottom of the treatment tank along with the insoluble carbonates of calcium and magnesium. Soft water passes up through the filters and drawn out from the out let.

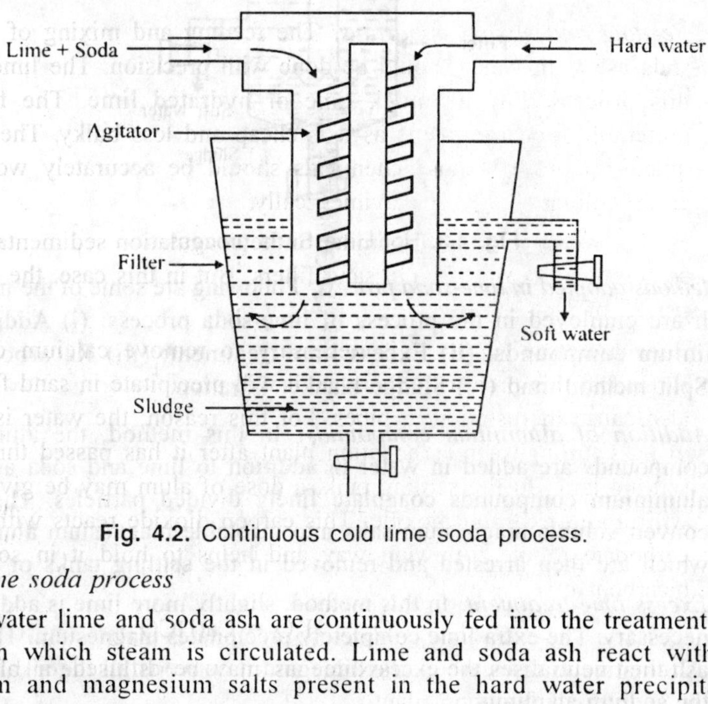

Fig. 4.2. Continuous cold lime soda process.

Hot lime soda process

Hard water lime and soda ash are continuously fed into the treatment tank through which steam is circulated. Lime and soda ash react with the calcium and magnesium salts present in the hard water precipitating respective carbonates. In this process, it is not necessary to add the

coagulating agent because at higher temperature, precipitates and suspended particles coagulate and settle to the bottom. Soft water moves upwards and continuously withdrawn. The sludge is removed from a separate outlet periodically (Fig. 4.3).

Advantages of Hot Lime Soda Process

At higher temperature dissolved gases like oxygen and carbon dioxide are expelled off. Since the precipitates coagulate and settle at higher temperature, it is not necessary to add the coagulating agents.

The process is faster than the cold lime soda process. Filtration is easier in hot condition. The hardness is reduced to a minimum level.

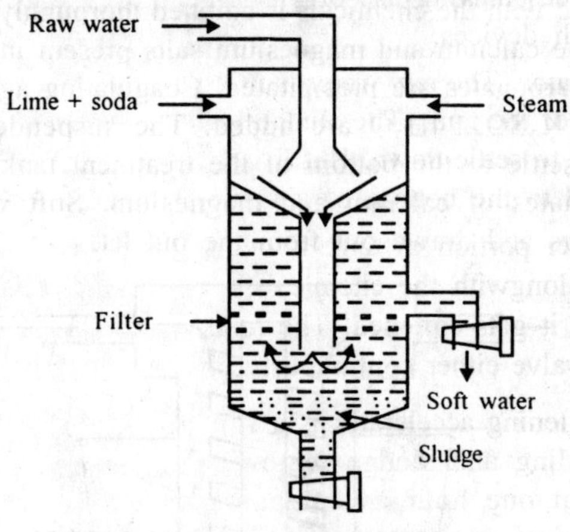

Fig. 4.3. Hot Lime Soda process.

Methods adopted in lime-soda process: Following are some of the methods which are employed in the process of lime-soda process: (i) Addition of aluminium compounds; (ii) Excess lime treatment; (iii) Recarbonation; (iv) Split method; and (iv) Sludge blanket filtration.

(1) *Addition of aluminium compounds:* In this method, the aluminium compounds are added in water in addition to lime and soda ash. The aluminium compounds coagulate finely divided particles. They also convert soluble magnesium salts into insoluble magnesium aluminates which are then arrested and removed in the settling tanks or filters.

(2) *Excess lime treatment:* In this method, slightly more lime is added than necessary. The extra lime completely precipitates magnesium. The soda ash then neutralises the excess lime and it converts all the alkalinity to the sodium alkalinity.

(3) *Recarbonation:* In this method, the raw water is treated with excess lime and the excess lime is then neutralised by the action of carbon dioxide.

(4) *Split method:* In this method, a part of raw water is given excessive treatment and its degree of hardness is lowered to the maximum possible extent. This excessively treated water is then added to the raw water to bring down the hardness of raw water. With the help of this method, it is possible to save a considerable quantity of chemicals.

(5) *Sludge blanket filtration:* In this method, the advantage is taken of the sludge collected at the bottom of coagulation sedimentation tank. The water is softened by filtration through this blanket of sludge. The various devices making use of this process are found out. Following are two such devices.

(a) *Accelerator:* The raw water is mixed with chemicals in the central portion of the tank. The sludge which is previously formed is not allowed to settle down. But it is kept suspension by a mixing fan attached to the bottom of a vertical rotating shaft.

Thus the lower portion of tank is occupied by gently moving sludge. The raw water alongwith the chemicals rises through this sludge blanket and in doing so, it gets softened. The removal of sludge can be done by sludge removal valve either continuously or intermittently.

The water softening accelerator is a very compact unit and it combines flocculation, settling and sludge removal. The detention period is also reduced to about one hour as compared to 3 or 4 hours in case of coagulation sedimentation tanks.

(b) *Spiractor:* The tank is in the shape of an inverted conical tank and one-half of its depth is filled with marble chips of size 0.10 mm to 0.20 mm. The pipes for raw water and chemicals are placed at the bottom of tank. The raw water as well as chemicals are forced under pressure through nozzles in a tangential direction. The raw water is thus forced upwards spirally. The softening of water is achieved by this spiral motion of water.

The velocity of flow is so adjusted that the marble chips are kept in suspension only and not carried away by water. The calcium carbonate precipitates and it accumulates around the marble chips. This increases the size of marble chips and consequently the volume occupied by them also increases.

The level of marble chips is maintained by periodical removal of marble chips from the bottom and by adding new chips from the top. The marble chips which are removed can be ground, washed and used again. Following are the *advantages* of this device:

(1) The detention period for the tank is only about 5 to 10 minutes.
(2) The wet sludge is not formed. But instead of it, the crystalline and hard granules are formed which are easy to handle.
(3) The tank combines flocculation, settling and sludge removal in one single unit.

Advantages of lime-soda process: Following are the *advantages* of the lime-soda process:

(1) The pH value of water treated by this process is increased which results in decrease in corrosion of the distribution system.
(2) When this process is adopted, less quantity of coagulant will be required.
(3) There is removal of iron and manganese also to some extent.
(4) There is reduction of total mineral content of water.
(5) There is likelihood of killing of pathogenic bacteria. This occurs when causticity caused by calcium hydroxide or sodium hydroxide of 20 to 50 p.p.m. is retained in the treated water for a period of about 4 to 5 hours.
(6) The process is economical.
(7) This process is most suitable for turbid and acidic waters where it will not be possible to adopt zeolite process.
(8) The whole process is easy and simple and it can be accommodated in the existing filter plant of any water supply scheme.

Disadvantages of lime-soda process: Following are the disadvantages of the lime-soda process:

(1) The large quantity of sludge formed during this process is to be disposed off by some suitable method. It can either be discharged directly into river or stream or municipal sewers or it can be used for raising the level of low-lying areas.
(2) The process requires skilled supervision for its successful working.
(3) If recarbonation is omitted, a thick layer of calcium carbonate will be deposited in the filtering media, distribution pipes, etc.
(4) The calcium carbonate is slightly soluble in water to the extent of about 30 mg/l and hence this process can remove hardness upto about 50 mg/l only. Thus the water of zero hardness cannot be produced by this

process. But as water of zero hardness is not required for public water supply, this disadvantage does not prove to be serious.

Zeolite process

This is also known as the *base-exchange* or *ion-exchange* process. The zeolites are compounds of aluminium, silica and soda. They have got the excellent property of interchanging base. They may be obtained from nature or may be prepared synthetically. The naturally available zeolite is green in colour and it is therefore known as the *green sand*. The exchange value of green sand is 7000 to 9000 gm of hardness per m^3 of zeolite (Fig. 4.4).

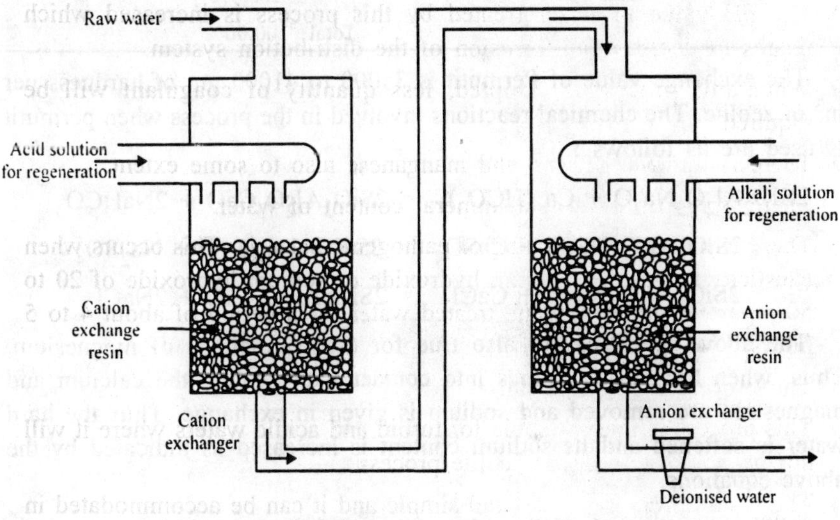

Fig. 4.4. Flow diagram of ion exchange process.

In this method certain mineral constituents of the soil i.e. hydrated silicates, possessed the property of base-exchange. Thus, in preparing base-exchange materials from solutions of sodium silicate and aluminium sulphate, the reversibility of the process could not be ascertained.

Thus, new methods were adopted in preparing a synthetic hydrated silicate of aluminium and sodium, known as the *Permutit*. Since then the process has steadily advanced and at present, several groups of materials of different character and manufacture having base-exchange properties are available.

The most common artificial zeolite is the *Permutit*. It is manufactured from felspar, kaolin, clay and soda. These chemicals are mixed in the required proportion and then the mixture is fused in a furnace. It is allowed to cool after attaining a certain degree of fusing.

The material thus formed is then crushed to form particles of diameter varies from 0.25 mm to 0.50 mm. It is then washed to remove all alkalies and impurities. The Permutit is white in colour and its chemical formula is $2SiO_2Al_2O_3 . Na_2O$. The chemical composition is as follows :

Alumina	22.00%
Silica	46.00%
Sodium oxide	13.60%
Water	18.40%
Total	100.00%

The exchange value of Permutit is 35000 to 41000 gm of hardness per m^3 of zeolite. The chemical reactions involved in the process when permutit is used are as follows :

$$2SiO_2Al_2O_3Na_2O + Ca\ (HCO_3)_2 = 2SiO_2Al_2O_3CaO + 2NaHCO_3$$

$$2SiO_2Al_2O_3Na_2O + CaSO_4 = 2SiO_2Al_2O_3CaO + Na_2SO_4$$

$$2SiO_2Al_2O_3Na_2O + CaCl_2 = 2SiO_2Al_2O_3CaO + 2NaCl$$

The above reactions are also true for the compounds of magnesium. Thus, when hard water comes into contact with zeolite, the calcium and magnesium are removed and sodium is given in exchange. Thus the hard water is softened and its sodium content is increased as indicated by the above equations.

After some interval of time, the sodium present in zeolite is exhausted. It is very easily regenerated by passing a solution of salt through the zeolite. The chemical reactions take place as follows :

$$2SiO_2Al_2O_3CaO + 2NaCl = 2SiO_2AlO_3Na_2O + CaCl_2$$

$$2SiO_2Al_2O_3MgO + 2NaCl = 2SiO_2Al_2O_3Na_2O + MgCl_2$$

Thus an exchange occurs again and salt of sodium is taken by zeolite. This process is known as the process of regeneration and in order to achieve the results as shown by above equations, the salt solution of required strength is prepared and it is allowed to trickle down over the bed of zeolite for a period of about 8 to 10 hours. The bed is then washed to remove calcium chloride or magnesium chloride for a period of about half an hour or so. The filter bed is then ready for use again.

Equipment required: The zeolite process is carried out in filters which are more or less similar to the pressure type filters except that the layer of sand is replaced by a layer of zeolite. The depth of zeolite layer varies from 750 mm to 1800 mm. The deeper the bed, the greater will be the capacity of filter.

The zeolite filters may be of downward flow type or of upward flow type. In the former case, the hard water is admitted from top of zeolite layer. It passes through zeolite bed and then through base material of gravel.

The softened water is collected by the underdrainage system and discharged for use by the outlet pipe. In the latter case, the reverse operations will take place. For a 750 mm deep bed of zeolite, the rate of flow is about 6000 litres per hour per m^2 of the filter area.

Advantages of zeolite process: Following are the advantages of the zeolite process.

(1) In this process, the sludge is not formed. Hence the problem of sludge disposal does not arise.

(2) The zeolite unit is compact in design. It can be easily operated and does not require any skilled supervision.

(3) There is no problem of the deposition of a layer of calcium carbonate in the distribution system.

(4) This process reduces hardness of water to zero. This advantage has got particular importance for the water to be used for boiler and certain textile industries.

(5) The water of any desired hardness can be prepared by adding softened water of zero hardness to the raw water.

(6) The process is completely automatic and highly skilled labour is not required.

(7) The process proves to be economical where salt is cheaply available.

(8) The load on zeolite process can be reduced by combining it with lime or aeration process. In such a case, the addition of lime or aeration preceeds zeolite process. In aerator lime zeolite system, the water is aerated first and it reaches to zeolite softener unit after passing through lime softener and filter.

Disadvantages of zeolite process: Following are the disadvantages of the zeolite process:

(1) The zeolite process cannot be adopted for highly turbid water. The suspended particles get deposited around zeolite particles and hence they cause obstruction to the working of zeolite.

(2) The zeolite process is unsuitable for water containing iron and manganese. This is due to the fact that iron zeolite or manganese zeolite formed during chemical reaction cannot be reconverted into sodium zeolite. The zeolite will thus be wasted in such a case.

(3) The zeolite unit should be carefully operated to avoid injury or damage to the equipment, quality of water and bed of zeolite.

(4) When the ion-exchange capacity of the zeolite has been exhausted, it has to be regenerated and at this time, it gives out a concentrated waste stream of the original contaminant.

Table 4.4 shows the comparison of lime-soda and zeolite processes of softening water supplies.

Table 4.4. Comparison between lime-soda and zeolite processes of softening water supplies.

Item	*Lime-soda process*	*Zeolite process*
Economy	Economical	Costly
Effect on bacteria	Can remove pathogenic bacteria	No effect on bacteria
Hardness of treated water	Can produce water upto hardness not less than 50 mg/l.	Water of zero hardness can be obtained
Hardness which can be treated	Excessively hard waters can be treated	Raw waters having hardness more than 800 mg/l cannot be treated
pH of treated water	It increases the pH value of treated water	It does not affect the pH value of treated water
Post treatment	Recarbonation is necessary after sedimentation and filtration	No such treatment is necessary
Removal of colour due to iron and manganese	Can remove such colour to a very small extent	Can remove such colour but proves very costly.
Size of plant	Bulky and large	Compact and small
Skilled supervision	Necessary	Not necessary
Sludge formation	Large quantity of sludge is formed.	No sludge is formed
Turbidity	Highly turbid and acidic waters can be treated.	Difficult to treat highly turbid waters.
Use	Public water supplies	Mostly for industrial water supplies

Demineralisation process

This process is also known as the deionisation process. It is similar to the previous process. But in this process, the hydrogen is exchanged for metallic ions.

The hard water is passed through a bed of resin or carbonaceous material in the hydrogen form. The chemical composition of such materials may be expressed as H_2Y where H represents hydrogen ions and Y represents the organic part of the substance. The chemical reactions involved in the process are as follows :

$$H_2Y + Ca(HCO_3)_2 = CaY + 2CO_2 + 2H_2O$$
$$H_2Y + CaSO_4 = CaY + H_2SO_4$$
$$H_2Y + CaCl_2 = CaY + 2HCl$$

The above reactions are also true for the compounds of magnesium. The treated water contains diluted carbonic acid, sulphuric acid or hydrochloric acid. These acids can be removed by mixing the treated water with the required proportion of alkaline water.

After some interval of time, the hydrogen content of the substance is exhausted. It is then regenerated by passing a solution of suitable strength of sulphuric acid or hydrochloric acid. The chemical reactions take place as follows:

$$CaY + H_2SO_4 = H_2Y + CaSO_4$$
$$CaY + 2HCl = H_2Y + CaCl_2$$

The equipment used in this process should be capable of resisting acidity and alkalinity. The demineralisation process is mainly used for preparing water to be used for the industrial purposes.

DESALINATION OF WATER

If the water contains 1000 to 35,000 mg lit^{-1} of dissolved salts, it is called brakish water. Sea water contains on an average 3.5% of dissolved salts. Brakish water has a peculiar salty taste and it is unfit for drinking purposes. If the water contains more dissolved salts than sea water, it is called brine. Desalination is the process of partial or complete removal of the dissolved salts from the sea water and brakish water. Following are some methods of desalination.

(1) *Flash evaporation:* It is a process of separating fresh water from the sea water. The sea water is subjected to the multistep flash evaporation. Flash evaporators consists of about 50 units. Each unit is

maintained at a slightly lower pressure and temperature than the previous unit. Saline water undergoes is passed through these units and a portion of water undergoes flash distillation, and the distillate is condensed. The residual water is heated under low pressure and subjected to flash distillation. As the saline water moves from one unit to the other, it is progressively desalinated, the residual water carries the salts removed as brine.

(2) *Freezing method:* When the saline water is cooled, ice crystals are formed leaving the salt in the mother liquor. The freezing point of water depends upon the concentration of the salt. The ice crystals thus formed are coated with brine solution. The ice crystals from ice brine slurry are separated by filtration. The ice crystals are then washed with fresh water to remove the contaminated brine. The ice crystals are then melted by the heat withdrawn from the incoming saline water.

(3) *Reverse Osmosis:* When two solutions of different concentrations are separated by a semi-permeable membrane, the solvent (water) flows from the dilute solution to the concentrated solution till the concentrations of two solution become equal. This process is known as osmosis. This flow of solvent from dilute solution to concentrated solution is due to osmotic pressure. If a hydrostatic pressure in excess of osmotic pressure is applied on the concentrated solution, the flow of solvent is reversed. i.e., the solvent moves from the concentrated side to the dilute side across the membrane. This process is known as reverse osmosis. The process of reverse osmosis can be used to separate pure water from the saline water (Fig. 4.5).

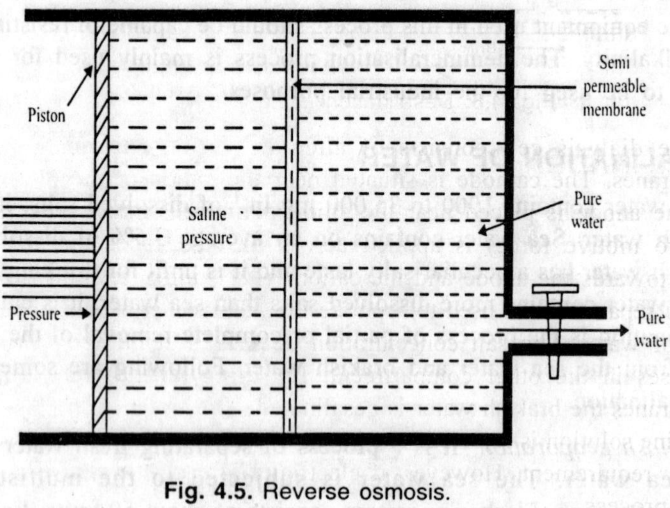

Fig. 4.5. Reverse osmosis.

Saline water is separated from pure water by a membrane which is permeable to water and not to salt. A pressure of about 15 to 40 Kg cm^{-2} is applied on to the saline solution. Pure water from the saline water passes through the membrane due to reverse osmosis, leaving behind the salts. By repeating the process, the water is desalinated.

By this process ionisable, non-ionisable, colloidal and high molecular weight organic compounds present in the water can be separated. The process has low maintenance cost and hence it is more economical.

(4) *Electrodialysis process :* The process of migration of ions present in the solution towards the oppositely charged electrodes under the influence of applied electric field is known as electrodialysis. This principle is used to drive away the dissolved minerals present in the saline water. For the electrodialysis process selectively permeable membranes are used (Fig. 4.6).

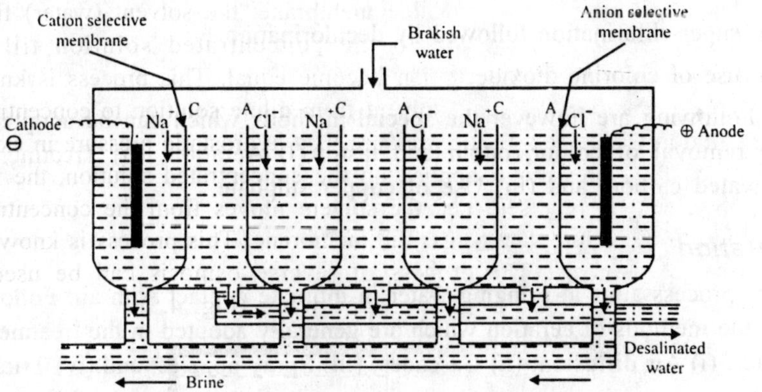

Fig. 4.6. Desalination using electrodialysis cell.

The dialysis cell consists of alternate cation and anion permeable membranes. The cathode is situated near the cation permeable membrane and the anode is placed near the anion permeable membrane. When emf (electro motive force) is applied across the electrodes the anions (Cl^-) move towards the anode and the cations (Na^+) move towards the cathode. Plastic separators are used to separate the membranes which also direct the flow of water. The salt concentration decreases in one compartment and increases in the other compartment. By using large number of pairs of membranes the brakish water is desalinated. The water of increased salinity i.e. brine solution is rejected. The process is not economical because of high energy requirement. However, if electricity is easily available it is a very good process.

MISCELLANEOUS METHODS OF WATER TREATMENT

The remaining miscellaneous methods which are adopted for the treatment of water under certain circumstances are

(1) Colour, odour and taste removal

(2) Iron and manganese removal

(3) Fluoridation.

Colour, Odour and Taste Removal

It may be mentioned that even pure water is not colourless and it is found to have a pale green-blue tint in large volumes. Some of the treatments methods to remove colour, odour and taste to certain extent are:

(1) coagulation followed by filtration

(2) pre-chlorination

(3) super-chlorination followed by dechlorination

(4) use of chlorine dioxide.

Following are however the special methods which can be adopted for the removal of colour, odour and taste: (i) Aeration; (ii) Treatment by activated carbon; and (iii) Use of copper sulphate.

Aeration

This process aims at bringing water in intimate contact with air. Following are the methods of aeration which are generally adopted in the treatment of water: (i) Air diffusion; (ii) Cascades; (iii) Spray nozzles; and (iv) Trickling beds.

(1) *Air diffusion:* In this method, the perforated pipes are installed at the bottom of tanks. The compressed air is blown through these pipes. The air bubbles while coming up from the bottom of tank come into close contact of water contained in the tank and aeration of water is thus achieved. The depth of aeration tanks is generally kept at about 2.5 m to 3 m and they work on the principle of continuous flow with minimum detention period of 15 minutes. The quantity of air consumed per 1000 litres of water varies from 0.3 to 0.6 m^3.

(2) *Cascade:* A cascade is a waterfall and a simple cascade consists of a series of three or four steps. These steps may be of concrete or metal. The water is allowed to fall through a height of about 1 to 3 metres and during this fall, it comes into close contact with air.

(3) *Spray nozzles:* In this method, the water is sprinkled in fine jets through nozzles to a height of about 2 m to 2.5 m. This method of aeration removes carbon dioxide to the extent of about 90 per cent or so. But it requires considerable heat of water for its working.

The nozzles are generally operated at pressure of 0.7 to 0.14 N/mm^2. The discharge through nozzle will depend upon its design and available head of water.

(4) *Trickling beds:* In this method, the beds of coke or slag are prepared. The beds are supported over perforated trays. The size of coke varies from 50 mm to 75 mm. The beds are arranged in vertical series. Usually three beds are placed one above the other. The depth of each bed is about 225 mm and the vertical distance between successive beds is about 100 mm to 150 mm.

The water is discharged through perforated pipes placed at the level of top bed and it is then allowed to trickle down from top bed to bottom bed. During the trickling process, the aeration of water occurs.

This method is found to be less effective than that of spray nozzles. But it gives better results than those obtained by the method of cascades.

Treatment by activated carbon

In this treatment, the activated carbon is used to remove colour, odour and taste from water. It is a form of charcoal or vegetable carbon. It is made by heating lignite, charcoal, paper mill waste, sawdust and similar carbonaceous materials in a closed vessel. It is then activated or oxidised by passing air or steam. This removes the hydrocarbons which might interfere with the adsorption of organic matter.

The activated carbon is available in various trade names such as Darco, Nuchar, etc. It is available in granular as well as powder form. The grains are of 6 mm size and below. Its weight is about 4 kN per m^3 and it is highly porous in structure. The usual dose of activated carbon varies from 5 to 20 p.p.m. Following are the *additional advantages* of this treatment:

(1) It aids the process of coagulation, if adopted before filtration of water.

(2) It reduces the chlorine demand of treated water.

(3) It removes the organic matter present in water.

(4) It removes taste due to excess chlorine, hydrogen sulphide, iron, manganese and phenol.

(5) Its overdose is harmless.

Methods of applications

The activated carbon can be applied either as a filter media or in a powder form. As a filter media, the layer of activated carbon is placed instead of sand layer in the rapid sand filters. As time passes, the carbon loses its adsorption quality. It then requires to be renewed or rejuvenated. The interval of time will depend upon the quality of water and may vary from one month to one year. The rejuvenation is carried out by forcing live steam through perforated pipes which are placed in the gravel bed of the filter. The process of rejuvenation of activated carbon can be carried out several times.

As a powder form, the activated carbon can be applied in various stages as follows :

(1) It can be added in raw water before any treatment takes place.

(2) It can be added along with the coagulant in mixing tank.

(3) A portion of activated carbon is added in the mixing tank and the remaining portion of it is added in water just before it enters the filter units. This is known as the *split method*.

(4) It can be added just at the point where water enters filters. It can be applied at constant rate or at varying rate. In the latter case, the rate is high when filter is washed and it becomes lower as filter gets clogged.

Use of copper sulphate

In this method, it is made of copper sulphate to remove colour, odour and taste from water. In addition, it also controls the growth of algae, bacteria, some types of aquatic weeds, etc.

Iron and Manganese Removal

The iron is present in practically all soils, gravels, sands and rocks. The rain, in percolating through soils and rocks, acquires iron in addition to other mineral constituents according to the character of the geological formation and it is observed that practically all waters, whether from surface or underground, contain at least some traces of iron. The manganese, usually in smaller amounts, may accompany iron in water. When their content in water exceeds 0.30 p.p.m., they become objectionable.

Fluoridation

It is found that a fluoride concentration of about 1 p.p.m. in water causes reduction in the cavities of teeth of young children. It also reduces the decaying and missing teeth. For such purposes, the fluoridation is practised.

The fluoride compounds which are usually adopted for the fluoridation of the public water supplies are sodium fluoride NaF, sodium hexafluorosilicate Na_2SiF_6, hexafluorosilicate H_2SiF_6, etc. The fluorides which may occur naturally in water are cryolite $AlF_3 3NaF$, fluorspar of calcium fluoride CaF_2 and fluorapatite $3Ca(PO_4)_2CaF_2$.

In general, it is found that cryolite and fluorspar occur in volcanic areas and fluorapatite is associated with phosphate deposits. The reaction of fluoride with teeth enamel is ionic and hence any fluoride compound that will give ionised fluoride in a dilute solution will give satisfactory results. It may also be noted that there is no difference in fluorides occurring naturally in water and those artificially added in water.

SEWAGE AND SEWAGE TREATMENT

The waste products of a society including the human excreta had been collected, carried 7 disposal of manually to safe point for disposal by sweepers. This primitive method of collecting and disposing of the society's wastes has now been modernised & replaced by a system in which these wastes are mixed with sufficient quantity of water and carried through closed conduits under the conditions of gravity flow.

This mixture of water & waste products, popularly called sewage, thus automatically flow to the place from where it is disposed off, after giving it suitable treatments. The treated sewage effluents may be disposal of either in a running body of water, such as stream, or may be used for irrigation of crops.

Characteristics of Sewage

Sewage although contains more than 99.9 per cent water, the small portion of solids it highly hazardous in nature and may cause pollution of stream, lake or harbour, cause nuisance, endanger water supplies. So it is important to know the characteristics of sewage which will give the idea of degree and method of treatment to be adopted for safe disposal. These characteristics are divided into three classes: (i) Physical; (ii) Chemical; and (iii) Bacteriological.

Physical characteristics

Fresh domestic sewage has a slightly soapy or oily odour and in appearance is cloudy. The solids of sewage are in solution and suspension and include both organic and inorganic matters. Suspended solids are those which can be filtered out on an asbestos mat and dried. They are relatively high inorganic matter. Dissolved solids are obtained by evaporating a filtered sample over water bath. Solids obtained after evaporation when ignited, the loss in weight is ascribed to the volatile solids, which are classed as organic matter; and the remaining or fixed solids are classified as inorganic.

Chemical characteristics

Sewage contains inorganic chemical compounds from the water supply and many complex organic materials derived from faeoes, urine and other organic wastes. It is normally alkaline when fresh but tends to acidity as it becomes stale. Complete treatment, however, will restore the alkalinity. It will contain nitrogenous compounds like urea, proteins, amino acids and carbonaceous compounds like soap, fats, carbohydrates etc. By treatment most of the organic matters get stabilised.

Stabilisation

By stabilisation is meant that organic matter has been broken down by bacterial action to simple inorganic substances that will decompose no further. The stabilisation may be brought about by anaerobic or aerobic bacteria. Anaerobic bacteria grow and liberate energy in the absence of free oxygen and obtain it from various compounds which they are to break down. But in anaerobic stabilisation reaction is slow and this process produces unpleasant odours.

Aerobic action takes place when free oxygen is available for bacteria. The oxygen may be available from atmosphere or sewage. In what is known as self-purification of stream, free oxygen is dissolved in the water and becomes available for aerobic bacteria. Nitrogenous compounds by gradual steps of oxidation are converted to nitrates. Those nitrates unlike free nitrogen are usable to the higher plant life. Thus this process is known as nitrogen fixation.

Ether-soluble matter

This is a determination of the fats and greases in a sample of sewage or sludge. The evaporated sample is extracted with ether, and the other is then evaporated to leave the grease behind. Fats and greases when excessive in amount cause scum formation in sedimentation tanks and are injurious to filters.

Disposal of Sewage

Sewage produced, collected, conveyed through sewers finally should be disposed off. It can be treated, or may be disposed off without treatment depending on conditions. This can be done by: (i) land disposal; and (ii) dilution technique.

Land disposal

Mainly there are three methods by which land disposal is done :

(1) *Broad irrigation:* This is also known as surface irrigation. In this process, sewage is allowed to flow over cultivable land, from which a part of sewage evaporates some percolates into the sub-surface and remaining part flows out and collected through surface drainage channel and disposed off.

(2) *Sub-irrigation:* Sewage is allowed to enter the land, through distributors. Sewage by the process of evaporation, filtration, biochemical action etc. gets purified like in previous method. Effluent thus purified can now be suitably disposed off.

(3) *Ridge and furrow method:* This is an improved method of sub-irrigation. Here the sewage is allowed to stand into furrows and crops are grown on the ridges. This way sewage does not get direct touch with the crops.

Sewage farming

It is interesting to note that sewage applying to land by above methods not only solves disposed problems, but also contributes land its fertilising values. This method of growing crops is known as sewage farming. Fertilising elements like nitrogen, phosphate, potash etc. from sewage gets entry to cultivable land through biochemical process.

Sewage sickness

If sewage is continuously applied to some land, the pores of the soil get clogged, preventing free circulation of air. This retards the aerobic action of the sewage also stops filtration of water. So the land looses the purifying capacity of sewage and known as sewage sick land. This condition may be stopped by: (i) applying pretreated sewage instead of raw sewage; (ii) giving rest to the land and using reserved land; and (iii) by rotation of crops.

Advantages of land disposal method

(1) Disposal is accompanied by natural treatment.

(2) In case of irrigation the fertilising value of sewage is returned to the land.

(3) The pollution of natural body of water is prevented.

(4) The method of cheap without requiring any installation of costly machineries involving high initial cost.

(5) Disposal of possible even without a nearby river with enough quantity of water.

Disadvantages

(1) This method requires huge quantity of land both for action as well as for reserve.

(2) In rainy season or wet climate the method is not suitable.

(3) Lands with clayey soil this method is not efficient.

(4) Where crop is grown, this method requires special attention to prevent the spread of diseases.

(5) If not properly supervised the land may become sewage sick.

Dilution technique

Disposal sewage or effluent into a body of water or water course is known as disposal by dilution. In this method we take advantage of the self-purification capacity of reviving water like stream, lake, sea etc. Prior to disposal, the sewage may be treated. The degree of treatment depends on the quantity and concentration of sewage and the dilution ratio.

The physical forces that are helpful in self-purification of water are dilution, dispersion, sedimentation and turbulence of water. By dilution the combined strength of any particular parameter like BOD, COD, chloride content etc. will be lower than that of the effluent quality.

Oxidation ponds

Oxidation ponds are also used for the disposal of sewage. They are shallow basins excavated in the ground which retain the sewage or waste-water for a certain desired period. Both settled or raw sewage can be applied and stored to these tanks.

During the process of storage sewage comes in contact with soil bacteria. Again under favourable climatic conditions namely warmth and sunshines there will be growths of algae. Bacterial decomposition of the sewage releases carbon dioxide, ammonia etc. Ammonia and other plant growing substances are used by algae. Algae again by the process of photosynthesis make the pond water to its oxygen saturation value creating favourable condition for decomposition of organic wastes in sewage by aerobic bacteria. The entire process is known as algal-bacterial symbiosis of stabilisation of wastes in oxidation pond.

The organic matter in the waste is oxidised by bacteria to carbon dioxide, ammonia, phosphorus compounds and micro nutrients.

Sewage Treatment

Sewage treatment is found necessary for the purpose of converting a raw harmful sewage into an acceptable final effluent and also to dispose of the solids removed in the process. It is therefore, essential first to determine the characteristics of the raw sewage and the required characteristics of the effluent, or the required degree of treatment, before proceeding with the design of the treatment plant.

Once these characteristics treatment requirements have been determined, it should not be assumed that they are fixed and final and not subjected to change. For example, pretreatment or elimination of industrial wastes can have a marked effect on the characteristics of a raw sewage. Even effluent requirement may be subjected to modification under certain conditions. A proper balance of influent characteristics, effluent requirements and degree of treatment should first be ascertained before fixing the design basis. To achieve this balance with optimum economy it will necessitate the use of rational analysis as well as other recognised tools of engineering.

Therefore, methods of treatment will depend on several considerations like, the characteristics of influent sewage, condition of available disposal points, effluent standards desired etc.

Classification of sewage treatment

Preliminary treatment

This can be defined as preparatory treatment. This will make sewage fit for further treatment. Screens, shredder or communitors, grit chamber and detritus tanks are included in this treatment. This removes about 10 per cent of total solids like floating matters, grit and detritus from domestic sewage.

Primary treatment

This is also known as simple treatment. In the first case the solids removed are mostly inorganic in nature. In this case attempts are made to remove the settleable organic matters. This is done with the help of plain sedimentation tanks. About 40 per cent of total solids gets removed. The BOD reduction varies from 30 to 40 per cent. But the dissolved solids including organic matters escapes along with the effluent.

Final or complete treatment

This is essentially a biological treatment. In this non-settleable and dissolved organic matters get removed. About 90 per cent of total solids and 85 per cent of BOD get removed from sewage after this treatment. Intermittent filters, trickling filter, activated sludge process etc. are included in this treatment.

Screening

Screening of sewage is done for removing floating matters like pieces of wood, charcoal, leaves etc. Sewage admitted to subsequent treatment units or pumping stations should be effectively screened to protect the machinery in the plant and to avoid difficulties and unsightly condition in subsequent stage of treatment.

Chemical precipitation

Chemical precipitation consists adding of such chemicals in sewage which by reaction with one another and also with the sewage constituents will produce a flocculent precipitate and accelerate sedimentation. This is quite similar process as coagulation adopted in water treatment. In this case also suspended solids, both colloidal and unsettleable can be settled easily by chemical precipitation. Sewage containing large quantities of substances that will need only a small amount of chemical to produce the required precipitate is the most favourable for this form of treatment.

The chemicals used for precipitation are lime, ferrous sulphate, ferrous chloride, copperas, sulphuric acid, ferric chloride, ferric sulphate etc. Alum though widely used as a coagulant in water treatment is not normally used in sewage precipitation the reason being it forms lighter floc not suited in sewage treatment. Ferric chloride forms heavy flocs and popularly used in treating sewage although it has the disadvantages of handling and storing difficulties.

Though the chemical reaction may be similar to water treatment process, because of the greater amount of suspended matter in sewage, the dosing usually are considerably greater. In order to keep the cost down, it is, therefore, even more important that optimum pH values be obtained.

In this process dry chemical is fed into a mixing device in which the chemical goes into solution or suspension of calculated strength. This solution is then fed into the sewage which is needed to be precipitated. The mixture flows into the flocculator, where it is stirred gently to induce good coagulation and flocculation. After flocculation the sewage flows to sedimentation tank or clarifier where sludge settles and gets separated.

Mixing period of above is kept about 1 minute, flocculation will take 15 to 20 minutes and for sedimentations detention period is usually kept for 2 hours.

By chemical treatment process dissolved solids are practically remain unaffected. The effluent however becomes quite clear as the turbidity is reduced considerably. With normal dosage of chemicals the suspended solids are removed by 60 to 80 per cent and BOD is reduced by 35 to 60 per cent.

Since chemical precipitation removes most of the suspended matter, it will produce much more sludge than plain sedimentation. The chemical itself forms the floc and add to the quantity of sludge. But as the chemical sludge contains only 90 per cent moisture as against 95 per cent in plain sedimentation the resulting volume may increase only marginally.

The process is particularly suitable where only seasonal treatment is required, as in a summer resort, or as a supplement to other methods of treatment, or for industrial wastes under variable conditions. The process is quickly adjustable to changing conditions and is especially suitable for industrial plants using chemical processes of manufacture. The advantages of the process can be summarised as given below :

(1) Where ordinary plain sedimentation is sufficient but low stream flow at times lowers the dilution factor and necessitates somewhat better treatment. As chemical dosing can be discontinued the process is flexible enough to meet this condition.

(2) As there is forced sedimentation, it is more efficient and space required for treatment unit is less.

(3) Degree of treatment is better than plain sedimentation.

(4) Many industrial wastes do not respond to biological treatment and others require very good clarification as a pre-treatment. In such cases this treatment is very useful.

(5) Colour in waste may be removed by this process.

Disadvantages of chemical precipitation :

(1) Operation cost is high due to high cost of chemicals.

(2) Skilled hands are required for operation of the processes.

(3) Additional provision has to be made for handling and storage of chemicals.

(4) Quantity of sludge accumulated is very high many times the treatment and disposal of them pose a problem.

Skimming tank

The objective of skimming tanks is the separation f om the waste-water of the lighter floating substances. The materials colle ted on the surface of skimming tanks, include oil, grease, soap, pieces of work, wood etc.

Skimming is required to remove the scum which otherwise will find its way to secondary treatment units and will interfere with the biological treatment processes, thus the efficiency of secondary treatment will be adversely affected.

Sludge accumulated gets digested and it is comparatively stable in nature. It is periodically cleaned and may be used as manure after further stabilisation.

Disposal of Septic tank effluents

The disposal of septic tank effluent may be done either by soil absorption or by biological filters. The construction of disposal system at the ground surface or below ground level will depend on sub-soil water level condition. When sub-soil water is within 180 cm from ground level the disposal units are constructed partly or fully above ground, otherwise sub-surface construction of this units are suitable.

In case the above mentioned conditions are not feasible and where the effluent has to be discharged into open drain it should be disinfected.

Imhoff tanks

This tank like the septic tank, combines sedimentation along with sludge digestion. There is, however, an important difference, digestion takes place in a different compartment from that where settlement occurs. Second difference is that it is found to be suitable for serving a population much bigger than that of a septic tank, normal range being 5,000 to 30,000 population.

Clarigester

It is a modified type of Imhoff tank. The sewage is fed in a settling tank with an interval cone like bottom. The inside of which acts as the digester hood. These are normally of 5 to 12 metres in diameter suitable for towns having population ranging between 10,000 to 25,000. It is a two storey tank as usual, but instead of hope bottoms, both the upper and lower chambers are flat. The sewage enters the sedimentation chamber through the inlet pipe and flows radially over a circumferential weir and is collected by an effluent channel.

The sludge is deposited on the bottom of the upper chamber. From there a gate permits the entry of sludge into lower chamber, but prevents sewage or gases to move upwards. A central vertical shaft rotates, which has arms attached at four different levels. The upper arms are provided with skimming blades for removal and collection of scum. The next sets of arm move the deposited sludge towards the central sludge door. The third set act as scum breaker arms. The lowest set is for raking and stirring the sludge.

Due to this stirring, mixing of sludge is better. This promotes efficient digestion. Gas collected by the process of digestion, may be utilised in heating the digestion chamber. As the bottom is not hopper, the tank depth is lesser. As the digestion is rapid, lesser space is required. More gas is released due to stirring. As sludge pipe is attached at the bottom for withdrawal of digested sludge.

SLUDGE TREATMENT AND DISPOSAL

Handling and disposal of sludge from biological waste-water treatment plants is an important problem and represents about half the cost of most sewage treatment plants. The concentration of solids in the primary sewage sludge is about 5%; the activated sludge contains less than 1% solids; and the sludge from trickling filters has about 2% solids. This means that the sludge is composed almost entirely of water and volume reduction is the key to economic disposal. In addition to reducing its high water content, the sludge must be stabilised so that its biological activity and tendency towards putrefaction are reduced drastically.

The common unit operations of sludge treatment and disposal involve concentration of thickening, digestion, conditioning, dewatering, oxidation and safe disposal. Fig. 4.7 represents the general flow sheet of sludge treatment techniques in approximately the order in which they could be applied in a treatment scheme.

Concentration

The purpose of concentration or thickening is to remove water from the sludge and reduce its volume as much as possible so that the sludge can be handled more efficiently. The common methods for thickening are gravity settling and flotation. In gravity thickeners the sludge is subjected to gentle agitation by means of a slow stirrer which enhances settling. In this manner the combined sludge from primary and secondary settlers can be thickened so as to contain 5–9% solids. Often the thickening of activated sludge is complicated by anaerobic action, particularly under warm conditions when the bacteria in the sludge decompose organic matter and release gases. This also creates settling and odour problems. The sludge can also be thickened by air flotation, particularly the secondary sludge, which keeps the system aerobic. The flotation technique can concentrate the sludge to bring it solids content to 4%.

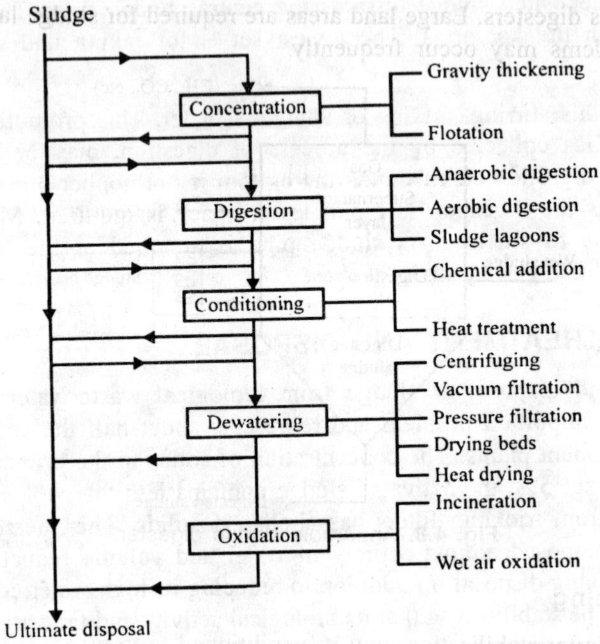

Fig. 4.7. Sequence of operations for sludge treatment (Arrows indicate possible flow paths).

Digestion

After concentration, the sludge is stabilised by digesting it under aerobic or anaerobic conditions. Anaerobic digestion is the most common method in which the organic content of the sludge decomposes to give mainly methane and carbon dioxide and at the same time the bound water is released from the sludge. Properly digested sludge is black with a faint smell of tar, and is stable. In a typical sludge digester, shown in Fig. 4.8, raw sludge is fed into the active digestion zone and gas lifts the sludge particles and other materials which form a supernatant layer on the top of the digestion zone. The gas is collected at the top and the digested sludge is withdrawn from the bottom. The normal detention period in standard digesters varies from 30 to 70 days depending upon the temperature conditions.

Aerobic digestion is also used, and it may be considered similar to extended aeration. The sludge is aerated in a tank for about 20 days at ambient temperatures. During the process the bacterial cells are destroyed and a substantial portion of the sludge is oxidised resulting in the reduction of the solid content by about 30%. Sometimes shallow lagoons are employed as digesters. Large land areas are required for sludge lagoons and odour problems may occur frequently.

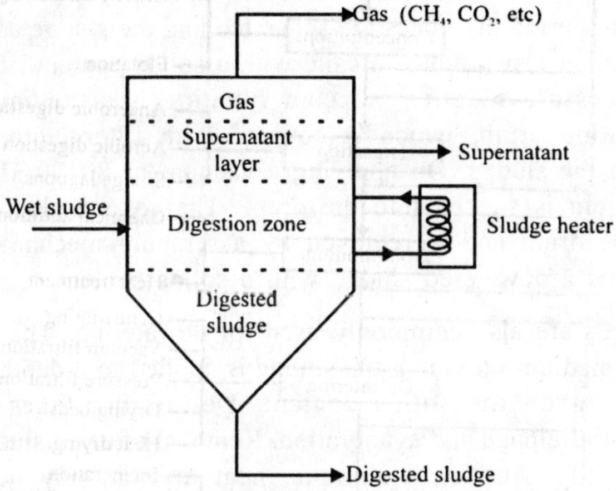

Fig. 4.8. Anaerobic sludge digester.

Conditioning

The sludge after stabilisation may be conditioned to improve its dewatering characteristics. This is done by adding chemicals like iron salts, alum, lime and polyelectrolytes. These chemicals bind the sludge particles together and encourage the release of absorbed water. Physical conditioning methods such as heat treatment are becoming popular. The sludge is heated under pressure and after a period of time the gel structure of the sludge breaks down so that the water as released. Heat treatment has the advantage of sterilising the sludge; at the same time the sludge is partially oxidised and completely stabilised.

Dewatering

The thickened sludge is dewatered for efficient handling and disposal. Dewatering is accomplished by mechanical methods, the most common being centrifugation and filtration, which includes pressure filtration and vacuum filtration. In centrifugation, conditioned sludge is added to a rotating bowl that separates the sludge into a cake and a dilute stream. The solid cake is transported within the bowl and is removed by a screw conveyor at one end of the bowl; the liquid is removed at the opposite end. Centrifugation is a compact method which requires careful control of process variables.

Filtration, using plate-and-frame pressure filters or rotary drum vacuum filters, is widely used for dewatering digested sludge. In pressure filtration, the sludge is pumped slowly with increasing pressure into filter plates supporting a cloth, which retains the solids. It is a bath process, and after the dewatering period the plates are separated and the sludge cake removed. Pressure filtration can produce a cake with a solids content of 25–50%. In contrast to pressure filtration, vaccum filtration is a continuous process where a rotating drum, which is covered with filter cloth, is partially submerged in the sludge. On applying a vacuum of 80–90 kPa inside the drum, the liquid is sucked into the drum. The cake is deposited on the outside of the drum and is removed by a scraping mechanism. Vacuum filtration yields a dewatered sludge with a solids content of about 25%.

Drying beds are also commonly used for dewatering. The bed consists of a filtering medium on which the sludge is applied to a depth of upto 250 mm, depending on the solids content. Dewatering takes place by a combination of drainage and evaporation. Removal of dried sludge is carried out mechanically. Another technique, heat drying, may be utilised in applications where the sludge is to be incinerated or when a saleable commodity can be produced. A major problem associated with this process is the control of gases and ash particles which may be generated in drying.

Oxidation

Before the final disposal, some sludges may be oxidised to reduce the organic content, with the consequent destruction of bacteria and a significant reduction in their volumes. Incineration and wet oxidation are the two common methods employed for sludge oxidation. Incineration is usually performed in a multiple hearth furnace, although fluidised beds or flash dryers may also be used. In the multiple hearth furnace the sludge passes downward through a series of hearths. Vapourisation of moisture occurs in the upper hearths, followed by incineration in the lower ones. The combined efficiency of evaporation and incineration in multiple hearth furnaces is about 55%. The fluidised bed system consists of a bed of sand fluidised by air. When the sludge is introduced, the sludge particles are dried almost instantly as they are dispersed, and are oxidised (Fig. 4.9).

Wet oxidation is a process in which the sludge is ground, mixed with air in stoichiometric proportions, and then subjected to high temperature reactor. Under the conditions of high temperature (260°C or more) and high pressure (8000–12000 kPa) there is no vapourisation of the sludge liquid. Initially, external heat is applied to start the oxidation of the organic matter in the sludge. The rate of oxidation increases until it reaches an equilibrium value. The products from the top of the reactor are passed through heat exchangers to recover heat which in turn is utilised by the incoming feed.

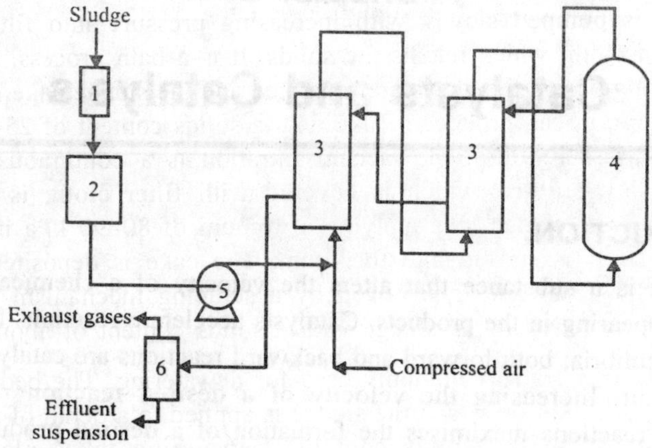

Fig. 4.9. Wet oxidation process flow diagram (1. Grinder; 2. Sludge storage tank; 3. Heat exchangers; 4. Reactor; 5. Sludge pump; 6. Separator).

Ultimate Sludge Disposal

Several methods are employed for the ultimate disposal of sludge. Wet digested sludge may be sprayed onto cropland where it functions as a fertiliser or fertiliser base. It may be lagooned; however, when the lagoons fill they must be abandoned. Dried sludge may be used as a landfill or a soil conditioner. Wet or partially dewatered sludge or ash from incineration may be transported from the shore to dumping grounds at sea.

Chapter 5

Catalysts and Catalysis

INTRODUCTION

A catalyst is a substance that alters the velocity of a chemical reaction without appearing in the products. Catalysts accelerate reactions but do not change equilibria; both forward and backward reactions are catalysed to the same extent. Increasing the velocity of a desired reaction relative to unwanted reactions maximises the formation of a desired product.

Catalysts are believed to function through an unstable chemical complex formed between the catalyst and reactant molecules. This complex reacts to produce new compounds with dissociation of the complex and regeneration of the catalyst which can then bring about the transformation of additional reactant molecules.

Chemical and Physical Aspects

The rate of a chemical reaction is proportional to a rate constant :

$$k = Ae^{-E/RT}$$

Because of the exponential form of the equation, even a small decrease in E, the activation energy, can bring about a large increase in velocity. The catalytic pathway is a series of three chemical reactions: formation of the complex (activated adsorption), surface reaction, and desorption.

Solid surface catalysts invariably adsorb reactant molecules, adsorption and chemical reaction occurring on active centres identified as certain types of lattice defects. The geometric factor in catalysis emphasises the significance of the reactant molecule's spacial arrangement and the catalyst crystalline lattice.

The activity and selectivity of a catalyst are affected by its pore structure, e.g. in a very slow reaction molecules can diffuse through a pore system to the centre of a catalyst particle before they react and the entire internal surface area will be used. The fraction of surface available in a given reaction can be calculated. The experimental rate constant, K_{exp}, is compared with :

$$K_d = (18/a^2)\ \rho B V_g D$$

where a = catalyst granule size, ρB = bulk density, V_g = porosity, and D = diffusion constant.

Various theories have been applied to catalysis, including quantum mechanical treatment of solids and the electron band, crystal field, resonance valence bond, and molecular orbital ligand field theories.

Techniques

The following factors are considered when choosing a catalyst for a specific reaction: selectivity, or the efficiency in catalysing a desired transformation; activity, or the overall conversion, expressed as the amount of reactant in contact with the catalyst under a given set of conditions converted to all products; stability, the ability to retain initial activity and selectivity over the catalyst's lifetime; physical suitability; regenerability; and cost.

Heterogeneous catalysts, in contrast to homogeneous catalysts, form a separate phase from reactants and products and are usually solids. They have a microporous structure and a very large internal surface area (it may reach 1000 m^2/g) but these colloidal catalysts are often inactivated by the presence of small amounts of poisons. Poisons act by combining with an active site in the catalyst surface and rendering it inactive.

Most poisons are strongly sorbed and react quantitatively with active sites to form stable inactive surface compounds, e.g. acidic catalysts are poisoned by basic nitrogen and alkali-metal ions, metal catalysts by sulphides, arsenic and lead compounds. Some poisons cause loss of selectivity, e.g. nickel and copper present in gas oils used in catalytic cracking. In some cases, poisons minimise a desired reaction while permitting an undesired reaction. Reactivation of a catalyst is sometimes accomplished by decomposition or desorption of the poison. In contrast, small amounts of promoters or co-catalysts may cause striking increases in catalytic activity.

In industrial use, catalysts lose activity and sometimes selectivity. In general, heterogeneous catalysts lose effectiveness owing to overheating and contamination. Contaminants affecting catalytic efficiency are metals entrained in feedstocks; oxygen, nitrogen, or sulphur compounds; and polynuclear aromatic compounds which form coke.

Preparation

Properties of a catalyst are determined by its preparation, with care taken to ensure duplication of characteristics. Colloid chemistry is important, including gel preparation (e.g. hydrous oxides and systems mainly of silica or alumina are amenable to gel formation), leaching selected components from solids (e.g. Raney nickel catalyst is prepared by leaching aluminium from a nickel-aluminium alloy by means of a caustic alkali), or decomposition of salts or hydrates with gas evolution. Frequently, the support is manufactured first by one of these techniques followed by impregnation with additional chemical components. The product may again require suitable treatment, e.g. precipitation, calcination, reduction (e.g. for hydrogenation catalysts such as nickel, cobalt, iron, or copper catalysts), etc.

Organic and Inorganic Chemicals

Important catalytic commercial processes include the Deacon process; platinum-catalysed oxidation of ammonia for nitric acid production; the Haber and Mittasch ammonia synthesis process; hydrogen production employing catalytic reactions such as reforming, shift, and methanation; hydrogenation techniques for hardening fat; synthesis of methanol from carbon monoxide and hydrogen; production of phthalic anhydride via oxidation of naphthalene; oxo reaction to produce aldehydes, which can be catalytically hydrogenated to the corresponding alcohols; catalytic dehydrogenation of butene; the catalytic cracking, reforming, or hydrodesulphurisation of petroleum; polymerisation for production of plastics and synthetic rubber; the production of polyurethane plastics.

Transition-Metal Complexes—Homogeneous Catalysi~

The distinction between homogeneous and heterogeneous catalysts has been disappearing as fundamental similarities have been recognised. However, certain transition-metal complexes are classified as homogeneous catalysts since they can be molecularly dispersed in solution, e.g., hydrocobalt tetracarbonyl, $HCo(CO)_4$. Some homogeneously catalysed reactions are hydrogenation, hydroformylation, double-bond migration, polymerisation, and oxidation.

Organometallic complex catalysts tend to be highly active, specific, and selective, and may resist poisoning better than heterogeneous systems. The catalytic centre may have two metal sites. Organometallic catalysts have a central transition-metal atom or ion bonded to molecules known as ligands which form a polyhedron around the metal. The nature of the metal and ligands critically influences the electronic structure of the complex. The coordination bonding properties of the metal are a key feature of the catalytic complexes. Important reactions of catalyst complexes are ligand exchange, oxidative addition, and the insertion reaction.

The most important industrial reactions are the Wacker process; vinyl acetate the oxo process; methanol carbonylation; and the Ziegler-Natta polymerisations.

The advantage of complex catalysts including the anchored complexes lies in the predictability of their properties, the ease of interpreting reaction mechanisms, and the consequent potential for extending the homogeneous catalysts to many new applications.

Oxidation Catalysis

Classical examples of heterogeneous catalysis include oxidation of CO over copper, of NH_3 over Pt, and of SO_2 over Pt or V_2O_5. Heterogeneous oxidation catalysis is similar to catalysis by metal complexes since both utilise a metal of variable valence.

Commercially important hydrocarbon-oxidation processes include oxidation of benzene to maleic anhydride and toluene to benzoic acid, naphthalene and xylenes to phthalic acids, ethylene and propylene to their respective epoxides. The catalytic ammoxidation of propylene, e.g., the Sohio process, is the successful process for the manufacture of acrylonitrile. Ethylene dichloride, the precursor of vinyl chloride, is manufactured by the oxidation of ethylene.

Catalysis Applied to Fuels

The concepts of acid and basic catalysis is applied to reactions of industrial importance, particularly hydrocarbon reactions used in large-scale petroleum refining.

Acid catalysts are employed in such diverse areas as Friedel-Crafts reactions and reactions in the presence of clays. The isomerisation of *n*-butane over an aluminium halide–hydrogen catalyst takes place only in the presence of olefins. In each case the active catalyst contains a proton. For example, hydrogen chloride plus aluminium trichloride give catalytic activity of acid catalysts is correlated with surface acidity for different reactions.

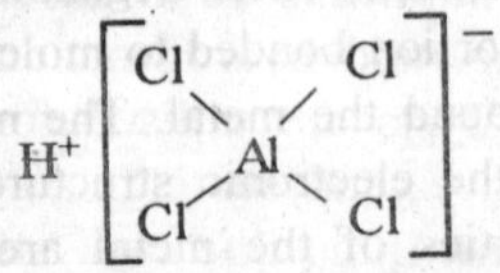

Basic catalysts are employed less than acid catalysts, but are used industrially for polymerisation of conjugated dienes as well as for the isomerisation of olefins, dehydrogenation of certain diolefinic materials to aromatics, alkylation of arylalkanes, and polymerisation of mono-olefins. Base catalysis gives different products than acid catalysis, e.g., ethylene alkylates toluene in the side-chain with a base catalyst whereas acid catalysis gives ring alkylation.

Zeolites (Molecular Sieves)

Crystalline zeolites are silicates with an open framework, regular structure, and apertures of molecular dimensions based on a tetrahedron of four oxygen ions surrounding a smaller silicon or aluminium ion. Important zeolite catalysts are known as X, Y (similar to the rare natural mineral, faujasite), erionite, and mordentite, and the newer ZSM-5 and ZSM-11. The most significant application of crystalline zeolites has been in catalytic cracking where catalysts containing 10–20% zeolites greatly improve selectivity and activity.

Dual-Function Catalysts

Reforming represented a significant advance in the utilisation of the dual-function catalysts, e.g., reforming with platinum catalysts is today the second largest industrial catalytic application. Another example of dual functionality is the conversion of C_{6}- cycloparaffins (cyclohexane and methylcyclopentane) over single- and dual-function catalysts for converting methylcyclopentane to benzene. For certain reactions, dual-function catalysts can accomplish more than passing the reactants through two reactors in sequence, each filled with a different single-function catalyst.

Multimetallic Catalysis

Bimetallic clustre catalysts are composed of atoms of two different metals in a state of high dispersion on a carrier. Dispersion refers to the fraction of atoms in a metal crystallite that exist on the surface, with the degree of dispersion increasing with decreasing crystallite size. These novel bimetallic systems offer a new range of catalytic properties and are not limited to combinations of metals that form bulk alloys. New metallic catalysts with high activity and unprecedented activity maintenance are employed in the reforming of petroleum naphthas to high-octane gasoline and in isomerisation, hydrocracking, and hydrogenation. In the petrochemicals field, improved Pd-Au catalysts are utilised for the synthesis of vinyl acetate, whereas more selective catalysts (e.g., Ag-Au, Cu-Au) are used for the partial oxidation of olefins. Bimetallic catalysts show increased activity, improved resistance to poisoning by coke accumulation or deposition, and optimise product selectivity, e.g., dehydrocyclisation and aromatic hydrogenation are possible with a minimum of cracking reactions (hydrogenolysis) while maintaining very high hydrogenation activity.

Synthetic Fuels

Synthetic fuels from coal, oil shale, and biomass are expected to become of enormous industrial importance with catalysis playing a key role in the conversion of these complex solids to more convenient gaseous and liquid forms, and in the removal of sulphur, nitrogen, and oxygen in addition to inorganic mineral matter.

Four general types of coal liquefaction processes are catalytic hydro-liquefaction, pyrolysis, solvent extraction, and indirect synthesis (Fischer-Tropsch). The mechanism of direct coal liquefaction is believed to consist of a primary thermal pyrolysis of coal, followed by catalytic hydrogenation of the reactive fragments or asphaltenes formed by pyrolysis. Several direct catalytic coal-hydroliquefaction processes use active cobalt–molybdenum-alumina catalysts developed for petroleum desulphurisation, including the H-Coal process (HRI), Synthoil (DOE), SRC and EDS (Exxon), and CCL (Gulf Oil), with others being researched.

In contrast to direct coal hydroliquefaction, in indirect liquefaction coal is gasified to a mixture of $CO + H_2$, followed by catalytic conversion to fuel products (Fischer-Tropsch process). The influence of the catalyst is profound, product distribution obtained varies considerably (Table 5.1). The SASOL process is in commercial operation in South Africa. A plant using the Mobil process is being built in New Zealand. Catalysis is effective in all three reactions involved in coal gasification, i.e., gasification, shift, and synthesis.

Methanation is used commercially to convert relatively small amounts of harmful carbon oxides to CH_4, whereas in ammonia synthesis, CO would especially interfere with catalytic conversions of gas mixtures. Methanation also will be used on a large scale in the manufacture of high heat values or pipeline gas, essentially methane. Chemicals are expected to be derived from the refining of coal hydro-liquefaction, especially aromatics, or from gas synthesis.

Table 5.1. Syngas process products.

Product	Process, % of product Fixed	SASOL-I Fluid	Mobil
Light gas C_1, C_2	7.6	20.0	1.3
LPG C_3, C_4	10.0	23.0	17.8
Gasoline C_5–C_{12}	22.5	39.0	80.9
Diesel C_{13}–C_{18}	15.0	5.0	0
Heavy oil C_{19+}	41.0	6.0	0
Oxygen compounds	3.9	7.0	0
Aromatics, % of gasoline		5	38.6

Fuel Cells

Catalysts, e.g., noble metals on electrically conducting supports, are used at both the anode and cathode in a fuel cell.

Environmental Control

Catalysts used for the abatement of noxious emissions were developed where the active component and substrate are supported on sturdy ceramic honeycomb structures, called monolithic supports. Platinum-group metal catalysts are used for nitric acid plant tail-gas cleanup. Platinum metals supported on SiO_2, Al_2O_3, and carbon are used as catalysts to eliminate organic fumes from air pollution aid for the safety. Hydrocarbon oxidation reactions are promoted via catalysts for emission control. Combustor systems are usually ceramic monolithic substrates coated with a catalytic agent. Catalysts are used in hydrogen sulphide conversion to sulphur.

About 60% of catalysts is used in the petroleum industry, and the remainder for major chemicals and chemical intermediates.

PHASE-TRANSFER CATALYSIS

Phase-transfer catalysis (PTC) is a technique by which reactions between substances located in different phases are brought about or accelerated. It is also known as extractive alkylation, catalytic two-phase reaction, and preparative ion-pair extraction. Typically, one or more of the reactants is an organic liquid or solid dissolved in a non-polar organic solvent and the coreactants are salts or alkali-metal hydroxides as solids or in aqueous solution. In principle, however, one reactant can be a gas, and the extracted species—normally an anion—can be a cation, a metal atom, or a neutral molecule. Catalysts used most extensively are quaternary ammonium or phosphonium salts, and crown ethers and cryptates. There are two distinct classes of reactions formed under PTC conditions :

(1) Reactions without added bases, e.g., displacement reactions such as $RX + Y^- \rightarrow RY + X^-$; oxidation reactions with oxidants such as $KMnO_4$, KO_2, $K_3[Fe(CN)_6]$, $K_2Cr_2O_7$, NaOCl; as well as some reduction reactions.

(2) Reactions in the presence of inorganic bases (conc. aqueous solutions or finely ground solids of NaOH, KOH, K_2CO_3, etc). These reactions include alkylations of weak C—H acidic compounds up to a pK_a limit of 22-25; alkylations of ambident anions; alkylations of OH, NH, and SH bonds; isomerisations and H—D exchange; additions across C=C and C=O bonds; β-eliminations; formation of carbenes by α-eliminations; hydrolysis and saponifications; Darzens reactions; Horner-Wittig reactions; nucleophilic aromatic substitutions; and many others.

Mechanism of Phase-Transfer Catalysis

A mechanistic picture of class (*1*) reactions shows that catalyst cation Q extracts the more lipophilic anion Y^- from the aqueous to the non-polar organic phase where it is present in the form of a poorly solvated ion pair $[Q^+Y^-]$. This then reacts rapidly with RX, and the newly formed ion pair $[Q^+ X^-]$ exchanges X^- for Y^- at the interphase for the next cycle. The rate of the reaction is proportional to the catalyst concentration and independent of the stirring speed above the minimum required for efficient mixing. Base-catalysed reactions are mechanistically diverse: They can be brought about by extraction of deprotonated substrate (Sub^-) from the aqueous phase, by deprotonation at the phase boundary and subsequent transport of $[Q^+Sub^-]$ into the depth of the organic phase, or by extraction of $[Q^+OH^-]$ as reacting base. In addition, many anions behave as bases under the virtually anhydrous condition, many anions behave as bases under the virtually anhydrous condition of PTC, e.g. F^-, Cl^-, $CH_3CO^-_2$.

Factors Influencing the Usefulness of Phase-Transfer Catalysis

The best solvents for PTC are non-miscible with water and non-hydroxylic, e.g., $CHCl_3$, hydrocarbons, chlorobenzene. The logarithms of extraction constants for symmetrical tetra-*n*-alkylammonium salts ($\log E_{QX}$) rise by ca

0.54 per added C atom, but the lipophilicity of phenyl and benzyl groups carrying ammonium salts is much lower than the number of C atoms might suggest. For example E_{QX} for tetra-*n*-butylammonium salts is about 140 times larger than E_{QX} for tetra-*n*-propylammonium salts of the same anion in the same solvent-water system. For practical application in mixtures of water-organic solvent, only ammonium and phosphonium salts containing 15 or more C atoms are sufficiently lipophilic unless a "salting-out effect" is operative as in the presence of concentrated sodium hydroxide. Crown ethers are sometimes as effective as the best onium salts in empirical catalyst comparisons; complexing agents advocated include 15-crown-5 and 18-crown-6 and their benzo derivatives or cryptates. Benzyltriethylammonium chloride is the most widely used catalyst under strongly basic conditions.

The successful extraction of the necessary anion into the organic phase is crucial for PTC. Often three anions compete for the catalyst cation: the one that is to react; the one formed in the reaction; and the one brought in originally with the catalyst.

Uses

Typical applications include displacement reactions; formation of esters; formation of ethers; alkylation of carbonions; Horner-Wittig reactions; oxidations using hypochlorite, permanganate, chromate or other oxidants; carbene reactions for preparing dihalocarbene adducts of olefins.

NEW ION-EXCHANGE RESIN CATALYSTS

Ever since catalysts were recognised as agents that speed up attainment of the reaction equilibrium, it has been customary to distinguish between homogenous and heterogenous catalysis. With the advent of ion-exchangers, this classification has become less neat. We expect the chemistry to remain that of homogenous catalysis with a 'free' catalyst species in a single fluid phase. Yet, mechanically, the system has all the attributes of heterogenous catalysis. The particulate nature of ion-exchange resins permits mechanical separation of catalyst and product mixture by filtration or decantation, thus eliminating distillation of product isolation.

Catalysis by ion-exchange resins lower the capital and processing costs by eliminating the steps required for catalyst removal in analogous homogenous process. Resins can often be used in hundreds of cycles without regeneration in a well designed process. It can work in both aqueous and non-aqueous systems. Resins having equivalent strength of strong acids can be handled without danger to personnel. The number of acid groups in contact with the equipment is very small, thus obviating the problems of corrosion. Agitation can further reduce the corrosion problems by minimising contact of the resin particles and the equipments.

The conclusion that an ion-exchange catalyst can at most produce same rate as dissolved electrolyte is still subject to revision. In fact the reactant might be preferentially sorbed by the IER and thus be attracted to just that portión of the system in which the catalyst ion is concentrated. If, so, a rate higher than in corresponding homogenous system could be achieved. Two reactions have acquired major commercial interest. The first is the production of ethers from olefins and alcohols for use as gasoline additives. The second is the inversion of sucrose to glucose and fructose, the latter with its sweetness, now being in great demand.

The major disadvantage of ion-exchange resins in catalysis is their relatively low thermal stability. For macroreticular resins, operation at 150°C for a long time may cause desulphonation. Research is on to develop IER having higher thermal stability.

Reactors used for IER Catalysed Reactions

Distillation column reactors (DCR)

DCR offer distinct advantages of exploiting the exothermicity of reactions such as in alkylations, etherifications, etc., for supplying heat required for distillation, lower mole ratios of reactants and yet get over the thermodynamic limitations encountered in simple reactors, lower resin inventory, etc., compared to common single stage reactors. The form of the catalyst is important. The standard size of the bead catalyst (0.5–2 mm) is not very suitable and techniques such as the use of tea bags have been considered. Resin catalyst in the form of random packings like raschig rings have also been made available.

It is necessary to have mechanical, thermal and chemical stability of the packings. Resin catalyst packings have also been moulded from resin particles and thermoplastic powder in a furnace for the hydrolysis of methyl acetate which is obtained as a by-product in the manufacture of polyvinyl alcohol from poly acetate and where the recycle of acetic acid to vinyl acetate is desirable. For making methylol, formaldehyde and methanol are reacted in the presence of a resin catalyst via reactive distillation by using multi reaction units and distillation tower. Nearly 100% conversion of methanol and formaldehyde was realised. Another example is the purification of phenol to make a better material, for instance, bisphenol-A of poly carbonate grade, where carbonyl compounds as impurities have to be reduced to vanishing levels. A continuous distillation column reactor containing nafion resin packings has been claimed as a versatile method to achieve the above objective. DCR has also been employed for the reaction of isomerisation of 2-hydrocarbyl-1-alkenes into 2-hydrocarbyl-2-alkenes using amberlyst-15 as a catalyst.

Ion Exchange Resin Catalysts in Production of Bisphenols

Generally bisphenols are made by reaction between phenols and carbonyl compounds in the presence of acidic catalyst. For polycarbonates, high purity bisphenols are required which can be advantageously made by using ion exchange resin as catalysts. For bisphenol production, ion exchange resins are not used directly. The cation-exchange resin is reacted with a mercaptoalkylamine and treated with p-toluene sulphonic acid. It is claimed that even after 500 hours, the conversion of phenol dropped from 11.6 to 10.3% recently. It has been claimed that by limiting the divinyl benzene content of the resin to 6% (wt %), long lasting activity of the catalyst can be obtained. The resin catalysts were used as packings in the reactor.

Transalkylations

Transalkylation reaction in the presence of nafion-H catalysts has been studied. Usually these reactions are carried out in the presence of $AlCl_3$ catalyst. However, it has been found that for trans-tert-butylation of aromatic compounds with 2,6-di(tert-butyl)p-cresol, more than molar equivalent of $AlCl_3$-nitromethane catalyst are required. On the other hand, 10% (w/w) of nafion-H affected smooth conversion to corresponding tert-butylated aromatic compounds and p-cresol. For transalkylation between tetrabromobisphenol-A and toluene in the presence of trifluoromethane sulphonic acid to get 2,6-dibromophenol, surprisingly, nafion-H did not catalyse the reaction effectively.

Reaction of Ethylene Carbonate with Methanol

Ethylene carbonate can be trans-esterified with methanol in the presence of ion-exchange resins containing sulphonic acid groups. Texaco Inc. has patented this process for producing ethylene glycol and dimethyl carbonate.

Higher Olefins to Secondary Alcohols

Acetate esters have been formed in high yields and at high rates using ion-exchange resin catalysts. These esters can be trans-esterified with methanol/n-butanol etc. to give the corresponding secondary alcohols which are also expected to be useful in making detergents.

Hydration of Acrolein and *A*-Methyl Styrene (AMS)

Hydration of acrolein gives hydropropanaldehyde via anti-markonikov addition which in turn can be hydrogenated to 1,3-propanediol.

Acetals and Ketals

The catalysis of acetal and ketal with cation-exchange resins have been reported. It has been found that the condensation of aq. formaldehyde with ethylene glycol in the presence of strongly acidic cation-exchange resin with continuous extraction of one or more of the products into benzene can lead to high conversions.

Dimerisation of Isoamylene and Cross-Dimerisation of Isoamylene with AMS

These reactions have been carried out in the presence of ion-exchange resins and acid-treated clays. In the case of isoamylene dimerisation, yields of more than 90% were achieved in the temperature range of 333–373K. In the cross-dimerisation reaction, with 1.2–2 mole ratio of isoamylene to AMS, 60% selectivity towards the cross-dimers, viz. 1,1,2,3,3-pentamethylindan and 3-ethyl-1,1,3-trimethylindan, was obtained at 353K. In both cases, ion-exchange resin was found to be a better catalyst as far as the activity was concerned.

Reactions with Polymeric Substances or Resulting in Polymeric Substances

Alkylation of poly (p-vinylphenol) with tert-butanol has been claimed in the presence of Amberlyst-15 as a catalyst. The preparation of phenolic polymers, based on dicyclopentadiene (DCPD), which are useful as thermosets, hardeners for epoxy resins, has been claimed using acidic ion-exchange resins such as Lewatit K-2611 as catalysts. The cracking of bisphenol-A in the presence of activated clay catalyst to give p-isopropenylphenol dimers, useful for eposy resins and polycarbonates has been reported. Ion-exchange resins can possibly serve the purpose with some added advantages.

Separation of Close-boiling Substances

The selectivity offered by the available cationic ion-exchange resins can be exploited to carry out difficult separations; further manipulations are possible through designing catalysts with pore sizes, acidity etc. The classic example of selective removal of isobutylene from C4 olefins/paraffins (and the corresponding example of isoamylenes from C5 olefins) should be cited as the difference in the reactivity of isobutylene and butene-1 is enormous and the selective etherification to give MTBE (ETBE) allows manufacture of not only pure MTBE, but also, through multi-stage reactors, raffinate with a negligible amount of isobutylene so that polymer grade butene-1 can be obtained by straight distillation. Further, the outgoing gases, containing the saturated amounts of methanol, can possibly go through a selective membrane to remove methanol so that purification of butene-1 of polymer grade is facilitated. MTBE can be cracked to give pure polymer grade isobutylene; similarly, methyl-tert-amyl ether can be cracked to give pure isoamylenes.

The efficacy of resin catalysts can be brought out in selective etherification of p-cresol can be in the presence of the close boiling 2,6-xylenol, phenol in the presence of o-chlorophenol, b-naphthol etc. with isobutylene. Here an added advantage is that the product often have different boiling points thereby further facilitating the separation. The ethers can be conveniently cracked to give the original phenol substance and isobutylene can be recycled.

Alkylation of Aromatic Amines

The alkylation of aromatic amines, such as anilines, toluene diamines, diphenyl amine etc., with olefins like isobutylene, diisobutylene, styrene, AMS, etc. is of considerable commercial importance. Thus dicumyl diphenylamine is sold as an anti-oxidant. These alkylation reactions are difficult and a variety of catalysts have been studied. Alkylation of diphenylamine with diisobutylene and a-methyl styrene in the presence of macroporous resin catalyst (e.g. amberlyst-15), was found to be unsuccessful even upto 393K. However, properly treated acid-activated clays are very useful catalysts. The lift of the catalyst, in this application, is short.

Some Engineering Aspects of the use of Cationic Ion-exchange Resins as Catalysts

It has been found that drying of ion-exchange resins by methanol percolation leads to lower moisture content as compared to that in other methods. The dissolved moisture content of organic reactants should be assessed and its possible impact on the rate of reaction be ascertained. Any impurity in the feed which can adversely affect the resin catalyst should be removed. Thus the residual acetonitrile in C4 olefins, after removal of butadiene via extractive distillation with acetonitrile as a solvent, can hydrolyse on the resin and give acetic acid and ammonia, and the later can neutralise the acidity of the catalyst and adversely affect the performance of the reactor. The diolefinic compounds can be selectively hydrogenated to mono-olefinic compounds with a supported Pd catalyst. Any ionisable inorganic compounds exchange with the acid sites and reduce the activity. It is reversible in nature, though.

Recent Developments of Ion-exchange Resin Catalysts

Above temperature of 125°C the commonly used sulphonated poly(styrene-DVB) resins undergo irreversible desulphonation and loose their activity. Resins with higher thermal stability can be obtained by carrying out the non-uniform sulphonation at higher temperature. High temperature favours the isomerisation of para-sulphonic acid groups to m-positions leading to increase in thermal stability. Increased thermal stability can also be achieved by use of oleum as a suphonating agent which leads to formation of SO_2 bridges during sulphonation.

It has been reported that acylation of polystyrene matrix before sulphonations leads to high activity and improves thermal stability upto 200°C. Sulphoalkylated styrene-DVB copolymers also offer thermal stability upto 200°C, but they are suitable only in aq. media due to their low acidity. One of the most attractive methods for further improving the activity and stability of the resin catalyst is the reaction of macroporous sulphonated polystyrene resin with elemental fluorine. The product was found to be much better as far as thermal stability and catalytic activity were concerned. Fast treatment with F_2 at room temperature produce resins with modified acid groups in the macrophase. These catalysts were more active especially in non-polar reactions such as alkylation of benzene.

The ion-exchange resins can be tailor-made for different industrially important reactions by partially exchanging the H^+ ions of the sulphonic acid resins by different species. Partially mercapto-exchange resins catalysts are used in the industrial synthesis of bisphenol-A. Partially Hg^{2+} or Ag^+ exchange resins are reported to accelerate the reactions involving acetylenic or olefinic compounds, respectively.

A trifunctional acid resin containing H^+, Hg^{2+}, Fe^{3+} was used as a catalyst for acetylene hydration. Moreover, metal exchange resins are sometimes used in the most reactive parts of column reactors to avoid hot spots and catalyst decomposition in exothermic reactions. Recently inorganic supported IER catalysts which have very good thermal and mechanical stability. The catalyst could be prepared with particle size in the range of 2-5mm, thus making it suitable for direct use in reactive distillation columns.

It offers low pressure drops, good reusability and activity. Catalyst was thermally stable upto 433°K. Limitation of this resin was the limited exchange capacity of the resin which was low at 0.4 m eq. H^+/gm of the resin as compared to 4.7–5.0 m eq. H^+/gm offered by other commonly used resin catalysts. Methylol was synthesised in an RDC using this catalyst to obtain upto 99% conversion of formaldehyde.

The other class of ion exchange resin with high thermal and chemical stability are perfluorosulphonic acid polymers, viz. nafion. These catalysts have higher acidity as compared to conventional resins. The main attraction for using the nafion catalyst lies in their chemical stability to strong oxidising reaction media like H_2O_2 where the conventional resins fail. Another attractive property of nafion lies in its ability to maintain acidity even when the resin is hydrated.

In hydration of ethylene oxide using nafion coated on Davison silica the conversion of ethylene oxide obtained was 94% with 94% selectivity of ethylene glycol whereas in the case of Amberlyst the corresponding figures were 19% and 67%, respectively. Their high price works against their possible large scale use in the industry. The cost of perfluorosulphonic acid can be reduced drastically by coating it onto inert support. The coating also helps in utilising maximum possible efficiency by making maximum number of acidic sites available. It has been observed that the rate of alkylation of phenol with 1-decene using 120 mesh powder of perfluorosulphonic acid resin proceeds at a rate seven times faster than that obtained with 30 mesh powder. The coating can be done in two ways: (i) coating with a polymer solution; and (ii) coating with polymer latex.

Coating a support from polymer solution, followed by heat treatment produces an improved support catalyst. Coating the support with an aqueous solution of the polymer in the sulphonyl fluoride form and subsequent conversion of sulphonyl fluoride functionality to sulphonic acid form is more convenient and allows more flexibility in production of supported catalyst. The supported perfluorosulphonic acid resin catalyst showed much higher activity and working life than the conventional Amberlyst. The use of supported perfluorinated sulphonic acid resins as the catalyst for oligomerisation of olefins, hydration of olefins, etherification of alcohols and hydrolysis of esters has been reported. The catalyst used was nafion supported on calcinated shot coke having hydrophobic surfaces and pore diameter greater than 1000 A°.

Allied Signal has claimed that directly fluorinated hydrocarbon ion exchange resin have higher acidity and can be used for alkylation of toluene with ethylene isobutane with isobutylene etc. which is currently made using liquid HF or H_2SO_4 as catalysts. Cationic ion exchange resin in the form of membranes has been used for the reaction of aqueous hydroxylamine sulphate and cyclohexanone to give the oxime. Pellicular resins, which consist of a thin layer of functional polymer grafted onto an inert core, are gaining importance as they can provide high surface area and larger particle size of the catalyst. An acidic catalyst has been prepared from porous silica having an area of 78 m^2/gm by polymerising potassium styrene sulphonate using an azo initiator. This catalyst was found to be comparable to amberlyst for alkylation of phenol with 1-dodecene.

Thus, the drive for clean, safe and selective processes will lead to more extensive adoption of cationic exchange resins as catalyst for small medium and large scale production of a vari .ᵗ ɔf speciality and bulk chemicals. There is scope for further improvement in the stability of the catalysts, particularly at temperatures higher than 403°K. Supported catalysts are likely to be commercialised (Table 5.2).

Table 5.2. Characteristics of some commercially available catalysts.

Name	*Type*	*Total capacity millieqiv. per gm.*
Cation exchangers		
	Phenolic	1.75
	Methylene	
	Sulphonic	
	Carboxylic	10.0
	Silicic acid	0.01–0.04
	Phenolic	3.25
	Methylene	
	Sulphonic	
Anion exchangers		
	Weak base	10.0
	Weak base	7.0
	Intermediate base	7.4
	Amphoteric	0.01

In particular, catalysts in the form of packings for distillation column reactors are likely to make an impact. The modeling and simulation of DCR's offer many challenges. The manufacture of MTBE, ETBE, and TAME will dominate the use of resin catalysts. Alternative catalysts based on acid-activated clays, pillared clays and supported heteropolyacids are expected to make a dent in some areas of application of resin catalysts.

ENZYME CATALYSED SELECTIVE OXIDATION REACTIONS

Enzymes are catalysts evolved in nature to speed up and co-ordinate the multitude of chemical reactions necessary to develop and maintain life. The variety of chemical transformations catalysed by enzymes make these catalysts a prime target for exploitation by the emerging biotechnological industries. Decades before the use of enzymes as synthetic reagents were seriously investigated, however, oxidative enzymatic reactions involving bioconversion or biotransformation by whole micro-organisms were being applied to the commercial production of steroid harmones. The development for such procedures has continued unabated despite interventions. The application of oxidative enzymes due, in part, to the nature of many oxidative enzymes, which for many years precluded their isolation in preparatively useful form. But, steady progress in science of enzyme isolation now promises that the handicap of inaccessibility of many oxidative enzymes may soon be reduced, if not eliminated.

Most of the oxidative reactions are generally catalysed by the class of enzymes called as "oxidoreductases". There are many biotechnological applications for oxidoreductases. Much of the industrial chemistry involves oxidation reactions, which offers significant potentialities for oxidoreductases. The use of oxidoreductases to carry out synthetic transformations is the major area of interest of practical significance in this regard are oxidoreductases mediated asymmetric synthesis of amino acids, steroids, pharmaceuticals, and of a host of specialty chemicals. Another area of majority importance is the current extensive use of oxidoreductases in clinical diagnosis and other analytical applications. Future development in the applications for oxidoreductases can be visualised in the areas as diverse as polymer synthesis, pollution control, and oxygenation of hydrocarbons.

Enzymatic Reaction—Advantages

Enzymes clearly are the most powerful and the efficient catalysts known, besides being most subtle and delicate. Nature makes lavish use of enzymes for synthesis of organic compounds with precision and efficiency. Hence, it would seem obvious to make use of these catalysts also in preparative organic chemistry.

Oxidative Enzyme Types

"Oxidoreductases" are the class of enzymes involved in biological oxidation reactions (Table 5.3). These can be grouped into variety of subclasses such as: (i) Dehydrogenases; (ii) oxidases; (iii) oxygenases; and (iv) peroxidases.

Table 5.3. Comparison between chemical and enzymatic/biological processes.

Criterion for comparison	*Chemical process*	*Enzymatic/biological process*
Catalyst	More or less active and selective, expensive if selective	Enzymes are highly active and selective (enantio and regio-selective)
	Expensive regeneration	Regeneration easy due to microbial growth
		Construction of new catalysts with genetic engineering
Reaction conditions	Mostly high temperature and pressure	Mainly at the temperature of 25°C and 1 bar pressure
		Biological demand of nutrients, biological optimisation, maintenance, mutation
Raw materials	Pure if poorly selective catalyst	Impure, inactive and diluted
Process	Often multistage process with recovery of intermediates	One stage possible without intermediatory product recovery
	Fast reaction rates at increase concentration levels and high yields	Mainly slow reaction rates, low concentrations
	Process may generate lot of toxic and hazardous waste	Eco-friendly process

Apart from the mechanistic diversity of the oxidoreductases they also require a stoichiometric amount of expensive amount of co-factors for catalytic activity. Some of the common co-factors for oxidoreductases are NAD, NAD(P), FAD. They also can be covalently bond or easily diffusible and effective regeneration of these co-factors is essential for successful application of enzymatic oxidation reaction for synthesis. Nicotinamide adenine dinucleotide (NAD) and it's analogous 2' phosphate bond (NADP) are involved as cofactors in many oxidation reactions. The nicotinamide ring system is redox active, accepting a hydride or two electrons and proton to form 1,4-dihydronicotinamide derivatives (NADH & NADPH). The reversible hydride transfer from a reduced substrate to NAD(P), and that from NAD(P) to an oxidised substrate is stereo selective and characteristic of individual enzymes.

The most prevalent co-enzymes apart from NAD(P) are the flavoproteins, which in many cases are firmly and specificially bound to the apoenzyme as definite prosthetic groups. The catalysis of redox reactions by flavoproteins is due to the alternatve oxidations and reductions of their flavin groups. The redox process involves the addition of two hydrides across the quinoid structure in the isoaloxazine nucleus of FAD and its subsequent regeneration. Flavin dinucleotide (FAD) can be differentiated from NAD by the very fact that the former completes its redox cycle while it is attached to one and the same enzyme protein and oxidised by other. NAD is therefore an intraenzymic redox carrier, and FAD an interenzymic redox carrier. The diversity of oxidoreductases is evident by the variety of subclasses it has. One of the most extensively used common subclass is the dehydrogenases.

Dehydrogenases

In this subclass the substrate is oxidised in absence of oxygen, but they require stoichiometric amount of co-factors such as NAD(P), FAD which serves as hydrogen acceptors.

$$SH_2 + A \longrightarrow S + AH + H^+$$

$$A = \text{Coenzyme. } SH_2 = \text{Substrate.}$$

These enzymes are often called as "anaerobic dehydrogenases", since they function in absence of oxygen. Dehydrogenases can be categorised on the basis of co-factor requirement as: (i) NAD(P) dependent dehydrogenases; and (ii) FAD dependent dehydrogenases. Some of the common dehydrogenases are enlisted below:

NAD(P) dehydrogenases	*FAD dehydrogenases*
Alcohol dehydrogenase	Succinate dehydrogenase
Glycerol dehydrogenase	D-amino acid dehydrogenase
β-hydroxy steroid dehydrogenase	Pyridoxine-5-dehydrogenase
Malate dehydrogenase	Acyl-co-A dehydrogenase
Glucose dehydrogenase	Nicotine dehydrogenase
Aldehyde dehydrogenase	
Alanine dehydrogenase	
Mannitol dehydrogenase	

Oxidases

The term oxidase is a recommended name for that oxidoreductases which utilises oxygen as an electron acceptor. More accurately an oxidase should be defined as an oxidoreductase that catalyses the reaction in which all the electrons taken from the donor are transferred to oxygen to form any of the reduction products (H_2O, H_2O_2).

Two electron transfer :

$2SH + O_2 \longrightarrow 2S + H_2O_2$ SH, SH_2 = Substrate

$SH_2 + O_2 \longrightarrow S + H_2O_2$

Four electron transfer :

$4SH + O_2 \longrightarrow 4S + 2H_2O$ SH, SH_2SH_4 = Substrate

$2SH_2 + O_2 \longrightarrow 2S + 2H_2O$

$SH_4 + O_2 \longrightarrow S + 2H_2O$

Thus oxidases catalyse two as well as four electron transfer reactions thereby reducing oxygen to hydrogen peroxide and water respectively. Some of the common oxidases are enlisted below:

Glycolate oxidase	Glucose oxidase
Pyranose oxidase	Cholesterol oxidase
Alcohol oxidase	Aldehyde oxidase
D-amino acid oxidase	Catechol oxidase

Oxygenases

The enzymes that incorporate oxygen atom from dioxygen molecule into the substrate are termed as "oxygenases". The oxygenases can further be classified into two categories: (i) Monoxygenases; and (ii) Dioxygenases.

Monoxygenases

These enzymes are characterised by their ability to introduce one of the atoms of molecular oxygen into an organic substrate, the other being incorporated into the molecule of water. Here oxygen plays dual role, of electron acceptor and oxygen transfer agent. Monoxygenases can further be classified into two classes depending on the source of electrons required to reduce the oxygen atom of dioxygen to water.

External monoxygenases. Electrons required are provided by the means of external reducing agent like NAD(P)H, $FADH_2$ and other reducing agents.

$$SH + O_2 + AH_2 \longrightarrow SOH + A + H_2O$$

$$SH = \text{Substrate} \qquad AH_2 = \text{Reducing agent}$$

Internal monoxygenases: Electrons required are provided by the substrate itself.

$$SH_2 + O_2 \longrightarrow SO + H_2O \qquad SH_2 = \text{Substrate}$$

Dioxygenases

Enzymes belonging to these classes incorporate both atoms of oxygen molecule into the substrate and as in monoxygenases, oxygen here also performs dual role. They can further be subclassed into intermolecular and intramolecular dioxygenases depending whether the two oxygen atoms are incorporated into separate substrates or into a single substrate molecule.

$$\text{Intramolecular: } S + O_2 \longrightarrow SO_2$$

$$\text{Intermolecular: } S + Co + O_2 \longrightarrow SO + CoO$$

$$S = \text{Substrate} \qquad Co = \text{Cosubstrate}$$

Some of the examples of oxygenases (mono and di) are enlisted below:

Monoxygenases	*Dioxygenases*
External	*Intramolecular*
Camphor-5-monoxygenase	Catechol-1,2-dioxygenase
Phenol-2-monoxygenase	Protocatechuate-2,3-catechuate
Cyclopentanone monoxygenase	Phthalate dioxygenase
Phenyl-4-monoxygenase	Cysteine dioxygenase
Cholesterol-7 monoxygenase	Sulphur dioxygenase
Alkane-1-monoxygenase	Benzene-1,2-dioxygenase
Internal	*Intermolecular*
Arginine-2-monoxygenase	Proline-2-oxoglutarate dioxygenase
Lysine-2-monoxygenase	Thymine-2-oxoglutarate dioxygenase
Lactate-2-monoxygenase	

Peroxidases

These are inclusive of the group of enzymes using hydrogen peroxide as oxidant with incorporation of one oxygen atom into the substrate but without the consumption of cofactor. The necessary reducing equivalents are derived from hydrogens of water :

$$SH + H_2O_2 \longrightarrow SOH + H_2O \quad SH = \text{Substrate}$$

Some of the common peroxides are mentioned in the following table:

Horseradish peroxidase	Catalase
Cytochrome C peroxidase	Chloroperoxidase
Lactoperoxidase	

Enzymatic Oxidation Reactions

Dehydrogenases

It is the largest class of oxidoreductases for which the most detailed mechanistic and substrate specificity data is available. The dehydrogenases oxidise the substrate in presence of suitable quantity of coenzyme and based on their coenzyme requirement they are classified into different subclasses. The applicability of dehydrogenases lies by the very fact that the enantiomerically pure and stereochemically predictable product can be synthesised because of the stereospecificity of most of dehydrogenases.

Dehydrogenases like L-phenylalanine dehydrogenase and D and L-hydroxyisocaproate dehydrogenases are of immense utility for the enzymatic production of L-phenylalanine which is one of the building blocks of the dipeptide sweetener, aspartame (L-α-aspartyl-L-phenylalanine methyl ester). It is about two hundred times sweeter than sucrose and has pleasant sweetness without bitter after taste, hence the sweetener is well received as diet sweetener. Apart from this is also useful for pharmaceutical purpose as food additive.

Enzymatic reaction involves the oxidation of D, L-phenyl lactic acid to phenyl pyruvic acid by the action of D and L-hydroxyisocaproate dehydrogenases of *Lactobacillus casei* and *Lactobacillus confusus* respectively and requires NAD as cofactor. The phenyl pyruvic acid so produced is subsequently converted to L-phenylalanine catalysed by phenyl alanine dehydrogenase requiring NADH as cofactor.

Alternative routes

Production of phenylalanine through phenylalanine lyase (PAL) as biocatalyst from transcinnamic acid was however hampered because of relatively low specific activity and instability of enzyme and sensitivity of the reaction to substrate inhibition. Taking note of the above impediments for commercial production a new route of L-phenylalanine production from phenylpyruvic acid was considered using phenyl dehydrogenase (PDH) but this route was negatively affected by instability of phenyl pyruvate and the enzyme was also subjected to substrate inhibition at high concentrations. Solution for such problem was sorted out by finding alternative cheap precursors that can be easily converted to phenyl pyruvate, followed by synthesis of L-phenylalanine via phenylalanine dehydrogenase activity. One of the alternative cheap precursor is D and L-phenyl lactate which can be oxidised by D and L-hydroxyisocaproate dehydrogenases to phenyl pyruvate. The problem of co-factor generation can be eliminated by employing formate dehydrogenase for co-factor recycling but now intrasequential cofactor regeneration can also be effectively employed.

Alternative enzymatic route

Dehydrogenases, like hydroxy steroid dehydrogenases accepts only substrates with limited structural differences and catalyse the reversible oxidoreduction of hydroxyl-keto groups of steroids. These dehydrogenases have high regio and stereo specificity, hence used in the synthesis of netural steroids and bile acids and also in analysis.

Hydroxy steroid dehydrogenases (HSDH) are useful in bile acid synthesis by regiospecific oxidation of cholic acid used as substrate. The reaction requires NAD(P)H as cofactor hence requires subsequent regeneration that can be done either enzymatically coupling of second reaction at the expense of cheaper substrates (formate, glucose-α-ketoglutarate) that acted as ultimate redox reagents. Oxidation was selective and occurs at specific positions (3, 7, 12) which involves oxidation of reversible hydroxy-keto group transformation. Immobilisation of the enzyme on the insoluble matrices stabilised the enzymes and also made it possible to recover at the end of the process.

Oxidases

Oxidases encompasses a group of enzymes in which oxygen acts as an electron acceptor and is subsequently reduced to any of its reduction products (superoxide radical, hydrogen peroxide and water). They include redox co-factors like quinoid structures as prosthetic groups, most frequently a flavin system. Oxidases are able to regenerate their redox center by direct transfer of electrons to dioxygen, a process oten referred to as "autoregeneration".

Oxidases that form an important group of oxidoreductases having applications in foods, pharmaceuticals and flavours. It is also of extensive utility in clinical diagnosis and other analytical applications. Some of the oxidative reactions catalysed by oxidases are summarised below:

One of the most prominent enzyme of commercial utility in the class of oxidases is glucose oxidase obtained from *Aspergillus niger*, a dimeric flavozyme (2FAD/molecule) with a molecular weight of 160 kDa. Glucose oxidase catalyses the oxidation of β-D-Glucose to D-glucono-lactone with the simultaneous evolution hydrogen peroxide, thereby reducing oxygen to hydrogen peroxide. Under aerobic conditions glucose oxidase autoregenerates.

This reaction has attracted the attention of analytical chemists due to its possible applicability in devising an amperometric glucose sensor for estimating clinically significant glucose levels using the archetypal and inexpensive redox enzyme glucose oxidase. Here the hydrogen peroxide produced in stoichiometric amounts with respect to oxidised glucose can be detected amperometrically or with a dye system.

Recently enzyme immobilised techniques with immobilisation of glucose polymer on electrically conducting polymer, polystyrene latex beads and even antigen-antibody reactions (immunological reactions) serves as an important tool for producing glucose biosensors. Glucose oxidase has high substrate specificity and almost exclusively accepts glucose hence this has not found much of synthetic utility. However it frequently used as technical enzyme in food industry in combination with its substrate to remove oxygen from vacuum sealed products. Immobilised glucose oxidase is useful for deoxygenation of fruit juices and beer.

Semi synthetic cephalosporins are increasingly being used in clinical practice due to their low toxicity and broad anti-bacterial spectrum. The key intermediate in the production of these cephalosporins is 7-amino cephalosporanic acid (7-ACA). Currently 7-ACA is produced by chemical method involving chemical deacylation of cephalosporin-C. The chemical process involves the use of hazardous and toxic chemicals so the focus shifted on eco-friendly and commercially feasible enzymatic alternatives.

The bioconversion of Ceph-C to 7-ACA involves a two step process: (i) Oxidative deamination and decarboxylation; and (ii) Enzymatic deacylation.

In the first step Ceph-C is oxidatively deaminated using immobilised D-amino oxidase isolated from the yeast *Trigonopsis variabilis* a flavozyme which is very specific for oxygen as acceptor, is converted to ketoadipyl 7-ACA. This is then spontaneously converted to glutaryl 7-ACA via a chemical route with hydrogen peroxide which is also a product of D-amino acid reaction. The second step involves enzymatic deacylation of glutaryl 7-ACA using immobilised glutaryl acylase isolated from recombinant *Escherichia coli*. Apart from environmental benefits this method also allows utilisation of raw material (Ceph-C) to be supplied as concentrated purified liquid instead of crystalline product thus resulting significant cost saving due to elimination of crystallisation process and thus resulting overall cost effectiveness.

Glycolate oxidase is flavin mono nucleotide (FMN) dependent peroxisomal enzyme that oxidises glycolic acid to glyoxalic acid with the subsequent release of hydrogen peroxide in the reaction. The hydrogen peroxide released has a tendency to react with the product, consequently the yield decreases. Catalase was included as co-catalyst to decompose the byproduct hydrogen peroxide, thus limiting peroxide dependent formate production and glycolate oxidase deactivation.

Addition of primary amine to the reaction mixture capable of forming hemiaminal or imine of glyoxylate resulted in the increase in the yield of glyoxylate. A synergistic effect was observed when using Tris or, EDA (ethylene diamine) along with catalase for the enzymatic oxidation of glycolic acid. Recently, immobilisation of these enzymes on the non-soluble support stabilises the enzyme against denaturation, resulting increased reaction rate, decreased reaction time, simplification of recovery and easy recycle of enzyme catalyst.

Glyoxalic acids serve as the useful synthon for vanillin production, which in turn is useful as flavour and fragrance ingredients as well as in manufacture of pharmaceuticals and agrochemicals. Manufacture of vanillin from glyoxalic acid assumes significance as the availability of vanillin from lignin is decreasing due to environmental concerns. As an alternative to chemical process of glyoxalic acid production biocatalytical route offers significant advantages, since it uses enzymes as catalysts and is relatively run under relatively mild conditions and few undesirable waste streams as compared to chemical process.

Pyranose-2-oxidase extracted from mycelia of polyporus obstusus catalyses the bioconversion of D-glucose to D-glucosone which was found to be key intermediate in the process for converting D-glucose to D-fructose. The reaction involves the enzymatic oxidation of D-glucose at C-2 to the corresponding ketose (D-glucosone) with the subsequent reduction of D-glucosone by catalytic hydrogenation to form D-fructose.

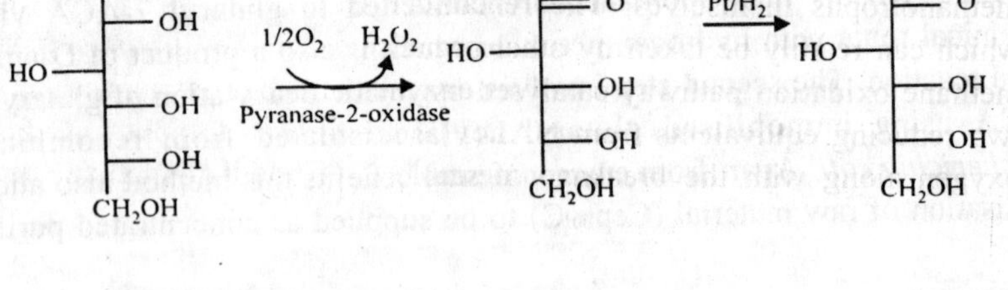

The oxidation of D-glucose to D-glucosone generates hydrogen peroxide that can be either utilised for the generation of propylene oxide through second enzymatic reaction coupled to glucose oxidation, or can be degraded by the use of catalyse.

Oxygenases

The oxygenases are most widely distributed of all enzymes, they incorporate molecular oxygen, and do so with high efficiency and selectivity. These enzymes convert alkalies to alcohols, olefins to epoxides, sulphides to sulphoxides. They also cleave aromatic rings or oxidise their substituents and hydroxylate polycyclic and aromatic hydrocarbons as well as variety of steroid derivatives.

Oxygenase catalysis is particularly intriguing since these involves selective direct oxyfunctionalisation of unactivated organic substrate and high degree of stereo and regio selectivity exhibited by them. Oxygenases have attracted much attention for industrial synthetic transformation, on both whole cell and isolated enzyme levels. Some of the oxidation reactions catalysed by oxygenases are enlisted below:

Monoxygenases are of immense academic and industrial interest not only because of regio and stereo selectivity exhibited during synthetic transformations involving oxygenation of substrates for practical applications, but also for providing information about unactivated hydrocarbon oxygenation, with long range goal of applying knowledge gained to future chemical catalytic systems. Some of the common oxygenases include methane mono oxygenase and Bayer villigeer monoxygenases (BVMO) useful in production of chiral syntheons like epoxide, sulphoxide, lactones.

Methane monoxygenase, a bacterial enzyme present in methanogenic bacteria is useful to oxidise methane produced in enormous quantities as the primary end product of anaerobic metabolism of methanogenic bacteria. The atmosphericegress of the large fraction of this methane that is generated in lakes, oceans and wet soils is blocked by the action of methanotrophic bacteria. It oxidises methane as their soul source of carbon and energy, part of the carbon from methane is assimilated as biomass by methanotrophs themselves. The remainder is released as carbon dioxide which can readily be taken by other organisms. The key initialising step of methane oxidation pathway catalysed by methane monoxygenase, in which two reducing equivalents from NAD(P)H are utilised to split O-O bond of oxygen along with the breakage of stable C-H bond of methane to form methanol in which one oxygen atom is incorporated into the substrate where as other oxygen atom is reduced to water.

$$CH_4 + O_2 + NAD(P)H \longrightarrow CH_3OH + H_2O + NAD(P)$$

Apart from oxidation of methane, methane monoxygenase also catalyses the adventitious oxidation of large number of hydrocarbons including saturated, unsaturated, linear, branched and cyclic hydrocabrons to approximately C_8 in size.

Bayer villiger oxidation of various substrates is a new emerging type of bioconversion. The enzymes which catalyse these reactions are termed as Bayer villiger monoxygenases. The BVMO are an interesting group of flavoproteins with considerable proven ability for yielding key chiral products (lactones and sulphoxides) of value in chemoenzymatic synthesis of wide range of useful compounds. BVMO are able to activate molecular oxygen for insertion into the cosubstrate thus, one of the oxygen is incorporated into the starting ketone, where as another one is converted to water.

Cyclohexanone monoxygenase is one of the most common BVMO which is thoroughly studied. Cyclohexanone performs an enantiodivergent biotransformation of the recemic ketone bicyclo hept-2-on 6-one to yield two homochiral lactones, these and equivalent biotransformations of various other unsubstituted bicyclic ketones have been used to yield chiral synthons that can themselves be incorporated into the chemoenzymatic synthesis of useful compounds as diverse as pheromones, multifidene and viridene, the antibiotic sarkomycin-A and the insect anti-feedant clerodin.

Dioxygenases, and microbes possessing dioxygenase activity have been studied as potential catalysts for transformations of aromatic hydrocarbons. Some of the dioxygenases such as catechol dioxygenases serve as a part of natue strategy for degrading aromatic molecules in the environment, they are found in soil bacteria and act in the last step of transforming precursors into aliphatic products.

The oxidation reaction involves the cleavage of aromatic molecules and the modes of cleavage may be intradiol or extradiol based on this there are two modes of dioxygenation : (i) Intradiol dioxygenation; and (ii) Extradiol dioxygenation.

The intradiol cleaving enzymes are represented by catechol 1,2-dioxygenase and protocatechuate 3,4-dioxygenase, where as extradiol cleaving enzymes are represented by catechol 2,3-dioxygenases. Enzymes like phthalate dioxygenase are involved in oxygenation of hydrocarbon rather than degradation of hydrocarbons in the soil. Phthalate dioxygenase belong to the group of enzymes called as "Rieske oxygenases" which catalyses the cis hydroxylation of phthalate.

Phthalate dioxygenase represents the other cis dihydroxylating enzymes (Benzoate 1,2-dioxygenase, toluene 2,3-deoxygenase) that catalyse the initial step of degradation of aromatic rings, the dihydrodiol serves as the precursor to catechols that are cleaved by catechol dioxygenases.

Peroxidases

Peroxidases are of immense synthetic potential since they can in principle mediate the same type of transformations as monoxygenases i.e. the incorporation of oxygen atom into the substrate but without consumption of cofactor. The necessary reducing equivalents are derived from hydrogen of hydrogen peroxide. More over peroxidases are generally extracellular enzymes and are considerably more robust than intracellular monoxygenases. One of the most common example belonging to this group is chloroperoxidase (CPO) produced from filamentous fungus caldariomyces fumago a heameprotein which catalyses the epoxidation of unfunctionalised *cis* alkene in presence of hydrogen peroxide to produce epoxides of variable yields.

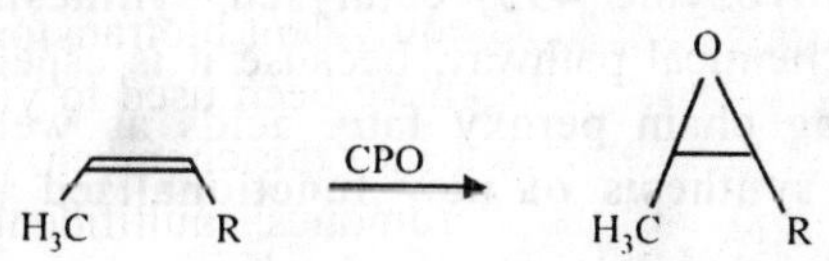

The epoxidation of alkenes catalysed by chloroperoxidase resembles closely to that of monoxygenases, except for the fact that the chloroperoxidase utilises molecular oxygen and a reducing agent such as NADH. Since chloroperoxidase is commercially available and has been cloned and over expressed it appears to be useful as "free monoxygenase" for synthetic applications. Certain epoxides with well defined absolute configuration may be obtained from halohydrins produced via chloroperoxidase catalysed addition of HOX to carbon carbon double bond.

Chloroperoxidase also catalyses the enantioselective asymmetric oxidation of sulphides using either hydrogen peroxide or chiral hydroperoxides as oxidation agent. In the latter case, enzyme is highly selective for the (R)-enantiomer of chiral hydroperoxides generating (R)-alcohols and unreacted (S)-hydroperoxides. In the oxidation of phenyl methyl sulphide using racemic 1-phenyl ethyl hydrogen peroxide as oxidant, the sulphide is selectively oxidised to (R)-sulphoxide with the concurrent formation of (S)-1-phenyl ethyl hydroperoxide and (R)-1-phenyl ethanol.

The sulphoxides with high enantiomeric excess were obtained even under low substrate concentration by this method. A drawback of this enzymatic process is that the undesired chemical oxidation of sulphide occurs, though it can be minimised considerably by controlling the concentration of substrates.

Others

Apart from oxidoreductases other enzymes like laccases and lipases also catalyse oxidation reaction of synthetic value. Lipase for example catalyses the reaction of fatty acids with hydrogen peroxide to generate peroxy acids for the epoxidation of the olefins. The reaction occurs by treating fatty acids with hydrogen peroxide (60% W/V) in presence of lipase in organic solvent using excess of hydrogen peroxide, besides simple peroxy fatty acids with unbranched saturated chain, those with internal hydroxy group or epoxy group has been prepared.

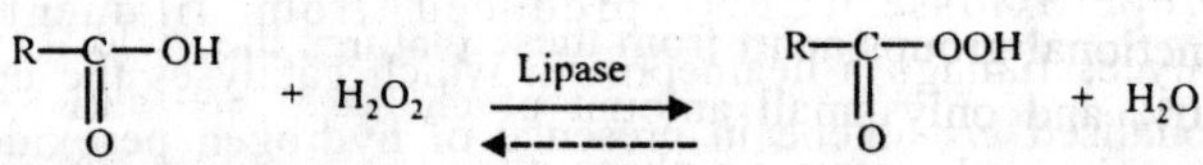

The lipase (Novozyme 435) catalysed synthesis of peroxy acids complements the chemical pathway, because it is especially useful for the preparation of long chain peroxy fatty acids as well as mild reaction conditions make synthesis of new functionalised peroxy fatty acids possible.

Peroxy fatty acids formed by the reaction of fatty acids with lipase/ H_2O_2 laccases on the other hand are employed in the selective oxidation of the methyl groups on the aromatic rings to the corresponding aldehydes using laccase/-$(NH_4)_2$-(2, 2'-azanobis (3-ethylbenzothiazoline-6-sulphonic acid) as the catalyst. Oxygen in the presence of this catalyst system is capable of converting aromatic methyl groups into formyl groups is of great importance with respect to the increasing interest in reactions carried under mild conditions.

Peroxybutyric acid
Peroxylauric acid
Peroxystearic acid
Peroxy-12-hydroxy stearic acid
Peroxy-9,10-epoxystearic acid
Peroxy-4-phenyl butyric acid
Peroxy-9, 10-dihydroxystearic acid

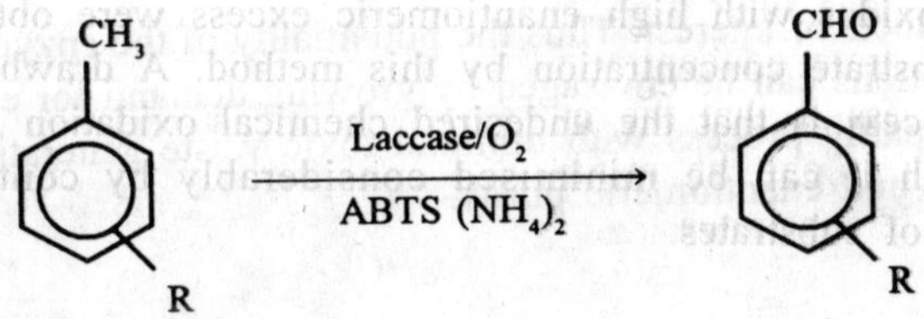

R = alkyl, alkoxy, nitro, halogen.

Selective oxidation involve lot of difficulties since the oxidation do not stop at the formation of products and the oxidation products are further subjected to oxidation more easily than the starting material. Once the methyl group is attacked it is further subjected to oxidation to carboxylic acids as the reaction is difficult to stop at the aldehyde level. The potentiality of this reaction lies in its broad applicability. It can be applied to methyl substituted aromatics such as alkoxytoluenes or nitrotoluenes, phenolic and benzylic hydroxyl groups require protection. The oxidation does not affect other functional groups apart from these features the yield was nearly found to be 90% and only small amount of chemical auxiliaries were applied, which remain in the aqueous phase after extraction of the aldehyde in the organic solvents hence very pure compounds were obtained. Representative examples of oxidation by laccase/ABTS-$(NH_4)_2$ catalyst.

Starting material	*Product obtained*	*Yield*
Toluene	Benzaldehyde	92%
p-Nitrotoluene	p-Nitrobenzaldehyde	98%
m-Chlorotoluene	m-Chlorobenzaldehyde	89%
3,4-dimethoxytoluene	3,4-dimethoxybenzaldehyde	90%

Conclusions

In the conclusion it can be stated that enzyme catalysed oxidation reactions offers significant potentiality. But, in present situation the use actually made of enzymes, left a lot to be desired. The deep discrepancy between promise and practice is a reflection of some kind of inherent difficulties like fragility (instability) of these enzymes and cofactor requirements, which can be circumvented to certain extent by the use of immobilised enzymes and by coupling second enzymatic reaction for cofactor regeneration, in some cases intrasequential regeneration is also employed. The use of oxidative enzymes like dehydrogenases is motivated by its stereospecificity resulting in enantiomerically pure and stereospecific product, oxidases on the other hand are useful in clinical diagnosis and other analytical applications, oxygenases and peroxidases used for generation of chiral synthons for obtaining products of interest. Thus the potentiality of the enzyme catalysed oxidation reactions can be envisaged by growing demand for economically viable, eco-friendly process with minimal toxic waste generating products of high yields and enantiomeric purity.

Chapter 6

Explosives and Propellants

INTRODUCTION

Propellants and explosives are chemical compounds or their mixtures that rapidly produce large volumes of hot gases when properly initiated. Propellants burn at relatively low rates measured in centimetres per second, whereas explosives detonate at rates of kilometres per second.

The energy liberated by explosives and propellants depends on the thermochemical properties of the reactants. As a rough rule of thumb, these material yield about 1000 cm^3 of gas and 4.2 kJ (1000 cal)/g. A comparison of the characteristics associated with propellant burning, explosive detonation, and the performance of conventional fuels is shown in Table 6.1. The greatest difference is the rate at which energy is evolved.

General Characteristics

Exothermic oxidation-reduction reactions provide the energy released in both propellant burning and explosive detonation. The reactions are either internal oxidation-reductions, as in the decomposition of nitroglycerine and pentaerythritol tetranitrate, or reactions between discrete oxidisers and fuels in heterogenous mixtures such as RDX and aluminium or ammonium perchlorate and a hydrocarbon fuel.

An activation energy of 125–250 kJ/mol (30–60 kcal/mol) is usually required to initiate the reaction, at which point the heat evolved is sufficient to cause the reaction to continue and become self-sustaining. Most explosives and propellants are organic compounds or mixtures of compounds that contain carbon, hydrogen, oxygen, and nitrogen. Metallic fuels such as aluminium may be added to increase the heat of reaction. Industrial dynamites have traditionally used nitroglycerine, nitrocellulose, and inorganic salts as sources of oxygen but are now increasingly formulated with ammonium nitrate as the oxygen source. Composite propellants commonly use ammonium perchlorate to supply the oxygen required. The most common gaseous products of the oxidation-reduction reactions are hydrogen, water, carbon monoxide, carbon dioxide, and nitrogen.

Table. 6.1. Characteristics of burning and detonation.

Characteristics	*Burning* *Fuel*	*Burning* *Propellant*	*Explosive detonation*
Typical material	coal-air	propellants	explosives
Linear reaction rate, m/s	10^{-6}	10^{-2}	$2\text{–}9 \times 10^{-3}$
Type of reactions	oxidation–reduction	oxidation–reduction	oxidation–reduction
Time for reaction completion, s	10^{-1}	10^{-3}	10^{-6}
Factor-controlling reaction rate	heat transfer	heat transfer	shock transfer
Energy output, J/g[a]	10^4	10^3	10^3
Power output, W/cm^2	10	10^3	10^9[b]
Most common initiation mode	heat	hot particles and gases	high temperature-high pressure shock waves
Pressure developed, MPa[c]	0.07–0.7	$0.7\text{–}7 \times 10^2$	$7 \times 10^3 - 7 \times 10^4$
Uses	source of heat and electricity	controlled gas pressure, guns, and rockets	brisance, blast, munitions, civil engineering

[a] To convert J to cal, divide by 4.184.

[b] This may be compared with the total United States electric-generating capacity of about 30×10^6 kW.

[c] To convert MPa to psi, multiply by 145.

Explosive detonation

Detonation proceed as a result of a reaction front moving in a direction normal to the surface of the explosive. However, detonation is a hydrodynamic phenomenon that differs in a fundamental sense from burning. Upon initiation, burning first occurs at an increasing rate for period of time up to several micro-seconds. A high pressure shock wave is formed which passes through the explosive at high velocity. As it does so, it causes exothermal decomposition of the explosive. The continued passage of the wave is supported by transfer of energy from the spent explosives to the unreacted explosive by shock compression. The rate of reaction depends on the rapid rate of transmission of a shock wave rather than on the relatively slow rate of heat transfer associated with propellant burning. The detonation rate is stable and constant and is primarily governed by the physical and chemical properties of the explosive, its geometry, degree of confinement, and particularly its density ρ, with which the detonation rate varies in a linear fashion for most explosives :

$$D_i = D_o + M(\rho_i - \rho_o)$$

where D_i = linear detonation rate dx/dt at density ρ_i, D_o = linear detonation rate at density ρ_o, and M = a constant characteristics of the explosive composition. Typical values of D_o at ρ_o = 1.0 g/cm^3 are about 5000–6000 m/s. Values of M are about 3000–4000 m/s.

Primary Explosives

Explosives are commonly categorised as primary, secondary or high explosives and propellants in order of decreasing sensitivity to energy output.

Primary, or initiator, explosives are the most sensitive to heat, friction, impact, shock, and electrostatic energy. They have been studied in considerable detail because of their almost unique capability, even when present in small quantities, to transform rapidly a low energy stimulus into a high intensity shock wave. Most recent evidence indicates that there is a minimum thickness or run-up distance before steady state detonation occurs and a gradual transition from an unreacted shock to a stable detonation.

Primary explosives are used to initiate the next element in a series which consists of explosives of increasing mass and decreasing sensitivity. They are arranged in sequence to amplify the input stimulus to an output level of sufficient intensity to maximise the probability of initiating the main charge. Overall energy intensification is about 10^7:1. Primary explosives are used in military detonators, commercial blasting caps, and in stab and percussion primers.

Mercury fulminate

Mercuric cyanate, [$Hg(ONC)_2$], is a gray-white powder, which explodes at 160°C, has a detonation rate 5.4 km/s at 4.2/ g/cm^3; is soluble to the extent of 0.1 g/100 H_2O at 20°C. It is prepared by the Chandelen process, essentially a large-scale laboratory process.

Lead azide

Lead azide ($Pb(N_3)_2$), is the primary explosive used in military detonators. Detonation rate 5.1 km/s at 4.8 g/cm^3. The azides are among the very few useful explosive compounds that do not contain oxygen. Lead azide is very stable at ambient and elevated temperatures, and has good flow characteristics. Since lead azide is less sensitive to ignition than mercury fulminate, a more readily ignitable material such as lead styphanate is often used as cover charge to ensure initiation. Lead azide is made in small batches buffered by the reaction solutions of lead nitrate or lead acetate with sodium azide :

$$Pb(NO_3)_2 + 2NaN_3 \rightarrow 2NaNO_3 + Pb(N_3)_2$$

All the phases of the manufacturing process are conducted by remote control in stainless-steel vessels using either distilled or demineralised water and filtered solutions. The overall precipitation time is about 60 min.

Silver azide

Silver azide, AgN_3, has received attention as a potential replacement for lead azide because it may be used in smaller quantities as an initiator and, therefore, offers the possibility of miniaturisation of fuse components. Silver azide requires somewhat less energy for initiation than lead azide and fires with a shorter time delay; detonation rate 6.8 km/s at 5.1 g/cm^3. It is less apt to hydrolyse and is more sensitive to heat than lead azide. It is incompatible with sulphur compounds, with tetrazene, and with some metals, including copper. Silver azide is made in the same manner as lead azide, except that silver nitrate is used in the reaction with sodium azide.

Lead styphnate

Lead styphnate, lead 2,4,6-trinitroresorcinate, is one of a number of compounds used in priming compositions to start the ignition-to-detonation process in the explosive sequence. Detonation rate 5.2 km/s at 2.9 g/cm^3. Lead styphnate monohydrate is precipitated as the basic salt from a mixture of solutions of magnesium styphnate and lead acetate followed by conversion to the normal form by acidification with dilute nitric acid.

Diazodinitrophenol (DDNP)

DDNP is an orange-yellow compound made by diazotising picramic acid with sodium nitrite and hydrochloric acid, washing the product with ice water, and recrystallising it from hot acetone, mp 157°C and detonation rate 6.9 km/s at 1.60 g/cm^3; it is somewhat less stable than lead azide. Since it is about equivalent to TNT in brisance, it is more effective for some purposes than lead azide and is used as an initiator in commercial blasting caps.

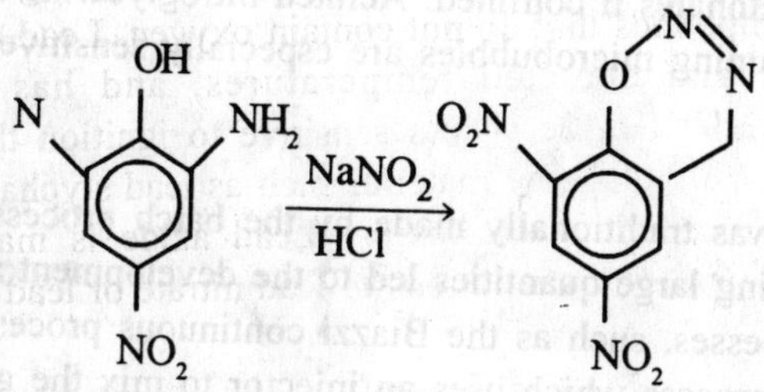

picramic acid DDNP

Tetrazene

Tetrazene is a pale-yellow crystalline explosive (explodes at 140–160°C) made by adding sodium nitrite to a solution of l-aminoguanidine hydrogen carbonate in dilute acetic acid at 30°C. It is used as a component in priming compositions.

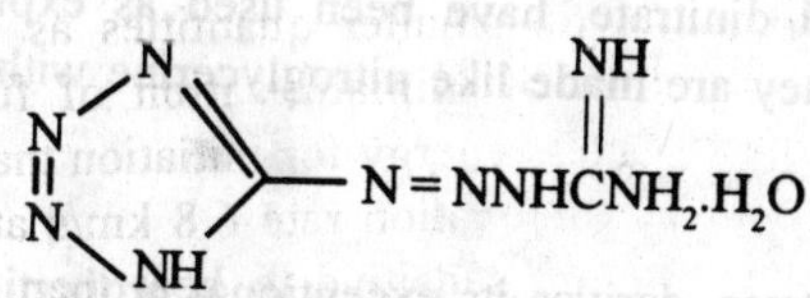

tetrazene

Secondary Explosives

Aliphatic nitrate esters

Aliphatic nitrate esters are among the most powerful explosives available. They are generally less stable than aromatic nitro compounds or nitramines because they tend to hydrolyse autocatalytically to form nitric and nitrous acids which further accelerate decomposition.

Nitroglycerine

Nitroglycerine, NG, glyceryl trinitrate, is primarily used as an explosive in dynamites and as a plasticiser for nitrocellulose in double- and multibase propellants. It is very sensitive to shock, impact, and friction and is employed only when desensitised with other liquids or absorbent solids or when compounded with nitrocellulose. It is readily soluble in many organic solvents, acting as a solvent for many explosives ingredients. Unconfined nitroglycerine burns without exploding if present in thin layers and in small quantities but detonates if confined. Aerated nitroglycerine and other liquid explosives containing microbubbles are especially sensitive to shock.

Manufacture

Nitroglycerine was traditionally made by the batch process. However, the hazard of handling large quantities led to the development and wide use of continuous processes, such as the Biazzi continuous process and the Nitro Nobel Injector process which uses an injector to mix the glycerol or other polyol with precooled nitration acid. In general, the methods used to make nitroglycerine are applicable to the manufacture of other liquid aliphatic polyhydroxy alcohol nitrates. Nitroglycerine is made from very pure glycerol to ensure stability of the product. Mixed acid (90% nitric acid and 25–30% oleum) is used in both batch and continuous processes.

Other glycol nitrates, e.g. ethylene glycol dinitrate (nitroglycol) and diethylene glycol dinitrate, have been used as explosive plasticisers for nitrocellulose. They are made like nitroglycerine with mixed-acid nitration.

Nitrocellulose

Nitrocellulose nitrate, derives its exceptional properties from the polymeric and fibrous structure of the cellulose from which it is made. Nitrocellulose provides mechanical strength as well as readily available energy to gun and rocket propellants. It is manufactured by nitration of cellulose with mixed acid.

Properties

D = 1.66 g/cm^3; detonation velocity = 7.3 km/s at 1.20 g/cm^3; hygroscopicity, heat of formation, heat of combustion and explosion point vary with nitrogen content. Dry nitrocellulose burns rapidly and furiously. It may detonate if present in large quantities or if confined and is dangerous to handle in the dry state since it is sensitive to friction, static electricity, impact and heat. Nitrocellulose is always shipped wet with water or alcohol. The higher the nitrogen content, the more sensitive it tends to be. Even nitrocellulose with 40% water detonates if confined and sufficiently activated.

Manufacture

The batch nitration process have included the pot process, the centrifugal process, the Thompson displacement process, and the Mechanical Dipper process. In the batch process, the raw materials proceed by gravity through the processing operations. The nitration of cellulose occurs very rapidly at first and the nitrocellulose is separated from spent nitrating acid. Semi-continuous nitration processes have been developed for military and industrial grades. Semi-continuous nitration uses a multiple-cascade system and a continuous wringing operation. The controlling factors in the nitration process are the rates of diffusion of the acid into the fibres, the composition of the mixed acid, and the temperature.

Nitramines

The four most important nitramines are: cyclotrimethylenetrinitramine (RDX); cyclotetramethylenetetranitramine (HMX); nitroguanidine (NQ); and 2,4,6-trinitrophenylmethylnitramine (tetryl). Tetryl has been increasingly replaced by RDX; both RDX and HMX are used as high energy explosives and may also be incorporated in high performance rocket propellants. Nitroguanidine is employed almost exclusively in gun propellants.

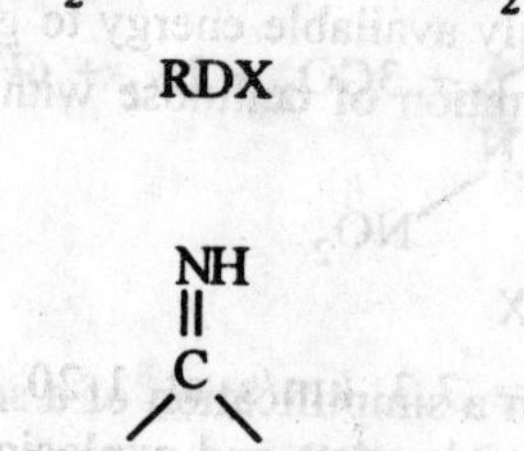

RDX

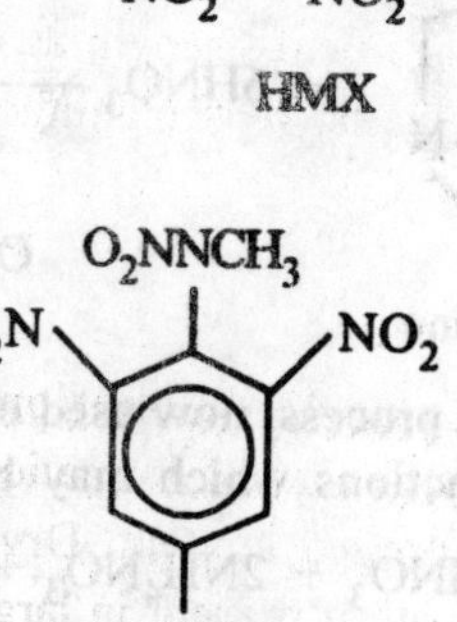

HMX

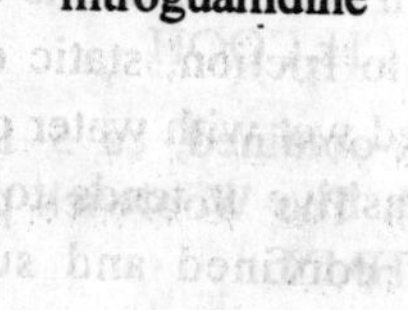

nitroguanidine

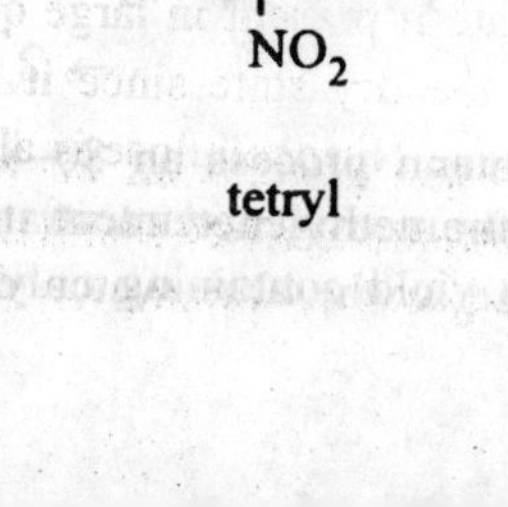

tetryl

RDX and HMX

The properties of RDX and HMX are quite similar; HMX has a higher density and a higher detonation rate, yields more energy per unit volume and has a higher melting point and higher explosion and cook-off temperatures.

Both are white, stable crystalline solids, less toxic than TNT and capable of being handled with no physiological effect if appropriate precautions are taken to assure cleanliness of operations. Both RDX and HMX detonate to form mostly gaseous, low molecular weight products with little intermediate formation of solids. The calculated mole fractions of detonation products of RDX is 3.00 H_2O, 3.00 N_2, 1.49 CO_2, and 0.02 CO. RDX has been stored for as long as ten months at 85°C without perceptible deterioration. At present, HMX is the highest-energy solid explosive produced on a large scale, primarily for military use. It exists in four polymorphic forms of which the beta form is the least sensitive and most stable and the type required for military use. The mole fractions of products of detonation of HMX in a calorimetric bomb are 3.68 N_2, 3.18 H_2, 1.92 CO_2, 1.06 CO, 0.97 C, 0.395 NH_3, and 0.30 H_2.

Both RDX and HMX are substantially desensitised by mixing with TNT to form cyclotols (with RDX) and octols (with HMX) or by coating with waxes, synthetic polymers, and elastomeric binders.

Manufacture

The two most common processes use hexamethylene-tetramine (hexamine) as starting material. The Woolwich or direct nitrolysis process used :

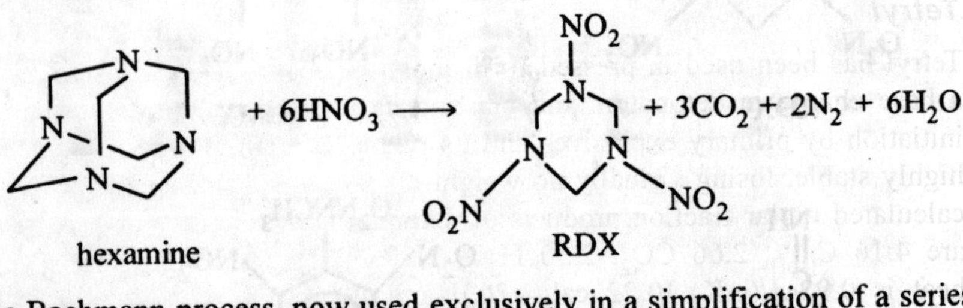

The Bachmann process, now used exclusively in a simplification of a series of complex reactions which may be summarised as follows:

$$C_6H_{12}N_4 + 4HNO_3 + 2NH_4NO_3 + 6(CH_3CO)_2O$$
$$\rightarrow 2\ RDX + 12\ CH_3COOH$$

In the Bachmann process an 80–84% yield is obtained, ca 5–10% of which is cyclotetramethylenetetranitramine (HMX). The Woolwich process gives a 70–75% yield containing only a trace of HMX.

A continuous process for medium-scale production of RDX based on the Woolwich process has also been developed. A modification of the Bachmann process used to make RDX with the same starting materials and in similar equipment is employed for the manufacture of HMX.

Nitroguanidine (NQ)

Nitroguanidine has been used to some extent as an industrial explosive but not as a military explosive because of its relatively low energy content and difficulty of initiation. It is stable and non-hygroscopic. Produced in the alpha crystalline form with a bulk density of about 0.2 g/cm^3, the crystals are needlelike and often hollow and are about 5 μm in diameter and 15 μm long.

Manufacture

Nitroguanidine may be made by several methods. In the typical Welland process, calcium cyanamide is first made from calcium carbonate and converted to cyanamide by acidification of a water slurry. The dimer dicyandiamide, which is then formed by filtration and evaporation of the filtrate, is fused with ammonium nitrate to form guanidine nitrate. Dehydration with 96% sulphuric acid gives nitroguanidine. In these processes, guanidine nitrate is the intermediate which is then dehydrated with sulphuric acid. In the Marquerol and Lorriette process, nitroguanidine is obtained directly in about 90% yield from dicyandiamide by reaction with sulphuric acid to from guanidine sulphate, followed by direct nitration with nitric acid.

Tetryl

Tetryl has been used in pressed form mostly as a booster explosive and as a base charge in detonators and blasting caps because of its sensitivity to initiation by primary explosives and its relatively high energy content. It is highly stable, losing virtually no weight on prolonged storage at 80°C. The calculated molar fraction products of detonation at a density of 1.70 g/cm^3 are 4.16 $C_{(sol)}$, 2.66 CO_2, 2.50 H_2O, 2.50 N_2, and 0.17 CO. Its specific heat is 0.92 J/(g.K) [0.22 cal/(g.°C)] and thermal conductivity 0.25 W/(m.K) [6×10^{-4} cal/(cm.s.°C)]. It is manufactured by a batch process in which dimethylanisidine dissolved in concentrated sulphuric acid is nitrated with mixed acid.

Nitroaromatics

The commonly used nitroaromatic explosives contain three NO_2 groups, generally in the 1, 3, and 5 positions. Aromatics are most often nitrated to the trinitro stage with mixed acid. Further nitration is difficult, and aromatics with four or more nitro groups attached to the ring tend to be relatively unstable. The most extensively used explosive is trinitrotoluene (TNT); however, hexanitrostilbene (HNS), hexanitroazobenzene (HNAB), and di- and triaminotrinitrobenzene (DATB and TATB) have found increasing application because of their low sensitivity to impact, shock, and friction and their excellent stability at elevated temperatures. Ammonium picrate (AP) has been used in armor-piercing gun projectiles because of its insensitivity to impact and shock.

Trinitrotoluene

2,4,6-trinitrotoluene (TNT) is very stable and may be stored indefinitely at temperate conditions without deterioration. TNT is non-hygroscopic and relatively insensitive to impact, friction, shock and electrostatic energy. It has been fired in high acceleration gun projectiles with reported prematured rates of less than one in a million (10^6). Bombs and projectiles are filled with steam-melted TNT by the casting process. Melted TNT also serves as the liquid carrier for RDX, HMX, aluminium, ammonium nitrate, and other high melting ingredients to form a wide range of castable slurries.

Manufacture

Trinitrotoluene is made by batch or continuous process. Toluene is nitrated in a three-stage operation by using increasing temperatures and mixed-acid concentrations to successively introduce nitro groups to form mononitrotoluene (MNT), dinitrotoluene (DNT), and trinitrotoluene (TNT). The steps used in each stage are similar and include acid mixing, addition of the oil, digesting the reaction to completion, cooling and settling the mix, and separating the oil from the acid. The TNT is purified by treatment with sodium sulphite solution.

Continuous processes used in many countries are based on nitration of toluene with an organic phase flowing countercurrent to the acid phase. A series of eight nitrator-separator reactors are used to conduct the nitration in six stages.

Picric acid and ammonium picrate

Picric acid (2,4,6-trinitrophenol) (PA) is of historic interest as the first modern high explosive to be used extensively as a burster in gun projectiles. It was first obtained by nitration of indigo, and used primarily as a fast dye for silk and wool. Picric acid has an energy content somewhat greater than that of TNT and a higher detonation rate. A large disadvantage is its tendency to form sensitive salts with calcium, lead, zinc, and other metals. Picric acid is no longer used as a military explosive. It can be made by gradually adding a mixture of phenol and sulphuric acid at 90–100°C to a nitration acid containing a small excess of nitric acid. The picric acid crystals are separated by centrifuging, washed, and dried.

Ammonium picrate (explosive D) (AP) is now used only where a high explosive is required that is particularly insensitive to shock. It has been used in a pressed form primarily as a burster in naval projectiles or armor penetration.

Metegang Methods

The least sensitive secondary explosives, of which the 94/6 prilled ammonium nitrate-fuel oil mixture is currently the most popular, require a relatively large booster. Other series of explosives of this type are the "slurry explosives" and "slurry blasting agents". The first slurry used commercially comprised essentially ammonium nitrate, coarse TNT, and water. Boosters are usually required for these relatively insensitive explosives. They comprise 0.1 to 1.0 pounds or more of cast 50/50 pentolite, cast TNT, or other water-insoluble material, usually cast, and water-insoluble explosives of very high brisance. Another early and still useful series of slurries substituted smokeless powder for TNT. Aluminium was added to these explosive-sensitised slurry types to increase strength. The most popular and economical commercial slurries in use today are the so-called blasting agents in which no explosive *per se* is used. Foremost among these are the aluminium-sensitised slurries. The higher-grade dynamites are also sometimes used, though inefficiently, as boosters for the slurry explosives and slurry blasting agents. Most military explosives also require boosters; a 10-g tetryl or pentolite charge usually satisfies the booster requirements of the secondary military explosives, illustrating the level of sensitivity of the explosives used by the armed forces. Artillery rounds may contain igniters, primary explosives, boosters, and secondary explosives, all assembled so that each element serves a definite function. Today cap-sensitive slurries, usually sensitised with fine aluminium, are rapidly coming into use for small-diameter, underground mining.

Detonation-wave velocities are measured by high-speed photography methods involving framing cameras capable of taking millions of frames per second, streak cameras with writing speeds of several miles per second, elaborate electronic timers called "pin-oscillographs," and centimetre-wave systems which use wave guide and doppler principles. Fig. 6.1 illustrates how the detonation velocity is measured by means of the streak camera. Since World War II, the above methods have largely replaced the older "Dautriche" and "Metegang" methods.

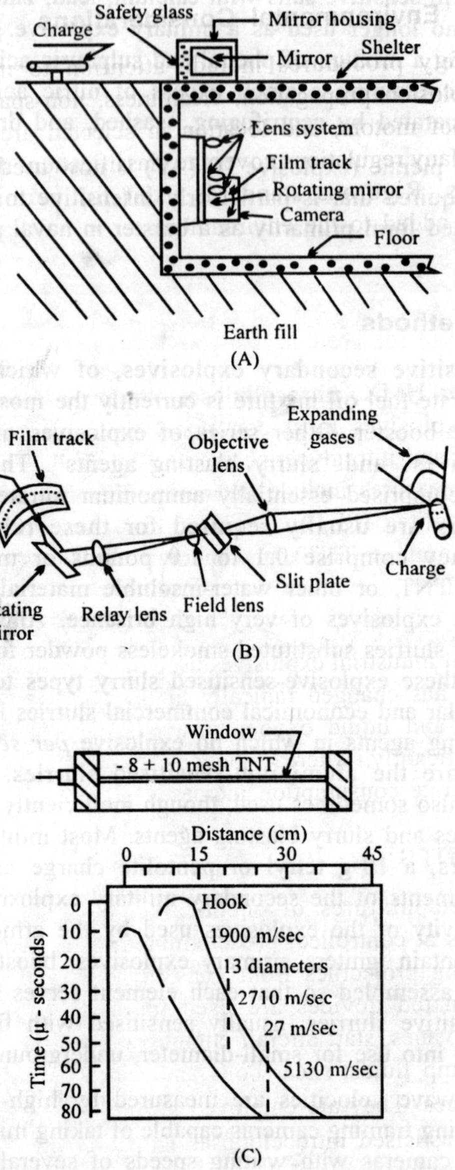

Fig. 6.1. Measurement of detonation velocity by means of the streak camera. (A) Arrangement in inside-out bombproof shelter. (B) Measurement of detonation velocity by means of the streak camera. (C) Streak camera "trace" illustrating detailed information made possible by following the detonation wave from the point of initiation until stable (steady state) detonation is established.

Safety and Environmental Considerations

Standard safety procedures include attention to intraline distances to minimise explosion propagation, cleanliness, non-sparking equipment and explosion-proof motors, and location in selected areas away from heavy population. Many regulations cover the classification, shipping, and handling of explosives. Rigorous standards exist for permissible quantities of contaminants added to the environment.

Uses

Military

TNT, RDX or HMX, nitrocellulose, and nitroglycerine are the single-component explosives most commonly used for military compositions. Nitrocellulose and nitroglycerine are used exclusively to make propellants. Other binary mixtures include the octols (HMX + TNT), cyclotols (RDX + TNT), pentolites (PETN + TNT), tetrytols (tetryl + TNT), amatols (ammonium nitrate + TNT), and picratols (ammonium picrate + TNT).

Industrial

The quantity of industrial explosives used in India continues to increase, but the type used has changed radically in the last twenty years. The use of black powder and liquid oxygen disappeared completely and dynamites declined significantly, whereas the consumption of dry and water-based ammonium nitrate consumption increased greatly.

PROPELLANTS

Propellants are mixtures of chemical compounds that produce large volumes of gas at controlled, predetermined rates. Their chief applications are in launching projectiles from guns, rockets, and for missile systems. Propellant-actuated devices are used to drive turbines, move pistons, operate rocket vanes, start aircraft engines, eject pilots, jettison stores from jet aircraft, pump fluids, shear bolts and wire, and act as sources of heat in special devices. They are employed in guns in the form of dense grains or sheets of plasticised nitrocellulose. Those that contain only plasticised nitrocellulose are known as single-base propellants; double-base propellants also contain a liquid explosive plasticiser such as nitroglycerine. Multi- or triple-base propellants incorporate a crystalline explosive such as nitroguanidine in the double-base formulation. Double- and triple-based nitrocellulose propellants are used in rockets as well as guns. Typical components of nitrocellulose propellants are listed in Table 6.2.

Table 6.2. Typical components of nitrocellulose propellants and their function.

Component	*Application*
Nitrocellulose	energetic polymeric binder
Nitroglycerine, metriol trinitrate, diethylene glycol dinitrate,	plasticisers
Triethylene glycol dinitrate, dinitrotoluene	energetic
Dimethyl, diethyl or dibutyl phthalates, triacetin	plasticisers fuels
Diphenylamine, diethyl centralite, 2-nitrodiphenylamine	stabilisers
Organic and inorganic salts of lead; e.g., lead stannate, lead stearate, lead salicylate	ballistic modifiers
Carbon black	opacifiers
Lead stearate, graphite, wax	lubricants
Potassium sulphate, potassium nitrate, cryolite (potassium aluminium fluoride)	flash reducers
Ammonium perchlorate, ammonium nitrate	oxidisers — inorganic
RDX, HMX, nitroguanidine and other nitramines	organic
aluminium	metallic fuels
	cross-linking catalysts
Lead carbonate tin	defouling agents

Polymer-based rocket propellants, referred to as composite propellants, contain a cross-linking polymer which acts as a viscoelastic matrix for holding a crystalline inorganic oxidiser such as ammonium perchlorate and for providing mechanical strength. Many other substances may be added including metallic fuels, plasticisers, extenders, and catalysts. Polymer-based composite propellants are too erosive to be used in guns because of the residues formed after repeated firings. Typical components of composite propellants are listed in Table 6.3.

Table 6.3. Typical components of composite rocket propellants.

Typical components	*Characteristics*
Binders	
Polysulphides	reactive group, mercaptyl (—SH), is cured by oxidation reactions; low solids loading capacity and relatively low performance; now mostly replaced by other binders
Polyurethanes polyethers	reactive group, hydroxyl (—OH), is cured with isocyanates; intermediate solids loading capacity and performance

(Contd...)

Typical components	*Characteristics*
Polyesters	
Polybutadienes copolymer, butadiene and acrylic acid	reactive group, carboxyl (—COOH) or hydroxyl (—OH), is cured with difunctional epoxides or aziridines; intermediate solids loading capacity and better performance than polyurethanes; less than adequate cure stability and mechanical characteristics
Terpolymers of butadiene, acrylic acid and acrylonitrile	superior physical properties and storage stability
Carboxy-terminated polybutadiene	cured with difunctional epoxides or aziridines; have very good solids loading capacity, high performance and good physical properties
Hydroxy-terminated polybutadiene	cured with diisocyanates; have very good solids loading and performance characteristics and good physical properties and storage stability.
Oxidisers	
Ammonium perchlorate	most commonly used oxidiser; it has a high density, permits a range of burning rates, but produces smoke in cold or humid atmosphere
Ammonium nitrate	used in special cases only, it is hygroscopic and undergoes phase changes, has a low burning rate and forms smokeless combustion products
High energy explosives (RDX-HMX)	have high energy and density; produce smokeless products; have a limited range of low burning rates
Fuels	
Aluminium	Almost commonly used; has a high density; produces an increase in specific impulse and smoky and erosive products of combustion
Metal hydrides	provide very high impulse, but generally inadequate stability, give smoky products, and have a low density
Ballistic modifiers	
Metal oxides	iron oxide most commonly used
Ferrocene derivatives other	permit a significant increase in burning rate coolants for low burning rate and various special types of ballistic modifiers
Modifiers for physical characteristics	
Plasticisers	improves physical properties at low temperatures, and processability; may vapourise or migrate; can increase energy if nitrated
Bonding agents	improve adhesion of binder to solids

Principle of the Rocket

Fig. 6.2. is a simplified illustration of how a rocket motor or engine works. Fig. 6.2a shows a chamber with a hole in one end in which gas is being generated rapidly enough to maintain pressure, P_c. Since the pressure is exerted equally in all directions, there will be a pressure imbalance forcing the chamber to the left because pressure is not exerted on the area where there is a hole on the right. The force, F, that will propel the rocket to the left is equal to the chamber pressure, P_c, times the area of the hole of throat, A_t.

$$F = A_t P_c$$

In an actual rocket this accounts for only part of the thrust. As shown schematically in Fig. 6.2b, an increase in thrust is obtained by acclerating the gas through the nozzle and expanding it to P_e, the pressure at the nozzle-exit plane. The new equation for force which is called thrust is:

$$F = A_t P_c C_f$$

where C_f is the thrust coefficient whose value depends on the ratio of P_c to P_e and the area of the throat to the area at the exit plane. Thus a nozzle-expansion cone is needed for high performance and is used on almost all rockets.

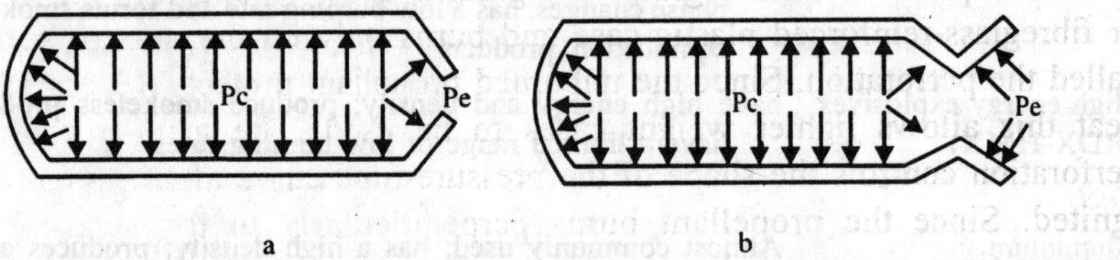

Fig. 6.2. Principle of rocket propulsion.

Since thrust is a function of motor design, it is not useful for comparing the performance of propellants. A much better criteria of rocket-propellant performance is specific impulse which can be defined as the thrust delivered per unit flow weight of propellant, which is the same as weight burned in unit time.

$$I_{sp} = \frac{1}{w}\int_{t_1}^{t_2} F dt$$

and the units are lb-sec (force)/lb of propellant. This is often abbreviated as sec. Measured I_{sp} values are determined by firing motors on elaborately instrumented thrust stands and converting the measured thrust according to the above equation.

It is time consuming and expensive to optimise propellant I_{sp} by firing numerous motors. Computer programs have been devised to calculate the I_{sp} for various propellant compositions. With a computer the calculations can be made rapidly, thus allowing many compositions to be investigated and the ratio of ingredients to be optimised. The results of the calculations are only as good as the thermodynamic data input to the computer program, but in most cases these data have been refined to the point that theoretical I_{sp}'s are dependable.

Types of Propellants

There are two general types of propellants in widespread use; these are solid propellants and liquid propellants. Both of these require an oxidiser and a fuel. If a single compound is used then it is called a monopropellant. There is also a combination of these two types known as a hybrid propellant in which a solid propellant fuel is used with a liquid oxidiser.

Solid propellants

A solid-propellant rocket motor is relatively simple in that it has very few moving parts (Fig. 6.3). The propellant in the motor is referred to as the grain. This is a carry-over from cannon powder terminology where each individual piece is called a grain. The grain is usually bonded to the metal- or fibreglass-reinforced plastic case and burns only on the interior surfacc called the perforation. Since the unburned propellant protects the case from heat this allows lighter weight cases to be used. The geometry of the perforation controls the shape of the pressure-time curve after the motor is ignited. Since the propellant burns perpendicularly to the surface, the pressure-time curve can be neutral (level), progressive, or regressive. A progressive curve leads to increasing acceleration, while a regressive curve can yield constant acceleration because of weight loss as the propellant burns. Various perforation shapes can be used including cylindrical, star, and cruciform. In some cases, end-burning grains are used without a perforation. The latter produce rather low thrust but long burning times.

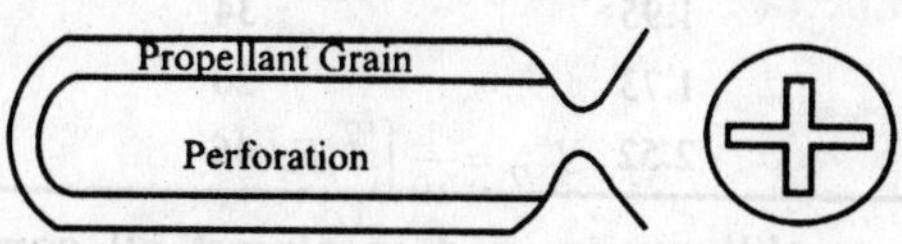

Fig. 6.3. Solid propellant rocket motor.

There are two general types of solid propellants, composite and double base. The composite propellants consist of a solid oxidiser and a metallic fuel intimately mixed in an clastomcric material which serves as both a binder and a fuel. Double base propellants use nitrocellulose plasticised with an energetic nitrato compound, and they may also contain a solid oxidiser and a metallic fuel. The double base propellants which contain a solid oxidiser and a metallic fuel are called composite-modified double base propellants.

Composite propellants

Binders for composite propellants are based on liquid polymers that contain functional groups which allow the polymer to be chain extended and crosslinked (cured) after they have been mixed with the solid oxidiser and fuel. When cured, the propellant exhibits rubbery properties. Table 6.4 gives information about the various common binders.

Table 6.4. Binders for composite propellants.

Polymer type	*Functional group*	*Curing agents*
Polybutadiene	COOH	Epoxides or aziridines
Polybutadiene	OH	Isocyanate
Polypropylene glycol	OH	Isocyanate
Polysulphide	SH	*p*-Quinone dioxime

The polybutadiene-type binders with the carboxyl functional group are most widely used in current solid-propellant rocket motors. The polyurethane binders based on polypropylene glycol have been used in motors such as the first stage Polaris. Polysulphide binders were used in some of the first successful composite propellants.

The common oxidisers used in composite propellants are ammonium perchlorate, ammonium nitrate, and potassium perchlorate. Some important properties of these oxidisers are shown in Table 6.5.

Table 6.5. Oxidisers for composite propellants.

Compound	*Density (g/cc)*	*Available O_2%*	*H_f kcal/100g*
NH_4ClO_4	1.95	34	–59
NH_4NO_3	1.73	20	–110
$KClO_4$	2.52	46	–32

Ammonium perchlorate is used in almost all composite propellants; ammonium nitrate and potassium perchlorate are used for specialised applications.

Metallic fuels are added to composite propellants primarily to increase the temperature of the gas which increases the specific impulse. While the metal oxides produced are usually liquids or solids which reduce specific impulse, the amount of heat released more than offsets this loss. The common metallic fuels are aluminium and beryllium. The addition of metals also increases propellant density which allows more propellant to be put in a given volume. Since many missile or space rocket motors have only a certain volume available, the denser propellants greatly improve performance.

Theoretical performance values for several typical composite propellants are shown in Table 6.6. It is clear from these values that the amount of oxidiser in the propellant has a large effect on I_{sp} and density (f) as seen by comparing 80 per cent and 86 per cent NH_4ClO_4. An even larger effect is noted when a metal fuel like Al or Be is added. While Al yields a propellant with a higher density, Be propellants have higher theoretical I_{sp}'s because Be evolves more heat on oxidation. The propellants containing metal hydrides which produce considerable low molecular weight gas products yield an even higher specific impulse but lower propellant density.

Table 6.6. Performance values for composite propellants.

	T_c(°K)	$I^{\circ\prime}_{sp}$(sec)	f(g/cc)
80% NH_4ClO_4, 20% PBD	2345	229.1	1.57
84% NH_4ClO_4, 16% PBD	2750	241.8	1.64
86% NH_4ClO_4, 14% PBD	2940	246.5	1.68
70% NH_4ClO_4, 16% Al, 14% PBD	3398	262.7	1.77
73% NH_4ClO_4, 13% Be, 14% PBD	3302	282.1	1.67
55% NH_4ClO_4, 25% BeH_2, 20% PBD	2650	306	1.26

Double-base propellants

Double-base propellants differ from composite propellants by having the oxidising and reducing functions contained in the same molecule. Nitrocellulose is swollen by a nitrato plasticiser such as nitroglycerine (glyceryl trinitrate) to give a rubbery gelled structure which, although it is not cross-linked, does have viscoelastic properties. The nitrocellulose desensitises the shock-sensitive nitroglycerine so that it can be used. Other plasticisers that are used for the nitrocellulose are trimethylol ethane trinitrate (TMETN), triethylene glycol dinitrate (TEGDN), and diethylene glycol dinitrate (DEGDN). Stabilisers such as *sym*-diethyl-diphenylurea (ethyl centralite) prevent the decomposition of the nitrate compounds and are a very important part of the formulation. The compositions of JPN, an old extruded propellant, and OV, made by the casting-powder process are shown below.

Table 6.7. Composition of JPN and OV propellants.

	JPN	*OV*
Nitrocellulose (13.25% N)	51.50	–
Nitrocellulose (12.5% N)	–	58.6
Nitroglycerine	43.00	24.2
Dimethyl phthalate	–	9.6
Diethyl phthalate	3.25	–
Potassium sulphate	1.25	–
Ethyl centralite	1.00	1.0
Dinitrotoluene	–	6.6
Carbon black (added)	0.20	0.1
Wax (added)	0.008	–

The addition of NH_4ClO_4 or HMX (cyclotetramethylene tetranitramine) or both, and Al increases both I_{sp} and density which can raise the theoretical I_{sp} to the 265 to 270 sec, and the density to 1.80 to 1.85 g/cc. The latter type of propellants are referred to as composite modified-double-base propellants.

Characterisation of solid propellants. Solid propellants must be characterised so that a process can be designed which is able to handle the uncured and cured propellants. In addition, physical and mechanical properties must be defined so that a satisfactory motor can be designed, and the consequent changes in the propellant needed to meet design requirements must be determined. Safety tests such as impact, friction, static-spark sensitivity, and thermal stability are conducted to determine whether the propellant can be handled safely. Rheological properties, including viscosity and shear stress versus shear rate as a function of time, assist in deciding whether the propellant will process satisfactorily. To ascertain whether the propellant grain in the motor can withstand the thermal and gravitational loads, stress, strain, and modulus are evaluated at various temperatures and strain rates under uniaxial, biaxial, and even triaxial conditions; stress-relaxation and strain-endurance values are also obtained. The ballistic properties evaluated include I_{sp}, burning rate, and ratio of burning surface to nozzle throat area (K_n) as a function of pressure, temperature sensitivity of pressure (π_k) and burning rate (σ_p); these are used to design the propellant grain and nozzle throat size.

Manufacture

Composite solid propellants are relatively easy to manufacture. The first step is to grind part of the oxidiser (usually ammonium perchlorate) to a particle size of 1 to 30μ, depending on the burning rate required. The finer the oxidiser the higher the burning rate. A premix containing the liquid polymer, the metallic fuel, and the curing agent is often made. Either this premix is added slowly to the vertical or horizontal Baker-Perkins mixer containing the oxidiser, or the oxidiser is added slowly to the mixer containing the premix. After mixing a sufficient time to blend the ingredients, and raising the temperature to 60°C or so to increase fluidity, the propellant is cast (poured) through a slit plate into the evacuated motor. The vacuum removes entrapped air. A mandrel installed in the motor forms the perforation in the grain. After heating for several days at a temperature above room temperature (usually between 45 and 60°C) the propellant is a tough viscoelastic solid which is bonded directly to the walls of the motor case. A liner which is usually made from the same polymer as the propellant binder covers the case wall for better adhesion of the propellant.

There are two main methods of manufacturing double-base-propellant rocket motors, the casting-powder process and the slurry casting process. In the casting-powder process, the solid ingredients, including the fibrous nitrocellulose, are mixed with a volatile solvent which yields a plastic mass. The plastic is consolidated in a blocking press, extruded, cut in small pieces and dried. The casting powder is placed in the motor and the casting solvent which includes the nitroglycerine or other nitroglycerine or other nitrato plasticiser is sucked into the evacuated motor to fill the voids. The motor is then heated to 40 to 60°C which causes the casting solvent to swell the casting powder and form a strong grain.

In the slurry casting process, a granular form of nitrocellulose is used. The plasticiser and solid ingredients including nitrocellulose, ammonium perchlorate, HMX, and aluminium can be added directly to the mixer. The mixture is stirred until it is homogeneous and then poured into the motor and cured. Since the plasticiser viscosity is low, it is sometimes desirable to predissolve some nitrocellulose in the plasticiser to increase its viscosity so that the solids do not settle.

Solid propellants have found wide application in rockets for both military and space applications because of their high reliability, low cost, and instant readiness. It is expected that they will continue to be used in many applications.

Liquid propellants

The liquid bipropellant system is used today for the largest rockets. In this system the liquid fuel and the liquid oxidiser are stored in separate tanks and are fed separately to the combustion chamber. The propellants are fed either by means of pumps or by pressurisation with an inert gas. In the case of the pum-fed rocket, the pumps are driven by gas turbines. The gas for these turbines is supplied by a generator which is actually a small rocket motor operating off a side stream of the main rocket propellants (Fig. 6.4) or, in some cases, a separate propellant system. In the largest rockets, thousands of pounds per second of propellants must be pumped to the engine, and the pump drives must develop thousands of horsepower. The propellants enter the thrust chamber through an injector which has a function similar to that of a carburetor in an internal combustion engine. The injector has to atomise and mix the propellant so that the oxidiser and fuel will be in the right proportions to vapourise and burn smoothly.

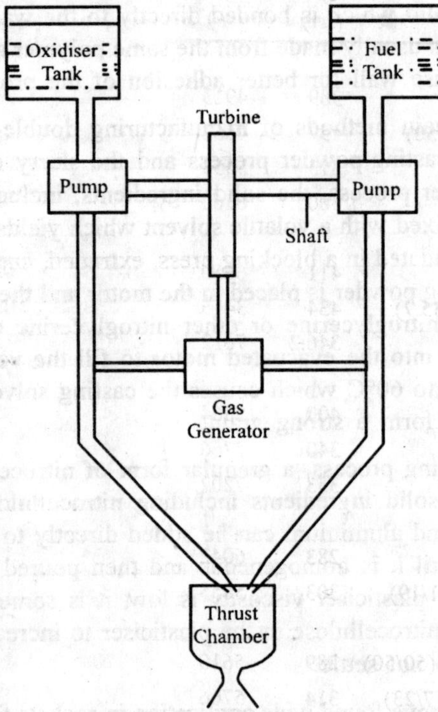

Fig. 6.4. Liquid bipropellant rocket engine.

Theoretical performance

Some typical liquid oxidisers are liquid oxygen (LOX), liquid fluorine, oxygen difluoride (OF_2), chlorine trifluoride (ClF_3) and nitrogen tetroxide (N_2O_4). Typical fuels are RP-1 (kerosene-like), liquid hydrogen, hydrazine (N_2H_4) monomethyl hydrazine (MMH), and unsymmetrical dimethylhydrazine (UDMH). Liquid propellant systems are usually classified as storable or cryogenic. Table 6.8 gives typical properties of several combinations of oxidisers and fuels. It is evident that the cryogenic propellants yield the highest I_{sp} values, however they also have the lowest densities, which is undesirable. Furthermore, they are more difficult to handle because of their low boiling points. They are, however, used in rockets. Such as the Saturn, for space applications because of the heavy payload requirements, and because instant readiness is not a problem. The storable propellants, which have higher densities, have lower I_{sp}, but they can be used in systems that require readiness.

Table 6.8. Properties of several common liquid bipropellants.

Oxidiser	*Fuel*	I_{sp}^{a} *(sec)*	T_c *(°F)*	*Boiling Pt (°C)*	*Density (g/cc)*	*Type*
LOX				–183		Cryogenic
	H_2	389	4913	–253	0.28	Cryogenic
	H_2-Be (50/50)	457	4662	–253	0.23	Cryogenic
	N_2H_4	313	5453	113	1.07	Storable
	RP-1	300	6160	370–510	1.02	Storable
F_2				–188		Cryogenic
	H_2	411	6650	–253	0.40	Cryogenic
	H_2-Li(64.3/35.7)	434	3415	–253	0.20	Cryogenic
	MMH	346.2	7612	87	1.25	Storable
OF_2						Cryogenic
	H_2	400	5864	–253	0.39	Cryogenic
	RP-1	340	7766	188–266	1.28	Storable
	MMH	353	7202	87	1.24	Storable
ClF_3						Storable
	MMH	283	6042	87	1.42	Storable
	N_2H_4-Be (81/19)	303	5855	113	1.43	Storable
N_2						Storable
	N_2H_4-UDMH (50/50)	289	5610	70	1.21	Storable
	MMH-Be (77/23)	314	5766	87	1.18	Storable
	N_2H_4	292	5406	113	1.22	Storable

Properties and manufacture

The liquid oxidisers and fuels that are in use today in operational and experimental rockets have been characterised rather thoroughly. The results of some of these studies and a brief description of the manufacturing methods are given below.

Cryogenic propellants

The ability to design and fabricate the large, high-performance rockets needed for space applications depends in large measure on advances in the engineering technology used in handling such intractable materials as liquid oxygen (b.p. – 183°C), liquid fluorine (b.p. – 188°C), and liquid hydrogen (b.p. – 253°C).

As indicated in Table 6.8 liquid hydrogen allows the highest I_{sp} values to be obtained. However, its low boiling point and low density (0.07 g/cc) make it difficult to handle and store. Also, its very low temperature can cause embrittlement of metals which results in the loss of impact strength. Storage tanks and transfer lines must be well insulated. Vacuum jackets, foamed plastics, and alternate layers of aluminium foil and glass-fibre mats have been successfully used.

Molecular hydrogen exists in two forms: *ortho*-hydrogen, wherein the nuclei of the two atoms are spinning in the same direction, and *para*-hydrogen, in which the nuclei spin in opposite directions. These two forms are in equilibrium with each other, and at room temperature the equilibrium mixture contains 75 per cent of the *ortho* form and 25 per cent of the *para* form. When cooled to the boiling point (–253°C), this mixture is unstable. Since the equilibrium concentration of *para*-hydrogen at this temperature is 99.8 per cent, the *ortho*-hydrogen will convert very slowly to the *para* form. The conversion from 25 to 90 per cent *para* takes about one month. This conversion is accompanied by the evolution of heat which is sufficient to cause severe evaporation losses of the liquid. In order to overcome this difficulty, a hydrous ferric oxide catalyst is used in connection with the liquefaction process to speed up the conversion so that a stable product may be obtained. Modern liquefiers can produce liquid hydrogen which is more than 99 per cent *para*-hydrogen.

Hydrogen is recovered as a by-product from petroleum refining and from the partial oxidation of fuel oil.

Liquid monopropellants are "unstable" liquid chemicals which, in the presence of a catalyst or a small amount of initiating energy, will decompose to give heat and large volumes of gas. Examples of such materials are hydrogen peroxide and hydrazine:

$$H_2O_2 \rightarrow 1/2\ O_2\ (g) + H_2O\ (g)$$

$$N_2H_4 \rightarrow N_2\ (g) + 2H_2\ (g)$$

In general, monopropellants exhibit low performance (at 300-pounds chamber pressure, exhausting to 1 atm, hydrogen peroxide has a specific impulse of 133 secs; hydrazine, 174 secs); materials containing enough energy to give a high theoretical specific impulse are usually too shock-sensitive to be useful in rocket work. The monopropellants are very useful as gas generators where cool temperatures are required or high I_{sp} is not a requirement.

Since fluorine is the most electronegative of all the elements, there are handling problems in addition to those associated with its low boiling point. Fluorine will violently attack nearly all oxidisable substances including metals. Fortunately, metal surfaces can be passivated by forming thereon a fluoride film which protects the metal from further attack. Many metals have been found to be suitable for rocket applications; they include "Monel", stainless steel, copper, brass, and aluminium. It is extremely important that all equipment be clean and free from any organic material to prevent hazardous conditions. Fluorine is produced by the electrolysis of hydrogen fluoride in fused potassium fluoride.

Liquid oxygen is highly reactive with most organic materials, but it is not corrosive like liquid fluorine. However, the surfaces which will be in contact with this oxidiser must be kept extremely free of any contamination. The low boiling point can also cause problems due to low-temperature embrittlement. Metals that can be used are "Monel", copper, bronze, brass, aluminium, and stainless steel; plastics which are suitable include Teflon and Kel-F. Liquid oxygen is produced by the fractionation of liquid air.

Storable propellants

Typical examples of storable liquid propellants are nitrogen tetroxide (N_2O_4), chlorine trifluoride (ClF_3), hydrazine (N_2H_4), and unsymmetrical dimethylhydrazine (UDMH). A brief description of the properties and production processes for these chemicals is given below:

Nitrogen tetroxide is very stable at room temperature, but at 150°C it begins to dissociate into nitric oxide and free oxygen. Upon cooling, nitrogen tetroxide is reformed. It is not corrosive to most common metals if the oxidiser is dry (< 0.1% H_2O); carbon, steels, aluminium, stainless steel, nickel, and inconel can be used. Under wet conditions 300-series stainless steel should be used. As a result of its good properties, nitrogen tetroxide is currently the most important operational earth-storable liquid oxidiser.

The catalyst oxidation of ammonia is the most important commercial method of preparing, N_2O_4. A mixture containing about 9 per cent ammonia in air is oxidised at pressures between 25 and 100 psig. The gaseous products containing nitric oxide (NO) and water are cooled and dried by scrubbing with cold nitric acid. The dry nitric oxide containing unreacted nitrogen from the air is further oxidised by the addition of air to form NO_2. Nitrogen dioxide forms an equilibrium mixture with its dimerised form, dinitrogen tetroxide, N_2O_4, and on cooling the equilibrium shifts to favour the formation of N_2O_4. The reaction sequence is illustrated by the following equations:

$$4NH_3 + 5O_2 \rightarrow 4NO + 6H_2O$$
$$2NO + O_2 \rightarrow 2NO_2$$
$$2NO_2 \rightarrow N_2O_4$$

Chlorine trifluoride is highly corrosive but is stable to shock, heat and electrical spark. Because it forms an inert fluoride film on some metals, it can be used with copper, brass, steel, aluminium, nickel, and Monel. Chlorine trifluoride has been prepared by the reaction of chlorine and fluorine at 280°C.

Hydrazine is stable to shock, heat, and cold. It begins to decompose at 160°C if no catalysts are present. It freezes at 1.4°C, but it contracts and does not decompose. It is compatible with stainless steel, nickel, aluminium, Teflon, and Kel-F.

Hydrazine is produced by the controlled oxidation of ammonia in the well-known Raschig synthesis. Sodium hypochlorite, formed by the chlorination of caustic soda, reacts with excess ammonia in a two-step process to form, successively, chloramine and aqueous hydrazine :

$$NH_3 + NaOCl \rightarrow NH_2Cl + NaOH$$
$$NH_2Cl + NaOH + NH_3 \rightarrow N_2H_4 + NaCl + H_2O$$

Hydrazine and water form an azeotrope which boils at 120.1°C. Anhydrous hydrazine is obtained by dehydration of the azeotrope with caustic, followed by a vacuum distillation. In another method, the azeotrope is broken by the addition of aniline as a third component, followed by a distillation in which the water-aniline azeotrope is taken overhead and anhydrous hydrazine is obtained as a bottoms product. Using this latter method hydrazine may be produced at 98 + per cent purity.

Unsymmetrical dimethyl hydrazine (UDMH) is neither shock- nor heat-sensitive, being stable to 249°C, its critical temperature. Furthermore, it is compatible with nickel, Monel, and stainless steel, but it does attack aluminium if water is present.

UDMH has been prepared by a modification of the Raschig process by reacting the chlorimine intermediate which dimethylamine instead of ammonia. It can also be produced commercially by nitrosation of dimethylamine to N-nitrosodimethylamine followed by reduction and purification. UDMH is usually used as a 50/50 mixture with hydrazine; in fact, this is the most common earth-storable fuel. Many applications have been found for liquid propellants because of their generally high performance. They have been widely used in space programs but have also found many military applications.

Selection

In addition to energy and burning-rate considerations, a propellant must meet other criteria including mechanical characteristics, stability, sensitivity, cost of manufacture, and uniformity of performance.

Burning Process

The mass rate of propellant burning at a given pressure and temperature depends on the amount of heat evolved during decomposition and the amount of heat transferred to the burning surfaces of the propellant from the hot gases above it. It is also influenced by the tangential velocity of the propellant gases and the radiation from the surrounding gases. Propellants burn in parallel layers so that the surface recedes in all directions normal to the original surface. The geometry of the grain on completion of burning is similar to its geometry at the start. Propellant burning at high gun pressures proceeds more smoothly and is less subject to erratic behaviour than burning at very low pressures because the conditions are appropriate for maximum energy transfer in minimum time. The burning rate at gun pressures usually varies somewhat less than the first power of the pressure. It changes more slowly at rocket pressures of 3.45–10.34 MPa (500–1500 psi), often to less than the square root of the pressure.

The operating pressure of the system has the predominant effect on the burning rate of propellants. Photographic evidence shows that increasing the pressure decreases the distance between the flame zone in the gas phase and the propellant surface. The rate of heat transfer to the propellant surface increases accordingly.

Burning-rate equations

Burning-rate equations have been developed to describe the performance of solid propellants based on the assumption that all the exposed propellant surfaces are ignited simultaneously and burn at the same linear rate.

$$r = a + bP \quad ...(6.1)$$

$$r = cP^n \quad ...(6.2)$$

where r is the linear burning rate, P is the pressure, n is the pressure exponent, and a, b, and c are constants that vary with temperature. Equation 6.1 is often used for propellants burning at high gun pressures, whereas equation 6.2 is associated with low pressure rocket systems.

Mechanism of burning of nitrocellulose-based propellants

Much of the information available on the burning process of nitrocellulose propellants is based on the decomposition of nitrate esters and the reaction of oxides of nitrogen with the products of decomposition. The three reaction zones identified are :

(1) The surface of the solid is a reaction zone ca 0.01 cm in thickness and ranging up to ca 600 K. Molecular bond breakage occurs in this zone, primarily the O—N bond in cellulose nitrate-nitroglycerine-type propellants.

(2) Above the foam zone is a non-luminous gaseous reaction zone, several thousands of a centimetre thick. This zone attains a temperature of ca 1500 K as a result of partial reaction among the material ejected from the surface. Aldehydes and alcohols are converted to smaller molecules nitrogen, water, carbon monoxide, carbon dioxide, and nitric oxide. About half the total heat evolved by the propellant is liberated in the fizz zone which is prominent at low pressures and disappears at high pressure.

(3) The reaction continues to completion and thermodynamic equilibrium is established in the flame zone. This zone, ca 0.001 cm thick, defines the flame temperatures of the propellant which may range from ca 1500 K for cool propellants to 3500 K for very hot ones. The nitric oxide reacts with the reaction products formed to produce carbon monoxide, carbon dioxide, hydrogen, water, nitrogen, and a small percentage of other molecules.

Mechanism of burning of composite propellants

A number of analytical models have been developed to permit estimations of the burning rates of composite propellants as a function of the factors that affect them most: caloric value, propellant temperature, chamber pressure, fuel-oxidant ratio, particle size distribution of the oxidiser, the presence of decomposition catalysts, and the incorporation of aluminium in powder, wire, or staple form.

The Beckstead-Derr-Price model (Fig. 6.5) considers both the gas-phase and condensed-phase reactions. It assumes heat release from the condensed phase, an oxidiser flame, a primary diffusion flame between the fuel and oxidiser decomposition products and a final diffusion flame between the fuel decomposition products and the products of the oxidiser flame. If aluminium is present, additional heat is subsequently produced at a comparatively large distance from the surface. Only small aluminium particles ignite and burn close enough to the surface to influence the propellant burn rate. The temperature of the surface is ca 500 to 1000°C compared with ca 300°C for double-base propellants.

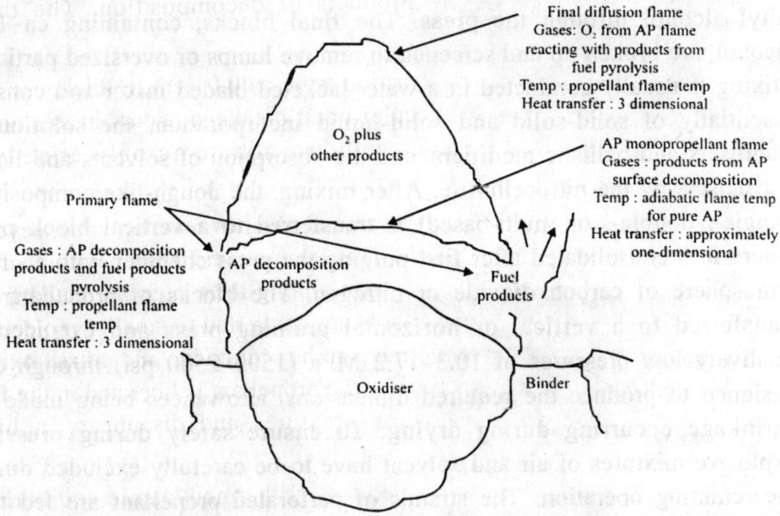

Fig. 6.5. The postulated flame structure for an AP composite propellant, showing the primary flame followed by the final diffusion flame. AP = ammonium perchlorate.

Manufacture

Nitrocellulose propellants

Nitrocellulose propellants are made with or without incorporation of a solvent as plasticiser by five processes: solvent extrusion, solvent emulsion, solventless extrusion, solventless rolling and casting. All require colloiding the nitrocellulose to eliminate its fibrous structure and cause it to burn predictably in parallel layers. Mechanical working of the ingredients contribute to plasticisation and uniformity of composition. Although the manufacture of extruded nitrocellulose propellants has been traditionally a batch process, continuous processes are being developed for single-, double-, and multi-base propellants.

Solvent extrusion batch process

Almost all standard gun propellants and small-webbed rocket propellant grains are made by the solvent extrusion process. Purified and blended, nitrocellulose of the required nitrogen content and wet with ca 30% water is received from the nitrocellulose plant. A charge is transferred to a double-acting hydraulic dehydration press and compressed at low pressure to remove some of the water. The remainder is removed by pumping 95% ethyl alcohol through the press. The final blocks, containing ca 18% alcohol, are broken up and screened to remove lumps or oversized particles. Mixing is usually conducted in a water-jacketed bladed mixer and consists essentially of solid-solid and solid-liquid incorporation, the solution of stabilisers and ballistic modifiers, and the absorption of solvents and liquid plasticisers by the nitrocellulose. After mixing, the dough-like composition (single-, double-, or multi-based) is transferred to a vertical block press where it is consolidated after first purging the press chamber with an inert atmosphere of carbon dioxide or nitrogen. The blocks of propellant are transferred to a vertical or horizontal graining press and extruded at relatively low pressures of 10.3–17.2 MPa (1500–2500 psi) through dies designed to produce the required dimensions, allowances being made for shrinkage occurring during drying. To ensure safety during pressing, explosive mixtures of air and solvent have to be carefully excluded during the remming operation. The strands of perforated propellant are fed to a mechanical cutter and sliced to specified lengths.

Continuous solvent extrusion process (for single-base propellants)

Continuous processes for making solvent propellant for cannon offer substantial reductions in labour, hazard, and pollutants. The main features in which the continuous process differs from the operation are thermal dehydration, water-wet nitrocellulose vacuum dried on a continuous vacuum belt filter followed by hot-air transfusion (80°C) to reduce the moisture to less than 2%; compounding, in which the alcohol-wet nitrocellulose is transferred from a surge feeder to a compounder by a continuous weigh-belt along with other ingredients of the composition, which are also weighed and added automatically. The loosely mixed paste is fed by a conveyor to a heavy duty reciprocating screw mixer that is temperature controlled and specially designed to thoroughly mix and work the paste by forcing it past pins in the mixer barrel and out through a die plate. As the plaste is extruded, it is cut into small pellets which are fed continuously to water-jacketed screw extruders and forced through multiple dies. The strands are cooled as they are extruded to facilitate cutting by an adjustable roll-type cutter. After the cut grains are screened to remove clusters and odd sizes, the solvent is removed in the solvent recovery–water dry systems where the propellant is treated first with hot inert gas and then hot water, then air dried until the moisture content is reduced from 12 to less than 0.8%.

Solventless extrusion process

The solventless process for making double-base propellant has been used for the manufacture of rocket-propellant grains with web thickness ca 1.35–2.0 cm and for thin-sheet mortar propellant. It is also used for making cannon propellant. In the water-slurry process for making solventless propellants, explosive and non-explosive liquid plasticisers and water-insoluble constituents are incorporated into nitrocellulose suspended as a slurry in a large volume of hot water. After removing the excess water, the resulting wet mass is partially dried and passed through heated rolls to thoroughly mix the colloidal composition into sheets of homogeneous propellant and remove any remaining water. These sheets may then be cut and rolled into scrolls for insertion in the press and extruded as propellant grains or cut up and used as a sheet or flake propellant.

Ball Powder

Ball powder is typically used in small-caliber weapons. It consists of spherically shaped or flattened ellipsoidal grains, ca 0.04–0.09 cm dia. The process permits the recovery and use of nitrocellulose from obsolete granular propellant.

Batch process

Water-wet fresh or extracted nitrocellulose is transferred as a slurry to a graining still. Calcium carbonate is added to neutralise any free acid released by the dissolved nitrocellulose. The required amount of ethyl acetate is added as well as other soluble components. The contents are heated to ca 70°C and agitated to dissolve the nitrocellulose and to form a lacquer. When of proper viscosity, a protective colloid is added to form an emulsion of nitrocellulose globules which prevents their coalescing. Sodium sulphate is added to form an osmotic-pressure differential between the water-laden nitrocellulose globules and the concentrated salt solution in the still, and water is extracted. Small spheres of dissolved nitrocellulose and the other soluble ingredients are formed, the ethyl acetate is distilled, and the remaining spherical particles are grained. After graining, the slurry is water-washed to remove adherent salt solution and then wet-screened into the required particle sizes.

Nitrocellulose-Based Cast Propellants

Cast nitrocellulose propellant is made by a two-step process. In the first stage, casting powder is produced by techniques that are almost identical to those used for the manufacture of conventional solvent-extruded small-grain gun propellants. The second stage consolidates the casting powder by filling the interstices of the granules with a fluid plasticiser that diffuses into the powder and causes swelling and ultimate coalescence of the granules into a monolithic grain. The plasticiser generally consists of a mixture of an explosive energy-producing liquid and an inert fluid. The process of consolidation is a physical one. No chemical reaction occurs and there is virtually no shrinkage during curing.

Polymer-based cast propellants

These propellants have an extensive range of performance characteristics, have excellent thermal and mechanical stability, and are relatively inexpensive. The manufacturing operations for making composite propellants are very similar, although a variety of polymeric binders may be used. A flow chart of a typical batch process is shown in Fig. 6.6.

Black powder

Black powder is mainly used as an igniter for nitrocellulose gun propellants and to some extent in safety blasting fuses, delay fuses, and in firecrackers. Potassium nitrate black powder is used for military applications. The slower-burning, less costly, and more hygroscopic sodium nitrate black powder is used industrially.

The critical relative humidity of black powder is 60%. It gains ca 2% moisture in 48 hours at 90% rh and 25°C. Ignitability decreases rapidly at ca 3 to 4% moisture. The structure of the granules also deteriorates during cycling through high humidity atmospheres. However, it can be stored satisfactorily for many years if dry. The hygroscopicity of black powder is caused by the carbon present and impurities in the potassium nitrate.

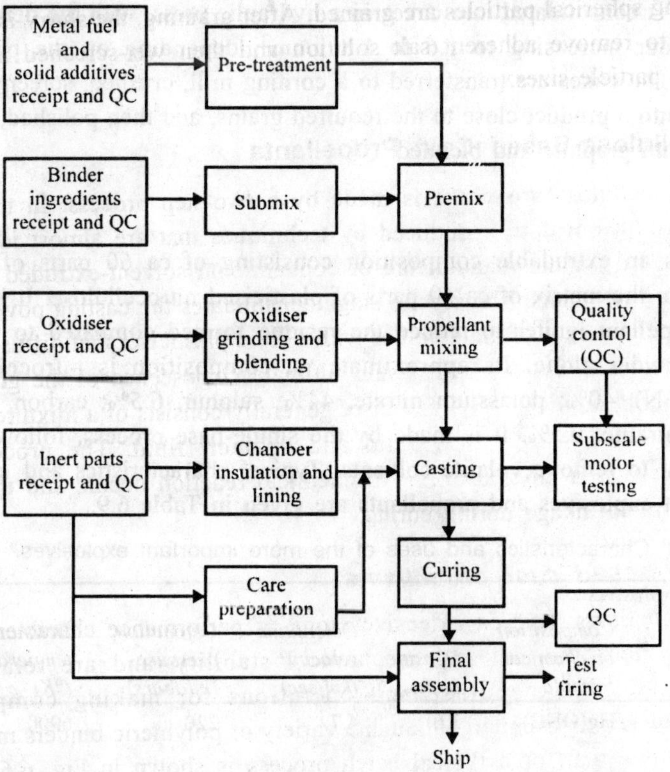

Fig. 6.6. Batch process for cast-composite polymer-based propellants.

Manufacture

Since black powder is a mechanical mixture, its performance is critically dependent on the degree of intimacy of the components in the product. A typical procedure for bringing the ingredients into maximum mutual contact is: dry potassium nitrate is pulverised in a ball mill and sulphur is milled into cellular charcoal to form a uniform mix in a separate ball mill. The nitrate and the sulphur-charcoal mix are screened and then loosely mixed in a tumbling machine. Magnetic separators may be used to ensure the absence of ferrous metals. The preliminary mix is transferred to an edge-runner wheel mill with large, heavy cast iron wheels. The magnitude of the gap between the pan and the wheels, a clearance needed for safety purposes, contributes to the density of the black powder granules obtained. Water is added to minimise dusting and improve incorporation of the nitrate into the charcoal. The milling operation requires ca 3 to 6 hour. The moist milled powders is consolidated into cakes in a hydraulic press, with the density of the powder increasing to 1.6 to 1.8 g/cm^3, depending on the pressure applied. The cakes are transferred to a corning mill, crushed, screened and formed into a product close to the required grains, and then polished, dried, mixed with graphite and blended.

Benite

Benite is an extrudable composition consisting of ca 60 parts of black powder in the matrix of ca 40 parts of plasticised nitrocellulose. It is used as a propellant igniter to reduce the residue formed compared to use of black powder alone. Its approximate wt composition is nitrocellulose (13.15% N), 40%; potassium nitrate, 44%; sulphur, 6.5%; carbon, 9.5%; ethyl centralite, 0.5%. It is made by the single-base process, followed by air-drying to remove volatile solvents. Typical characteristics and uses of important explosives and propellants are given in Table 6.9.

Table 6.9. Characteristics and uses of the more important explosives.

Primary explosives

Name	*Composition or chemical formula*	*Density (g/cc)*	*Detonation velocity[a] (km/sec)*	*Detonation pressure (kilobars)*	*Detonation temperature[a] (°K)*
Mercury fulminate	$Hg(ONC)_2$	3.6	4.7	220	6900
Lead azide	PbN_6	4.0	5.1	250	5600
Lead styphnate	$C_6H(NO_2)_3O_2Pb$	2.5	4.8	150	–
Nitromannite (Mannitol hexanitrate)	$C_6H_{58}(ONO_2)_6$	1.73	8.3	300	6000
Dinitrodiazophenol (DDNP)	$C_6H_2N_4O_5$	1.5	6.6	160	–

Secondary high explosives

Name	*Composition or chemical formula*	*Density (g/cc)*	*Detonation velocity (km/sec)*	*Available energy (kcal/g)*
Ammonia gelatin	30–90% grades same as straight gelatins except	1.2–1.5	4–6.5	0.75–1.15

(Contd...)

Name	Composition or chemical formula	Density (g/cc)	Detonation velocity (km/sec)	Available energy (kcal/g)
dynamites	for some NG and $NaNO_3$ replacement by NH_4NO_3			
Semigelatin dynamite	15–20% NG 1–2% DNT oil, AN-SN dope	1.2	3.5–5 (depends on diameter)	0.9
Prilled AN-oil	94/6 NH_4NO_3/oil	0.8–0.9	1.5–4	0.81–0.83
Slurry explosives	TNT 17–40	1.4–2.0	5–8	0.7–1.8
TNT-SE	Oxidiser* 30–65			
	H_2O 12–25			
	Al 0–20			
	Other 0.3- 1.5			
Smokeless powder SE	SP 20–40	1.35–1.9	4–7	0.65–1.7
	Oxidiser* 30–60			
	H_2O 3–25			
	Al 0.20			
	Other 0.3–10			

[a] Most important properties of detonators.

* AN, SN, perchlorates, etc.

Sensitivity	Major characteristics	Uses
Very high	Best primary explosive for single-component (fuse) detonators; easily detonated by flame, spark, heat, or friction; easily dead-pressed.	In fuse caps (mixed with $KClO_3$); propellant primer; in fuses for shells; small arms cartridges caps.
Very high (higher than NG; less than mercury fulminate)	Powerful detonator but requires strong igniters, e.g., lead styphnate.	Primary explosive in composition (EB) caps; military fuses.
Exceedingly high	Extremely sensitive to sparks, static electricity; explodes rapidly on ignition; good thermal stability.	Igniter in composition caps, military fuses; very satisfactory detonator explosive for fast ignition.
Very high (greater than NG; less than lead azide)	Stronger and more brisant than NG, RDX, PETN.	In composition caps and fuses.
Very high (less than lead azide)	Does not dead-press. About 3/4 as strong as TNT.	In composition caps and fuses.

(Contd...)

Sensitivity	Major characteristics	Uses
High	More economical; only slightly less brisant than straight gelatin; exhibits low-order detonation with threshold priming and high pressures.	General small and large diameter blasting in hard rock and under water.
High	Stringy, plastic; easily loaded in "uppers;" economical; high strength; moderate brisance.	Popular small diameter metal-mining explosive.
Low (requires booster)	One of the cheapest sources of explosive energy available today; flammable and will explode when ignited under strong confinement; no water resistance; adaptable to to-it-yourself operations.	Open-pit and underground blasting where dry conditions prevail; most adaptable to soft, easy shooting.
Low (requires booster)	Gel or thick pea-soup consistency; capable of detonation at high pressures, excellent water resistance.	Large diameter, open-pit, small diameter underground, oil well, submarine, water-filled boreholes, deep-water bombs.
Low (requires boosters)	Generally similar to TNT slurry.	Large diameters, open-pit blasting.

Name	Composition or chemical formula		Density (g/cc)	Detonation velocity (km/sec)	Available energy (kcal/g)
Al-SE	Al	1.0-10	1.1–1.5	2–5	0.7–1.3
Slurry blasting agents	Al	0–35	1.1–1.6	2–6	0.7–2.0
	Oxidisers*	50–80			
	H_2O	4–18			
	Other	0.2–10			
Nitrostarch powders	Nitrostarch in place of NG		1.2	4–5	0.8–1.0
Composition B	40/59/1 TNT/RDX/wax		1.7	7.8	1.1
Composition B-3	40/60 TNT/RDX		1.73	7.9	1.15
Haleite or EDNA	$(CH_2NHNO_2)_2$		1.6 (pressed)	7.9	1.2
Ammonium picrate (Explosive D)	$(ONH_4)C_6H_2(NO_2)_3$		1.56 (pressed)	6.6	0.7

(Contd...)

Name	Composition or chemical formula	Density (g/cc)	Detonation velocity (km/sec)	Available energy (kcal/g)
Nitrostarch	Mixtures of various nitro esters of starch	1.4 (pressed)	6.4	0.95
Tetryl	$(NO_2)_3C_6H_2CH_3N_2NO_2$	1.45 (pressed)	7.0	0.95
PETN (penta-erythritol tetranitrate)	$C(CH_2ONO_2)_4$	1.6 (pressed)	7.92	1.31
Pentolite	50/50 TNT/PETN	1.63 (cast)	7.7	1.1
Trinitrotoluene (TNT)	$CH_3C_6H_2(NO_2)$	1.59 (cast)	6.9	0.9
		1.45 (pressed)	6.9	–
		1.03 (Pelletol)	5.1	–
		0.8 (grained)	4.2	0.8
Amatols	50/50 AN/TNT	1.55 (cast)	5–6.5 (depending on diameter)	0.95
	80/20 AN/TNT	1.0 (loose)	4 (large diameter)	0.93
		1.45 (pressed)	5.6 (large diameter)	

Sensitivity	Major characteristics	Uses
From cap sensitive to very low–depends on aluminium fineness.	Gelatin; no explosive ingredients.	Small diameter, underground and general blasting.
Low (requires boosters)	Gelatin to thick or thin pea-soup consistency.	Large diameter, underwater, wet- and dry-hole blasting, large bombs.
Moderately high, but less than dynamites	Good "fumes"; fair water resistance; powerful; economical.	Small diameter blasting.
Average	Very high brisance.	Bursting charge and special weapons.
High	Very high brisance.	Experimental standard.
High	High brisance; less sensitive RDX and PETN.	In Ednatols for bursting charges.

(Contd...)

Sensitivity	*Major characteristics*	*Uses*
Very low	Insensitive to shock and friction; melts with decomposition; shells filled with high-pressure pressing.	Armour-piercing shells.
High	Highly inflammable white powder	Demolition blocks and Trojan blasting explosives.
High	Very sensitive; rapidly reacting; easily pressed with 1–2% graphite; high brisance.	Booster; base charge in caps; in tetrytols for bursting charges.
High	Very powerful and sensitive (more sensitive than RDX, less than NG).	In Primacord fuse; base charge in caps.
Moderate	High pressure or brisance; primacord sensitive	Booster and special weapons; commercial booster for prilled AN-fuel oil and slurry explosives.
Low	Easily melted and cast; suitable liquid for slurrying with other explosives; easily pressed into blocks; completely waterproof.	Military; "Nitropel" TNT used in slurry explosives and in filling annulus between charge and bore-hole in water filled holes; in amatols.
Low	Insensitive; hygroscopic, not waterproof; less brisant but stronger than TNT; 50/50 can be cast; 80/20 either pressed or granulated.	Military; oil well shooting; quarrying; dry-hole booster for very low-sensitive types.

Name	*Composition or chemical formula*	*Density (g/cc)*	*Detonation velocity (km/sec)*	*Available energy (kcal/g)*
Dinitrotoluene (DNT)	$CH_3C_6H_3(NO_2)_3$	1.28 (liquid)	5	0.7
		0.8 (granular solid)	2–3.5 (depending on diameter)	
Nitromethane (NM)	CH_3NO_2	1.12	6.2	
Cyclonite (RDX)	$C_3H_6N_6O_6$	1.2 (loose)	6.8	1.32
		1.6 (pressed)	8.0	
HMX	$C_4H_8N_8O_8$	1.89	9.1	1.35

(Contd...)

Name	*Composition of chemical formula*	*Density (g/cc)*	*Detonation velocity (km/sec)*	*Available energy (kcal/g)*
HBX	Mixtures of RDX, TNT, aluminium, and wax	1.78	7.5	1.5
Plastic explosives (compositions A,C, C-2, C-3, C-4)	Waxed RDX	1.45–1.6	8.0	1.1–1.3
PBX 9404	94/3/HMX/binder/ nitrocellulose	1.84	8.8	1.3
Nitroglycerine (NG)	$C_3H_5(ONO_2)_3$	1.59	7.8	1.41
Ethylene glycol dinitrate (EGDN)	$C_2H_4(ONO_2)_2$	1.48	7.4	1.43
Straight dynamites[b]	20–60% NG, in balanced SN dope 20% grade ≡ 20% NG, etc.	1.3	4–6	0.55–0.85[b]
Ammonia dynamites (and permissibles)	As above except NH_4NO_3 replaces part of NG and $NaNO_3$	0.8–1.2[b]	1.5–5.5 Depends on AN particle size, NG content.	0.7–0.9[b]
Blasting gelatin	92/8/NG/nitrocotton ("Solidified" NG contains some wood pump to minimise low-order detonation).	1.55 (1.45)	7.5 (7.2)	1.45 (1.4)

[b] Depends on grade.

Sensitivity	*Major characteristics*	*Uses*
Very low	Reddish brown or yellow liquid.	Sensitiser in "Nitramons"; 60/ 40 NG/DNT in oil well shooting; up to 20% in TNT bursting charges; in FNH (flashless) propellant; 6% in small-arms ammunition (with guncotton).
Moderate	Clear, watery liquid	Special demolition, experimental studies of liquid explosives.

(Contd...)

Sensitivity	*Major characteristics*	*Uses*
High	High thermal stability in solid state; expressively sensitive in pure state; 1.65 times as strong as low density TNT; 1.45 times as strong as cast TNT.	Major ingredients in plastic explosives; one of most brisant explosives in cast TNT (composition B); base charge in caps.
High	Better than RDX in all respects.	Same as RDX.
Average	Very powerful.	Underwater explosive.
Moderate	Plastic, easily moulded or pressed	Specialised military demolition.
Moderate	Plastic bonded.	Specialised military demolition.
Very high (almost a primary explosive)	Oily, toxic liquid; volatile above 50°C; gelatinised by nitrocotton; exhibits low-order detonation with threshold priming.	Shooting oils wells; main explosive in dynamites; used in double-base powders.
Very high	Closely resembles NG; more volatile, toxic, slightly stronger but less brisant (owing to lower density).	Used in solution with NG as freezing point depressant.
High	Cheesy, plastic substance; packed in paper cartridges; may be slit and tamped in borehold for greatest blasting effect; fired by detonator as are all dynamites; heat, friction, shock, and flame sensitive.	Ditching, stumping, other uses where high propagation-by-influence "sensitiveness" is required.
High	Cheaper than comparable grade straight dynamites; must be waterproofed by special additives.	General small and large dynamite blasting; permissible (some grades).
High	Strongest, most brisant dynamite; completely waterproof; exhibits low-order detonation with threshold priming and under high pressures.	Oil well and submarine blasting, tunnel drilling, demolition.

(Contd...)

Propellants

Name	*Composition or chemical formula*	*Sensitivity*
Colloidal nitrocellulose (N.C.) Powders	Pyrocotton: cellulose nitrate with 12.6%N. Guncotton: cellulose nitrate with 13.3% N.	Low
Double-base powders	60–80% Nitrocellulose. 20–40% Nitroglycerine.	Moderate
Cordite	65% N.C., 30% NG, 5% vaseline	Low
FNH (Flashless non-hydroscopic powders)	Either straight N.C. or double-base powders with addition of coolants, etc., to prevent muzzle flash, and decrease water adsorption.	Low
Albanite; DINA powder	Di(2-nitrooxyethyl)nitramine.	Low
Rocket powder (solventless powder)	Nitrocellulose plasticised with about 50% NG, plus stabilisers and potassium salts.	Low
Chemical propellants[c]	Hydrogen perioxide, 80–90% H_2O_2 plus Ca, Na, or K permanganate (solid or aqueous solution). Hydrazine hydrate plus methyl alcohol. Fuming nitric acid-aniline. Mixed acid-monoethylaniline. Liquid oxygen-kerosene.	
Black powder	75% KNO_3 (or $NaNO_3$), 15% charcoal, 10% sulphur.	High

[c] Many new rocket propellants have been described; the best ones are under security classifications.

Sensitivity	*Major characteristics*	*Uses*
High	Jelly-like substance; powerful, waterproof; exhibits low-order detonation under threshold primering and high pressure.	In hard rock; mud-capping demolition; submarine blasting.

Major characteristics	*Important uses*
Burning rate controlled by graining, hygroscopic, smokeless flame, with intense flash; gelatinised with alcohol-ethel.	Combined with stabilisers and modifiers to make smokeless powders for artillery, small arms, and sporting ammunition.
Pyrocotton and guncotton are usually blended to secure an average of 13.15% N.	Dry guncotton in fibre form is used in primers fired by an electric current.
Very rapid burning rate, controllable by surface area; more powerful and more readily ignitable than straight N.C. powders; causes erosion of gun bores; can be detonated and is subject to DDT.[d]	Propellant for mortars and sporting ammunition; not used by U.S. armed forced as cannon powder because of bore erosion.

(Contd...)

Major characteristics	*Important uses*
Gelatinised with acetone.	Propellant for large caliber naval guns (English).
Like other smokeless powders, but can be rolled into sheets; flash reduced by DNT, potassium salts, etc.	Propellant for small armor-piercing rockets such as the "Bazooka" (NG base); for naval ammunition (NG base).
Better flashless powder than FNH powders.	Naval ammunition.
Very rapid, uniform burning rate; can be made with thick section since no solvent need be removed.	For rockets up to 4.5 inches.
Catalytic decomposition into water and O_2 releases about 1000 Btu per pound. Supplies oxygen to burn petroleum fuel.	For driving turbines on submarines: V-2 rocket-fuel pumps; jet motors; launching device for ram-jets.
Rapid combustion; fuel and oxidiser are both liquid.	For torpedo turbine drives.
Rapid rate of reaction generates heat and gases.	For launching device for ram-jets; jet motors.
As above.	As above.
	Rocket motors.
Cheap; excellent "heaving action," persistent smoky flame; very sensitive to friction, spark, and heat; hygroscopic.	Time (delay) fuses for blasting and shell; in igniter and primer assemblies for propellants; pyrotechnics; $NaNO_3$ powder, in commercial black powder and for practice bombs and saluting charges; (it is being discontinued as blasting charge).

[d] DDT = deflagration to detonation transition.

Chapter 7

Industrial Gases

INTRODUCTION

Those of the gases which are required in large quantities for industrial purposes are termed as industrial gases. The most important and largely used industrial gases are: (i) Hydrogen, (ii) Oxygen, (iii) Nitrogen, (iv) Carbon dioxide, (v) Acetylene, (vi) Ethylene, and (vii) Helium. Industrial gases have a large variety of essential functions: some are used as raw materials for the manufacture of other chemicals; some are essential medicaments like oxygen and helium. Some of the gases in the liquid and solid states are used for creating cold, largely by absorbing heat via evaporation, or by performing work, or by melting. The most modern application of industrial gases is in the cryogenic techniques.

HYDROGEN

From Hydrocarbons and Steam (Steam Reformer Process)

Process

The largest quantities of hydrogen are manufactured by catalytically reacting hydrocarbons and steam, to yield hydrogen and carbon oxides, followed by the water-gas shift reaction. The most commonly used raw material is natural gas, although other natural and refinery hydrocarbons, including naphtha cuts may be used. The following description is based on propane as a raw material.

Reaction

$$CH_3CH_2CH_3 + 3H_2O \rightarrow 3CO + 7H_2$$

$$3CO + 3H_2O \rightarrow 3CO_2 + 3H_2$$

$$CH_3CH_2CH_3 + 6H_2O \rightarrow 3CO_2 + 10H_2$$

Commercial propane, obtained from either natural-gasoline plants or oil refineries, contains small amounts of organic sulphur compounds which are removed before processing. The propane from storage passes in the form of vapours through a heater at a temperature of about 370°C. The hot gases then pass over a bauxite or metallic oxide catalyst which converts the sulphur compounds (mercaptans, organic sulphides, and carbonyl sulphide) to hydrogen sulphide. After cooling, the gases are scrubbed with aqueous sodium hydroxide and water to remove the soluble sulphides. When natural gas is used as the process feed, sulphur is removed by passing the gas through drums containing activated carbon (Fig. 7.1).

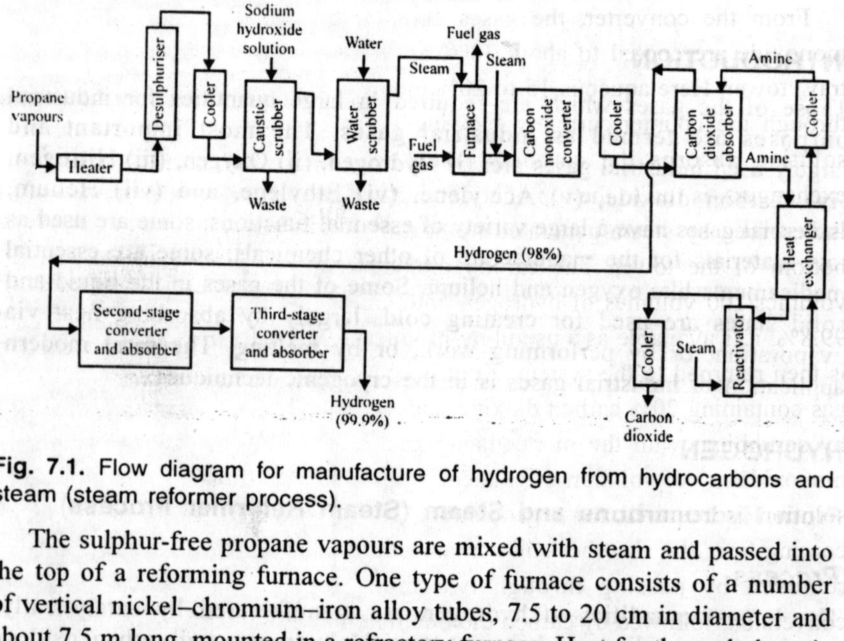

Fig. 7.1. Flow diagram for manufacture of hydrogen from hydrocarbons and steam (steam reformer process).

The sulphur-free propane vapours are mixed with steam and passed into the top of a reforming furnace. One type of furnace consists of a number of vertical nickel–chromium–iron alloy tubes, 7.5 to 20 cm in diameter and about 7.5 m long, mounted in a refractory furnace. Heat for the endothermic reaction is supplied by multiple horizontal burners located at various levels, with the flue gases passing upward counter-current to the process gas. At temperatures of 760 to 980°C, the propane gas passes down the tubes over a supported nickel catalyst at a space velocity of about 600 volumes/hour per volume of catalyst. The propane is converted to hydrogen, carbon monoxide, and carbon dioxide, with a trace of a methane remaining in the mixture.

Operating pressure in the reformer may be as high as 600 psi (4.2 MPa). Although the steam-hydrocarbon reaction is favoured by low pressure, there are compensating economic advantages in operating at high pressure. Accordingly, new plants usually are designed for high-pressure operation.

In any event, the reformed gases are cooled to about 370°C by mixing with steam, and then passed into the first-stage carbon monoxide converter containing an iron oxide catalyst promoted with chromium oxide. Here the exothermic conversion reaction (water-gas shift) takes place at a temperature of about 425°C and a space velocity of 100 (or greater) volumes of gas/volume of catalyst per hour. Both this catalyst and the nickel reforming catalyst are rugged and have a normal life of 1 year or more.

From the converter, the gases containing a small amount of carbon monoxide are cooled to about 38°C and passed into a packed (or bubble-tray) tower. Here aqueous 15 to 20% monoethanolamine is circulated down through the counter-currently blowing gas (Girbotol process). The amine solution absorbs the carbon dioxide and, after passing through heat exchangers, is run to the top of a reactivating tower. Here the carbon dioxide is desorbed by steam generated by heating the solution in a reboiler at the bottom of the tower. The carbon dioxide removed amounts to about 30 volumes/100 volumes of hydrogen and, since it is recovered at a purity of 99.8%, it is available as a useful by-product. The regenerated amine solution is then returned to the system. At atmospheric pressure and 38°C hydrogen gas containing 20% carbon dioxide may be purified to 0.1% carbon dioxide by scrubbing, with the monoethanolamine absorbing 15 to 30 m^3 carbon monoxide/m^3 of solution circulated. Approximately 0.12 kg steam/litre of solution is required for regeneration. The carbon dioxide-free hydrogen coming from the absorber still contains about 1% carbon monoxide. This is removed by passing through two more stages of carbon monoxide conversion, followed by carbon dioxide removal. From the last absorber, the purified hydrogen analysis better than 99.9% pure. The gas may be compressed to about 150 psi (1 MPa) and charged into storage tanks. Many new plants require only one stage of shift conversion. A highly active low-temperature catalyst is used.

Shift converter gas may also be purified by absorption of carbon dioxide in hot carbonate solutions. In one modification (Vetrocoke process), the potassium carbonate solution is activated with arsenic trioxide. Small amounts of carbon oxides may be removed by conversion to methane by passing the mixture over a nickel catalyst at 315 to 425°C.

From Hydrocarbons by Partial Oxidation

Process

Synthesis gas, a mixture of hydrogen and carbon monoxide, may be made by partially burning any hydrocarbon feed (natural gas through fuel oil) with oxygen. The carbon monoxide is then reacted with water vapour in a shift converter to form more hydrogen (Fig. 7.2).

Reaction

$$\underset{\text{Fuel oil}}{C_nH_{2n-2}} + mO_2 \rightarrow xCO_2 + n - xCO + n - 1 - yH_2 + yH_2O$$

$$CO + H_2O \rightarrow CO_2 + H_2$$

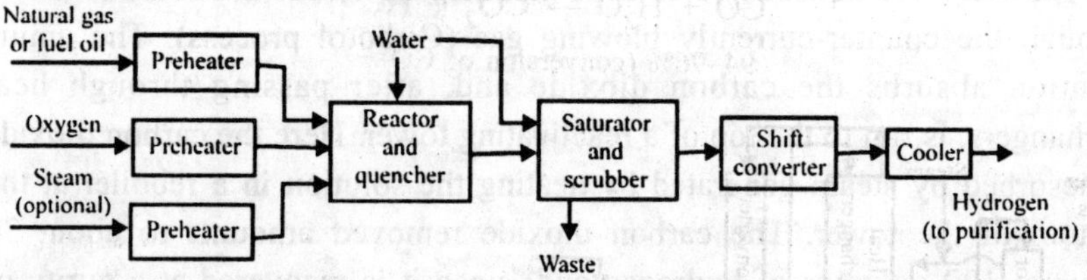

Fig. 7.2. Flow diagram for manufacture of industrial gases from hydrocarbons by partial oxidation.

Natural gas, for example, is preheated to 650°C and sent to a refractory-lined pressure vessel where it reacts with preheated oxygen. Hydrocarbon feed rate varies from 1000 to 3000 volumes/unit volume of reactor space. The oxygen/carbon ratio is held between 1.0 and 1.2 to hold carbon formation below 14 kg/million ft^3 (0.5 kg/1000 m^3) of gas. Customary operating pressure is 200 to 500 psi (1.4 to 3.5 MPa).

The reaction products are removed continuously from the reaction zone. In addition to the major products, carbon monoxide and hydrogen, the burner effluent contains 5 mole % water, less than 2 mole % carbon dioxide, less than 0.1 mole % oxygen and about 0.5% unreacted methane. The effluent gases are quenched rapidly, in less than 1 sec., from 1425°C to 530 or 815°C. In a somewhat similar process used largely in Europe, the partial combustion is moderated by the presence of nitrogen, steam, or carbon dioxide; the quench is carried out in a waste-heat boiler.

In either case the gases may be scrubbed (to remove carbon particles) and saturated, and then passed to a shift converter, where the carbon monoxide reacts with steam to produce carbon dioxide and hydrogen. The hydrogen may be purified by conventional means.

From Water Gas and Steam

Process

Hydrogen is produced along with carbon dioxide by catalytically reacting water gas (40% carbon monoxide–50% hydrogen) with steam at elevated temperatures. The carbon dioxide is removed by scrubbing the gas; relatively pure hydrogen remains. The process is particularly adaptable to low-purity hydrogen containing nitrogen and carbon monoxide for the synthesis of ammonia and methanol, respectively (Fig. 7.3).

Reaction

$$C\ (\text{amorph}) + H_2O \rightarrow CO + H_2\ (\text{water gas})$$

$$CO + H_2O \rightarrow CO_2 + H_2$$

94–96% (conversion of CO)

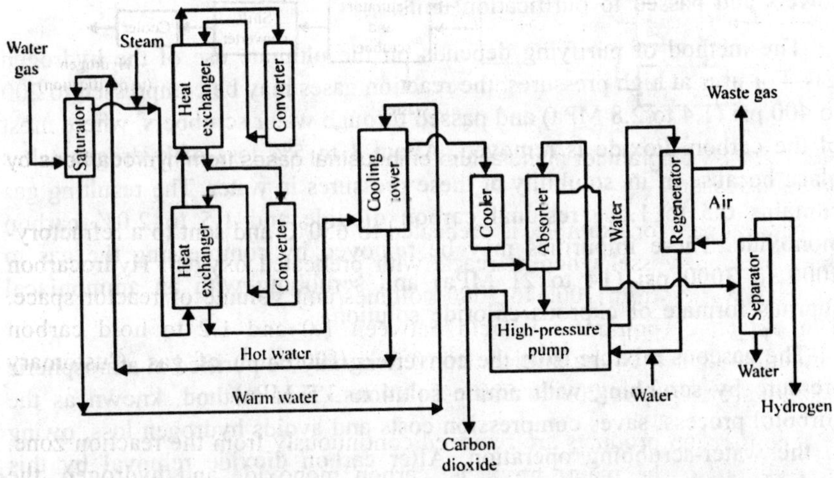

Fig. 7.3. Flow diagram for manufacture of industrial gases from hydrocarbons by partial oxidation.

Water gas, also called blue gas, is produced by the reaction of steam on incandescent coke or coal at a temperature of 1000°C or higher. Analysis of the gas approximates 40% carbon monoxide, 50% hydrogen, 5% carbon dioxide, and 5% nitrogen and methane. The gas from a holder is run into a saturator, where it contacts hot water and is heated to 75 to 85°C. The saturated water gas (1 volume) is then mixed with steam (3 volumes) and passed into a two-stage catalytic converter, where carbon monoxide is reacted with water vapour. The first stage operates at a temperature of 425 to 480°C, and the second at 370 to 400°C, with heat exchangers between the stages. Two stages are employed because of the exothermic character of the reaction and the decreased conversion at higher temperatures. By using this two-stage procedure on the so-called water-gas shift reaction, the major part of the conversion takes place with a relatively small amount of catalyst, whereas the balance is effected at a lower temperature conducive to high overall yield. The catalyst commonly used is iron oxide promoted with chromium oxide, and is required in amounts of 180 to 270 kg/1000 ft^3 (6425 to 9600 kg/1000 m^3) of water gas converted per hour. The life of the catalyst is long; it is ordinarily not necessary to remove sulphur compounds from the water gas, because the catalyst is sulphur-resistant and converts sulphur compounds into hydrogen sulphide which is removed with the carbon dioxide.

The exit gases, containing about 64% hydrogen, 31% carbon dioxide, 4% nitrogen and methane, and 1% carbon monoxide, are cooled in water towers and passed to purification units.

The method of purifying depends on the ultimate use of the hydrogen gas. For uses at high pressures, the reaction gases may be compressed to 200 to 400 psi (1.4 to 2.8 MPa) and passed through water scrubbers, where most of the carbon dioxide is removed. About 3 to 5% loss of hydrogen takes place because of its solubility at these pressures in water. The resulting gas contains 0.5 to 1.5% residual carbon dioxide and 1.5 to 2.0% carbon monoxide. These impurities may be removed by compressing the gas to 2000 to 3000 psi (14 to 21 MPa) and scrubbing with an ammoniacal cuprous formate or cuprous chloride solution.

The gaseous mixture from the converters may be purified at atmospheric pressure by scrubbing with amine solutions. This method, known as the Girbotol process, saves compression costs and avoids hydrogen loss, owing to the water-scrubbing operation. After carbon dioxide removal by this method, the gases analyse about 97.2 mole % hydrogen, 1.2% carbon monoxide, 1.2% nitrogen, 0.3% methane, and 0.1% carbon dioxide. Pure hydrogen may be obtained by starting with a lower-nitrogen-content water gas and catalytically removing the carbon monoxide.

The composition of the starting water gas is varied to suit the end use destined for the hydrogen. For ammonia synthesis, a semi-water gas is generated so that after conversion it contains the proper ratio of nitrogen to hydrogen. A low-nitrogen water gas is produced for methanol synthesis, which requires a ratio of 2 moles hydrogen to 1 mole carbon monoxide in the starting gases.

Carbon monoxide may also be liquefied by refrigeration and thus separated from the water gas produced by the first reaction shown, leaving fairly pure hydrogen. In one such process, the refrigeration is by liquid air after preliminary cooling by an ammonia refrigeration system; in another, the cooling is obtained by expansion of the hydrogen from a pressure of 20 atm (2.1 MPa) to a lower one by doing work against a piston. The liquid carbon monoxide is then easily separated from the still gaseous hydrogen.

The steam-water gas process is practically obsolete, but is described here for its historical interest, since it was the dominant process for hydrogen production at one time.

Steam-Iron Process

This process produces hydrogen by the reaction of steam at high temperature on reduced iron oxide to yield hydrogen, and then re-reducing the iron oxide with a reducing gas such as water gas or producer gas in a cyclic operation, this is process is obsolete.

From Water by Electrolysis

Direct current is passed between iron or nickel-plated iron electrodes in a solution consisting of 10 to 25% caustic soda or potash. Only distilled water is added to the electrolyte, since the solute is not consumed. Operating voltage is about 2 V; current density about 650 A/m^2. Electricity required is 140 or more kWh/1000 ft^3 of hydrogen (4950 kWh or 18 GJ/1000 m^3) plus one-half that volume of oxygen (if the oxygen is recovered). Where low-cost electricity is available this process is attractive, and it is one of the ways of producing heavy water.

Hydrogen is also obtained as a by-product in other electrochemical processes, although it is not always considered economical to recover it. Typical examples are the electrolysis of sodium chloride or potassium chloride solutions for the manufacture of chlorine, sodium or potassium hydroxide, sodium chlorate, and sodium perchlorate.

Thermal Decomposition of Hydrocarbons

Hydrogen is obtained as a by-product in the manufacture of carbon black by the thermal decomposition of natural gas. A brick checker-work chamber is heated to 1090 to 1370°C, and natural gas is passed through. After the carbon black is separated, the gases contain 70 to 85% hydrogen. By scrubbing to remove carbon oxides, hydrogen pure enough for ammonia synthesis is obtained. This process is one of the cheapest sources of low purity hydrogen.

Catalytic Reforming of Petroleum Stocks

Considerable hydrogen, 400 to 800 ft^3 (11 to 22 m^3)/barrel of feedstock, is produced by petroleum refiners in catalytic reforming operations. Hydrogen-rich gas from the reforming operation contains hydrogen sulphide and low-molecular-weight hydrocarbons. Purification involves only a caustic wash (to remove hydrogen sulphide) followed by refrigeration to condense most of the water and higher hydrocarbons. The resulting gas is usually further dried by passing it over a solid adsorbent (e.g., activated alumina) and may be washed with liquid nitrogen to remove more hydrocarbons. Although the product (95% hydrogen) is suitable for ammonia synthesis and was formerly so used, present refining processes have such large requirements of hydrogen (for hydro-desulphurisation and other uses) that all or practically all by-product hydrogen is used on-site and does not reach the market.

From Methanol and Steam

Hydrogen of higher than 98% purity may be manufactured by the catalytic reaction of methanol and steam at 260°C, according to the equation :

$$CH_3OH + H_2O \rightarrow CO_2 + 3H_2$$

The exothermic reaction at moderately low temperatures may be conducted at up to 300 psi (2.1 MPa), so that the hydrogen can be delivered at elevated pressures. For 1000 standard ft^3 (28 m^3) of hydrogen plus 18 kg of pure by-product carbon dioxide, approximately 20 litres of methanol and 15 litres of fuel are required. The ease and simplicity of fabricating the plant, handling raw materials, and purifying the hydrogen make this process particularly practical for portable plants. The present cost of methanol precludes the widespread or large-scale use of this process.

From Ammonia by Dissociation

For metal treating and hydrogen welding, a source of mixed gas (75% hydrogen–25% nitrogen) is the catalytic cracking of ammonia. In the process liquid ammonia is vapourised, preheated by hot gases from the reaction chamber, and then passed into an electrically heated reactor. In the presence of a supported nickel oxide catalyst, the ammonia may be cracked at pressures up to 20 psig (350 kPa) and temperatures ranging from 900 to 980°C. The product gases may be scrubbed with water to remove traces to unconverted ammonia, and then dehydrated to yield a dry gas.

Since each kilogram of liquid ammonia yield 2.8 m^3 of dissociated gas containing approximately 2.1 m^3 of hydrogen and 0.7 m^3 of nitrogen, it is claimed that a single 150-lb (68-kg) cylinder of liquid ammonia can replace 33 cylinders of hydrogen. Pure nitrogen may also be obtained by burning out the hydrogen at a cost comparable to or perhaps lower than cylinder nitrogen. Although it is not economically feasible to produce pure hydrogen at a cost comparable to or perhaps lower than cylinder nitrogen. Although it is not economically feasible to produce pure hydrogen by this process, the cracked gas is quite satisfactory for many uses, particularly in the steel industry in the annealing process.

High-purity hydrogen

Extremely pure hydrogen (99.9%) can be produced by diffusing less pure gas through palladium films. At about 315°C, molecular hydrogen dissociates to atoms on palladium; the atoms then diffuse through the metal and recombine on the other side. The liquefaction process, described in outline below, also yields extremely pure hydrogen.

Liquid hydrogen

The steam reformer process is usually the source of hydrogen in liquefied hydrogen plants. After the removal of impurities as described, further purification is obtained by refrigeration and adsorption; feed to the liquefier must be quite pure to prevent blocking of the lines and valves by condensable impurities.

The final liquefier section depends on the ortho-para shift of the hydrogen molecule. Hydrogen can exist in two forms, depending on the rotation of the two atoms in the molecule. In orthohydrogen, they spin in the same direction, in parahydrogen in opposite directions. The conversion temperature is also very low, –70°C, above which Joule-Thompson expansion raises rather than lowers the temperature.

Normal hydrogen is only 25% para at atmospheric temperature; at liquid hydrogen temperature it is 99.7% para, and the heat generated by the conversion is about 11 + % of the total heat removed during liquefaction. If normal liquid hydrogen were stored, as much as two-thirds of it would be vapourised as conversion took place. Therefore the liquefier cycle includes rapid conversion over a ferric hydroxide catalyst. This takes place in two steps, at liquid nitrogen and liquid hydrogen temperatures, in the presence of liquid refrigeration. Thus liquid hydrogen is over 95% para. Purity may exceed 99.99%.

Uses

Ammonia, Methanol, Refinery, and other chemicals. The manufacture of hydrogen is largely an integral part of other chemical manufacturing processes, the production of ammonia and methanol. In these operations it is manufactured and used as a mixture with other gases, nitrogen in the case of ammonia synthesis and carbon monoxide in the case of methanol synthesis. Nevertheless, the manufacturing processes are essentially identical, except that the gas mixtures must be purified to produce pure (95 to 100%) hydrogen.

The steam reforming process is by far the most important; next in importance is the hydrocarbon partial oxidation process. Natural gas, liquefied petroleum gases, and propane may all be used as feedstocks, according to availability and cost. An important advantage of the partial oxidation process lies in its versatility with respect to feedstocks, particularly its adaptability to fuel oil. Of the minor manufacturing processes, ammonia dissociation (cracking) is most widely used. Several companies furnish push-button plants for producing the nitrogen-hydrogen mixture that results; the nitrogen content is not detrimental in such applications as annealing of steel, powder metallurgy, and atomic-hydrogen welding.

Large quantities of hydrogen are produced as a by-product of catalytic reforming of petroleum stocks. Some of this is used in the refinery for hydrotreating, some for hydrocracking processes, and some for ammonia manufacture. Considerable amounts are burned for fuel or flared.

Other uses include hydrogenation of edible oils to produce shortenings and margarines, many of which are low in cholesterol. Hydrogen is used in large electric generators to reduce windage losses and heat. Liquefied hydrogen is used as a rocket fuel; it has become the commercial source of the ultrapure hydrogen required in some segments of the electronics industry.

CARBON DIOXIDE

From Steam and Natural Gas

Process

Most (more than 60%) of merchant carbon dioxide is a by-product of the production of ammonia, which in turn is produced by the reaction of nitrogen and hydrogen. Hydrogen for this reaction is produced in several ways, but primarily by the steam reforming process summarised in the above equation. Most newer producers of carbon dioxide make use of this by-product as their source of product.

Reaction

$$CH_4 + 2H_2O \rightarrow 4H_2 + CO_2$$

From flue or kiln gases

Burning of fuel : $C + O_2 \rightarrow CO_2$

Burning of limestone : $CaCO_3 + \text{heat} \rightarrow CaO + CO_2$

Process

All carbonaceous fuels, in burning, produce carbon dioxide along with oxides of other ingredients of the fuel, such as sulphur dioxide. When limestone is calcined to make lime, carbon dioxide is evolved and may be recovered. These sources, once important, are now relatively minor, at least so far as carbon dioxide reaching the market is concerned.

From Fermentation

Process

The fermentation of grain to produce ethyl alcohol also produces a stoichiometric quantity of carbon dioxide. Since the advent of purely chemical methods of producing ethanol, this source has become relatively unimportant. Similar fermentations to produce beer, wine, and liquor also produce carbon dioxide, but the carbon dioxide is either usually not recovered or is used by the breweries themselves.

Reaction

$$C_6H_{12}O_6 \xrightarrow{\text{Yeast}} 2C_2H_5OH + 2CO_2$$

From Sodium Carbonate and Phosphoric Acid

Process

In plants manufacturing sodium phosphates by the reaction of phosphoric acid with sodium carbonate, pure carbon dioxide is available as a by-product, and a small amount is thus recovered for sale.

Reaction

$$Na_2CO_3 + H_3PO_4 \rightarrow Na_2HPO_4 + CO_2 + H_2O$$

Recovery of Carbon Dioxide

Methods for recovering carbon dioxide in pure form depend primarily on whether it is in dilute and impure form, as from combustion or lime kilns, or in practically pure form, as in certain types of natural gas or as a by-product of hydrogen production, in which case it needs only to be liquefied or solidified. Recovery of carbon dioxide from flue gases will illustrate the former.

From Flue or Kiln Gases by Absorption

Process

Coke is burned under specially designed boilers in a Carborundum-lined furnace. Combustion is controlled by drafts (about 0.6 m^3/kg of coke), so that the flue gases contain 17 to 18% carbon dioxide. At the same time steam (125 to 150 psi) is generated in the boilers to furnish power for the pumps and compressors and heat for the lye boiler. The hot (232°C) flue gases, containing oxygen, carbon monoxide, nitrogen, dust, and some sulphur and organic compounds, in addition to 18% carbon dioxide, are passed through a heat exchanger and economiser. Here the temperature of the gases is reduced to about 120°C, and the excess heat is taken up by counter-currently flowing strong lye. The gases are scrubbed in limestone-packed towers with water from the coolers to remove sulphur dioxide and dust and to reduce the temperature to about 38°C. The dilute carbon dioxide gas is then fed into the bottom of coke-filled absorption towers, where it passes counter-current to an aqueous solution of sodium carbonate (sp gr. 1.12) called weak lye. After absorbing the carbon dioxide, the solution of bicarbonate, known as strong lye, is pumped through heat exchangers operating at 104 to 120°C, where the solution is heated before entry into the lye boiler. The absorbers, operating at about 30°C, remove all but about 9% of the carbon dioxide in flue gas, which is then released to the atmosphere (Fig. 7.4).

The preheated solution of sodium bicarbonate is heated with exhaust steam (from the compressors) in the lye boiler. Here, at a temperature of about 118°C, the bicarbonate is decomposed into sodium carbonate (weak lye), which is returned through the heat exchanger to the absorbers for reuse. The liberated gas, consisting of 99.8% carbon dioxide, escapes through the rectifying section of the lye boiler, where it warms the entering strong-lye solution. The gas passes through a water cooler, where the temperature is further lowered, and the moisture in the saturated gas is condensed and returned to the weak lye. The cooled gas is collected in a gas holder.

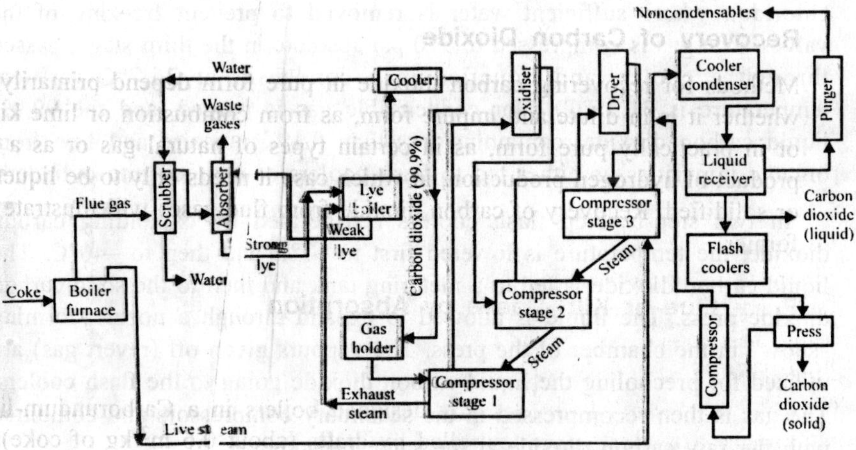

Fig. 7.4. Flow diagram for manufacture of carbon dioxide from flue or kiln gases by absorption.

In place of sodium carbonate, other absorbents, such as aqueous potassium carbonate, monoethanolamine (10 to 20%), and triethanolamine may be used in a similar manner.

Before or during liquefaction, the concentrated gas is purified by removal of organic impurities that would affect its taste and odour, and the gas is dried. Various methods or so-called cycles of liquefaction are employed, such as the precooling, binary, ternary, bleeder, or flush-cooling cycles. These operate on the raw gas or on a combination of raw and revert gases (return or back-blow gas from the solid carbon dioxide press), using two- or three-stage compression and ammonia refrigeration, carbon dioxide flash cooling, or a combination of both.

In the ternary cycle, the carbon dioxide from the gas holder is cooled to 4°C and raised to a pressure of 75 psi absolute in the first stage of the primary compressor. The compressors, usually two sets, are driven by live steam from the coke boiler, and the exhaust steam from them is used in the lye boiler.

From the first stage, the gas is passed through a purification system consisting of an oil separator and a scrubber, containing potassium permanganate or dichromate solution, which oxidises organic impurities. The gas from the scrubbers, raised in temperature to about 10°C, enters the second compression stage and is discharged at a pressure of about 350 psi absolute. The temperature is lowered by water coolers to 4°C, at which point some of the water and vapourised lubricating oil (such as glycerine) are condensed. The gas passes through a desiccant drying tower (calcium chloride), where sufficient water is removed to prevent freezing of the valves. The gas is compressed to 970 psi absolute in the third stage, passed through a cooler, and liquefied in a condenser at 26.7°C (critical temperature is 31.35°C). Non-condensable gas is purged, and the 99.9% liquid carbon dioxide, containing less than 0.1% moisture and free from organic impurities, is either fed to flash coolers or charged into cylinders.

In two step-by-step flash coolers refrigerated by expanding carbon dioxide, the temperature is lowered first to –7°C and then to –40°C. The liquid carbon dioxide is fed to a metering tank and then to the solid carbon dioxide press. The liquid is allowed to expand through a nozzle, forming "snow" in the chamber of the press. The vapours given off (revert gas) are utilised for precooling the liquid carbon dioxide going to the flash coolers. The gas is then recompressed in the secondary compressors and combined with the raw carbon dioxide at the same pressure.

The "snow" formed in the press is converted to a 100-kg, 50 × 50 × 25 cm block by lowering the upper ram under 2500 psi (17.5 MPa) until the block reaches a 25-cm thickness. Then the lower ram is dropped, and the block is ejected. After quartering by band saws, 25-kg, 25-cm cubical blocks are wrapped in kraft paper and either stored or shipped in insulated cars.

Uses

Refrigeration, carbonated beverages, chemical, inerting and pressure.

Properties

Colourless, odourless, non-combustible gas liquefiable to a heavy, volatile, colourless liquid, which may be compressed into a white snowlike solid. Mol. wt. 44.01, mp –56.6 (5.2 atm)°C, bp sublimes at –78.5°C, sp. gr., Gas 1.53 (air = 1), Liquid 1.101 (–37°C), Solid 1.56 (–79°C). Soluble in water (179.7 cm^3/100 g at 0°C, 90.1 cm^3/100 g at 20°C), acids, and alkalies. Threshold limit value (ppm by volume) 5000.

OXYGEN

From Air by Liquefaction (Modified Linde-Frankl Cycle)

Process

Commercial oxygen is produced in two grades: high purity, 99.5%, and low purity, 90 to 98%, although the latter is becoming less and less important since it is practically as cheap with modern technology to produce the better grades. The process consists of the liquefaction and subsequent fractionation of air. Several cycles have been proposed. These vary in methods of air compression, purification, and refrigeration, as well as in the design of heat-exchange, rectifying, evaporating, and condensing equipment. The pressures of the various cycles range from about 60 to 3000 psi (0.4 to 21 MPa).

Air consists of an invariable mixture of gases, in per cent by volume: nitrogen-78, oxygen-21, argon-0.94, hydrogen-0.01; and smaller amounts of neon, helium, krypton, and xenon. It also contains variable concentrations of carbon dioxide (0.03 to 0.07), water vapour (0.01 to 0.02), and hydrocarbon gases such as acetylene and methane, as well as local pollutants.

In the modified Linde-Frankl low-pressure cycle, reversing exchangers are used to lower the temperature of the incoming air which flows counter current to the outgoing gas. Critical cooling is obtained by the Joule-Thomson effect, that is, the cooling effect produced by the expansion of compressed air with no external work [1 kg of air at 3000 psi (21 MPa) and 10°C, when expanded to 75 psi (520 kPa), produces a cooling effect of 10 kcal (42 kJ)].

Air is filtered and compressed to 77 psi (530 kPa) in a two-stage, steam-driven centrifugal compressor, and then passed through a catalytic oxidation chamber. Here traces of hydrocarbons are converted to carbon dioxide and water. After cooling, the hydrocarbon-free air is passed through a water separator and then to automatic three-way reversing valves which select the proper air channels through the exchangers. The exchangers, which function both to cool and to purify the air, have a concentric, triple-tube construction in which the oxygen product follows the same path through the centre tube, while waste nitrogen and incoming air switch channels in the two outer annuli. This periodic switching (about every 3 min) permits the carbon dioxide and water vapour, which condense from the incoming high-pressure air, to be evaporated and purged out of the system by the low-pressure waste nitrogen (Fig. 7.5).

The high-pressure air entering the exchangers is saturated with water at about 32°C and 77 psi (530 kPa). It also contains approximately 0.03 mole % carbon dioxide. As the air is cooled, it deposits first water, then ice, and finally carbon dioxide snow. Leaving the top of the last exchangers through automatic check valves and at a temperature of –168°C, the cold, purified air passes through activated carbon filters which remove any remaining carbonaceous impurities. The air stream is then divided into three portions; part of the air, called the "unbalanced" stream, is returned to an extra jacket at the top of the exchangers, where it is warmed to about –84°C, passed through a heat exchanger against waste nitrogen, and then blended with a second portion of the air stream to give a mix temperature of –153°C. The air stream may be passed through silica-gel dryers and then fed into a turbo-expander.

The temperature unbalance system of diverting part of the air stream introduces a small amount of refrigeration into the reversing exchangers to compensate for heat losses. It also gives the necessary cold-end temperature difference (2°C) to permit constant low exchanger-outlet temperatures and insures complete purging of impurities, thereby giving excellent air purification. The silica-gel dryer is used, when starting up the cycle, to remove water contained in as yet unpurified air. The turbo expander supplies the refrigeration for the whole unit. In it the air at –153°C and 77 psi (530 kPa) is expanded through nozzles against a high-speed turbine connected to an electric generator. The air leaves the turbine at its dew point, –187°C, and 7.5 psi (52 kPa), and is fed into the upper section of a double-fractionating column.

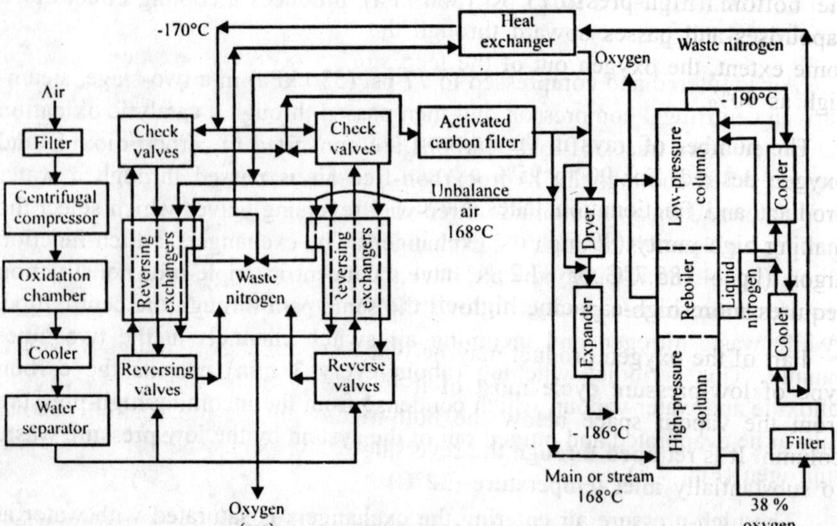

Fig. 7.5. Flow diagram for manufacture of oxygen from air by liquefaction (modified Linde-Frankl cycle).

The third and main portion of the air stream, at a temperature of –168°C, enters the bottom of the double column. In the high-pressure section, operating at 71.5 psi (500 kPa), the air is condensed, and preliminary fractionation takes place. The more volatile nitrogen (bp –195.8°C or

77.3K) rises to the top of the high-pressure section, where it is condensed in a condenser-reboiler by heat exchange with boiling oxygen in the low-pressure section of the column. The nitrogen, about 94% pure, serves as reflux to the trays in the high-pressure section and is then withdrawn from the column. After passing through a subcooler, where the temperature is reduced by heat exchange with the overhead waste-nitrogen stream from the low-pressure section, the liquid nitrogen at about 71 psi (500 kPa) is expanded through a valve near the top of the low-pressure column to 7 psi (49 kPa). Here it serves as reflux to the top tray of this section, giving up some of its oxygen and forming the waste-nitrogen exhaust.

At the base of the column, so-called rich liquid air, containing about 38% of the less volatile oxygen (bp – 183°C or 90.1 K), is removed and passed through a glass-cloth filter to remove any traces of precipitated carbon dioxide. It is then cooled in a subcooler against the waste nitrogen gas and fed through an expansion valve at an intermediate point in the low-pressure (7-psi, 49-kPa) section of the column. Here it is further fractionated to give oxygen of desired purity and waste nitrogen. Reboiler heat is furnished at the bottom of the top section by the condensation of nitrogen at the top of the bottom (high-pressure) section. The nitrogen in the top section vapourises and passes upward through the liquid reflux, which scrubs, to some extent, the oxygen out of the feed stream to give an oxygen yield as high as 90%.

The number of trays in the column depends on the purity and yield of oxygen desired. For 90 to 95% oxygen, most of the argon remains in the product, and fractionation takes place between oxygen and nitrogen. For making high-purity (99.5%) oxygen, fractionation must take place between argon (bp – 185.7°C or 87.2 K) and oxygen; this is more difficult and requires more high-capacity, high-efficiency trays.

Part of the oxygen product may be withdrawn as liquid, although in this type of low-pressure cycle most of it is withdrawn as vapour at –180°C from the vapour space below the bottom tray in the top section of the column. It is returned through the reversing exchangers, where it is warmed to substantially inlet temperature (32°C).

The waste nitrogen leaves the top of the column at –191°C and is used to subcool both the nitrogen reflux and the rich air feed. It is then returned through the heat exchanger, where it cools the unbalance air and passes to the main reversing exchangers. It enters these at a temperature of –170°C and a pressure at 7 psi (49 kPa) and is reheated to within 4°C of the entering air. After the impurities of carbon dioxide and water released by the incoming air are removed, the waste nitrogen is exhausted into the atmosphere.

The process part of the plant is enclosed by a steel section surrounding the various pieces of equipment. All the void spaces are filled with mineral wool, thus affording the required insulation. The units are of necessity particularly tight and leak-free and are usually constructed of non-ferrous metals such as copper, aluminium, and high-chrome-nickel alloys.

In some modern plants, the air is not purified before liquefaction. Rather, most of the impurities are frozen out during liquefaction and are rejected from the exchangers when the flow is reversed. Any hydrocarbons remaining in the liquid are removed by passing the liquid product from the high-pressure column through a bed of silica gel before it is fed to the low-pressure column.

Conventional Linde-Frankl Cycle

The other low-pressure process for making oxygen, the conventional Linde-Frankl cycle, differs essentially from the foregoing in its use of "cold accumulators" for purification and heat transfer. These are pressure vessels (operating at about 70 psi, 480 kPa) packed with aluminium spirals in which both waste-nitrogen and oxygen product cool the incoming air by a periodic reversal of the flow. Since the oxygen product purges the impurities, this process produces only low-purity oxygen. A second difference is in the use of a small amount (4 to 5%) of high-pressure (3000-psi, 21-MPa) supplementary air to make up refrigeration deficiencies. The supplementary air also aids in purging the accumulators, being used in place of the unbalance air stream employed in the previously described process.

Other Cycles

Other methods for producing oxygen from air operate at moderate to high pressure and, in general, follow the basic Linde or Claude cycles. Most plants rely on chemical methods for removing carbon dioxide, and some employ chemicals also for water removal. Alkalies such as sodium or potassium hydroxide may be used for both, or sodium hydroxide for carbon dioxide removal and silica gel for water. Water is often separated from the air stream, however, by freezing in an exchanger and removing with waste nitrogen or warm refrigerant gas.

In the Linde cycle with auxiliary refrigeration, the air is compressed as high as 3000 psi (21 MPa), scrubbed with a sodium hydroxide solution, and cooled in alternating heat exchangers against product oxygen and nitrogen.

At –15°C most of the water is frozen out, and the air is further cooled to –46°C in a second pair of alternating exchangers against refrigerated ammonia. The cold purified air is then fractionated in a double column to yield oxygen and nitrogen.

Claude Cycle

The basic Claude cycle operates at a lower pressure (about 400 psi, 2800 kPa) than the Linde cycle, but is similar in the method of air purification. The novel feature is passing up to 80% of the main air stream through an expansion machine. By this adiabatic expansion, the power consumption per unit of air is reduced by about 35%. After the air has expanded to do work, it is combined with the rest of the air, which has been allowed to expand for self-refrigeration, and passed into a fractionator.

One modification of the Claude cycle used in large plants consists of compressing filtered air, scrubbing with caustic, and drying with activated alumina or silica gel. While the air is cooling in exchangers against outgoing oxygen and nitrogen, part of it is taken off and put through a reciprocating expansion engine. The other part of the air, after completing the heat-exchange cooling, is expanded from 750 psi (5175 kPa) operating pressure to column pressure. It is combined with the air from the expander and fractionated in a double column to produce liquid oxygen, which is discharged at cylinder-filling pressure.

A further modification is the Heylandt cycle, which claims a still lower power consumption (30% less) by using high pressure (3000 psi, 21 MPa) and introducing about 60% of the air to the expander at room temperature (rather than at 88°C as in the Claude cycle). The remaining 40% is throttled.

By using any of these cycles or various modifications, it is possible to produce liquid or gaseous oxygen ranging in purity from about 90 to 99.5%. The liquid is withdrawn directly from the column, or, if gaseous oxygen is desired, it may be stored in a gas holder and then compressed prior to cylinder filling. The nitrogen, produced as a co-product, may be recovered or vented to the atmosphere. The rare gases, especially argon, may also be separated by additional fractionation. Plants separating these gases obtain an argon fraction (boiling between oxygen and nitrogen) and may process the nitrogen gas to obtain a neon-helium fraction (lower boiling than nitrogen). By special processing of liquid oxygen, a krypton-xenon fraction (higher boiling than oxygen) may be obtained.

Oxygen is also produced as a co-product with hydrogen in the electrolysis of water; only about 0.3% or less of the oxygen produced is obtained in this way.

Uses

Steel, welding, acetylene, missiles, medicine, chemical process industries, pulp and paper, water treatment etc.

Properties

Colourless, odourless, tasteless gas liquefiable at –190°C into a slightly bluish liquid. Mol. wt. 32.00, mp –218.8°C, bp –183°C, sp. gr., Gas 1.1053 (air = 1) Liquid 1.14 (–183°C), Solid 1.426 (–252.5°C).

Most plants manufacturing oxygen also recover nitrogen, used in liquid form as a refrigerant and as a gas for blanketing sensitive chemical reactions and similar uses. Many plants also recover the rare gases argon, krypton, and xenon. The purified gases are widely used for fluorescent and display lighting. Argon in particular has gained a mass market, chiefly in gas-shielded, arc-welding processes. Although nitrogen is often considered an inert gas, many metals react with it at elevated temperatures, including especially the so-called refractory metals such as titanium and zirconium. In these cases argon must be used in all stages of their production and finishing.

ACETYLENE

From Calcium Carbide

Process

Calcium carbide is formed when lime (essentially free of phosphates and magnesium carbonate) and coke (with low ash content), mixed in a ratio of 60:40, are heated to a temperature of 2000 to 2100°C in an electric furnace. The reaction is : $CaO + 3C \rightarrow CaC_2 + CO$. The carbon monoxide may be recovered and used as a fuel in lime kilns or as a raw material for chemical synthesis. The liquid calcium carbide is allowed to cool and solidify for 24 to 48 hours and is then crushed and screened for size. The grinding operations are carried out in an atmosphere of nitrogen to prevent explosion from acetylene generated by moisture in the air. The approximate requirements per metric ton of carbide are: lime (991 kg), coke (683 kg), electrode paste (17 to 20 kg), electricity (3200 kWh), and labour (5 man-h). The carbide so produced is about 80% calcium carbide; 1 kg of carbide yields about 0.29 m^3 of acetylene (Fig. 7.6).

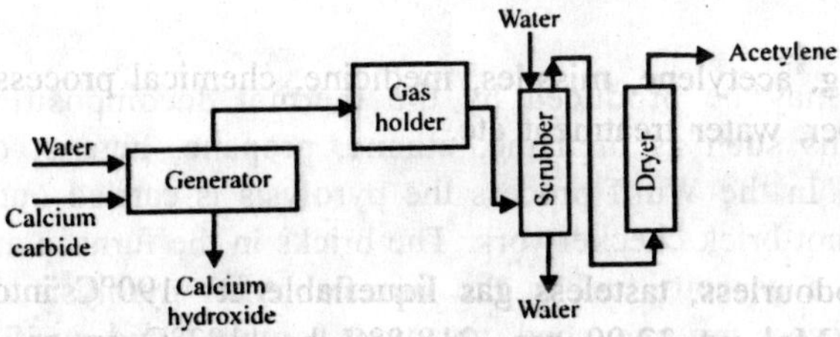

Fig. 7.6. Flow diagram formanufacture of acetylene from calcium carbide.

There are two standard methods for the production of acetylene from calcium carbide and water: the wet process and the dry process. The wet process consists of adding the carbide to a relatively large quantity of water, releasing acetylene gas while the calcium hydrate residue is discharged in the form of a lime slurry containing approximately 90% water.

Reaction

$$CaC_2 + 2H_2O \rightarrow CH{\equiv}CH + Ca(OH)_2$$

93–95% yield

In order to eliminate the waste of calcium hydrate, the dry process was originated in which a limited amount of water (1:1 by weight) is added to calcium carbide in a generator. The heat of reaction (1475 kcal/m^3 acetylene generated) is employed to vapourise the excess water over the chemical equivalent, leaving a substantially dry calcium hydrate suitable for reuse as a lime source. The temperature must be carefully controlled, since acetylene polymerises to form benzene at 600°C and violently decomposes at 780°C, and air mixtures may explode at 480°C. Generators of both types are usually designed to operate below 150°C and 15 psi (100 kPa).

The crude gas (containing traces of hydrogen sulphide, ammonia, and phosphine) from the generator is either scrubbed with water and caustic soda solution or led to a purifier where the impurities are absorbed by the use of iron oxide or active chlorine compounds. The dry gas is fed to cylinders or manufacturing units.

From Paraffin Hydrocarbons by Pyrolysis (Wulff Process)

Reaction (with Butane Feed):

$$C_4H_{10} \xrightarrow{\text{Steam}} CH \equiv CH + C_2H_4 + CO + H_2$$

$$C_2H_4 \longrightarrow CH \equiv CH + H_2$$

$$2CH_4 \longrightarrow CH \equiv CH + 3H_2$$

25-50% yield

Process

Acetylene may be produced by the thermal decomposition of certain hydrocarbons such as methane, ethane, propane, butane, ethylene, and natural gas. In the Wulff process the pyrolysis is carried out in a furnace containing hot brick checkerwork. The bricks in the furnace are first heated by burning a gaseous fuel in the furnace. After pyrolysis the acetylene-containing gases are quenched, and then scrubbed and purified to recover acetylene (Fig. 7.7).

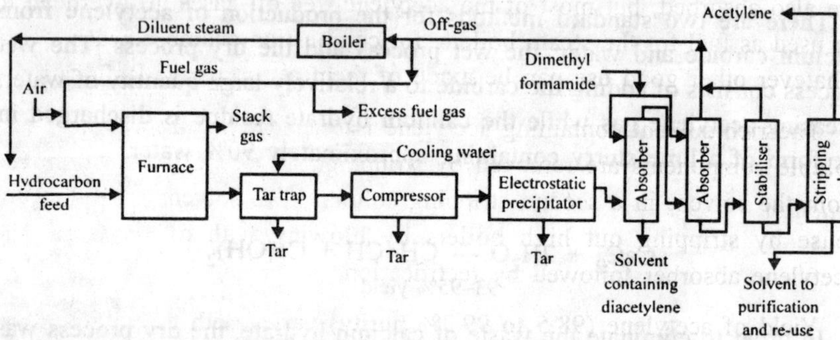

Fig. 7.7. Flow diagram for manufacture of acetylene from paraffin hydrocarbons by pyrolysis (Wulff process).

The Wulff regenerative furnace is essentially a rectangular steel box filled with refractory brick checkerwork. The furnace is operated on a 4-min cycle in which the checkerwork is heated for 1 min and the feed gas pyrolysed for 1 min. The same sequence of operation is then done in the reverse direction through the furnace. To facilitate reversal of the gas flow, fuel gas burners and hydrocarbon feed pipes are located on each side of the combustion chamber. To allow continuous flow of cracked gas to the purification train, two furnaces are usually operated on staggered cycles.

In usual operation off-gas from the acetylene recovery system is used as fuel for heating the combustion chamber. The volume of off-gas is much more than is required for fuel, so it may either be recycled to the furnace (recycle process) or used as a raw material for some other operation (once-through process).

Before the hydrocarbon feed is sent to the chamber, it is diluted with steam (up to 1:8 ratio). The feed is carried through the chamber at sub-atmospheric pressure by virtue of a large vacuum pump (see flow diagram). In this way residence time may be reduced to as little as 0.03 sec. Cracked gas leaves the chamber at about 370°C (maximum temperature in the furnace just after the heating cycle approaches 1315°C). It is then further quenched in a tar trap, where steam and various tars are removed. The gas is compressed to atmospheric pressure, passed through a knock-out drum and electrostatic precipitator, and sent to the recovery system.

Usually diacetylene and acetylene are separated by absorption in dimethylformamide. By proper adjustment of solvent ratio and temperature, diacetylene may be removed in the first scrubbing column. In the acetylene absorber small quantities of ethylene, carbon dioxide, and higher acetylenes are also absorbed, but most of the acetylene-free off-gases go overhead to be used as fuel for the steam boilers, for combustion chamber heating, and whatever other good use can be made of them at the particular location.

The rich solvent containing acetylene is sent to a stabiliser, where less soluble components are removed by stripping. Acetylene is then removed from the solvent in a second stripping column. The solvent is readied for reuse by stripping out high boilers by blowing with off-gas from the acetylene absorber followed by rectification.

Yield of acetylene (98.5 to 99.3% purity) varies with the hydrocarbon feed stock used. Average yields for the once-through process are 22.5 kg/100 kg methane, 38.8 kg/100 kg ethane, and 35.5 kg/100 propane. The off-gas is principally ethylene, carbon monoxide, hydrogen, and methane.

From Natural Gas by Partial Oxidation (Sachsse Process)

Process

Acetylene may be produced from a wide variety of hydrocarbon feed stocks (natural gas, LPG, naphtha, fuel oil, even crude oil) by high-temperature cracking. Heat for the cracking operation is developed by partial oxidation of the feed stock with oxygen. The heat evolved cracks the excess hydrocarbon to acetylene. After rapid quenching with water, the acetylene is separated from the gas stream by absorption in a suitable solvent. The process herein described is the basic Sachsse process using natural gas as raw material (Fig. 7.8).

Reaction

$$CH_4 + 2O_2 \rightarrow CO_2 + 2H_2O$$

$$2CH_4 \rightarrow C_2H_2 + 3H_2$$

Methane conversion—90–95%

30% yield

Methane (natural gas) is partially oxidised with oxygen (from a separate unit). The heat evolved cracks the excess methane to acetylene. After rapid quenching with water, the acetylene is separated from the gas stream by absorption in a suitable solvent.

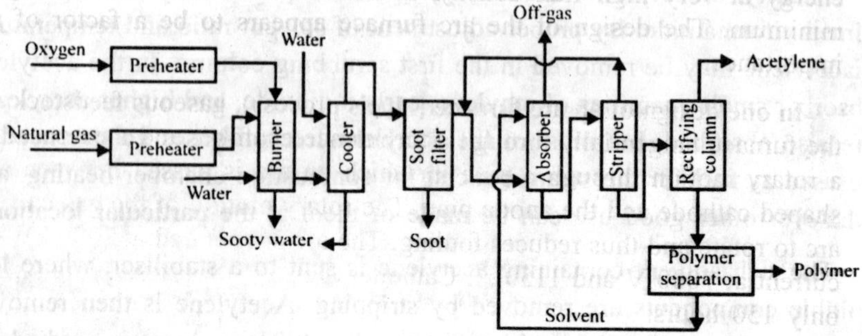

Fig. 7.8. Flow diagram for manufacture of natural gas by partial oxidation (Sachsse Process) from acetylene.

Natural gas (1 mole) and oxygen (0.65 moles 95% O_2) are preheated separately to 510°C and fed to a specially designed burner.

The converter is a vertical cylindrical unit built in three sections: mixing chamber, flame room, and quench chamber. After rapid and thorough mixing of the oxygen and methane in the mixing chamber, the gases are fed to the flame room through ports in a burner block designed to prevent back-travel or blowoff. The heat of combustion heats the gases to 1550°C to allow cracking of the excess methane to acetylene. Residence time is 0.001 to 0.01 sec. By rapid quenching of the resulting gases with water to 40°C, decomposition of the acetylene is prevented. The cooled effluent gases contain (on the dry basis) 8% acetylene, 54% hydrogen, 26% carbon monoxide, 5% methane, 4% carbon dioxide, and 3% nitrogen and higher acetylenes. These gases are run to a filter where carbon black (26 kg/metric ton of acetylene produced) is removed on a moving bed of coke. Clean gas leaving the filter is compressed to 150 psi (1 MPa) and fed to an absorption unit, where acetylene is separated and purified in a manner similar to that described for the Wulff process. Acetylene of 99.5% or higher purity is produced. A variation of the Sachsse process in wide use is the BASF process.

Several other partial oxidation processes have been developed and are available for commercial use. The chief differences among processes are in burner design, hydrocarbon feed, recovery solvent, and number of reaction stages. Two-step processes use stoichiometric fuel-oxygen mixtures in the first step and add the raw material to be cracked to acetylene in a second step.

From Hydrocarbons by the Arc Process

Several processes have been developed in which hydrocarbon cracking is carried out in an electric arc. In these processes the electric arc provides energy at very high flux density, so that reaction time can be kept at a minimum. The design of the arc furnace appears to be a factor of major importance.

In one design of an arc furnace (Huels process), gaseous feedstock enters the furnace tangentially through a turbulence chamber and then passes with a rotary motion through a pipe in which an arc is passed between a bell-shaped cathode and the anode pipe. The rotary motion of the gas causes the arc to rotate and thus reduces fouling. The arc is operated at 8000 kW direct current at 7000 V and 1150 A. Cathodes are said to last 800 hours, anodes only 150 hours.

Feed to the arc is fresh hydrocarbon and recycle gas. The reaction effluent gases are quenched and purified. In one pass through the furnace, a weight yield of 35% purified acetylene may be obtained, along with 17% weight yield of ethylene and 10% weight yield of carbon black, plus hydrogen and minor products.

Uses

Vinyl chloride monomer, Acrylates, Acetylenic chemicals, Vinyl acetate, Chlorinated solvents.

Properties

Colourless, flammable gas; odourless, when pure, but ordinarily has a garlick-like odour because of impurities :

Mol. wt. 26.04, fp –80.8°C, Sp gr 0.618 (32°C/4) bp –84°C.

One pound of acetylene is equivalent to 14.5 ft^3 (1 kg = 0.9 m^3). Soluble in acetone (2500 ml gas/100 g, 30,000 ml gas / 100 g at 12 atm), ethanol (600 ml/100 g), water (100 ml/100 g), and liquid ammonia at room temperature. Flash point gas, Ignition temperature 644°C (pure), Vapour density, (air = 1), 0.91, Explosive limits (% by volume in air), Lower 2.5, Upper 82, Threshold limit value (ppm) 1000. Acetylene is produced on the spot where needed by means of portable generators utilising calcium carbide.

RARE-EARTH ELEMENTS

The rare earths comprise a group of 17 elements in the periodic table and have similar properties in aqueous solutions. All are metals in the elemental state and all form salts that are strong electrolytes when dissolved in water. They ionise in this medium to give triply charged ions and, because of the high charge on these ions, react strongly with water dipoles to form a tight sheath of water molecules about them. Other ions in aqueous solutions only contact this sheath, giving rise to the similar properties of rare-earth cations in water.

The group consists of the following elements: scandium (^{21}Sc), yttrium (^{39}Y), lanthanum (^{57}La), all of which appear in the Group IIIB of the periodic table, and cerium (^{58}Ce), praseodymium (^{59}Pr), neodymium (^{60}Nd), promethium (^{61}Pm), samarium (^{62}Sm), europium (^{63}Eu), gadolinium (^{64}Gd), terbium (^{65}Tb), dysprosium (^{66}Dy), holmium (^{6}Ho), erbium (^{68}Er), thulium (^{69}Tm), ytterbium (^{70}Yb), and lutetium (^{71}Lu).

Occurrence

The rare earths are widely distributed in low concentrations throughout the earth's crust. They occur as mixtures in many massive rock formations, e.g., basalts, granites, gneisses, shales, and silicate rocks, in which they are present in amounts of 10–300 ppm. They also occur in ca 160 discrete minerals, most of which are rare, but in which the rare-earth content, expressed as R_2O_3 or Ln_2O_3 (REO), can be as high as 60% (REO is an abbreviation used in industry for rare-earth oxide content). Ln_2O_3 is used when Y_2O_3 is not present in the ore or mineral. Approximately ten of these occur in sufficient quantities that they may furnish some REO to commerce, but more than 95% of the RFO occurs in three minerals: monazite and bastnasite for the light rare earths, and xenotime for yttrium and the heavy rare earths.

Properties

Physical

The rare-earth metals alloy with most metals to form intermetallic compounds and occasionally solid solution. In rare-earth-rich alloys, other elements can change the properties of the pure metal by drastically lowering (or, more rarely, raising) the melting point by 200–300°C in some cases. Alloying with other elements can make the rare earth either pyrophoric or corrosion-resistant.

Chemical

The chlorides, bromides, nitrates, bromates, and perchlorate salts are soluble in water and, when their aqueous solutions evaporate, they precipitate as hydrated crystalline salts. The acetates, iodates, and iodides are somewhat less soluble. The sulphates are sparingly soluble and are unique in that they become less soluble with increasing temperature. The oxides, sulphides, fluorides, carbonates, oxalates, and phosphates are insoluble in water. The oxalate, which is important in the recovery of highly pure Ln, can be calcined directly to the oxide.

Anhydrous rare-earth salts usually cannot be prepared by evaporating the water of crystallisation.

The lanthanides can form hydrides of any composition up to LnH_4. Small amounts of hydrogen dissolve interstitially but, with increasing amounts of hydrogen, a second phase (LnH_2) appears. These alloys are metallic in their properties and loss hydrogen at relatively high temperatures.

The rare earths form many compounds with organic ligands; some of which are water-soluble. Chelating agents, which are organic molecules that engulf rare-earth ions and displace some of the water of hydration, form complexes with rare-earth ions that have a much wider range in their formation constants than do those of ordinary mineral-acid salts.

Production and Processing

Monazite and bastnasite are the minerals used commercially to supply most of the rare-earth chemicals. Monazite is a brown, dense phosphate mineral. There are extensive deposits of enriched monazite sands on beaches in many parts of the world, e.g., along the southwest coast of India and the east coast of Brazil; in the uplands of Australia, South Africa, the erstwhile USSR; and in the United States in Idaho, South Carolina, and Florida. Such deposits are dredged, pulverised if necessary, and further enriched by flotation methods. Sometimes they are also subjected to cross-belt magnetic separation since they are weakly magnetic.

Liquid-liquid extraction

As the industrial demand for individual rare earths increased, some of the rare-earth companies developed liquid-liquid extraction systems. An organic stream immiscible with water is flowed counter-current to the aqueous stream containing the rare-earth mixture. The organic phase may absorb neutral rare-earth molecules by complexing with them, e.g., as does tributyl phosphate, or it may contain a complexing molecule or anion added to it for the same purpose. Also, complexing ligands may be added to the aqueous phase so as to effect a synergistic equilibrium between the rare-earth ions and the complexant. If the various equilibria between the organic and aqueous phases have overall equilibrium constants sufficiently different for various rare earths, separation occurs.

Health and Safety

The rare earths are considered only slightly toxic according to the Hodge-Sterner classification system and can be handled safely with ordinary care.

Uses

The rare earths are used in gasoline-cracking catalysts; in carbon arcs; as additives to steel and cast irons; as polishing compounds; to make optical glass; as both colourants and decolourants for glass, depending upon the rare earth element used; and in magnets. In the electronic area, the most important industrial rare-earth compounds are garnet-based materials, e.g., yttrium iron garnet and gadolium gallium garnet, used in microwave devices, memory storage and oxygen sensors (Y_2O_3-stabilised ZrO_2). In nuclear reactors, the rare earths are used in the form of oxides as absorbers of neutrons. They are also used as additions to superalloys, as hydrogen-storing materials and as synthetic gems. Some of the important rare earth elements are discussed as under.

HELIUM

Helium (He), inert element of atomic number 2; first element in the noble gas group of the periodic table. Atomic weight 4.0026; valence 0 (not known to combine with any other element). Helium nuclei are alpha particles. Most important isotope is helium-3.

Properties

Colourless, non-combustible gas; odourless and tasteless; liquefies at 4.2°K to form helium I; at 2.2°K there is a transition (lambda) point at which helium II is formed. bp. –268.9°C (1 atm); fp. –272.2°C (25 atm) lowest of any substance. Density 0.272.2°C. Very slightly soluble in water; insoluble in alcohol. Rate of diffusion through solids is 3 times that of air. Helium is an asphyxiant gas.

Preparation

From natural gas, by liquefaction of all other components, followed by purification over activated charcoal.

Uses

To pressurise rocket fuels; welding; inert atmospheres (growing germanium and silicon crystals); inflation of weather and research balloons; heat-transfer medium; leak detection; chromatography; cryogenic research; magnetohydrodynamics; luminous signs; geological dating; aerodynamic research; lasers; diving and space vehicle breathing equipment. Possible future uses include coolant for nuclear fusion power plants and in superconducting electric systems.

NEON

Neon (Ne), inert element of atomic number 10, noble gas group of the periodic system. Atomic weight 20.179. It has three stable isotopes.

Properties

Colourless, odourless, tasteless gas; does not form compounds, but ionises in electric discharge tubes. Liquefies at –245.92°C; f.p. –248.6°C; sp. gr. 0.6964 (air = 1); sp. vol. 11.96 cu ft/lb (21°C, 1 atm). Noncombustible; nontoxic. Slightly soluble in water. An asphyxiant gas.

Preparation

By fractional distillation of liquid air, it constitutes 0.0012% of normal air.

Uses

(Gas) Luminescent electric tubes and photo-electric bulbs; electronic industry, high-voltage indicators; lasers. (Liquid) cryogenic research.

ARGON

Argon (Ar), non-metallic element of atomic number 18, in noble gas group of the Periodic System. Atomic weight 39.948; occurs in air to extent of 0.94% by volume.

Properties

Colourless, odourless, tasteless, monotomic gas; it is not known to combine chemically with any element, but forms a stable clathrate with beta-hydroquinone; fp. –189.3°C; bp. –185.8°C; sp. gr. 1.38 (air = 1); sp. vol. 9.7 cu ft/lb (21.1°C, 1 atm). Slightly soluble in water. Non-combustible; an asphyxiant gas.

Preparation

(1) By fractional distillation of liquid air.

(2) By the treatment of atmospheric nitrogen with certain metals such as magnesium and calcium to form nitrides.

(3) Recovery from natural gas oxidation bottoms steam in ammonia plant.

(4) Originally formed by radioactive decay of potassium 40.

Methods of Purification

(1) Highly purified argon is obtained by passing the gas through a bed of titanium at 850°C.

(2) Synthetic zeolite molecular sieves separate oxygen from argon to give high purity gas.

Uses

Inert gas shield in arc welding; furnace brazing; plasma jet torches (with hydrogen); electric and specialised light bulbs (neon, fluorescent, sodium vapour, etc.); titanium and zirconium refining; flushing molten metals (steel) to remove dissolved gases; in Geiger-counting tubes, lasers; inert gas or atmosphere in miscellaneous applications; decarbonisation of stainless steel (AOD process).

KRYPTON

Krypton (Kr), inert element of atomic number 36, noble gas group of the periodic table. Atomic weight 83.80; valence 2 (possibly others); has six stable isotopes and a number of artificially radioactive forms.

Properties

Colourless, odourless gas; bp. –152.9°C (1 atm); fp. –157.1°C; sp. gr. 2.818 (air = 1); sp. vol. 4.61 cu ft/lb (21°C, 1 atm); only slightly soluble in water. Known to combine with fluorine, at liquid nitrogen temperature, by means of electric discharges or ionising radiation, to make KrF_2 and KrF_4. These compounds decompose at room temperature. Non-toxic; non-combustible.

Preparation

By fractional distillation of liquid air. Air contains 0.000108% of krypton, by volume.

Uses

Incandescent bulbs and fluorescent light tubes; lasers; high-speed photography.

Note: Solid krypton exists at cryogenic temperatures, as a white, crystalline substance, m.p. 116°K.

XENON

Xenon (Xe), inert element of atomic number 54; noble gas group of the periodic table. Atomic weight 131.30. Valences 2, 4, 6, 8. There are 9 stable isotopes.

Properties

Colourless, odourless gas or liquid. Gas (at STP) has density of 5.8971 g/l (air = 1.29 g/l) and dielectric constant (25°C), 1 atm of 1.0012. Liquid has b.p. of –108.12°C at 1 mm; density (at b.p.) 2.987 g/cc. Liquefaction temp. –106.9°C. Non-combustible; non-toxic. Chemically unreactive, but not completely inert.

Preparation

Fractional distillation of liquid air.

Uses

Luminescent tubes; flash lamps in photography; fluorimetry; lasers; tracer studies; anesthesia.

RADON

Radon (Rn) gaseous radioactive element. Atomic number 86; noble gas group of Periodic table. Atomic weight 222.02; valences 2, 4 (6); 18 radioactive isotopes, all short-lived. The 222 isotope has a half-life of 3.8 days, emits alpha radiation.

Properties

Colourless gas; density 9.72 g/1 (0°C); soluble in water; can be condensed to colourless transparent liquid (b.p. –61.8°C) and to an opaque, glowing solid. The heaviest gas known.

Preparation

Radioactive decay of radium. Radon is obtained by bubbling air through a radium salt solution and collecting the gas plus air.

Uses

Medicine (cancer treatment); tracer in leak detection; flow rate measurement; radiography; chemical research.

SULPHUR DIOXIDE

Properties

Colourless gas or liquid with sharp pungent odour. Soluble in water, alcohol, and ether. Forms sulphurous acid H_2SO_3. Sp. gr. 1.4337, liquid at 0°C; fp. –76.1°C; bp. –10°C; vapour pressure 3.2 atmosphere at 20°C; refractive index (liquid) 1.410 (n 24/D). An outstanding oxidising and reducing agent. Non-combustible.

Preparation

(1) By roasting pyrites in special furnaces. The gas is readily liquefied by cooling with ice and salt, or at a pressure of 3 atm.

(2) By purifying and compressing sulphur dioxide gas from smelting operations.

(3) By burning sulphur.

Toxic by inhalation; strong irritant to eyes and mucous membranes, especially under pressure. Dangerous air contaminant and constituent of smog. Tolerance, 2 ppm in air. Not permitted in meats and other sources of vitamin B_1.

Uses

Chemicals (sulphuric acid, salt cake, sulphites, hydrosulphites of potassium and sodium, thiosulphates, alum from shale, recovery of volatile substances); sulphite paper pulp; ore and metal refining; soybean protein; intermediates; solvent extraction of lubricating oils; bleaching agent for oils and starch; sulphonation of oils; disinfecting and fumigating; food additive (inhibition of browing, of enzyme-catalysed reactions, bacterial growth); reducing agent; anti-oxidant.

CARBON MONOXIDE

Properties

Colourless gas or liquid; practically odourless. Burns with a violet flame. Slightly soluble in water; soluble in alcohol and benzene. Sp. gr. 0.96716 (air = 1.0); b.p. −190°C; solidification point −207°C; specific volume 13.8 cu ft/lb (21.1°C). Autoignition temp. (liquid) 1128°F (609°C). Classed as an inorganic compound.

Preparation

(1) Obtained almost pure by placing a mixture of oxygen and carbon dioxide in contact with in-adescent graphite, coke or anthracite.

(2) Action of steam on hot coke or coal (water gas) or on natural gas (synthesis gas). In the latter case, carbon dioxide is removed by absorption in amine solutions, and the hydrogen and carbon monoxide separated in a low-temperature unit.

(3) By-product in chemical reactions.

(4) Combustion of organic compounds with limited amount of oxygen as in automobile cylinders.

(5) Dehydration of formic acid.

Highly toxic by inhalation; highly flammable, dangerous fire and explosion risk; flammable limits in air, 12 to 75% by volume. Tolerance 50 ppm (industrial workrooms); U.S. standard, 35 ppm.

Note: CO has an affinity for blood haemoglobin over 200 times that of oxygen. A major air pollutant.

Uses

Organic synthesis (methanol, ethylene, isocyanates, aldehydes, acrylates; phosgene); fuels (gaseous); metallurgy (special steels, reducing oxides, nickel refining); zinc white pigments.

NITROUS OXIDE

Properties

Colourless, sweet-tasting gas; non-combustible; sp. gr. gas, 1.52 referred to air; liquid, 1.22 (−89°C); fp. −90.8°C; bp. −88.5°C; soluble in alcohol, ether, and concentrated sulphuric acid; slightly soluble in water. An asphyxiant gas.

Preparation

Thermal decomposition of ammonium nitrate; controlled reduction of nitrites or nitrates. Supports combustion; can form explosive mixtures with air. Narcotic in high concentrations.

Uses

Anaesthetic in density and surgery; propellant gas in food aerosols; leak detection.

Chapter 8

Industrial Carbon

INTRODUCTION

Carbon is an extremely versatile and useful element. Its applications grow each year. "The possibility of the expansion of carbon industries is very great, for they are intimately bound up with our highly complex life of the present day." This expansion has been realised. The usefulness of carbon is apparent only when we single out the many phases of our everyday living where this adaptable element is essential: in the black pigment of the ink of our books, magazines, and newspapers; in carbon paper; in pencils; as the black colour in many paints, automobile finishes, and shoe blacking; as a strengthening and toughening constituent of rubber tires, tubes, and other rubber goods; and as an essential element in the construction of a very large amount of electrical and nuclear equipment, from the household vacuum cleaner to the largest dynamos and nuclear reactors. The carbon arc is used in the production of visible and ultra-violet radiation in an increasingly large number of industrial processes dependent on photochemical reactions.

Non-fabricated industrial carbon is represented by lampblack, carbon black, activated carbon, graphite, and industrial diamonds. The first three are examples of amorphous carbon. Lampblack is soot formed by the incomplete burning of carbonaceous solids or liquids. It is gradually being replaced in some uses by carbon black, the most important of these amorphous forms and also the product of incomplete combustion. Activated carbon is amorphous carbon that has been treated with steam and heat until it has a very great affinity for adsorbing many materials. Graphite is a soft, crystalline modification of carbon that differs greatly in properties from amorphous carbon and from diamond. Industrial diamonds, both natural and synthetic, are used for drill points, special tools, glass cutters, wire-drawing dies, diamond saws, and many other applications, where this hardest of all substances is essential.

CARBON AND ARTIFICIAL GRAPHITE

Structure, Terminology, History and Applications

Carbon occurs widely in its elemental form as crystalline or amorphous solids. Coal, lignite, and gilsonite are examples of amorphous forms; graphite and diamond are crystalline forms. Carbon forms chemical bonds with other elements, but is also capable of forming compounds in which carbon atoms are bound to carbon atoms and, as such, is the chemical elements that is the basis of organic chemistry. Carbon and graphite can be manufactured in a wide variety of products with exceptional electrical, thermal, and physical properties. They are a unique family of materials in that properties of the products can be controlled by changes in the manufacturing processes and by raw materials selection.

Elemental carbon exists in nature as two crystalline allotropes: diamond and graphite. There are six electrons in the carbon atom with four electrons in the outer shell available for chemical bonding. Three of the four electrons form strong covalent bonds with adjacent in-plane carbon atoms. The fourth electron forms a less-strong bond of the van der Waals type between the planes. The weak forces between planes account for such properties of graphite as good electrical conductivity, lubricity, and the ability to form interstitial compounds.

Although the terms carbon and graphite are used interchangeably frequently, the two are not synonymous. The terms carbon, manufactured carbon, amorphous carbon, or baked carbon refer to products that results from the process of mixing solid particulates (such as petroleum coke, carbon blacks, or anthracite coal), called fillers, with binder materials of coal tar or petroleum pitch, forming these mixtures by moulding or extrusion, and baking the mixtures in furnaces at temperatures from 800–1400°C. The filler is material that makes up the body of the finished product. Green carbon refers to form carbonaceous material that has not been baked.

Graphite, also called synthetic or artificial graphite, electrographite, manufactured graphite, or graphitised carbon, refers to a carbon product that has been further heat-treated at a temperature exceeding 2400°C and preferably 2800–3000°C.

With the development of nuclear and aerospace technologies, several new forms of carbon and graphite are being produced commercially. Products that are deposited on a heated graphite substrate by vapour-phase decomposition of gaseous hydrocarbons, usually methane, at 1800–2300°C, are termed pyrolytic carbons. Chemical vapour deposited (CVD) carbon or pyrocarbon are other terms used to designate pyrolytic carbons. Pyrolytic graphite is a product resulting from high temperature annealing and has a crystallite interlayer spacing similar to that of ideal graphite.

The term polymeric carbon is a generic term for products that result when high polymers with some degree of cross-linking are heated in an inert atmosphere. Vitreous or glassy carbon prepared by carbonisation of cellulose and phenolic or polyfurfuryl resins is an example of a polymeric carbon.

There is no standard industry-wide system for designating the various grades of carbon and graphite that are commercially available. The data cited in the following sections are representative average values for commercially available materials.

The most significant event in the development of the carbon and graphite industry was the invention of an electric furnace capable of reaching approximately 3000°C, the temperature necessary for graphitisation. This permitted the development of improved products which were used in the production of alkalies, chlorine, aluminium, calcium, and silicon carbide, and for electric furnace production of steel and ferro-alloys. Further, a new application for graphite was found when it was used as a moderator in the first man-made self-sustaining nuclear chain reaction. This nuclear application and subsequent use in the developing aerospace industries opened new fields of research and new markets for carbon and graphite. Carbon and graphite fibres are an example of a new form and a new industry. A list of main applications is shown in Table 8.1.

Table 8.1. Applications for manufactured carbon and graphite.

Aerospace	*Metallurgical*
Nozzles	electric furnace electrodes for
nose cones	the production of iron and
motor cases	steel, ferro-alloys, and
leading edges	non-ferrous metals
Control vanes	furnace linings for blast
blast tubes	furnaces, ferro-alloy
exit cones	furnaces, and cupolas
Thermal insulation	aluminium pot liners and

(Contd...)

Aerospace	Metallurgical
Chemical	extrusion tables
Heat exchangers and centrifugal pumps	run-out troughs for molten iron from blast furnaces and cupolas
Electrolytic anodes for the production of chlorine, aluminium, and other electrochemical products	metal fluxing and inoculation
	tubes for aluminium and ferrous furnaces
Electric furnace electrodes for making elemental phosphorus	ingot moulds for steel, iron, copper, and brass
Activated carbon	extrusion dies for copper and aluminium
porous carbon and graphite reaction towers and accessories	
Electrical	*Nuclear*
Brushes for electrical motors and generators	moderators
	reflectors
anodes, grids, and baffles for mercury are power rectifiers	thermal columns
	shields
electronic tube anodes and parts	control rods
telephone equipment products	fuel elements
rheostat disks and plates	*Other*
welding and gouging carbons	motion picture projector carbons
electrodes in fuel cells and batteries	turbine and compressor packing and seal
Contacts for circuit breakers and relays	rings spectroscopic electrodes and powders for spectrographic analyses
Electric discharge machining	structural members in applications requiring high strength-to-weight ratios

Activated Carbon

Activated carbon, a micro-crystalline, non-graphitic form of carbon, has been processed to develop internal porosity. Commercial grades of activated carbon are designated as either gas-phase or liquid-phase adsorbents. Liquid-phase carbons are generally powdered or granular in form; gas-phase, vapour-adsorbent carbons are hard granules or hard, relatively dust-free pellets. Activated carbons are used widely to remove impurities from liquids and gases and to recover valuable substances from gas streams.

Surface area is the most important physical property of activated carbon. For specific applications, the surface area available for adsorption depends upon the molecular size of the adsorbate and the pore diameter of the activated carbon. Generally, liquid-phase carbons are characterised as having majority of pores of 3 nm dia and larger, whereas most of the pores of gas-phase adsorbents are 3 nm in diameter and smaller. The most important chemical properties of activated carbon are the ash content, ash composition, and pH of the carbon.

For economic reasons, lignite, coal, bones, wood, peat, and paper-mill waste (lignin) are most often used for the manufacture of liquid-phase or decolourising carbons, and coconut shells, coal, and petroleum residues are used for the manufacture of gas-adsorbent carbons.

Activation of the raw material is accomplished by two basic processes, depending upon the starting material and whether a low or high density, powdered or granular carbon is desired :

(1) Chemical activation depends upon the action of inorganic chemical compounds, either naturally present or added to the raw material, to degrade or dehydrate the organic molecules during carbonisation or calcination.

(2) Gas activation depends upon selective oxidation of the carbonaceous matter with air at low temperature, or steam, carbon dioxide, or flue gas at high temperature. The oxidation is preceded usually by a primary carbonisation of the raw material.

Antipollution laws have increased the sales of activated carbon for control of air and water pollution. Pollution-control regulations also affect the manufacture of activated carbons. Chemical activating agents have a large potential for the emission of corrosive acid gases during the activation process. If the emissions from chemical activation processes cannot be economically controlled, alternative selective oxidation methods will dominate. Selection of raw materials favouring low-sulphur materials will depend upon the local standards for permissible emission levels of sulphur oxides.

Regeneration of activated carbons is a major factor in the cost effectiveness of the use of carbon. Gas-phase activated carbons used for recovery of valuable process gases and solvents are responsible for the continued cost effectiveness of many industrial processes, particularly solvent-based fibre and tape manufacture, dry cleaning, and rotogravure printing.

To avoid contamination and possible loss of properties, gas-phase activated carbon should not be exposed to vapours or moisture. Although activated carbon is not an easily combustible material, stored carbon should be kept away from heat, electricity, and flames. Use of activated carbon in protective masks does not protect the wearer from an oxygen-depleted atmosphere. An asphyxiation hazard also exists inside activated-carbon vessels because of adsorption of oxygen from the air onto the carbon. Proper ventilation and breathing equipment should be used by persons entering activated-carbon vessels.

Approximately 60% of the activated carbon manufactured for liquid-phase applications is used in powdered form. The two principal uses of powdered activated carbon are removal from solution of colour, odour, taste, or other objectionable impurities such as those causing foaming or retarding crystallisation, and concentration or recovery from solution of a solute. Miscellaneous uses for liquid-phase carbons include laboratory uses, reaction catalysis, and aquarium water filters.

The largest single application for gas-phase carbon is in gasoline-vapour emission-control canisters on automobiles. It also is used in purification and separation of gases, in air conditioning systems to remove industrial odours and irritants as well as body, tobacco, and cooking odours from building air, in nuclear-reactor systems to adsorb radioactive gases in carrier or coolant gases and from emergency exhaust systems, for recovery of volatile organic compounds from process air streams, in cigarette filters to adsorb some of the harmful components of tobacco, and as the catalyst in the manufacturing of phosgene.

Baked and Graphitised Carbon

The raw materials used in the production of manufactured carbon and graphite largely determine the ultimate properties and practical applications of the finished products. This dependence can be attributed to the nature of carbonisation and graphitisation processes. Throughout the entire process of the thermal conversion of organic materials to carbon and graphite, the natural chemical driving forces cause the growth of larger and larger fused-ring aromatic systems, and ultimately result in the formation of the stable hexagonal carbon network of graphite. Differences in the final materials depend upon the ease and extent of completion of these overall chemical and physical ordering processes.

In order to produce useful carbon and graphite bodies, filler and binder materials are mixed, formed by moulding or extrusion, and finally baked, or baked and graphitised to yield the desired shaped-carbon or graphite bodies. More than 30 different raw materials are used in the manufacture of carbon and graphite products. The primary materials, in terms of tonnage consumption, are the petroleum coke or anthracite coal fillers and the coal-tar or petroleum-pitch binders and impregnants. Other materials, called additives, are often included to improve processing conditions or to modify certain properties in the finished products.

Coke and pitch purchased by carbon companies from suppliers must meet rigid specifications. Some materials, such as pitch and natural graphite, may be used as received from suppliers; other materials, such as raw coke and anthracite, require calcining, a thermal treatment to temperatures above 1200°C. Calcining consists of heating raw filler to remove volatiles and to shrink the filler to produce a strong, dense particle.

Processing of Baked and Graphitised Carbon

In the first step of artificial-graphite production, the run-of-kiln coke is crushed, sized and milled to prepare it for the subsequent processing steps. The degree to which the coke is broken down depends on the grade of graphite to be made. Roll crushers commonly are used to reduce the incoming coke to particles which are classified in a screening operation.

Generally, the guiding principle in designing carbon mixes is the selection of the particle sizes, the flour content, and their relative proportions in such a way that the intergranular void space is minimised. If this condition is met, the volume remaining for binder pitch and the volatile matter generated in baking are also minimised. The volatile evolution is often responsible for structural and property deterioration in the graphite product. In practice, most carbon mixes are developed empirically with the aim of minimising binder demand and making use of all the coke passed through the first step of the system. Typically, a coarse-grained mix may contain a large particle (e.g., 6 nm dia), a small particle half this size or smaller, and flour. In this formulation, approximately 25 kg of binder pitch would be used for each 100 kg of coke.

The manufacturing process begins with the mixing operation, the purpose of which is to blend the coke filler materials and to melt and distribute the pitch binder over the surfaces of the filler grains and structural integrity of the graphite. Thus, the more uniform the binder distribution is throughout the filler components, the greater the likelihood for a structurally sound product.

Following the mixing operation, the hot mix must be cooled to a temperature slightly above the softening point of the binder pitch at which point it is ready to be charged into an extrusion press or mould. One purpose of the forming operation is to compress the mix into a dense mass so that the pitch-coated filler particles and flour are in intimate contact. Another purpose of the step is to produce a shape and size as near that of the finished product as possible.

The next stage is the baking operation during which the product is fired to 800–1000°C. One function of this step is to convert the thermoplastic pitch binder to solid coke. Another function of baking is to remove most of the shrinkage in the product associated with pyrolysis of the pitch binder at a slow heating rate. This procedure avoids cracking during subsequent graphitisation where very fast firing rates are used. The conversion of pitch to coke is accompanied by marked physical and chemical changes in the binder phase, which if conducted too rapidly, can lead to serious quality deficiencies in the finished product. For this reason, baking is regarded as the most critical operation in the production of carbon and graphite.

In some applications, the baked product is taken directly to the graphitising facility for heat treatment to 3000°C. However, for many high performance applications of graphite, the properties of stock processed in this way are inadequate. The method used to improve those properties is impregnation with coal-tar or petroleum pitches. Upon carbonisation the impregnant deposits additional coke in the open pores of the baked stock, thereby improving properties of the graphite product. Many nuclear and aerospace graphites are multiply pitch-treated to achieve the greatest possible assurance of high performance.

Graphitisation, an electrical heat treatment of the product to ca 3000°C, causes the carbon atoms in the petroleum coke filler and pitch coke binder to orient into the graphite lattice configuration. This ordering process produces graphite with the intermetallic properties that make it useful in many applications.

In the temperature range of 1500–2000°C, most petroleum cokes undergo an irreversible volume increase known as puffing. This effect has been associated with thermal removal of sulphur from coke and increases with increasing sulphur content. Because of the recent emphasis on the use of low sulphur fuels, many of the sweet crudes that have been used as coker feeds are now being processed as fuels. Desulphurisation of the sour crudes available for coking is possible but expensive. The result is an upward trend in the sulphur content of many petroleum cokes, leading to greater criticality in heating rate in the puffing-temperature range during graphitisation.

Properties of Manufactured Graphite

The graphite crystal, the fundamental building block for manufactured graphite, is one of the most anisotropic bodies known. Properties of graphite single crystals illustrating this anisotropy are shown in Table 8.2. Anisotropy is the direct result of the layered structure with extremely strong carbon-carbon bonds in the basal plane and weak bonds between planes. The anisotropy of the single crystal is carried over in the properties of commercial graphite, although not nearly to the same degree. By the selection of raw materials and processing conditions, graphites can be manufactured with a wide range of properties and degree of anisotropy.

Table 8.2. Room-temperature properties of graphite crystals.

Property	*Value in basal plane*	*Value across basal plane*
Resistivity, Ω.m	40×10^{-4}	ca 6000×10^{-4}
Elastic modulus[a], TPa	0.965	0.034
Tensile strength (est)[a], TPa	0.096	0.034
Thermal conductivity, W/(m.K)	ca 2000	10
Thermal expansion, °C^{-1}	-0.5×10^{-6}	27×10^{-6}

[a] To convert TPa to psi, multiply by 1.45×10^{8}.

Manufactured graphite is a composite of coke aggregate (filler particles), binder carbon, and pores. Most graphites have a porosity of 20–30%, although special graphites can be made that have porosity well outside this range. Manufactured graphite is a highly refractory material that has been thermally stabilised to as high as 3000°C. At atmospheric pressure, graphite has no melting point but sublimes at 3850°C, the triple point being approximately 3850°C and 12.2 MPa (120 atm). The strength of graphite increases with temperature to 2200–2500°C; above 2200°C graphite becomes plastic and exhibits viscoelastic creep under load. Graphite is a valuable structural material because it has high resistance to thermal shock, does not melt, and possesses structural strength at temperatures well above the melting points of most metals and alloys. For many applications of graphite, one or more of the following characteristics is important: density, elastic modulus, mechanical strength, electrical and thermal conductivity, and thermal expansion.

Other Forms of Carbon and Graphite

Carbon and graphite foams have been produced from resinous foams of phenolic or urethane base by careful pyrolysis to preserve the foamed cell structure in the carbonised state. The foams are attractive as high temperature insulating packaging material in the aerospace field and as insulation for high temperature furnaces. Variations of the resinous-based foams include the syntactic foams where cellular polymers or hollow carbon spheres comprise the major volume of the material bonded and carbonised in a resin matrix.

The carbon commercially produced by the chemical vapour deposition (CVD) process is referred to as pyrolytic graphite. The material is not true graphite in the crystallographic sense, and wide variations in properties occur as a result of deposition methods and conditions. Commercial applications for pyrolytic graphite include rocket nozzle parts, nose cones, laboratory ware, and pipe liners for smoking tobacco. Pyrolytic graphite coated on surfaces or infiltrated into porous materials also is used in other applications such as nuclear-fuel particles and prosthetic devices.

A special form of pyrolytic graphite is produced by annealing under pressure at temperatures above 3000°C. This pressure-annealed pyrolytic graphite exhibits the theoretical density of single-crystal graphite; and although the material is polycrystalline, the properties of the material are close to single-crystal properties. The highly reflective, flat faces of pressure-annealed pyrolytic graphite have made the material valuable as an x-ray monochromator.

When carbon is produced from certain non-graphitisable carbonaceous materials, the material resembles a black glass in appearance and brittleness; hence the terms glassy carbon or vitreous carbon. Non-graphitising carbons are obtained from polymers that have some degree of cross-linking, and it is believed that the presence of these cross-linkages inhibits the formation of crystallites during subsequent heat treatments. By pyrolysing polymers such as cellulosics, phenol-formaldehyde resins, and poly(furfuryl alcohol) under closely controlled conditions, glassy carbon is produced. These carbons are composed of random crystallites, of the order of 5.0 nm across, and are not significantly altered by ordinary graphitisation heat treatment to 2700°C.

The properties of glassy carbon are quite different from those of conventional carbons and graphites. The density is low (1.4–1.5 g/cm^3), but the porosity and permeability also are quite low. The extreme chemical inertness of glassy carbon, along with its impermeability, makes it a useful material for chemical laboratory glassware, crucibles, and other vessels. It has found use as a susceptor for epitaxial growth of silicon crystals and as crucibles for growth of single crystals.

Carbon and graphite have been produced for many years in roughly spherical shape in small sizes (1–2 mm dia). Globular carbon of this type was used in telephone receivers in the past. Carbon of nearly spherical shape can be produced by mechanically tumbling or rolling a mixture of finely divided carbon or graphite with a resinous binder, screening to select the desired size balls, and carefully carbonising the selected material. Usage of these nearly spherical carbon balls has been limited to specialised applications where flowability was at a premium or uniform contact area was important. Shim particles for use in nuclear reactor fuel applications are examples.

A new type of carbon sphere, usually hollow, has been made from pitch. The spheres are of uniform, very nearly spherical shape and are graphitisable. The particle density is usually 0.20–0.25 g/cm^3, and consequently, attention has been given to application in syntactic carbon foam for high temperature insulation and for lightweight composite structures. The spheres can be activated to produce an adsorptive material of high flow-through capacity.

CARBON BLACK

Carbon black is an important member of the family of industrial carbons. Its various uses depend on chemical composition, pigment properties, state of subdivision, adsorption activity, and other colloidal properties. The basic proccss for manufacturing carbon black has been known since antiquity. The combustion of fuels with insufficient air produces a black smoke containing extremely small carbon-black particles which, when separated from the combustion gases, comprise a fluffy powder of intense blackness (Fig. 8.1). The term carbon black refers to a wide range of such products made by partial combustion or thermal decomposition of hydrocarbons in the vapour phase, in contrast to cokes and chars that are formed by the pyrolysis of solids. Whether partial combustion or thermal decomposition methods are used, the basic reaction is represented by:

$$C_xH_y \xrightarrow{\Delta} xC + \frac{y}{2} H_2$$

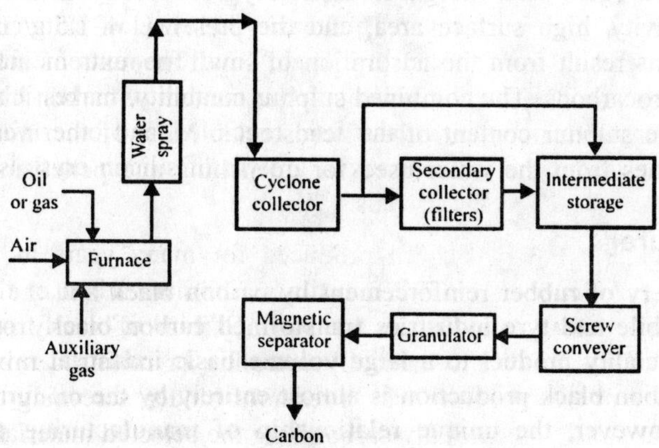

Fig. 8.1. Flow diagram for manufacture of carbon black from oil or natural gas.

Carbon exists in two crystalline forms and numerous so-called amorphous, less-ordered forms. The crystalline forms are diamond and graphite, and the less-ordered forms are mainly cokes and chars. All forms of industrial carbons other than diamond and graphite, including carbon black, can be classified as amorphous carbons characterised by degenerate or imperfect graphitic structures.

Carbon blacks differ in particle size or surface area, average aggregate mass, particle and aggregate mass distributions, morphology or structure, and chemical composition. The form of these products, loose or pelleted, is another feature of some special grades. The ultimate colloidal units of carbon black, or the smallest dispersible entities in the elastomer, plastic, and fluid systems are called aggregates, fused assemblies of particles. The particle size is related to the surface area of the aggregates. The oil-furnace process produces carbon blacks with particle diameters of 10–250 nm and the thermal black process, 120–500 nm.

A primary aggregate is further characterised by its size, the volume of carbon comprising the aggregate, and its morphology. Aggregates have a variety of structural forms. Some are clustered like a bunch of grapes, whereas others are more open, branched, and filamentous. The three most important properties used to identify and classify carbon blacks are surface area, structure, and tinting strength.

Most commercial rubber-grade carbon blacks contain over 97% elemental carbon. A few special pigment grades have carbon contents below 90%. In addition to chemically combined surface oxygen, carbon blacks contain varying amounts of moisture, solvent-extractable hydrocarbons, sulphur, hydrogen, and inorganic salts. Hygroscopicity is increased by surface activity, high surface area, and the presence of salts. Extractable hydrocarbons result from the adsorption of small amounts of incompletely burned hydrocarbons. The combined sulphur content of carbon black has its origin in the sulphur content of the feedstocks. Most of the inorganic salt content comes from the water used for quenching and pelletising.

Manufacture

The discovery of rubber reinforcement by carbon black and the growth of the automobile and tyre industries transformed carbon black from a small volume speciality product to a large volume basic industrial raw material. Today's carbon black production is almost entirely by the oil-furnace black process. However, the unique relationship of manufacturing process to special performance features of products has prevented the total abandonment of the older processes. The thermal, lamp black, channel black, and acetylene black processes account for less than 4% of the world's production. The easy sea and land transportation of feedstocks for the oil furnace process has made possible worldwide manufacturing facilities convenient to major consumers. The remarkable success of the oil-furnace process was due to its improved yield performance in the range of 50–70%, higher capacities, and its ability to produce a wide range of products.

The feedstocks used by the carbon-black industry are viscous, residual aromatic hydrocarbons consisting of branched polynuclear aromatic types mixed with smaller quantities of paraffins and unsaturates. Preferred feedstocks are high in aromaticity, free of coke or other gritty materials, and contain low levels of asphaltenes, sulphur, and alkali metals. As a result of the introduction of more efficient petroleum cracking catalysts, decant oil from gasoline production has now become the most important feedstock, displacing residual thermal tars and catalytic cycle-stock extracts.

The reactor feeds are preheated feedstock, preheated air, and gas. Carbon-black properties, yield, and production rate can be adjusted by controlling reactor feed variables. The carbon-containing combustion products are quenched with water sprays and pass through heat exchangers that preheat the primary combustion air. The production stream is again cooled by a secondary water quench in vertical towers. Carbon black in light fluffy form is separated in bag filters and conveyed to micropulverisers discharging into a surge tank. From the surge tank, the carbon black is fed to dry drums for dry pelletisation or pin-type wet pelletisers, followed by dryers, to produce pelleted products. Some of the special carbon-black grades used for pigmentation are lightly densified and shipped as "loose" or "fluffy" products.

Most of the carbon black used by the rubber industry is wet pelletised. Dry pelletisation is preferred for some rubber applications and for pigmentation. In the dry process the slightly compacted loose black is conveyed to large rotating steel drums whose rolling motion causes a steady increase in bulk density and the formation of pellets. A fraction of the pelletised product is recirculated and added to the loose black at the drum entry to increase the density of the bed and provide seed material to increase the formation rate.

At least two different types of reactors are required to make all the furnace grades for the rubber industry: one for the reinforcing and another for the less reinforcing grades. Additional types are used to produce special pigment grades. During the last decade, reinforcing furnace-black reactors have undergone substantial design changes to give higher gas velocities and turbulence, more rapid mixing of reactant gases and feedstock, and increased capacity. Some of the latest reactor designs have refractory-lined metal construction, heat-recovery chambers for preheating combustion air, increased thermal efficiency, more efficient feedstock atomisation, and improved quenching. The industry has also adopted improved reactor and burner designs that provided more rapid and uniform feedstock atomisation, more rapid mixing, and shorter residence times. Higher capacities and increases of 6–8% in yields for products rated at equivalent roadwear levels have been reported. On account of the demands of the automobile industry for lower rolling resistance tyres have resulted in the development of new carbon-black grades for tyre treads. These grades are used to make lower hysteresis tread compounds which reduce tyre rolling resistance without sacrificing tread-wear performance. The new tread grades have broader aggregate size distributions and lower tinting strengths and give evidence of higher surface activity.

Carbon black manufacturing processes

The high temperature decomposition of hydrocarbons in the absence of air or flames is the basis for the manufacture of thermal blacks and acetylene black. Thermal black is made by a strongly endothermic reaction requiring a large heat-energy input, whereas the acetylene-black decomposition reaction is strongly exothermic. Thermal-black production has been drastically curtailed in recent years, and thermal black now has only limited use as a functional filler for specially engineered rubber and plastic products.

Acetylene black is made by a continuous decomposition process at 800–1000°C in water-cooled metal retorts lined with refractory. The process is started by burning acetylene and air to heat the retort to reaction temperature, followed by shutting off the air supply to allow the acetylene to decompose to carbon and hydrogen in the absence of air. A major use for acetylene black is in dry-cell batteries because it contributes low electrical resistance and high capacity. In rubber it gives electrically conductive properties to heater pads, heater tapes, antistatic belt drives, conveyor belts, and shoe soles. It also is used in electrically conductive plastics. Many of these electrical applications are now also served by newly developed electrically conductive grades of furnace blacks. Some applications of acetylene black in rubber depend on its contribution to improved thermal conductivity, such as rubber curing bags for tyre manufacture.

Early lampblack processes used large open shallow pans from 0.5 to 2m in diameter and 16m deep for burning various oils in an enclosure with a restricted air supply. The smoke from the burning pans was allowed to pass at low velocities through settling chambers from which it was cleared by motor-driven ploughs. Today lampblack substitutes are made by the furnace process. Traditional lampblack manufacture is still practised, but the quantities produced are small.

The following trends in carbon-black manufacture can be expected; increasing use of by-product pyrolysis tars from ethylene manufacture for feedstocks; replacement of natural gas by liquid hydrocarbons for the primary heat source; increased, sensible-heat recovery from the combustion stream; increased use of tail gas for fuel in the primary combustion zone; use of higher sulphur content feedstocks; and increasing use of automatic computer control. Carbon-black consumption by the rubber industry has grown rapidly from the time of the discovery of its outstanding reinforcing properties.

Carbon is a relatively stable, unreactive element, insoluble in organic or other solvents. Carbon in the form of char has been used in the pharmaceutical industry for many years. Recently, vitreous and pyrolytic carbons have been found to be biocompatible and useful for artificial heart valves and other medical applications. There is no evidence of carbon-black toxicity in humans despite the fact that it contains trace amounts of some polynuclear aromatic compounds known to be carcinogenic.

Owing to the low thermal conductivity of bulk carbon black, storage fires may go undetected for some time. Fires may be controlled by purging with carbon dioxide. When this is done, respirators must be used to avoid breathing carbon monoxide. Even in the absence of a fire, an air line or adequate respirator should be used when entering any confined carbon-black storage facilities to avoid breathing possible toxic gases or low oxygen atmosphere.

The combustion of fuels can cause serious problems of atmospheric pollution from the emission of particulate material, sulphur compounds, nitric oxides, hydrocarbons, and other gases. Because of the intense blackness of carbon black, it cannot be allowed to escape to the atmosphere even in minute quantities. Huge bag filters for complete recovery are put into operation.

Uses

The general effect of different carbon blacks on rubber properties is dominated mainly by surface area and aggregation or structure. High surface area and small particle size impart higher levels of reinforcement as reflected in tensile strength, tear resistance, and resistance to abrasive wear with resulting higher hysteresis and poorer dynamic performance. Higher structure gives improved extrusion behaviour, higher stock viscosities, improved green strength, and higher modulus values. Rubber property changes produced by carbon-black addition depend on loading. Tensile strength and abrasion resistance increase with increased loading to an optimum and then decrease. The optimum is normally in the range of 40–60 phr (parts per hundred or rubber). Increasing the loading level to 35–80 phr produces linear increases in hardness and modulus. The magnitude of these changes increases with structure.

About 8–9% of the total carbon black production consists of special industrial carbon grades for non-rubber applications (special blacks). Of increasing importance in recent years have been applications in plastics where special carbon blacks are required to improve weathering resistance, or to impart antistatic and electrically conductive properties. About 35% of the special blacks is used in printing inks, 8% in paints and lacquers, 33% in plastics, 1% in the paper, and 23% in miscellaneous applications. News inks account for most of the printing-ink market. For this use, special N300 series (HAF) grades containing 6% mineral oil are made to give rapid dispersion. Medium and high colour blacks, the most expensive grades, are used in enamels, lacquers, and plastics for their intense jetness.

The low thermal conductivity of carbon black makes it an excellent high temperature insulating material. For high temperature insulation up to 3000°C, it must be maintained in an inert atmosphere to prevent oxidation. Thermal-grade carbon black has been most widely used. Carbon black also is a source of pure carbon both for ore reduction and carburising. Carbon brushes and electrodes are fabricated from carbon black. It is used as well as a pigment in the cement industry, in linoleum, leather coatings, polishes, and plastic tile.

DIAMOND (NATURAL)

Diamond, the high pressure allotrope of carbon, changes to graphite, the high temperature allotrope of carbon, when heated above 1500°C *in vacuo*. Diamond is metastable and chemically inert at moderate temperatures. Although diamond is the hardest known substance, it is quite brittle and is readily crushed to grit and powder, the form in which approximately 75% of all industrial diamond is used. These properties make diamond of ever-increasing importance in industry and technology as fixed abrasive in saws, drills, and grinding wheels. In the form of single crystals (stones), it is used in turning and boring tools, wire, dies, indenters, tools to shape and dress conventional abrasive wheels, and in drill crowns. Those diamonds of near flawless quality having colours of blue, blue-white, water-white, pink, green, or a slight tinge of yellow are of greatest value as gems. Approximately 20% of mined diamonds are classified as gems.

Diamonds are found in ancient volcanic pipes in relatively soft, dark, basic perioditite rock called blue ground or kimberlite. As a result of erosion of the volcanic pipes, diamonds are also recovered from alluvial gravels and marine terraces.

Diamond is composed of the single element carbon and has a cubic crystal structure of space group O_h^7. Diamonds are classified into four groups based on their optical and electrical properties as types 1a, 1b, 11a, and 11b. Typical physical properties of natural diamond are listed in Table 8.3.

Table 8.3. Physical properties of natural diamond.

Property	*Value*
Density,[a] g/cm^3, 25°C	3.51524 ± 0.00005
thermal conductivity,	
W/(cm.K), 20°C	
Type 1	9
Type 11a	26
Maximum at 190°C	
Type 1	24
Type 11a	120
Thermal expansion	
coefficient at 20°C	$0.8 \pm 0.1 \times 10^{-6}$
Specific heat (constant volume)	6.184
at 20°C, J/(mol.K)[b]	
Refractive index at 546.1 nm (Hg green),	2.4237
μm	
Dielectric constant, at 0–3 kHz, 27°C	5.58 ± 0.03
optical transparency	
Type 11a diamond	225 nm to 2.5 μm > 6 μm
Type 1 diamond	340 nm to 2.5 μm > 10 μm
Resistivity, 20°C, Ω.cm	
Type 1 and most Type 11a	$> 10^{16}$
Type 11b	$10–10^3$
Hardness, Mohs scale (scratch hardness)	10
Knoop scale (indention hardness), kgf/mm^2[c]	5,700–10,400
compressive strength, GPa[d]	

(Contd.

Property	Value
average (flawless octahedral diamond)	8.68
maximum	16.5
Modulus of elasticity, GPa[d]	1160
Young's modulus, GPa[d]	1160

[a] Average of 35 diamonds.
[b] To convert J to cal, divide by 4.184.
[c] To convert kgf/mm^2 to psi, multiply by 1422.
[d] To convert GPa to psi, multiply by 145,000.

DIAMOND (SYNTHETIC)

Catalysed Synthesis

In this process, a mixture of carbon (e.g., graphite) and catalyst metal is heated high enough to be melted while the system is at a pressure high enough for diamond to be stable. Graphite is then dissolved by the metal, and diamond is produced from it. Effective catalysts are Cr, Mn, Fe, Co, Ni, Ru, Rh, Pd, Os, Ir, Pt, and Ta, and their alloys and compounds. If the metal is not molten, graphite is obtained instead of diamond even at pressures high enough to produce diamond. (The exception is tantalum which does not have to be molten to be effective).

Many kinds of apparatus have been devised for simultaneously producing the high pressures and temperatures necessary for diamond synthesis. An early, successful design is the belt apparatus, in which two opposed, conical punches, made of cemented tungsten carbide and carried in strong steel binding rings, are driven into the ends of a short, tapered chamber that also is made of cemented tungsten carbide supported by strong steel rings. A compressible gasket, constructed in a sandwich fashion of stone, usually pyrophyllite, and steel cones, seals the annular gap between punch and chamber, distributes stress, provides lateral support for the punch, and permits axial movement of the punches to compress the chamber contents. The reaction zone, usually a cylinder, is buried in pyrophyllite stone in the chamber. The pyrophyllite, a good thermal and electrical insulator, is easily machined and transmits pressure fairly well. The reaction zone is heated electrically with a heavy current.

Size, shape, colour, and impurities are dependent on the conditions of synthesis. Lower temperatures favour dark-coloured, less pure crystals; higher temperatures promote paler, purer crystals. Low pressures (about 5 GPa or 50 kbar) favour the development of the cube faces, whereas higher pressures produce octahedral faces. Nucleation and growth rates increase rapidly as the process pressure is raised above the diamond graphite equilibrium pressure.

If diamond seed crystals are placed in the active diamond-growing zone of a typical graphite-catalyst metal apparatus, new diamond usually forms on the seed crystals. However. the new growth tends to be uneven in the thickness and quality, with gaps or inclusions of foreign material. Such defects probably appear because the main driving force for the nucleation and growth under these conditions is the Gibbs free energy difference between diamond and graphite, which is a function of pressure, temperature, and composition; none of these variables can be sufficiently controlled. However, excellent growth can be obtained if pressure and composition are held relatively constant while the change of composition with temperature is employed as a driving force.

Direct Graphite-to-Diamond Process

In this process, diamond forms from graphite without a catalyst. The refractory nature of carbon demands a fairly high temperature (2500–3500 K) for sufficient atomic mobility for the transformation, and the high temperature in turn demands high pressure (above 12 GPa; 120 kbar) for diamond stability. The combination of high temperature and pressure may be achieved statically, or dynamically using shock waves. During the course of experimentation on this process, a new form of diamond with a hexagonal (wurtzitic) structure was discovered.

A diamond prepared by the direct conversion of well-crystallised graphite, at pressures of about 13 GPa (130 kbar), shows certain unusual reflections in the x-ray-diffraction patterns. They could be explained by assuming a hexagonal diamond structure (related to wurtzite) with $a = 0.252$ and $c = 0.412$ nm, space group $P6_3/mmc$- D^4_{6h} with 4 atoms per unit cell. The calculated density would be 3.51 g/cm^3, the same as for ordinary cubic diamond, and the distances between nearest-neighbour carbon atoms would be the same in both hexagonal and cubic diamond, 0.154 nm.

Synthetic Diamond Industry

Soon after the first successful diamond syntheses by the catalyst process, small batches of crystals were prepared in the laboratory. When mounted in abrasive wheels, these first synthetic diamonds performed better than comparable natural diamonds for shaping of ceramics and cemented tungsten carbide. A pilot plant for producing synthetic diamond was established, the efficiency of the operation was increased, production costs declined, and product performance was improved.

With the exception of the rare, natural type 11b, diamond is normally a good electric insulator. However, semiconducting diamonds are prepared by adding small amounts of boron, beryllium or aluminium to the growing mixture, or by diffusing boron into crystals at high pressures and temperatures. Some natural diamonds known as carbonado or ballas occur as tough, polycrystalline masses. The production of synthetic sintered diamond masses of comparable excellent mechanical properties has only been achieved recently. Natural single-diamond crystals and carbonado can now be replaced in many industrial uses by sintered-diamond tool blanks.

NATURAL GRAPHITE

Natural graphite, the mineral form of graphitic carbon, occurs worldwide. It differs from the carbon of coal and of diamond in its predominantly lamellar hexagonal crystal structure. The ore usually contains associated silicate minerals that vary in kind and amount with the source. Except for technical terminology, the name natural graphite is seldom used. It may be simply termed graphite or any of several common names such as plumbago, black lead, silver lead, carburet of iron, potelot, crayon noir, carbo-minerals, and reissblei. The macrophysical form depends on geological genesis, whereas the properties depend on both the macrophysical form and associated mineral suite. The commercial value depends on specific characteristics such as form, percentage, and kind of mineral suite, and availability. Graphite occurs in widely distributed places as flakes, lumps and cryptocrystalline masses referred to commercially as amorphous graphite.

Parallel layers of condensed planar C_6-rings constitute the graphite crystallite. Each carbon atom joins to three neighbouring carbon atoms at 120°C angles in the palne of the layer. The C—C distance is 0.1414 nm (this bond is 0.1397 nm in benzene); the width of each C_6-ring is 0.246 nm. The hexagonal form of graphite (Fig. 8.2) contains the most common stacking order, i.e., ABABAB... . Grinding graphite to particle sizes smaller than ca 0.1 μm reduces the crystallite size to < 20 nm, at which size two-dimensional ordering of carbon (char) replaces the three-dimensional ordering of graphite. This is depicted in the Franklin-Bacon curve (Fig. 8.3). The weak van der Waals forces that pin the carbons in adjoining layers are strong enough to hold the three-dimensional ordering of graphite at 0.3354 nm distance, but at 0.344 nm the structure is completely disoriented in the third dimension because the forces are too weak. This accounts in part for the marked anisotropic properties of graphite crystal. Table 8.4. lists some of the physical properties of natural graphite.

Table 8.4. Physical properties of natural graphite.

Property	Value
Density[a], g/cm³	
calculated	2.265
experimental, pure	ca 2.25
Compressibility, GPa^{-1}[b]	
at low pressures	4.5×10^{-2}
at high pressures	2×10^{-2}
average	3.1×10^{-2}
Shear modulus, GPa[c]	2.3
Young's modulus, GPa[c]	1.13×10^5
Heat of vapourisation, kJ/mol[d]	711
Sublimation point, K	4000–4015
Triple point, K	
graphite-liquid-gas, 10.1 MPa[c]	3900 ± 50
graphite-diamond-liquid, 12-13 GPa[c]	4100–4200
surface energy, J/cm^2[d]	ca 1.2×10^{-5}

[a] The difference between the calculated and experimental values of density is caused by dislocations and imperfections.

[b] To convert GPa^{-1} to cm^2/dyn, multiply by 10^{-10}.

[c] To convert GPa to psi, multiply by 145,000.

[d] To convert J to cal, divide by 4.184.

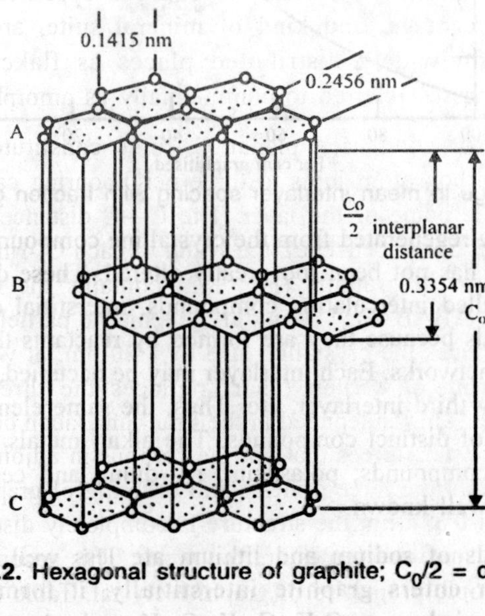

Fig. 8.2. Hexagonal structure of graphite; $C_0/2$ = d-spacing.

Graphite burns slowly in air above 450°C, the rate increasing with temperature and exposed area. The particle size and shape govern the ignition temperature. Flake graphites generally resist oxidation better than granular graphites. Above 800°C graphite reacts with water vapour, carbon monoxide, and carbon dioxide. Chlorine has a negligible effect on graphite, and nitrogen none. Many metals and metal oxides form carbides above 1500°C. These reactions occur with the carbon atom and destroy the graphitic structure. A series of compounds in which the graphite structure is retained, known as graphite compounds, consists of two general kinds: crystalline and covalent compounds.

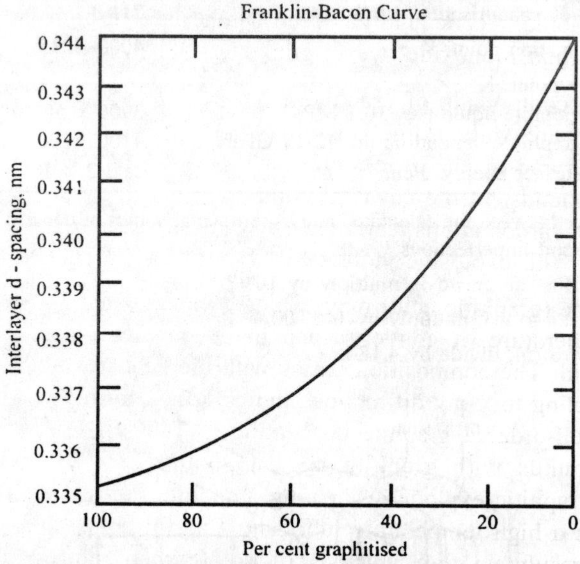

Fig. 8.3. Change in mean interlayer spacing with fraction graphitised.

Graphite can be regenerated from the crystalline compounds because the graphitic structure has not been too greatly altered. These dark crystalline compounds are called intercalation compounds, interstitial compounds, or lamellar compounds because they are formed by reactants that fit between the planar carbon networks. Each interlayer may be occupied, or every other interlayer, or every third interlayer, etc. Thus, the same element, or group, can form a series of distinct compounds. The alkali metals form a variety of such addition compounds; potassium, r 'dium, and cesium reactions with graphite are well known.

The Compounds of sodium and lithium are less well known. When potassium vapour enters graphite interstitially, it forms a series of intercalation compounds, e.g., C_8K, $C_{24}K$, $C_{26}K$, and others, depending on the sequence of interstitial layers filled. Compounds of graphite with alkali metals or ammonia are electron donors. Compounds with the halogens (except fluorine), metal halides, and sulphuric acid (graphite sulphate) are electron acceptors.

Graphite sulphate forms when graphite is warmed in concentrated sulphuric acid containing a small quantity of an oxidising agent such as concentrated nitric acid. The graphite swells and becomes blue. Bromine vapour forms C_8Br by direct addition to well-oriented graphite. Some intercalated crystalline compounds find their way into commerce as catalysts for chemical synthesis. Graphite-$FeCl_3$.KCl has been used in the synthesis of ammonia. Acetic acid has been synthesised in high yield from methanol and carbon monoxide over a graphite-$RhCl_3$-I_2 compound.

The covalent compounds of graphite differ markedly from the crystal compounds. They are white or lightly coloured electrical insulators, have ill-defined formulae, and occur in only one form, unlike the series typical of the crystal compounds. In the covalent compounds, the carbon network is deformed and the carbon atoms rearrange tetrahedrally as in diamond. Often they are formed with explosive violence.

Graphite oxide (graphitic acid) may explode when heated above 200°C. Below this temperature, it converts to a black powder once known as pyrographitic acid. The composition varies with the heat treatment and the end point; according to x-ray diffraction studies, it is a form of carbon that reconverts to well-ordered graphite on heating to 1800°C. Fluorine forms covalent compounds with graphite, e.g., C_4F and CF. These fluorine compounds of graphitic explode on heating. Graphite fluoride continues to be of interest as a high temperature lubricant. Careful temperature control at 627 ± 3°C results in the synthesis of poly(carbon monofluoride), a compound that remains stable in air to ca 600°C and is a superior lubricant under extreme conditions of high temperatures, heavy load, and oxidising conditions.

Graphite's many useful properties give rise to a wide variety of products; unctuous, dry lubricant; marks readily, writing and drafting pencils; combination of lubricity and electrical conductivity, motor and generator brushes; excellent weathering properties and inertness, industrial paint pigment; solubility in molten iron, carbon raiser for steel; poorly wet by most metals and alloys, foundry-mould facings; and burns slowly, conducts heat, and retains strength over a large temperature range, refractories such as crucibles, retorts, and stopper heads for steel ladles.

Some additional properties of interest include hydrophobicity, formation of water-in-oil emulsions, negative charge, low photoelectric sensitivity, strong diamagnetism, and infrared absorbent. Some minor uses of graphite include coating smokeless powder and gunpowder grains to control burning rate and to prevent static sparking from friction between grains, roofing granules, packings, brake linings, gaskets, stove polish, static eliminator, polish for tea leaves and coffee beans, pipe-joint compounds, boiler compounds, wire drawing, welding-rod coatings, catalysts, oil-well drilling muds, lock lubrication, coatings for eight-track tape cartridges, mechanical mounts in cassettes, mercury and silver dry cells, exfoliated flake for gaskets and packings, aircraft disk brakes, catalyst pellet production, O-rings and oil seals, and interior and exterior coating for cathode-ray tubes.

Chapter 9

Sulphur and Sulphuric Acid

INTRODUCTION

One of the most important basic raw materials in the chemical process industries is sulphur. It exists in nature both in the free state and combined in ores such as pyrite (FeS_2). It is also an important constituent of petroleum and natural gas (as H_2S). Various industrial and academic research groups are developing new uses for sulphur. Among some of the new uses being studies are :

(1) As an additive for asphalt

(2) Sulphur concretes and mortars

(3) Plant and soil treatment

(4) Sulphur-alkali metal batteries

(5) Foamed sulphur insulation

Sulphur is an active element which combines directly with most of the known elements. It can exist in both positive and negative oxidation states and can form ionic as well as covalent and coordinate covalent compounds. The uses of sulphur are limited primarily to the manufacture of sulphur compounds. However, large quantities of elemental sulphur are used in the vulcanisation of rubber, in lime-sulphur sprays to destroy plant parasites, in the manufacture of artificial fertiliser and certain types of cements and electric insulators, in certain ointments and medicinals and in the manufacture of gunpowder and matches. Sulphur compounds are used in the manufacture of chemicals, textiles, soaps, fertilisers, leather, plastics, refrigerants, bleaching agents, drugs, dyes, paints, paper and other products.

The oxides of sulphur which have been characterised have the formulae SO, S_2O_3, SO_2, SO_3, S_2O_7 and SO_4. Sulphur dioxide, SO_2 and sulphur trioxide, SO_3, are of far greater importance than the others. Sulphur dioxide can act as an oxidising agent and as a reducing agent. It reacts with water giving an acidic solution (often called sulphurous acid) and bisulphite (HSO_3) and sulphite (SO_3^{2-}) ions. The dioxide is used as a refrigerant gas, as a disinfectant and preservative, as a bleaching agent and in the refining of petroleum products. Its major use, however, is in the manufacture of sulphur trioxide and sulphuric acid. Sulphur trioxide is used primarily in the preparation of sulphuric acid and sulphonic acids.

Although salts (or esters) of all the oxy acids are known, in many cases the free acids themselves have not been isolated because of their instability. Sulphurous acid is not actually known as a pure substance. Sulphuric acid (H_2SO_4) is a colourless, viscous liquid with the melting point 10.31°C (50.56°F). It is a strong acid in water and reacts with most metals in either the dilute or concentrated form. The concentrated acid is a strong oxidising agent, especially at elevated temperatures. Pyrosulphuric acid ($H_2S_2O_7$) is an excellent sulphonating agent and loses sulphur trioxide on being heated. It also reacts vigorously with water, liberating a considerable quantity of heat. The persulphuric acids (monoperoxysulphuric acid, H_2SO_5, called Caro's acid and diperoxysulphuric acid, $H_2S_2O_8$, called Marshall's acid) are known as the acids and salts. Sulphonic acids are known as the esters and halides.

Sulphuric acid, H_2SO_4, is a bulk chemical of immense value. The consumption of sulphuric acid has many times been cited as an indicator of the general state of a nation's economy, and although many other indicators (such as energy consumption) might today be regarded as more important, sulphuric acid consumption still follows general economic trends and is, by far, the largest volume chemical.

Sulphuric acid, once called 'oil of vitriol', has been an important item of commerce for at least 250 years, and has been known and used since the Middle Ages, when small quantities were obtained in glass vessels when sulphur was burned with saltpeter in a moist atmosphere.

SULPHUR

Process of Manufacture

From hydrogen sulphide by oxidation

The hydrogen sulphide present in "sour" natural gas or in various petroleum-refinery streams may be separated from the base fuel by standard extraction or saponification methods. In the most common process, the hydrogen sulphide is stripped from the gas with an ethanolamine solution and later liberated from the solvent.

The concentrated hydrogen sulphide may then be converted to sulphur by oxidation. Details of the process differ from plant to plant, depending on the source of the hydrogen sulphide and other factors. The following description refers to a process commonly used by petroleum refineries, in which sulphur recovery plants are almost always necessary owing to pollution control laws.

Reaction

$$2H_2S + 3O_2 \rightarrow 2SO_2 + 2H_2O$$

$$2H_2S + SO_2 \rightarrow 3S + 2H_2O$$

$$\text{Overall: } 2H_2S + O_2 \rightarrow 2S + 2H_2O$$

In the Claus process, (Fig. 9.1) one-third of the hydrogen sulphide feed to the plant is burned at 1000°C in a pressurised boiler, where 80% of the total heat of reaction is removed by generating steam. The hydrogen sulphide is converted to sulphur dioxide by the first reaction shown above; the second reaction (Claus reaction) produces sulphur.

After the removal of the molten sulphur in a condenser, the remaining gases, now 2 parts hydrogen sulphide and 1 part sulphur dioxide, are passed through a series of two or three bauxite or alumina catalyst beds at 200-260°C, with sulphur removal between each converter stage. Overhead conversion is 92–95% with two catalytic stages, 95-96% with three and 96-97% with four.

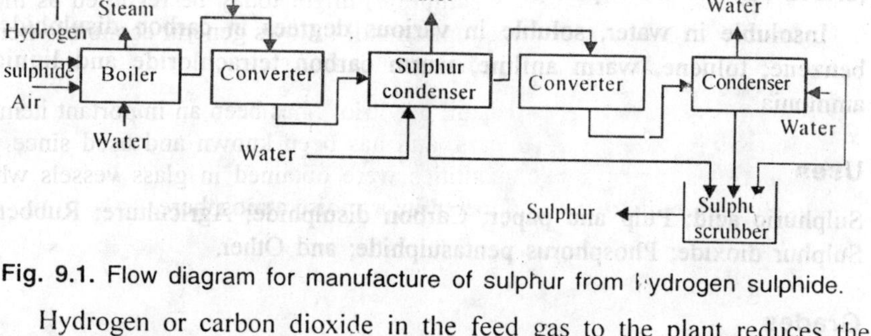

Fig. 9.1. Flow diagram for manufacture of sulphur from hydrogen sulphide.

Hydrogen or carbon dioxide in the feed gas to the plant reduces the sulphur yield by 0.25–2.5%, by allowing the formation of carbonyl sulphide (COS) and carbon disulphide (CS_2). The composition of the tail gas from the final converter is about one-third sulphur dioxide and hydrogen sulphide, one-third carbonyl sulphide and carbon disulphide, and one-third sulphur. Although this tail gas may be burned in an incinerator to yield sulphur dioxide, which is then vented to the atmosphere, pollution regulations normally do not permit this, and the tail gases must be treated. In one treatment process (Beavon process) all sulphur compounds in the Claus tail gases are reconverted to hydrogen sulphide by hydrogenation in a converter containing a cobalt molybdate catalyst. No hydrogen needs to be added, since the Claus tail gas already contains sufficient, ca. 2.5%. The gas containing hydrogen sulphide from the condenser that follows the converter may then be treated by any appropriate process to remove the hydrogen sulphide. Another treatment (Stratford process) uses sodium carbonate solution to react with the hydrogen sulphide, forming sodium hydrosulphide (NaHS) which in turn is oxidised to sulphur by sodium vanadate in solution. The finely divided sulphur froth is skimmed off, washed and dried. The effluent gas is now low enough in hydrogen sulphide to be vented.

Although what is described above refers to off-gases from petroleum refineries, the hydrogen sulphide content of other gaseous sources, such as sour natural gas, can be treated by similar methods. As a rule, differences between processes are in heat-recovery methods, or in the concentration of hydrogen sulphide in the raw material.

Properties

Sulphur (brimstone). Pale-yellow, odourless, brittle solid which commonly occurs in two crystalline forms. There is also an allotropic form known as plastic sulphur, which reverts to the crystalline form on standing.

Atomic wt. 32.06, mp 112.8°C, Specific gravity 2.07, Flash point (closed up) 207°C, Explosive limits Lower 30 mg/litre.

Insoluble in water, soluble in various degrees in carbon disulphide, benzene, toluene, warm aniline, warm carbon tetrachloride and liquid ammonia.

Uses

Sulphuric acid; Pulp and paper; Carbon disulphide; Agriculture; Rubber; Sulphur dioxide; Phosphorus pentasulphide; and Other.

Grades

Crude (run-of-mine) "bright" (99.5%) and "dark" (up to 1% carbonaceous material); both types are free of arsenic, selenium, and tellurium. Refined (99.8%) in the form of rock, lumps, coarse granular pieces, cones and sticks; also cast solid in wooden barrels. High purity (99.97%). Sublimed (flowers of sulphur). Flour, ground refined or crude sulphur in various mesh sizes. Rubber-markers', a ground special grade. Insoluble (85% insoluble in carbon disulphide) for special uses.

Packing

In bulk, bags and barrels. Molten sulphur in tankers, barges, tank cars and tank trucks.

In the 18th and 19th centuries, sulphuric acid was almost entirely produced by the Chamber process, in which oxides of nitrogen (as nitrosyl compounds) are used as homogenous catalysts for the oxidation of sulphur dioxide. The product made by this process is of rather low concentration (typically 60 Baume, or about 77–78 wt% H_2SO_4), not high enough for many commercial applications. During the first part of the 20th century, the chamber process was gradually replaced by the Contact process, and today the former is considered obsolete. Primary impetus for development of the contact process came from a need for high strength acid and oleum to make synthetic dyes and organic chemicals. The contact process, employing platinum catalysts began to be used on a large scale for this purpose late in the 19th century. Its development accelerated during World War I to provide concentrated mixtures of sulphuric acid and nitric acid for explosives production.

Essentially all sulphuric acid today is manufactured by the Contact process. The technology is very mature, although important alterations in the detailed arrangement of the conversion and adsorption stages were introduced recently to increase the sulphur dioxide conversion efficiency, primarily in the interest of environmental protection. Some plants are now designed for a sulphur dioxide conversion efficiency exceeding 99.8%. Nevertheless, the basic principle of the process remains the same today as when it was first introduced in the 1930s.

SULPHUR DIOXIDE

Sulphur dioxide, SO_2, is a colourless, non-flammable gas with strong suffocating odour. It condenses at –10°C and ordinary pressure to a colourless liquid with a density of 1.5. When mixed with oxygen and passed over red hot platinum, it is converted to sulphur trioxide. With water it forms sulphurous acid, H_2SO_3. Sulphur dioxide bleaches vegetable colours and is intensely irritating to the eyes and respiratory tract. Sulphur dioxide is supplied compressed in cylinders and is used in preserving fruits, vegetables, etc. Sulphur dioxide is also used as a disinfectant in breweries and food factories, bleaching textile fibres, straw, wicker ware, gelatin, glue and beet sugars. Liquid sulphur dioxide is used as a solvent.

SULPHUR TRIOXIDE

Sulphur trioxide SO_3 (mol. wt. 80.07) is a colourless liquid (at room temperature and atmospheric pressure) that fumes in air. Trace amounts of water or sulphuric acid can catalyse the formation of polymers. The three trimorphic phases—alpha, beta and gamma—have melting points of 62.3, 32.5 and 16.8°C, respectively. The alpha-form is the stable modification, while the beta- and gamma–forms are metastable. The alpha and beta forms melt to give liquid gamma SO_3.

Sulphur trioxide may be prepared in the laboratory by heating fuming sulphuric acid and collecting the sublimate in a cooled receiver. If the vapour is condensed above 27°C, the gamma form is obtained as a liquid. Below 27°C and in the presence of a trace of moisture, a mixture of all three forms is obtained, which can be separated by fractional distillation.

Absolutely dry SO_3 is not corrosive to metals and shows no acid reaction. On exposure to air, it absorbs moisture rapidly, emitting dense white fumes. It combines with water with explosive violence (heat of dilution 504 cal/gm) forming sulphuric acid. Due to this avidity for water, SO_3, chars many organic substances. As a general rule sulphur trioxide reacts very briskly with all organic compounds and the organic compound may be sulphonated, oxidised (decomposed) with the release of water, or dehydrated.

Sulphur trioxide is a strong oxidising agent and at the same time it is one of the strongest known Lewis acids. The bifunctional activity is apparent, for example, in the reaction with hydrogen halides. With hydrogen fluoride, sulphur trioxide reacts as a Lewis acid to form fluorosulphuric acid. Hydrogen bromide and hydrogen iodide, on the other hand, are oxidised under the same conditions to the respective free halogens, while the sulphur trioxide is reduced to give sulphur dioxide or even hydrogen sulphide.

Sulphur trioxide reacts at 50–150°C with elemental sulphur to give sulphur dioxide. This reaction is used industrially for the production of pure sulphur dioxide (using coke or pulverised coal at elevated temperatures).

Under absolutely dry conditions, sulphur trioxide is unreactive toward most metals. With metal oxides it reacts at moderately high temperature to form the corresponding metal sulphates. Due to sulphur trioxide's importance as an industrial intermediate in the production of sulphuric acid, the thermodynamics and kinetics of its generation by the oxidation of sulphur dioxide have been extensively studied, as has been the reverse reaction. Pure sulphur trioxide is extremely resistant to thermal decomposition because of kinetic inhibition, even at elevated temperatures where thermodynamic equilibrium is shifted heavily towards $SO_2 + O_2$. However, certain catalytically active substances are able to increase the rate of equilibration substantially. In the presence of metals such as platinum or of metal oxides and sulphates (e.g., of iron, copper and vanadium) the decomposition approaches equilibrium at temperatures above 700°C.

In principle, sulphur trioxide has the same toxic effects as sulphuric acid. Inhalation of the gas itself or the sulphuric acid mist formed when it comes in contact with humid air causes irritation, burning and degeneration of the tissue of moist skin, eyes and mucous membranes, especially in the respiratory tract. This happens relatively rapidly at concentrations more than 10 ppm by volume.

When storing and handling liquid sulphur trioxide, it is extremely important to prevent polymcrisation, because any solid formed is very inconvenient to remove. Sulphur trioxide reacts explosively with water, so the vessels must never be cleaned with water. Concentrated sulphuric acid is required for dissolving solid deposits of sulphur trioxide. In order to maintain sulphur trioxide in the liquid state during storage, tanks and piping must be kept at about 30°C by supplementary heating. Addition of a stabiliser also effectively inhibits the formation of solid polymers at temperatures above the natural solidification point of sulphur trioxide. A number of inorganic and organic compounds are used or have been recommended as stabilisers, but the extensive patent literature on the subject proves no indication of their mode of action.

Liquid sulphur trioxide is among the most dangerous of all industrial chemicals, and strict safety regulations govern its handling. In Germany, for example, every act of transporting liquid sulphur trioxide over public roads requires a special permit, which can be withdrawn at any time.

Pure sulphur trioxide is used in organic synthesis for sulphonation reactions, including the manufacture of chlorosulphonic acid, thionyl chloride, aminosulphonic acid, dimethyl sulphate and sulphamide. The reaction is difficult to control if the sulphur trioxide is introduced in liquid form, so it is preferentially supplied as a gas (pure or diluted with an inert gas), as a solution in liquid sulphur dioxide or some other solvent that is inert under the reaction conditions or as an addition compound with an organic base. Sulphur trioxide is also used in the manufacture of explosives.

Properties

Physical properties

Pure sulphuric acid is a colourless, water white, odourless, slightly viscous (dense), only, very corrosive liquid. It has a melting point of 10.4°C, a boiling pointing of 279.6°C and a density of about 1.84. It decomposes at 340°C into sulphur trioxide and water. It can be mixed with water in any ratio. Aqueous sulphuric acid solutions are defined by their H_2SO_4 content in weight-per cent terms. Anhydrous (100%) sulphuric acid is even today sometimes referred to as 'monohydrate', which simply means that it is the monohydrate of sulphur trioxide. Sulphuric acid has a great affinity for water, abstracting it from the air and also from many organic substances, and hence it chars sugar, wood, etc. It is miscible with water and alcohol with the generation of much heat and with contraction in volume. When diluting, the acid should be added to the diluent. Sulphuric acid will dissolve any quantity of sulphur trioxide, forming 'oleum' (fuming sulphuric acid). The concentration of oleum is expressed in weight-per cent of dissolved SO_3 ('free SO_3') in 100% H_2SO_4.

The physical properties of sulphuric acid and oleum are dependent on the H_2SO_4 and SO_3 concentrations, the temperature and the pressure. At constant temperature, the density of sulphuric acid increases steeply with rising H_2SO_4 concentration, reaching a maximum of about 98%. From there upto a concentration of 100% the density decreases slightly, but it rises again in the oleum range up to a concentration of about 60% free SO_3. On account of the clear relationship between density and concentration at defined temperatures in the lower concentration range, density measurement provides a quick method for determining concentrations upto about 93% H_2SO_4. Hydrometers used for this purpose were formerly calibrated in 'degrees Baume' (Be) and for that reason sulphuric acid concentration was often and sometimes still is, expressed in Be. In USA, the Baume scale is calculated utilising the formula: Degree Be = 145 – (145/sp.gr). In Europe, the Baume scale is calculated using 144.3 as the constant.

Higher concentrations are not covered because of the great difficulty in differentiating between acid concentrations in the range of 93–100% H_2SO_4 by specific gravity (density) measurements.

Sulphuric acid with free SO_3, is designated in commerce as 'oleum' (fuming H_2SO_4). Available grades contain upto about 80% free SO_3. It is colourless or slightly coloured, viscous liquid, emitting choking fumes of sulphur trioxide. It is extremely corrosive and should be handled with great care.

Chemical properties

Sulphuric acid is a strong acid with characteristic hydroscopic and oxidising properties. Like the sulphate ion, sulphuric acid is chemically and thermally very stable. The dehydrating effect of concentrated sulphuric acid is due to the formation of hydrates. Several hydrates have been identified in the solid state and these explain the irregular variation of some of the physical properties of sulphuric acid with concentration, such as its freezing temperature.

Pure sulphuric acid is ionised to only a small extent. Due to this the electrical conductivity of sulphuric acid solution has its lowest value at 100% H_2SO_4. When pure sulphuric acid is diluted with water, dissociation occurs increasingly and the conductivity rises accordingly. At higher water content the second stage of dissociation becomes increasingly important. However, on account of the diminishing total concentration of sulphuric acid, the conductivity reaches a maximum at about 30% H_2SO_4 (the exact value depends on the temperature) and decreases steeply down to 0 wt % H_2SO_4. Dilute sulphuric acid is the preferred electrolyte for industrial metal electrowinning and electroplating plants on account of its high conductivity and the chemical stability of the sulphate ion. To take advantage of the electrical conductivity maximum, sulphuric acid of about 33% concentration is used in lead storage batteries.

Dilute sulphuric acid is a strong dibasic acid and dissolves all base metals forming respective metal sulphates and bisulphates (hydrogen sulphates) with the release of hydrogen. Hot, concentrated sulphuric acid has an oxidising effect, reacting with precious metals and with carbon, phosphorus and sulphur by which it is reduced to sulphur dioxide. A very important property of sulphuric acid is, its ability to decompose the salts of most other acids. Industrially important examples are given below:

(1) Production of sodium sulphate and hydrogen chloride from sodium chloride.

(2) Decomposition of sulphites to sulphur dioxide.

(3) Decomposition of phosphate rock (natural calcium phosphates) to phosphoric acid and calcium sulphate.

The reactions of concentrated sulphuric acid with organic compounds are frequently dominated by its oxidising and hygroscopic properties, e.g., carbohydrates are decomposed to the point of carbonisation. Organic condensation reactions in which water is eliminated are promoted by sulphuric acid because it effectively removes the water as soon as it is formed. Sulphuric acid is, therefore, frequently used in industry for this purpose. It also exercises a catalytic effect on certain reactions involving organic compounds.

Sulphuric acid is oxidised both by hydrogen peroxide and anodically to diperoxysulphuric acid $H_2S_2O_8$, and the unstable monoperoxysulphuric acid (Caro's acid), H_2SO_5. Since it is a strong oxidant, Caro's acid can oxidise sulphur dioxide to sulphuric acid, a property that has been exploited in pollution control for sulphuric acid plants. Nitrogen oxides ($NO + NO_2$) react with sulphuric acid at concentrations above 70 wt% H_2SO_4 to give nitrosyl hydrogen sulphate, $NOHSO_4$.

Manufacture

Catalysts

Sulphuric acid is manufactured by chamber process and contact process. However, the chamber process, i.e., nitrogen oxide process is disadvantaged due to lower product concentration of around 78% compared to 98% concentration possible by the contact process.

Initially, platinum was used as catalyst in the contact process and remained the predominant catalyst until the 1930s. However, vanadium pentoxide eventually succeeded in replacing the platinum catalyst because of its insensitivity to catalyst poisons and its considerably lower cost.

With the development of vanadium pentoxide catalyst together with increasing demand for concentrated sulphuric acid, the proportionate share of world sulphuric acid output produced by chamber process declined steadily. Nonetheless, the nitrogen oxide process has continued to be the object of interest and a certain amount of development work, especially for the processing of gases with extremely low sulphur dioxide content.

In 1936–37, Lurgi introduced the wet contact process for converting moist sulphur dioxide-containing gases over a vanadium catalyst. This made it possible to process hot gases from the combustion of hydrogen sulphide in coking plants directly to sulphuric acid. In succeeding years, a number of factors affected the development of the contact process. First, the raw material basis of the industry changed progressively from mainly roaster gases to sulphur combustion gases containing higher concentrations of sulphur dioxide. Second, plant capacities increased as a result of rapid rise in the consumption of sulphuric acid by the fertiliser industry. These and other factors provided a stimulus for the introduction of improvements in the individual process steps and in the design of associated equipment (e.g., the shift to tray converters from tube converters).

In 1960, Bayer patented the so-called double-catalysis process, and the first plant using this process, built by Lurgi, started up in 1964. By incorporating a preliminary SO_3 absorption step ahead of the final catalytic stages, this improved contact process permitted a decisive increase in overall SO_2 conversion, thus reducing SO_2 emissions substantially. Because the essential difference between this process and the ordinary contact process is in the number of absorption stages, it is referred to as the 'double-absorption' process. However, this process did not substantially change the nature of the process or the process equipment. In the 1970s the principal industrial countries introduced more stringent regulations for environmental protection, which made the use of the double-absorption process more or less mandatory in new plants. Nevertheless, the conventional contact process continues to be used in countries where environmental regulations are less exacting.

In the 1970s and 1980s, the increased value of energy and production of sulphuric acid from a variety of waste products, including off-gases and spent sulphuric acid, led to a number of process and equipment modifications. In recent years also, the main thrust of development in the contact process is toward increasing the recovery and utilisation of the very substantial amount of process heat. Indeed, a large, modern sulphuric acid plant may be looked upon not just as a chemical plant but also as a thermal power plant.

Contact process

Today, the contact process is the most widely used process for the manufacture of sulphuric acid throughout the world. Sulphuric acid may be produced by the contact process from a wide range of sulphur-bearing raw materials by several different process variants, depending largely on the raw material used. In some cases, sulphuric acid is made as a by-product of other operations, primarily as an economical or convenient means of minimising air pollution or disposing of unwanted by-products.

The raw materials used to make sulphuric acid are elemental sulphur, spent (diluted and contaminated) sulphuric acid, and hydrogen sulphide. Elemental sulphur is by far the most widely used. Till 1970s iron pyrite and related compounds were the predominant raw materials. A large amount of sulphuric acid is also produced as a by-product of non-ferrous metal smelting, i.e., roasting sulphide ores of copper, lead, molybdenum, nickel, zinc or others. Moderately concentrated sulphuric acid is also produced by concentration and purification of spent or waste sulphuric acid. Reprocessing of sulphuric acids generated in large quantities in many processes and recycling such regenerated acid to the user is becoming increasingly important from an environmental production standpoint.

The sulphuric acid manufacturing process may be divided into the following distinctive steps:

(1) Generation of sulphur dioxide gas.
(2) Catalytic oxidation of sulphur dioxide to sulphur trioxide.
(3) Absorbing of sulphur trioxide to form sulphuric acid.

The chemical reactions may be represented as follows:

(1) $S + O_2 \longrightarrow SO_2$

(2) $SO_2 + \frac{1}{2} O_2 \longrightarrow SO_3$ $\Delta H = -99.0$ KJ

(3) $SO_3 + H_2O \xrightarrow{\text{Catalyst}} H_2SO_4$ $\Delta H = -132.5$ KJ

All the three reactions are exothermic. In addition to the above mentioned process steps, drying of sulphur dioxide gas and cooling of sulphuric acid (after SO_3 absorption) are important. The gas drying stage is not applicable to a plant of wet catalysis type which are relatively uncommon. Almost without exception, the contact process plants operate under essentially atmospheric pressure; compression is required only for driving the gases through the plant.

Where elemental sulphur or hydrogen sulphide is used as raw material, considerable heat is evolved during the initial combustion. Additional heat is generated in the next two steps. In such plants, most of the heat is typically used to produce steam utilised either for heating requirements in other processes or to generate power via turbines. In many cases, large plants of this type are essentially co-producers of steam or power and sulphuric acid; products which have significant commercial value. Where spent acid is used as raw material, it usually is decomposed in furnaces fired by gas, oil, or other fuels (sometimes H_2S or sulphur) and the high temperature gas from such furnaces can also generate steam or power.

Generation of sulphur dioxide gas

In all types of contact plants, the first step is to produce a reasonably continuous, contaminant-free gas stream containing appreciable sulphur dioxide and some oxygen. This gas is preferably dry, but plants can be designed to handle wet gas directly from H_2S combustion. This requires careful design of equipment to minimise mist formation in the condensation – absorption section of the plant. If the initial oxygen concentration of the process gas is low, additional air or oxygen must be added prior to or during catalytic oxidation to ensure that there is an excess over stoichiometric needs for conversion of SO_2 to SO_3.

In general, SO_2 gas derived from metallic sulphides, spent acids or gypsum anhydride is purified before drying by cold, i.e., wet, gas purification. Various equipment combinations including humidification towers, reverse jet scrubbers, packed gas cooling towers, impingement tray columns and electrostatic precipitators are used to clean the gas. Plants that produce good quality elemental sulphur or H_2S gas generally have no facilities for purifying SO_2.

Before the advent of relatively pure Frasch or recovered sulphur, however, hot gas purification was frequently used in which the SO_2 gas stream was passed through beds of granular solids to filter out fine dust particles just prior to its entering the converter. Sulphur shipped as a solid frequently becomes contaminated with dirt and scale during shipping and handling. In places where solid sulphur is still handled, molten sulphur is frequently filtered prior to use as an alternative to, or in combination with, hot gas purification.

The air used for combustion must be dried. In the usual arrangement, filtered air from the atmosphere is drawn through a drying tower by the main blower. Since most of the plants use sulphur burning to generate SO_2 gas, it is described in detail here. Spent acid or H_2S burning and ore roasting, sintering or smelting are also used to some extent and require specialised equipments.

For sulphur burning, there is a trend toward very large single-train plants. Due to this the usual practice is to use horizontal, brick-lined combustion chambers with dried air and atomised molten sulphur introduced at one end. Atomisation typically is accomplished either by pressure spray nozzles or by mechanically driven spinning cups. Because the degree of atomisation is a key factor in producing efficient combustion, sulphur nozzle pressures are typically 2.76 Mpa (150 psi) or higher. Sulphur burners are typically designed as proprietary items by companies specialising in acid plant design and construction. Sulphur burners are normally operated at moderate pressures, in the other temperatures and concentrations are in similar proportion. SO_2 concentrations in the gas stream range from 4–14 vol %.

The gases are dried, generally in counter-current with fairly concentrated sulphuric acid in irrigated packed towers. The residual water content of the gases after drying corresponds theoretically to the partial pressure of water vapour above the drying-tower acid at the prevailing temperature and concentration. To achieve high drying efficiency, the temperature of the irrigation acid is normally maintained at 50–60°C (by cooling).

At high flame temperatures, small amounts of nitrogen react with oxygen to form nitrogen oxides, NO_x, primarily nitric oxide, NO. Ultimately, some of them form nitrosylsulphuric acid, which ends up either as trace amounts in product acids or, in considerably higher concentrations, as condensed acid collected at mist eliminators.

Conversion of SO_2 to SO_3 and its absorption

The catalytic oxidation of SO_2 to SO_3 is highly exothermic and, equilibrium becomes increasingly unfavourable for SO_2 formation as temperature increases above 410–430°C. Unfortunately, this is about the minimum temperature level required for typical commercial catalysts to function. Consequently, catalytic reactors (called as converters) are typically designed as multistage adiabatic units and have gas cooling between each stage.

To improve equilibrium or driving force for the conversion, the sulphuric acid industry has attempted one or a combination of the following methods—design modifications—increasing concentration of SO_2 in the process gas stream, increasing concentration of oxygen in the process gas stream by air dilution or oxygen enrichment, increasing the number of catalyst beds, removing the SO_3 product by interpass absorption—known as the double absorption process, lowering catalytic converter inlet operating temperature, i.e., using better catalysts, and increasing the catalytic converter operating pressure (pressure plants).

In early years, the contact process frequently employed only two or three catalyst stages (passes) to obtain overall SO_3 conversions of approximately 95–96%. Later four-pass converters were used to obtain conversions from 97% to slightly better than 98%.

For sulphur-burning plants, this typically resulted in sulphur dioxide stack emissions of 1500–2000 ppm. The air pollution requirements led to the adoption of the double contact or double absorption process, which provides overall conversions better than 99.7%. Most industrialised nations have emission standards that cannot be achieved without utilising double absorption or tail-gas scrubbers.

Single absorption sulphur-burning plants

These plants were standard in the industry for many years. These single absorption plants used either relatively low strength (about 8 vol. %) SO_2

gas without air dilution, or air dilution designs and higher (about 10 vol. %) inlet gas strength. Air dilution was a common design option using additional dry air, instead of heat exchangers, to cool the process gas entering the last one or two converter passes.

The additional air improved conversion at the final converter pass by increasing oxygen concentration and reducing equivalent sulphur dioxide concentration of the process gas. Its chief advantage was reduced investment over designs with heat exchangers for inter-pass cooling.

Fig. 9.2 shows a typical four-pass, single absorption sulphur-burning plants without air dilution. Plants of this design burn sulphur to generate a process gas stream of about 7.5–9.0% SO_2. Typical converter operating conditions are listed in Table 9.1.

Table 9.1. Converter conditions for a single absorption plant.*

Converter pass temperature, °C	*1*	*2*	*3*	*4*
Inlet	410-445	430-450	430-435	425-430
Outlet	595	500	450	430-435
DT	150-185	50-70	15-20	5

* Using 8% SO_2, inlet gas.

The scheme shown produces a full range of products including 66 deg. Be and 98.5% acid and oleum. If 66 deg. Be acid product is not needed, a common pump tank is usually used for both the drying and absorbing towers. If oleum production is not needed, the equipment within the dashed box is omitted and process gas enters the absorbing tower directly from the economiser.

Most of the heat of combustion from the sulphur is removed in a boiler, which reduces the process gas temperature to the desired converter inlet temperature. Typically, the inlet temperature (Table 9.1) to the first converter pass is dictated by catalyst performance, catalyst bed depth and process gas strength. Standard, i.e., sodium- or potassium-promoted, vanadium pentoxide catalysts do not have sustained catalytic activity at temperatures of more than 400–410°C, although fresh catalysts may have an initial reaction ignition temperature as low as 385°C. Such low ignition temperatures cannot be sustained by conventional catalysts. The catalyst ignition temperature is the temperature below which substantial catalytic conversion (approaching equilibrium) cannot be sustained in any given bed or pass. Newer catalysts promoted with cesium, have a considerably lower sustainable ignition temperature (about 375°C) and have proved useful in special situations.

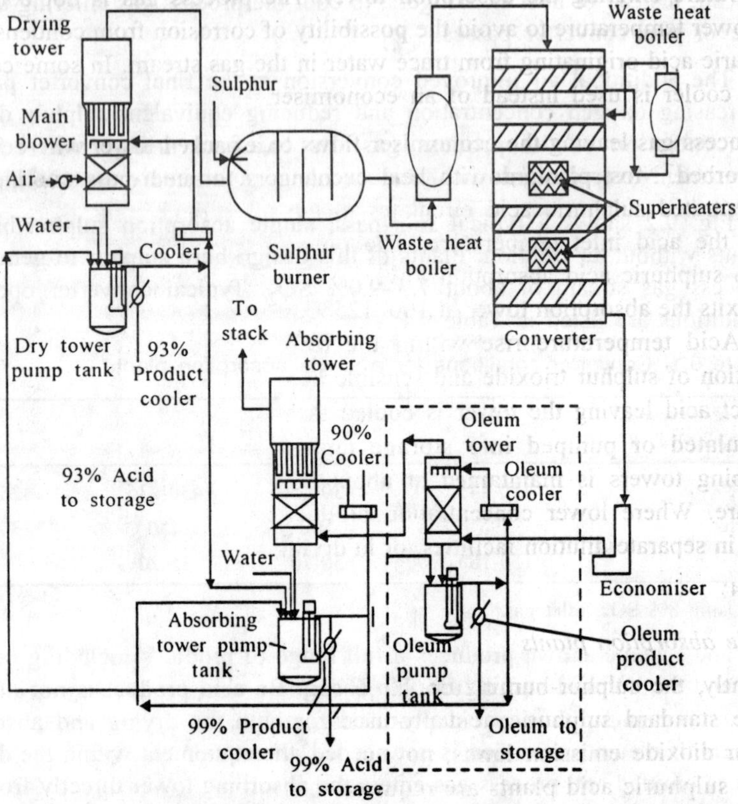

Fig. 9.2. Flow sheet for a single absorption sulphur-burning plant.

Sulphur dioxide gas is catalytically oxidised to sulphur trioxide in a fixed bed reactor (converter) which operates adiabatically in each catalyst pass. The heat of reaction raises the process gas temperature in the first pass to approximately 600°C (Table 9.1). The temperature of hot gas existing the first pass is then lowered to the desired second pass inlet temperature (430–450°C) by removing the heat of reaction in a steam super-heater or second boiler. To obtain optimum conversion, the heats of reaction from succeeding converter passes are removed by super-heaters or air dilution. The temperature rise of the process gas is almost directly proportional to the SO_2 converted in each pass, even though SO_2 and O_2 concentrations can vary widely. Gas leaving the converter is normally cooled to 180–250°C using boiler feedwater in an economiser. This increases overall plant energy recovery and improves SO_3 absorption by lowering the process gas temperature entering the absorption tower. The process gas is not cooled to a lower temperature to avoid the possibility of corrosion from condensing sulphuric acid originating from trace water in the gas stream. In some case, a gas cooler is used instead of an economiser.

Process gas leaving the economiser flows to a packed tower where SO_3 is absorbed. Most plants do not produce oleum and need only one tower. Concentrated sulphuric acid circulates in the tower and cools the gas to about the acid inlet temperature. The typical acid inlet temperature for 98.5% sulphuric acid absorption towers is 70–80°C. The 98.5% sulphuric acid exits the absorption tower at 100–125°C, depending on acid circulation rate. Acid temperature rise within the tower comes from the heat of hydration of sulphur trioxide and sensible heat of the process gas. The hot product acid leaving the tower is cooled in heat exchangers before being recirculated or pumped into storage tanks. Acid circulated over SO_3 absorbing towers is maintained at about 98.5% to minimise its vapour pressure. Where lower concentration product acid is desired, it is made either in separate dilution facilities, or in drying towers operated at 93–95% H_2SO_4.

Double absorption plants

Currently, the sulphur-burning double-absorption process is considered to be the standard sulphuric acid production process for conforming with sulphur dioxide emission limits, now in force in most countries. In USA, newer sulphuric acid plants are required to limit SO_2 stack emissions to 2 kg of SO_2 per metric ton of 100% acid produced. This is equivalent to a sulphur dioxide conversion efficiency of 99.7%. Acid plants used as pollution control devices, for example those associated with smelters, have different regulations. This high conversion efficiency is not achievable by single absorption plants using available catalysts, but it can be attained in double absorption plants when the catalyst is not seriously degraded.

A typical double absorption plant design uses intermediate SO_3 absorption after the second or, more commonly, the third converter pass. This is called the 3+1 configuration. As of the mid-1990s, newer double absorption plants usually contained a total of four catalyst passes in a 3+1 configuration. Plants having five passes in a 3 + 2 configuration have also been built. The Fig. 9.3 presents a typical flow diagram for a double absorption sulphur-burning plant. Typical converter temperatures in a double absorption converter are given in the Table 9.2.

Table 9.2. Converter conditions for a 3+1 double absorption plant.*

Converter pass temperature, °C	*1*	*2*	*3*	*4*
Inlet	415-420	430-445	430-445	425-430
Outlet	600-610	530	470	450
DT	185	90	30	20

* Using 11.5% SO_2, inlet gas.

Approximately 90–95% of total sulphur trioxide produced by the double absorption process is absorbed in the inter-pass absorption tower. The sulphur trioxide produced in subsequent converter passes is absorbed in the final absorbing tower. Inter-pass absorbing tower operation is similar to an absorbing tower in a single absorption plant. In both cases, acid irrigation rates are designed so that acid temperature exiting the tower is 100–125°C. The smaller amount of sulphur trioxide absorbed in the final absorption tower of double absorption plants typically raises its acid temperature to only upto 105°C. Inter-pass and final absorbing towers of double absorption plants are very similar in size because tower diameter is dependent on total gas throughput, not sulphur trioxide concentration.

Another major difference between single and double absorption processes is that, after inter-pass absorption, the process gas is reheated from approximately 80°C to approximately 425°C before re-entering the converter. This re-heating is accomplished in gas-to-gas heat exchangers (Fig. 9.3), using some of the heat from the initial converter passes. The gas-to-gas heat exchangers are a primary cost item. All other plant operations are very similar to the corresponding single absorption processes.

Assuming the feed gas contains 10 vol. % SO_2, the 99.7% conversion efficiency of double absorption plant corresponds to a sulphur dioxide concentration in the tail gas of about 400 ppm SO_2. Sulphur burning double-absorption plants are today designed for production capacities of 3000 T/day or more of 100% sulphuric acid in a single stream.

Wet-catalysis processes

Wet-catalysis processes differ from other contact sulphuric acid processes in that the feed gas still contains moisture when it comes into contact with the catalyst.

Sulphur trioxide formed by catalytic oxidation of sulphur dioxide reacts instantly with the moisture, to produce sulphuric acid, in the vapour phase, to an extent determined by the temperature. Liquid acid is subsequently formed by condensation of the H_2SO_4 vapours and not by the absorption of SO_2 in concentrated sulphuric acid, as in a contact process based on dry gases. The concentration of the product acid depends on the H_2O/SO_3 ratio in the catalytically converted gases as well as on the condensation temperature. The wet-catalysis process is especially suitable for processing the wet, dust-free gases obtained in the combustion of hydrogen sulphide-containing off-gases, which need only be cooled to the converter inlet temperature of about 440°C. Also their capacities are small—the largest one is of 250 TPD. A sulphur dioxide conversion of about 97.5% is achieved in the wet-catalysis process with a fourth converter.

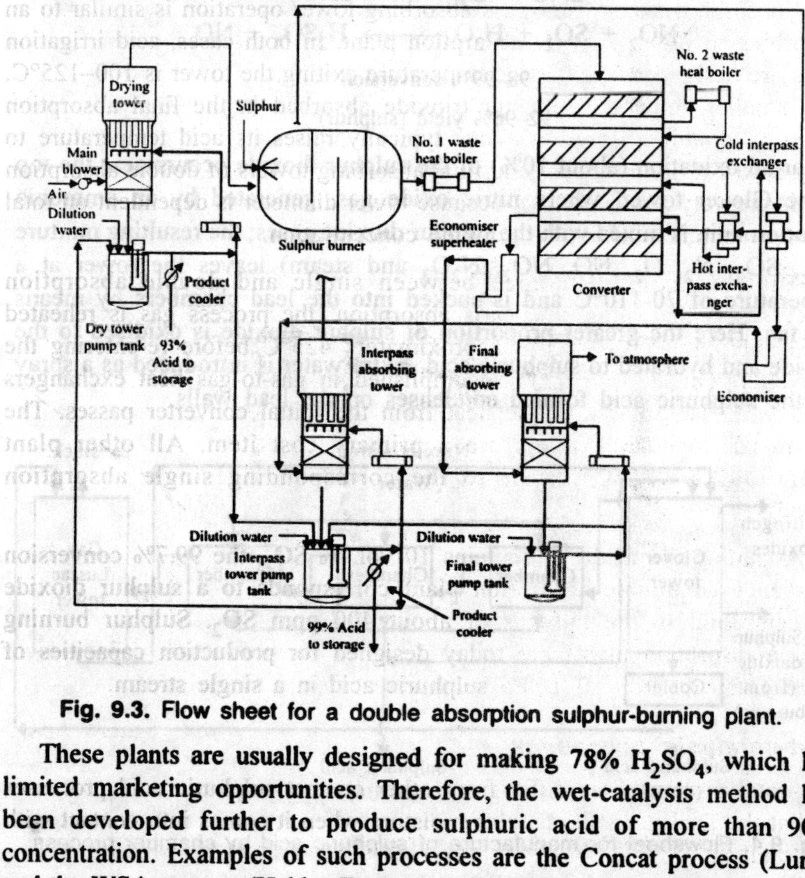

Fig. 9.3. Flow sheet for a double absorption sulphur-burning plant.

These plants are usually designed for making 78% H_2SO_4, which has limited marketing opportunities. Therefore, the wet-catalysis method has been developed further to produce sulphuric acid of more than 90% concentration. Examples of such processes are the Concat process (Lurgi) and the WSA process (Haldor Topsoe). Another special development in this area is the wet-dry contact process with intermediate condensation.

Chamber process

Fig. 9.4 depicts the flow diagram of chamber process for the manufacture of sulphuric acid. Hot sulphur dioxide gases (425 to 600 °C) from the burners (7 to 9% sulphur dioxide and 9 to 12% O_2) are introduced into the base of a Glover tower. The gases pass counter-current to a cool stream of nitrous vitriol (72.8% H_2SO_4), which has been diluted from 60°Bé (77.7% H_2SO_4) acid obtained from the Gay-Lussac towers. The hot (100 to 140°C) Glover or "tower" acid (60° Bé) is cooled; part is used again in the Gay-Lussac tower and the rest is sent to storage.

Reaction

$$2NO + O_2 \longrightarrow 2NO_2$$

$$NO_2 + SO_2 + H_2O \longrightarrow H_2SO_4 + NO$$

98–99% conversion

92–96% yield (sulphur)

Partial oxidation (about 10%) of the sulphur dioxide occurs near the top of the Glover tower, where nitric oxide gas, generated by an ammonia oxidation unit, is mixed with the sulphur dioxide gases; the resulting mixture (SO_2, SO_3, N_2, O_2, NO, NO_2, N_2O_3 and steam) leaves the tower at a temperature of 70-110°C and is sucked into the lead chambers by means of a fan. Here the greater proportion of sulphur dioxide is oxidised to the trioxide and hydrated to sulphuric acid. Extra water is introduced as a spray and the sulphuric acid formed condenses on the lead walls.

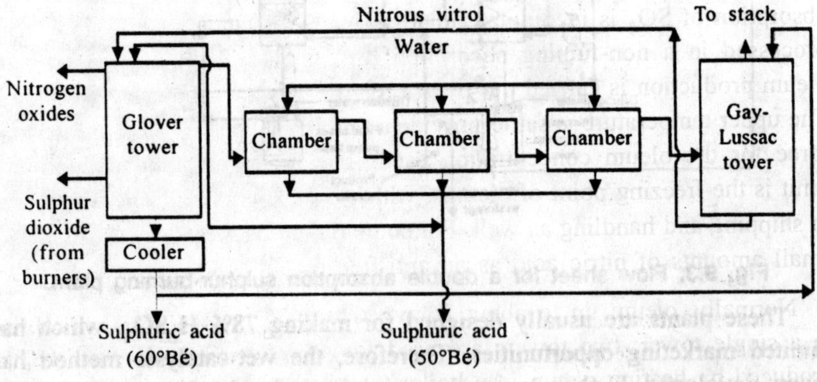

Fig. 9.4. Flowsheet for manufacture of sulphuric acid by chamber process.

This chamber acid (62-68% H_2SO_4), as it is called, is pumped to the Glover tower for concentration and nitrogen oxide removal. The gases leaving the chambers contain little sulphur dioxide but are led into the bottom of the Gay-Lussac tower counter-current to cold (35–40°C) 60°Bé (77.7%) sulphuric acid which strips nitrogen oxides from the gases, forming the nitrous vitriol containing 1–2.5% N_2O_3 by weight. The nitrous vitriol is pumped to the top of the Glover tower. The stripped gases are led to a stack and discharged to the atmosphere.

The number of lead chambers varies from 3–12 and the size depends on their design. Box-like chambers require 0.4 to 0.75 m^3 space/kg sulphur burned per 24 hour, whereas externally water-cooled truncated cones require only 0.2 to 0.3 m^3. With good operation the average chamber set lasts more than 20 years.

The chamber process produces directly a weaker acid of sp. gr. 1.53 to 1.71 (50 to 60 Bé). Up to sp. gr. 1.84 (66 Bé) acid may be obtained by subsequent concentration.

The chamber proccss is practically obsolete. It is described here for its historical interest and because many of the terms used in the industry (tower acid, etc.) and standard concentrations, still used for pricing, are derived from it.

MANUFACTURE OF OLEUM

To produce oleum, i.e., fuming sulphuric acid, sulphur trioxide is absorbed in one or more special absorption towers irrigated by recirculated oleum. Because of oleum vapour pressure limitations the amount of SO_3 absorbed from the process gas is typically limited to less than 70%. Since the absorption of SO_3 is incomplete, the gas leaving the oleum tower must be processed in a non-fuming absorption tower. The absorption of SO_3 for oleum production is carried out over a relatively narrow temperature range. The upper temperature is set to provide a reasonably partial pressure driving force for the oleum concentration used. The lower practical temperature limit is the freezing point of oleums, which is high enough to be a problem in shipping and handling as well. For some oleum uses it is practical to add small amounts of nitric acid as an anti-freeze.

Normally, oleum up to about 35 wt % free SO_3 content can be made in a single tower; two towers are used for 40 wt % SO_3. Liquid SO_3 is produced by heating oleum in a boiler to generate SO_3 gas, which is then condensed. Oleums containing more than 40 wt % SO_3 are usually produced by mixing SO_3 with low concentration oleum.

Equipment

Due to the highly corrosive nature of the gases and sulphuric acid product, a lot of attention has been given to various equipments used in their manufacture and also their materials of construction. The important equipment include—absorption and drying towers, acid coolers, catalytic converters, gas-gas heat exchangers, mist eliminators, oleum equipment.

Tail gas scrubbers are sometimes used on single absorption plants to meet SO_2 emission requirements, most frequently as an add-on to an existing plant, rather than on a new plant. Ammonia scrubbing is most popular, but to achieve good economics the ammonia value must be recovered as a usable product, typically ammonium sulphate for fertiliser use. A number of other tail gas scrubbing processes are available, including use of hydrogen peroxide, sodium hydroxide, lime and soda ash. Other tail gas processes include active carbon for wet oxidation of SO_2, molecular sieve adsorbents, and the adsorption and subsequent release of SO_2 from a sodium bisulphite solution.

Small amounts of sulphuric acid mist or aerosol are always formed in sulphuric acid plants whenever gas streams are cooled, or SO_3 and H_2O react, below the sulphuric acid dew point. The dew point varies with gas composition and pressure but typically is 80-170°C. Higher and lower dew point temperatures are possible depending on the SO_3 concentration and moisture content of the gas.

Such mists are objectionable because of both corrosion in the process and stack emissions. More recently sulphuric acid mists have been satisfactorily controlled by passing streams through equipment containing beds or mats of small diameter glass or teflon fibres. Such units are called mist eliminators. Use of this type of equipment has been a significant factors in making the double absorption process economical and in reducing stack emissions of acid mist to tolerably low levels. Coalescing demister pads have been used in some single absorption plants instead of packed fibre beds to remove mist from the stack gas. For sub-micrometre particle collection these devices are not as efficient as packed fibre beds. Nevertheless, they have been used in some plants to obtain nearly invisible emissions. Successful use of a coalescing demister pad requires careful control of plant operating conditions to minimise mist formation.

Catalysts

Of all substances tested for catalytic activity toward sulphur dioxide oxidation, only vanadium compounds, platinum and iron oxide have proven to be technically satisfactory. Today, vanadium pentoxide is used almost exclusively. Commercial sulphuric acid catalysts typically consist of vanadium and potassium salts supported on silica, usually diatomaceous earth. Catalyst pellets are available in various formulations, shapes and sizes depending on the manufacturer and the particular converter pass in which they are to be used. Commercial catalysts contain 4-9 wt% vanadium pentoxide, V_2O_5, as the active component, together with alkali-metal (usually potassium) sulphate promoters. Under operating conditions these form the liquid melt in which the oxidation of sulphur dioxide is thought actually to take place. Some catalysts also contain sodium sulphate to reduce the melting point. The carrier material is silica in the form of diatomaceous earth, silica gel or zeolites. In recent years, cesium-doped catalysts have also been developed and installed in various facilities. Cesium sulphate as a promoter reduces the melting point of the active components, resulting in significantly lower temperature limits for sustainable stable activity.

Solid cylindrical extrudates or pellets ranging from 4–10 mm diameter were used in the past. But now, ring-shaped catalysts (or starring) or variations, including rings having longitudinal ribs, are almost exclusively used primarily as a means of saving energy via reduced gas pressure drop. The various ring-shaped pellets also have greater resistance to dust fouling. Ring catalysts also have somewhat higher activity per unit volume than pellet catalysts. An important property of the vanadium catalyst is the low temperature limit at which stable operation is possible under fixed gas conditions. This temperature is about 410–430°C for a conventional catalyst and about 380–390°C for a cesium-doped catalyst.

The average service life quoted by most catalyst producers is about ten years. Service life is generally determined not so much by progressive loss of activity as by catalyst losses incurred when filling and emptying the reactor and during routine screening. Depending on the dust load of the gas entering the converter, the size and shape of the catalyst grains and the properties of the active melt, dust will accumulate in the catalyst bed over a period of time. This dust eventually increases the gas-pressure drop through the catalyst bed and reduces both gas throughput and SO_2 conversion efficiency. Therefore, the catalyst must be screened from time to time. The first converter-pass catalyst pellets are screened at every significant turnaround, typically every 12–24 months. Second-pass catalyst pellets need screening less frequently because the first converter-pass catalyst bed acts as a fitter for the rest of the converter. Typical screening losses range from 10–15% of the catalyst bed per screening. Screening losses depend on screen mesh size and catalyst hardness, as well as on screening rate.

In contrast to platinum, vanadium catalyst is largely insensitive to catalyst poisons. Catalyst ageing is a combination of a loss of catalytically active material from catalyst pellets and irreversible changes within the pellets. Catalyst aging is accelerated by increasing temperatures and temperature cycling. Exposure to moisture above the dew point is not detrimental to the catalyst. But exposure to moisture at temperatures below the dew point produces irreversible damage. Prolonged exposure to moisture reduces the vanadium to the + 3 oxidation state, which is very difficult to reoxidise under converter conditions. Moisture also damages the binders that hold together the silica support. This reduces catalyst hardness resulting in higher than normal screening losses.

Catalytic converters of different engineering firms vary in design. Stainless steel converters are frequently preferred, although carbon steel converters are also used. Stainless steel designs generally use all-welded interior construction, including stainless screens for catalyst supports. Traditional carbon steel converter designs use steel shells, sometimes partly or fully brick-lined, with cast iron and alloy internals. Essentially all designs use horizontal catalyst beds arranged one over another with gas flowing down through the catalyst. A relatively new innovation in stainless steel converter design uses structurally shaped supports grids and division plates, giving improved resistance to temperature differentials during start-up and a higher strength design requiring less metal.

The design of sulphuric acid manufacturing plants has evolved to a great extent. In addition to different variations of the contact process, several special plant designs have also been installed—for example, energy-efficient plants, cement plants (cement and sulphuric acid from calcium sulphate), oxygen enriched processes. Sulphuric acid concentrators are also designed to concentrate the large quantities of dilute H_2SO_4 produced as a by-product in many chemical processes.

STORAGE AND HANDLING

Sulphuric acid, oleums and liquid SO_3 are very corrosive and dangerous chemicals. Carbon steel is used in concentrated sulphuric acid storage tanks because in quiescent and low temperature conditions its corrosion rate is acceptable. Carbon steel is not suitable for handling sulphuric acid in concentrations between 80-90% or less than 68% even under quiescent conditions, unless passivating agents are present. Steel tank cars, often lined to minimise iron contamination are usually employed for high concentrations of sulphuric acid. Bottom outlets or valves are not allowed, nor are internal steam coils. Tank contents must be unloaded via standpipe. Using air pressure to unload is not recommended for safety reasons, but if air pressure is used, gauge pressure should be held at less than 0.21 Mpa (30 psi). General handling precautions should be observed and equipment appropriate for exposure conditions should be worn. The acid should be handled only in areas having sufficient ventilation to prevent irritation. Acid containers must be kept closed, water must not enter containers. An emptied container retains vapour and product residue. Thus all labeled safeguards must be observed until the container is cleaned, reconditioned or destroyed. Drums, if not safe-venting, should be periodically vented to prevent accumulations of hydrogen in dismantling lines and equipment, it should always be assumed that a spray of acid may occur and suitable precautions should be taken. Tightening flange bolts on pipe filled with acid is dangerous because of the possibility of mechanical failures.

In case of a spill or leak, keep peoplc away and up wind of the spill. If it is necessary to enter the spill area, self-contained breathing apparatus and full protective clothing must be worn. The area should be diked using sand or earth to contain the spill the acid removed by vacuum truck and the spill area flushed with water. Washings should be neutralised with lime or soda ash and pollution control authorities notified of any runoff into streams or sewers and of any air pollution incidents. Safety showers with deluge heads, protected against freezing, should be readily available at appropriate locations in any plant producing or using sulphuric acid.

SPECIFICATIONS AND STANDARDS

Typical specifications for several common types or grades of sulphuric acid are shown in Table 9.3.

Table 9.3. Typical sulphuric acid specifications.

Property	*Acid type*			
	Electrolyte[a, b] class 1[f]	*Technical[c] class 1[d]*	*Food chemicals codex*	*Technical[b] 66Be (93%)*
H_2SO_4, wt %	93.2	93.0		93.2
$Sp^2.\ gr^2.$, 15.5/15.5	1.8354	1.8347		1.835 - 1.837
Non-volatiles[e], %	0.03	0.025		0.02 - 0.03
As[e], ppm	1.0	2.5	3	
SO_2[e], ppm	40		40[f]	40 - 80

(Contd...)

Property	*Electrolyte*[a, b] *class 1*[f]	*Acid type Technical*[c] *class 1*[d]	*Food chemicals codex*	*Technical*[b] *66Be (93%)*
Iron[2b], ppm	50		200	50 - 100
Heavy metals[e], ppm			20	
Nitrate[i], ppm	5.0		10	5 - 20
Colour[i]	per test			100 - 200 APHA

[a]—Limits are also specified for platinum, organics, copper, zinc, antimony, selenium, nickel, manganese, ammonium and chloride.

[b]—Typical industry sulphuric acid.

[c]—The other classes of lower strength acids are included.

[d]—One other class of lower strength acid is included.

[e]—Value given is maximum.

[f]—Value is for reducing substances as SO_2.

Similar limits are generally used for other sulphuric acid concentrations, with the exception of turbidity values for high strength acids (and oleum) and SO_2 and nitrate values in oleums. Because iron sulphate is relatively insoluble in concentrated acids, the turbidities of 98–99% H_2SO_4 and oleum may be higher than shown, even at acceptable total iron concentrations. Sulphur dioxide concentrations in oleum are rarely specified or measured, but typical values are considerably higher than in acids of less than or equal to 99 wt % concentrations. A number of different grades of sulphuric acid are produced for specialised uses, such as reagent grade, food grade and electrolyte grade. In addition some producers offer special premium priced grades that contain little or no turbidity and colour, or in some cases a maximum iron concentrations of 10 ppm.

Certain objectionable elements such as arsenic, lead, mercury and selenium are not commonly specified for technical grade acid, but some producers attempt to hold each of these at less than one ppm, or less than 2-5 ppm in the case of lead, to minimise possible problems. Selenium is not usually present except at a few metallurgical-type plants or at plants using volcanic sulphur as raw material.

Indian specifications for various grades of sulphuric acid are given by IS-266. They are listed in the Tables 9.4, 9.5 and 9.6. The density composition tables for aqueous solutions of sulphuric acid are given in IS-4048. The safety code is given by IS-4262, while the limits for gaseous emissions in sulphuric acid industry are given by IS-8635.

Metallurgical (smelter) plants and spent acid decomposition plants usually produce acid of good (low) colour because the SO_3 feed gases are extensively purified prior to use. In some cases, however, and particularly at lead smelters, sufficient amount of organic flotation agents are volatilised from sulphide ores to form brown or black acid. Such acid can be used in many applications, particularly for fertiliser production, without significant problems.

APPLICATIONS

Sulphuric acid is one of the most widely used industrial chemicals. But most of its uses can be considered as indirect, because it functions as a reagent rather than as an ingredient. Surprisingly little of it appears in end products, and most ends up as spent acid or some type of sulphate waste. A number of products incorporate the sulphur of sulphuric acid, but nearly all of them are low-volume, speciality items.

Table 9.4. Requirements for technical grade sulphuric acid.

Property	*Value*
Specific gravity at 25°C, min.	1.828
Sulphuric acid, % by wt.. min.	93.0
Residue on ignition, % by wt., max.	0.2
Iron (as Fe), % by wt., max.	0.05
Heavy metals (as Pb), % by wt., max.	0.005

Table 9.5. Requirements for battery grade sulphuric acid (concentrated and dilute).

Property	*Value*	
	Concentrated	*Dilute*
Specific gravity at 25°C	1.8340	1.21
Sulphuric acid, % by wt, min.	95.0	29.7
Residue on ignition, % by wt, max.	0.06	0.02
Iron (as Fe), % by wt, max.	0.002	0.0006
Chlorides (as Cl), % by wt, max.	0.001	0.0003
Arsenic (as AS_2O_3), % by wt, max	0.0003	0.0001
Oxidisable impurities, as SO_2 and organic matter	To satisfy permanganate and charring test	To satisfy permanganate test
Selenium, % by wt, max.	0.002	0.0006
Manganese, % by wt, max.	0.0001	0.00003

(Contd...)

Property	*Value*	
	Concentrated	*Dilute*
Copper, % by wt, max.	0.003	0.001
Zinc, % by wt., max.	0.003	0.001
Nitrate, nitrite and ammonia as nitrogen (N), % by wt, max.	0.003	0.001

Table 9.6. Requirements for pure grade and analytical reagent grade sulphuric acid.

Property	*Value*	
	Pure	*Analytical reagent*
Specific gravity at 25°C	1.8340	1.8360
Sulphuric acid, % by wt, min.	95.0	96.0
Residue on ignition, % by wt, max.	0.01	0.0025
Iron (as Fe), % by wt, max.	0.001	0.0001
Chloride (as Cl), % by wt, max.	0.0035	0.0003
Heavy metals (as Pb), % by wt, max.	0.002	0.0002
Arsenic (as AS_2O_3), % by wt, max.	0.0005	0.00001
Oxidisable impurities as SO_2 % by wt, max.	0.004	0.0005
Nitrate, nitrite and ammonia as nitrogen (N), % by wt, max.	0.003	-
Nitrate (as NO_3), % by wt, max.	-	0.00002
Ammonia (NH_3), % by wt., max.	-	0.0005

Sulphuric acid is used in the manufacture of fertilisers, explosives, dyestuffs, other acids, parchment paper, glue, in purification of petroleum and pickling of metal. Formerly, dilute sulphuric acid was used in the treatment of gastric hypoacidity and the concentrated acid was used as a topical caustic.

Indirect Uses

The largest single consumer of sulphuric acid by far is the fertiliser industry. Most goes into the production of phosphoric acid, which in turn is used to manufacture such fertiliser materials as triple superphosphate and mono- and diammonium phosphates. Lesser amounts are used for producing superphosphate and ammonium sulphate. About 60% of the sulphuric acid produced is utilised in fertiliser manufacture.

Substantial quantities of sulphuric acid are used as an acidic dehydrating reaction medium in organic chemical and petrochemical processes involving such reactions as nitration, condensation and dehydration, as well as in oil refining, in which it is used for refining, alkylation, and purification of crude oil distillates.

In the inorganic chemical industry sulphuric acid is used notably in the production of TiO_2 pigments, hydrochloric acid, and hydrofluoric acid. In the metal processing industry, sulphuric acid is used for pickling and descaling steel, for leaching copper, uranium and vanadium ores in hydrometallurgical ore processing and in the preparation of electrolytic baths for nonferrous-metal purification and plating. Certain wood pulping processes in the paper industry require sulphuric acid, as do some textile and chemical fibre processes and leather tanning.

Direct Uses

Under certain conditions, sulphuric acid is occasionally used directly in agriculture for rehabilitating extremely alkaline soils. However, this is not a very important use in volume terms. Probably the largest use of sulphuric acid in which the sulphur becomes incorporated in the final product is organic sulphonation, particularly for the production of detergents. Other minor organic chemicals and pharmaceuticals are also made by sulphonation. One of the most familiar consumer products containing sulphuric acid—the lead-acid battery—accounts for only a tiny fraction of total sulphuric acid consumption.

Sulphuric acid has several desirable properties that lead to its use in a wide variety of applications. It typically is less costly than any other acid, it can be readily handled in steel or common alloys at normal commercial concentrations, and it is available and readily handled at concentrations more than 100 wt % (oleum).

CONTROL OF SULPHUR DIOXIDE POLLUTION DUE TO THE MANUFACTURE OF SULPHURIC ACID

In India, sulphuric acid capacity during the past five years has increased from about 5 million to about 6 million ton per year. When pollution control of sulphur dioxide is considered, the sulphuric acid plants were the first target. Rightly so, since at six million tons output, the traditional contact process would emit 94,000 tons of sulphur dioxide which result in about 144,000 tons of acid rain! This has been reduced to less than one-fourth of sulphur dioxide emission by the well established Double Catalysis, Double Absorption (DCDA) process.

With co-generation of power, sulphuric acid will reduce SO_2 emissions resulting from the corresponding generation of electric power, the sulphuric acid manufacture can no longe be viewed as a polluter of environment with sulphur dioxide emissions as in the past.

Modern sulphuric acid plants are viewed as power generating plants with sulphuric acid as by-product. For example, if the total capacity of six million tons of sulphuric acid in India was with best power generation capability, it would provide additional 180 MW to the grid. If credit is given to the saving of corresponding SO_2 emission for the power generated, the co-generation of power from sulphuric acid plants is attractive not only from economies of cheap power source, but also from lower SO_2, pollution due to corresponding power generation. For example, 180 MW would emit 80 MT/Day of SO_2 in air as permissible by Pollution Control Board. The corresponding emission from acid plants will be only 30 MT/Day of SO_2 generating the same amount of power.

Hence it has become imperative that all future sulphuric acid plants be equipped with maximum co-generation of power. This would make sulphuric acid production with no sulphur dioxide pollution and offer a sizeable reduction in SO_2 pollution for the generation of electric power.

Latest Trends Achievements in Emissions of SO_2 per ton of Sulphuric Acid

Pollution Control Board is to become more stringent in SO_2 emissions from sulphuric acid plants. The MINAS (Minimum National Standard) provides for 4 kg of SO_2 emission per ton of acid produced, the State Pollution Control Board insists on a figure of 3 kg of SO_2 per ton of acid produced. Well operated DCDA plants give an average of 2 kg SO_2 per ton of acid. However, if alkali scrubbing system is kept in operation continuously, this can be brought down to 0.5 kg SO_2 per ton of acid produced.

A well designed alkali scrubber has to take into account the high load on the system during the unsteady state of start-up operation after short stoppage due to maintenance problems or power failure or after a prolonged annual stoppage for tower inspection and routine maintenance.

At present there are two types of scrubbing systems which are found most suitable. One is the TCA (Turbulent Contact Absorber) offered by Nissan of Japan and the second is the Ventury-cum-Fixed Bed Absorber offered by Lurgi of Germany. Both are indigenously marketed and can be supplied on turn-key basis. One of the main drawbacks in this system is the additional running cost of providing power and alkali.

Since alkali scrubber cannot be economically viable on a continuous basis, a modified DCDA System using five passes instead of conventional four passes is found more attractive. In the conventional process, the fourth pass which operates at 410-430°C is found inadequate to convert the last traces of SO_2 to SO_3 due to thermodynamic and equilibrium restrictions.

The new modified DCDA System employs one more, namely, the fifth bed. Allowing a cooling between 4th and 5th passes and introducing additional air to cool gases to 360°C, the approach to chemical equilibrium is favoured. In addition, use of low kindling catalysts will guarantee the stock analysis below 100 ppm without the use of alkali scrubber. This claim is yet to be proved on a plant in India.

Chapter 10

Hydrochloric Acid and Miscellaneous Inorganic Chemicals

INTRODUCTION

Hydrochloric acid, although not manufactured in such large quantities as sulphuric acid, is an important heavy chemical. Manufacturing techniques have changed and improved in recent years, and new procedures are now employed, such as the burning of chlorine in hydrogen.

Hydrogen chloride (HCl) is a gas at ordinary temperature and pressure. Aqueous solutions of it are known as *hydrochloric acid* or, if the HCl in solution is of the commercial grade, as *muriatic acid.* The common acids of commerce are 18°Bé (1.142 sp. gr.) or 27.9% HCl, 20°Bé (1.160 sp. gr.) or 31.5% HCl, and 22°Bé (1.179 sp. gr.) or 35.2% HCl. Anhydrous HCl is available in steel cylinders at a very considerable increase in cost, because of the cylinder expense involved.

The largest users of hydrochloric acid are the metal, chemical, food, and petroleum industries. The major use of hydrochloric acid is in steel pickling (surface treatment to remove mill scale). Hydrochloric acid has taken over this market because it reacts faster than sulphuric with mill scale, less base metal is attacked by it, the pickled steel has a better surface for subsequent coating or plating operations, and much smaller quantities of waste pickle liquor are produced.

HYDROCHLORIC ACID

Manufacture of Hydrochloric Acid from Organic Chlorinations

Mostly hydrochloric acid production results as a by-product of various organic chlorination reactions, such as the chlorination of benzene to chlorobenzene. Other similar sources include the cracking of ethylene dichloride to yield vinyl chloride, the catalytic cracking of 1,1,2,2-tetrachloroethane to trichloroethylene, and the production of linear alkylbenzene. The recovery of magnesium from seawater yields some by-product hydrochloric acid. (Fig. 10.1)

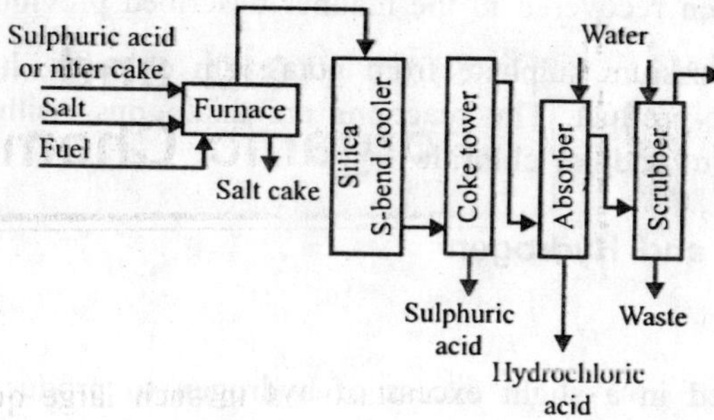

Fig. 10.1. Flow diagram for manufacture of hydrochloric acid from organic chlorinations.

Process

Salt and sulphuric acid, sp gr 1.7 (slight excess), or salt and an equivalent amount of niter cake are charged to a furnace equipped with a rake agitator (Mannheim furnace), where the reacting mass is slowly heated to a temperature just below fusion (843°C). Hydrogen chloride is evolved and led through a cooling and condensing system to the absorbers. Salt cake (crude sodium sulphate) is continuously discharged from the periphery of the furnace.

Reaction

$$NaCl + H_2SO_4 \rightarrow HCl + NaHSO_4$$

$$NaCl + NaHSO_4 \rightarrow HCl + Na_2SO_4$$

98% yield

The combustion gases, containing hydrogen chloride (approximately 30% HCl), leave the furnace at about 840°C, pass into silica bends (coolers) externally cooled with water, and leave the coolers at about 38°C. The cooled gases then pass through a coke-packed tower, in which sulphuric acid mist and any solid particles present are removed, and thence to another series of silica bends, where the hydrogen chloride is absorbed by water to produce hydrochloric acid. Exhaust gases from the absorber are scrubbed with water and discharged to the atmosphere. The more recently designed plants have made use of impervious graphite and structural plastics for absorbers and coolers.

A few plants use the Laury furnace, a horizontal rotary kiln, in which the reacting mass comes into contact with fuel combustion gases. These gases, containing as little as 5% hydrogen chloride, leave the furnace at 300°C, and are cooled and then recovered in the manner described previously.

In making potassium sulphate from potassium chloride, hydrochloric acid results as co-product. The reactions are analogous, with potassium chloride in place of sodium chloride.

From Chlorine and Hydrogen

Process

Chlorine is burned in a slight excess of hydrogen to produce hydrogen chloride. Several types of burners are used: for example, the silica burner, the ceramic-lined burner, the graphite burner, and the water-jacketed steel burner. The last-mentioned burner cannot of course be used with wet gases. (Fig. 10.2).

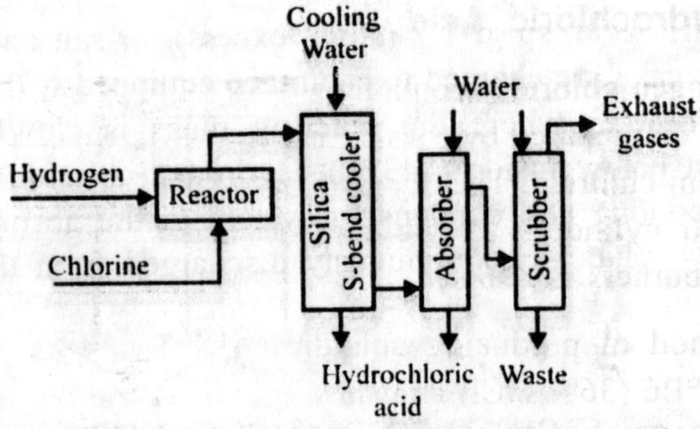

Fig. 10.2. Flow diagram for manufacture of hydrochloric acid from chlorine and hydrogen.

Reaction

$$H_2 + Cl_2 \rightarrow 2HCl$$

90–99% yield

A submerged combustion burner, operating beneath a surface of muriatic acid, has been developed. The burner gases, practically pure hydrogen chloride, are cooled, absorbed, and scrubbed in a system essentially the same as that used in the salt process. The purifying coke tower, however, is omitted, and in some systems (where the gas concentration approaches 100% HCl) the scrubber may be omitted. Strong hydrochloric acid (35%) is removed directly from the bottom of the cooler by means of a trap, and weak acid (28%) leaves the bottom of the absorber. Recently installed absorbers have been built of tantalum and structural carbon, as well as silica. This process readily gives a water-white product.

The concentration of hydrogen chloride in burner gases generally depends on the degree of chlorine utilisation. With 0.03% or less chlorine in the product gases, hydrogen chloride concentration may be as high as 98.5%. Where 0.1% chlorine is allowable, hydrogen chloride concentration may reach 99 to 99.5%. Manufacturers often insist on 100% chlorine utilisation, in which case exit gases may be only 90% hydrogen chloride because of the 5 to 10% excess hydrogen used.

Hydrochloric acid is also made by burning chlorine in methane or water gas, according to the equation :

$$2Cl_2 + CH_4 \ \text{air} \longrightarrow 4HCl + CO_2$$

The recovery system is the same as that used in the chlorine-hydrogen process.

Anhydrous Hydrochloric Acid

Anhydrous hydrogen chloride (containing 99.8% HCl) is available in steel cylinders. It may be prepared by passing hot (260°C) burner gases over cold anhydrous calcium chloride. The dry hydrogen chloride is then compressed and charged into cylinders. A yield of 99% of the hydrogen chloride supplied by the burners is realised.

Another method of producing anhydrous hydrogen chloride involves distillation of 22°Bé (36% HCl) acid in a graphite stripping column under 20 psig (136 kPa). The overhead gases (97% HCl and 3% water) are cooled in a brine condenser to –12°C. A liquid containing 50% hydrogen chloride condenses; the anhydrous hydrogen chloride passes through a mist trap and is ready for use. The condensate and stripping column bottoms are recycled to appropriate places in the acid recovery plant.

Uses

Metal treating, chemicals, oil well acidising, food processing etc. In metal treating the largest use is in the pickling of steel. It is superior to sulphuric acid for this use; it leaves a cleaner surface, reacts more slowly with the basis metal and, perhaps most important, the spent pickle liquor can be recovered and recycled, whereas spent pickle liquor from sulphuric acid pickling is an annoying waste product. In chemical processing, techniques have been developed for using hydrochloric acid in place of chlorine for organic chlorinations (oxyhydrochlorination).

In oil-well acidising, it increases the permeability of wells by dissolving part of the limestone and dolomite in the formations.

In food processing, it is used as a modifier for food starch, as a hydrolytic agent in the manufacture of sodium glutamate and gelatin, for conversion of corn starch to syrup, and for adjusting pH in brewing.

Properties

Hydrochloric acid is a solution of hydrogen chloride in water. Pure hydrogen chloride is a colourless, pungent poisonous gas at ordinary temperatures and pressures.

Mol. wt. 36.46, sp. gr. (0°C) 1.268 (air = 1), Density (0°C) 1.639 g/litre, mp –112°C, bp –35°C.

It is liquefiable under a pressure of 82 atm (8.2 MPa) at 51°C. At atmospheric pressure, the liquefied gas boils at –85°C and has a specific gravity of 1.194. Easily soluble in ethanol and ether, as well as in water. A constant-boiling mixture of hydrogen chloride and water contains 20.24% hydrogen chloride and boils at 110°C at 760 mm pressure. Threshold limit value 5 ppm, 7 mg/m^3. Other possibilities for alleviating the hydrochloric acid surfeit are the use of hydrogen chloride in place of chlorine as a chlorinating agent, and replacement of sulphuric acid in steel pickling.

The production of hydrochloric acid is considerably less than that of sulphuric, phosphoric, and nitric acids. It is commonly used for its acid characteristics where its volatility, non-oxidising characteristics, high acidity per unit weight, and ease of neutralisation are advantageous. Its chief drawbacks are cost and extreme corrosiveness to metals.

HYDROFLUORIC ACID

From Fluorspar and Sulphuric Acid

Process

By treating fluorspar (calcium fluoride) with concentrated sulphuric acid in a furnace, hydrogen fluoride gas is evolved, leaving a residue of calcium sulphate. After being cleaned of dust and sulphuric acid content, the gas is condensed as 98 to 98.5% hydrogen fluoride. Distillation increases its strength to 99.9 to 99.95%. Aqueous grades may be made by dilution with water (Fig. 10.3).

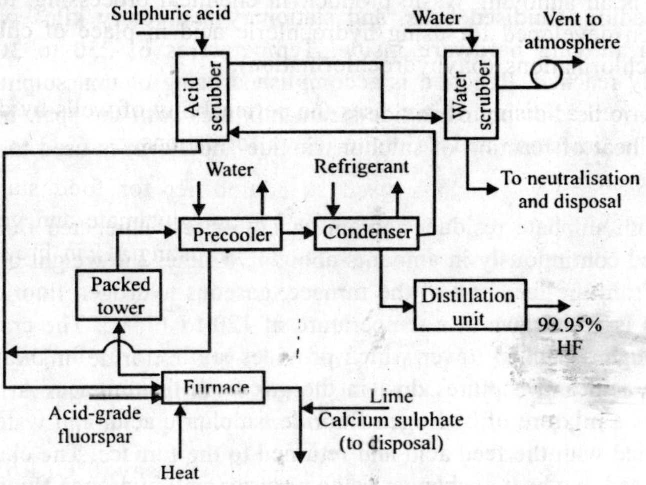

Fig. 10.3. Flow diagram for manufacture of hydrofluoric acid from fluorspar and sulphuric acid.

Reaction

$$CaF_2 + H_2SO_4 \rightarrow 2HF + CaSO_4$$

85 to 95% yield

Crude fluorspar, fluorite, as it comes from domestic mines, varies in calcium fluoride content from about 50 to 90%. To be suitable for acid production, the ore must be upgraded. This is usually done by flotation and results in an acid-grade fluorspar containing about 98% calcium fluoride, 1% silica , 0.03% sulphur and less than 0.1% moisture.

The finely powdered fluorspar is withdrawn from storage silos and transferred by air conveyors to steel conical-bottom hoppers. From here it is fed continuously along with concentrated (93 to 99%) sulphuric acid into a rotating furnace in a refractory-lined shroud. The space between the shroud and the furnace is fired with oil or gas to provide heat for the endothermic reaction. The cylindrical steel furnaces have been constructed in a variety of dimensions, up to 3.7 m in diameter and 25 m long, having a metal thickness of 5 cm. The production capacity of these furnaces ranges up to 55 metric tons/day.

The raw materials are fed into the furnace at a ratio of 1 mole of fluorspar to 1.1 to 1.3 moles of acid. Several methods have been proposed to keep the reacting mass of fluorspar and sulphuric acid flowing freely enough to allow complete reaction. They include two-stage reaction system with kneading, fluidised beds, and stationary and rotary kilns with fixed flights or moving hardware inside. Temperatures of 250 to 300°C are commonly reached. Reaction is accomplished using oleum, sulphuric acid, or sulphuric acid-sulphur trioxide-steam mixtures with the spar. In the last case, the heat of reaction of sulphur trioxide and steam is used to maintain the temperature.

Calcium sulphate residue, containing 1% or less unreacted fluorspar, is discharged continuously in amounts about 1.75 times the weight of the spar charge. From the feed end of the furnace, gaseous hydrogen fluoride (70 to 75% HF) is withdrawn at a temperature of 120 to 175°C. The crude gases pass through a packed tower which provides surface area for the reflux of sulphuric acid and captures dust in the gas from the furnace. A precooler condenses a mixture of hydrogen fluoride, sulphuric acid, and water, which is combined with the feed acid and returned to the furnace. The cleaned gas is condensed in a heat exchanger using a refrigerant; hydrogen fluoride boils at 19°C at atmospheric pressure. Non condensables pass through a packed tower, using the feed acid to absorb any uncondensed hydrogen fluoride. This gas passes through another tower in which water is the absorbent to remove silicon tetrafluoride (SiF_4) and any remaining hydrogen fluoride. Motive power for gas movement is usually provided by an induced draft blower.

The 98.5% acid from the condenser is distilled to raise its assay to 99.9 to 99.95%. The main impurities removed by distillation are sulphuric acid, sulphur dioxide, and water.

Variations are used in several plants. For instance, the furnace or retort may be a stationary, horizontal steel shell in which an alloy-steel conveyor propels the reacting fluorspar and product calcium sulphate through the unit. In one plant the fluorspar and 99% sulphuric acid are mixed in a premixer or kneader, where the initial and highly corrosive stage of the reaction is carried out. The partially reacted mass is then discharged to the rotary kiln. It is claimed that the premixing is so thorough that excess acid is not required and capital costs are reduced.

Solid and liquid particulates may be removed by such techniques as cyclones, scrubbing with sulphuric acid, coke boxes, and demisters. The gases are cooled to 20 to 30°C by sulphuric acid scrubbing and indirect heat exchangers. The resultant acid can be 98% pure, containing as impurities sulphuric acid, water, sulphur dioxide, and silicon tetrafluoride; this acid is suitable for cryolite manufacture without refining. The low boilers are readily fractionated as described.

The vent gas from the last cooler will contain by-product silicon tetrafluoride and hydrogen fluoride in a mole ratio of 1:2. On absorption by water in a packed tower, fluosilicic acid is formed according to :

$$SiF_4 + 2HF \rightarrow H_2SiF_6$$

This by-product may be converted to the sodium salt for use in the fluoridation of water supplies; more often it is neutralised with lime and discarded with the calcium sulphate. Despite many efforts to utilise the calcium sulphate, it is normally dumped as landfill.

Uses

Fluorocarbons; aluminium industry; atomic energy; alkylation; stainless steel; etc.

Properties

Anhydrous hydrofluoric acid is a colourless, fuming, corrosive liquid, which causes extremely serious burns on skin contact. Vapours are very irritating to the eyes and mucous membranes.

Mol. wt 20.01, Sp. gr. 0.988 (13.6°C), mp –83°C, bp 19.4°C, Threshold limit value 3 ppm, 2mg/m^3. Although the formula of hydrogen fluoride is sometimes written H_2F_2, there seems to be no real justification for the doubled formula. Various polymerised species, up to H_6F_6, are in equilibrium. Safety is an important item around a hydrofluoric acid plant because of the great hazards to personnel involved in handling it. Waste disposal also poses a special problem.

FLUORINE

The element is far too reactive to occur naturally, but it is found as fluorides in certain minerals, e.g. fluorspar or fluorite, CaF_2, and cryolite, Na_3AlF_6, and fluo-apatite, $CaF_2.3Ca_2(PO_4)_2$. Traces of fluorides are found in teeth and bones and in the ash of plants.

Preparation

An electrolytic method is employed, but considerable difficulty is encountered because of the very great reactivity of fluorine. The apparatus

(Fig. 10.4) must be kept completely free of moisture, oil, and grease, because fluorine attacks these materials vigorously. In the modern process an electrolyte of composition KF.2HF (i.e. a mixture of KHF_2 and HF) is used at about 100°C. The anode is made of special non-graphitic carbon impregnated with copper so that it is not readily attacked by fluorine, and the cathode is of steel. The cell itself is made of nickel steel, which is fairly resistant to attack, and a diaphragm is fitted to ensure complete separation of the products, since fluorine and hydrogen combine explosively.

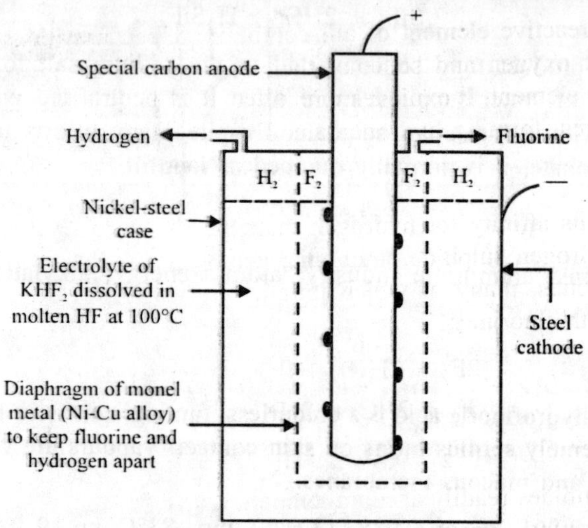

Fig. 10.4. Preparation of fluorine.

The electrolyte may be regarded as a solution of potassium fluoride in anhydrous hydrogen fluoride; it contains, therefore, a high concentration of potassium and fluoride ions and a much lower concentration of hydrogen ions. The changes at the electrodes are :

$$\text{At the anode:} \quad F^- \rightarrow F + e \text{ and } 2F \rightarrow F_2 \uparrow$$

$$\text{At the cathode:} \quad H^+ + e \rightarrow H \text{ and } 2H \rightarrow H_2 \uparrow$$

Thus the overall effect of the electrolysis is to decompose hydrogen fluoride into its elements, and the cell has to be topped up with more anhydrous hydrogen fluoride from time to time. The fluorine liberated at the anode can be freed from traces of hydrogen fluoride by passing it over anhydrous sodium fluoride :

$$NaF + HF = NaHF_2$$

Properties

It is a greenish-yellow diatomic gas, slightly paler in colour than chlorine, which it resembles in smell. It melts at –220°C and boils at –188°C. Its low critical temperature, –129°C, makes it impossible to store liquid fluorine at ordinary temperatures, but it can be transported in bulk as a liquid by using liquid nitrogen as a coolant.

Reactions

It is the most reactive element of all, combining directly with every other element except oxygen and some of the noble gases. It is also the most electronegative element. It explodes with hydrogen, even in the dark at very low temperatures, forming hydrogen fluoride :

$$H_2 + F_2 = 2HF$$

So great is its affinity for hydrogen that fluorine also reacts vigorously with water, hydrogen sulphide, ammonia, and hydrogen chloride and with organic compounds, many of which inflame spontaneously when brought into contact with fluorine :

$$2F_2 + 2H_2O = 4HF + O_2$$
$$F_2 + 2HCl = 2HF + Cl_2$$
$$4F_2 + CH_4 = 4HF + CF_4$$

Fluorine combines readily at room temperature with silicon, phosphorus, and sulphur, forming fluorides in which these elements show their highest covalency, e.g., SiF_4, PF_5 and SF_6. It also reacts with carbon, particularly when heated, giving carbon tetrafluoride, and with boron giving boron trifluoride. The alkali and alkaline earth metals ignite spontaneously in fluorine at room temperature and most other metals, including platinum, react with it when heated, but some metals such as copper and nickel show some resistance to attack probably because they form impervious fluoride layers on their surfaces which hinder further action.

When fluorine is passed through a very dilute solution of sodium hydroxide some fluorine monoxide is formed as a colourless gas (pale yellow), b.p. –146°C; it is the only oxide of fluorine stable at room temperature :

$$2F_2 + 2NaOH = 2NaF + F_2O\uparrow + H_2O$$

Uses of Fluorine and Its Compounds

Fluorine is required in great quantities for making uranium hexafluoride, UF_6, which is the only volatile compound of uranium. Fluorine compounds such as CCl_2F_2 and CCl_3F are being increasingly used as refrigerants because of their inertness, stability, non-toxicity, and thermodynamic efficiency, and also as propellents in aerosols. Boron trifluoride and hydrogen fluoride are used as catalysts in the petroleum industry, and cobalt fluoride and bromine trifluoride as fluorinating agents. Other fluorides are used as disinfectants in the brewing industry and for preventing rot in timber. The fluorocarbon, tetrafluoroethylene, C_2F_4, can be polymerised into a plastic, which is renowned for its chemical inertness; it is already important as an electrical insulator and as a lubricant. Sulphur hexafluoride is used as an electrical insulator in gas-filled high voltage cables. Molten cryolite is used as a solvent for alumina in the manufacture of aluminium. The mineral was the original source, but now-a-days a synthetic product is used.

Fluorosilicic acid, H_2SiF_6, is used for fluoridation of drinking water. By adding fluorides artificially in this way so that their concentration is only a few parts per million it has been found possible to reduce dental decay, particularly in children, presumably by strengthening the fluoride-containing enamel which protects the teeth. Toothpastes containing fluorides (e.g. stannous fluoride) are claimed to have the same beneficial effect.

ALUMINA (ALUMINIUM OXIDE)

The mineral corundum is natural aluminium oxide, and emery, ruby, and sapphire are impure crystalline varieties. The mixed mineral bauxite is a hydrated aluminium oxide.

Properties

Vary according to the method of preparation. White powder, balls or lumps of various mesh. Sp. gr. 3.4–4.0; m.p. 2030°C; insoluble in water; difficultly soluble in mineral acids and strong alkali. Non-combustible; non-toxic.

Preparation

(1) Leaching of bauxite with caustic soda followed by precipitation of a hydrated aluminium oxide by hydrolysis and seeding of the solution. The alumina hydrate is then washed, filtered and calcined to remove water and obtain the anhydrous oxide.

(2) Coal mine waste-waters are used to obtain aluminium sulphate, which is then reduced to alumina. It is toxic by inhalation of dust.

Manufacture of Alumina

The Bayer process, the most economical method of manufacture, takes advantage of the reaction of aluminium trihydroxide and aluminium oxide hydroxide with aqueous caustic soda to form sodium aluminate :

$$Al\ (OH)_3 + NaOH \rightleftharpoons NaAlO_2 + 2\ H_2O$$

$$AlO(OH) + NaOH \rightleftharpoons NaAlO_2 + H_2O$$

The reaction equilibria move to the right with increases in caustic soda concentration and temperature. The following operations are performed:

(1) Dissolution of the alumina at high temperature.

(2) Separation and washing of the insoluble impurities of bauxite (red muds) to recover the soluble alumina and caustic soda.

(3) Partial hydrolysis of sodium aluminate at a lower temperature to precipitate aluminium trihydroxide.

(4) Regeneration of the solutions for recycle to step (1) by evaporation of the water introduced by the washings.

(5) Transformation of the trihydroxide to anhydrous alumina by calcination at 1450 K. Fig. 10.5 shows the flow sheet of the Bayer process.

Variations of lime-soda sintering process are used in treatment of high-silica bauxites. Numerous sintering and acid-extraction processes have been investigated for kaolin and other clays, as well as from dawsonite and coal ash.

Electrolysis of Alumina

Nearly all aluminium is produced by the electrolysis of alumina (Al_2O_3) dissolved in a molten cryolite-based bath, the Hall and Héroult process. The aluminium is deposited onto a carbon cathode which also serves as a melt container. Simultaneously, oxygen is deposited on and consumes the cell's carbon anode(s).

Cryolite ionises to form sodium (Na^+) and hexafluoroaluminate (AlF_6^{3-}) ions. The latter dissociates to form tetrafluoroaluminate (AlF_4) and fluoride (F^-) ions. Alumina dissolves at low concentrations by forming oxyfluoride ions with a 2:1 ratio of aluminium to oxygen ($Al_2OF_{2\,n}^{(4-2n)}$); at higher alumina concentrations, oxyfluoride ions with a 1:1 ratio of aluminium to oxygen ($Al_2OF_{2n}^{(4-2n)}$) are formed. Cells are generally operated with a 2–6 wt % Al_2O_3 in the electrolyte. Saturation ranges between 7–12% Al_2O_3, depending upon electroyte composition. Ion transport measurements indicate that Na^+ ions carry most of the current; however, aluminium is deposited. Most probably, a charge transfer occurs at the cathode interface and hexafluoroaluminate ions are discharged, forming aluminium and F^- ions to neutralise the charge of the current carrying Na^+ ions :

$$12\ Na^+ + 4AlF_6^{3-} + 12\ e \rightarrow 12\ (Na^+ + F^-) + 4Al + 12\ F^-$$

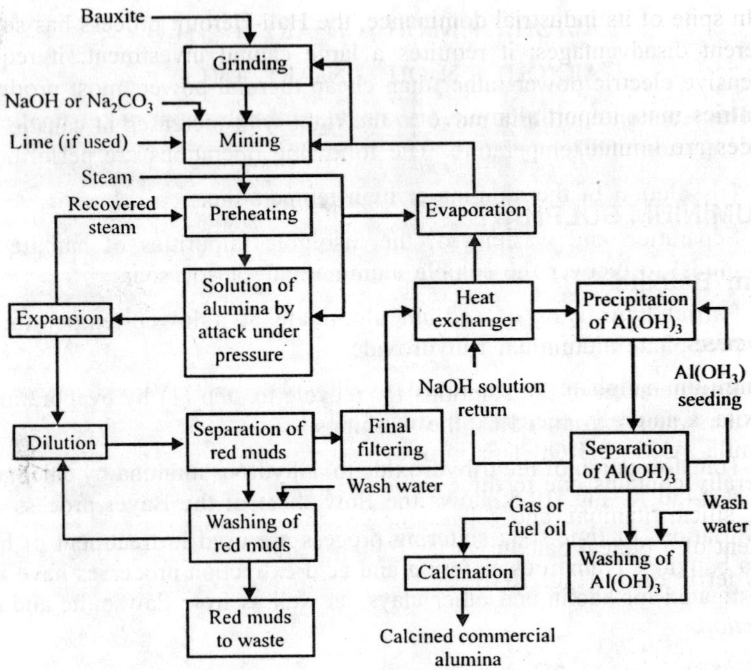

Fig. 10.5. Flow diagram for manufacture of alumina oxide by Bayer process.

Oxyfluoride ions discharge on the anode, forming carbon dioxide and AlF_6^{3-} ions :

$$3\ Al_2O_2\ F_4^2 + 3C + 24\ F^- \rightarrow 3\ CO_2 + 6AlF_6^{3-} + 12\ e$$

The addition of anode and cathode equations, plus solution of alumina gives the overall reaction :

$$2\ Al_2O_3 + 3C \rightarrow 4\ Al + 3\ CO_2$$

According to Faraday's law, one Faraday (268.A·h) theoretically should deposit one gram-equivalent to aluminium (8.994 g). In practice, only 85–95% of this amount is obtained. A modern alumina-smelting cell consists of a rectangular steel sheet lined with refractory insulation surrounding an inner lining of baked carbon which can withstand the combined corrosive action of molten fluorides and molten aluminium. Steel collector bars are joined to the carbon cathode at the bottom to conduct electric current from the cell. Current enters through prebaked carbon anodes or through a continuous Soderberg anode.

In spite of its industrial dominance, the Hall-Héroult process has several inherent disadvantages; it requires a large capital investment; it requires expensive electric power rather than cheap thermal power; most producing countries must import alumina or bauxite; and petroleum coke supplies for anodes are limited.

ALUMINIUM SULPHATE

From Bauxite

Process

Aluminium sulphate is produced by the reaction of sulphuric acid on bauxite, a naturally occurring hydrated alumina ore having the approximate formula $Al_2O_3.2H_2O$. The mined material varies in composition and generally contains one to three molecules of water plus impurities such as iron, silica, titanium, and selenium. The soluble aluminium oxide (Al_2O_3) content of a typical bauxite varies from 52 to 57% and it may contain 1 to 10% ferric oxide (Fig. 10.6).

Reaction

$$Al_2O_3 \cdot 2H_2O + 3H_2SO_4 \rightarrow Al_2(SO_4)_3 + 5H_2O$$

92% yield

The crude bauxite ore is ground to a fine powder (80% passes 200 mesh) and charged into open lead-lined steel reaction tanks. Sulphuric acid (sp. gr. 1.7) is added, and the raw materials are thoroughly agitated using paddle agitators, hot air, or live steam. The reaction mixture is kept at a temperature of 105 to 110°C by the live steam or lead steam coils. An excess of bauxite is fed to the reactor, so that there is an excess of 0.1 to 0.2% of soluble aluminium oxide. From 15 to 20 hours are required to complete the reaction. At the end of this time, a reducing material is added to the reaction mixture to reduce the iron (ferric sulphate) to a colourless ferrous condition. Barium sulphide in the form of black ash is commonly utilised, although sodium sulphide, hydrogen sulphide, sodium bisulphite, or sulphur dioxide may also be used. If the operation is performed in batches, the charge is allowed to settle in settling tanks. Flaked glue or some similar coagulable substance is generally added to remove the finely divided suspended material remaining in the supernatant liquid. This liquid is drawn off, and the residue is washed several times. The washings are combined with the decanted liquor, which is then sent to concentrators.

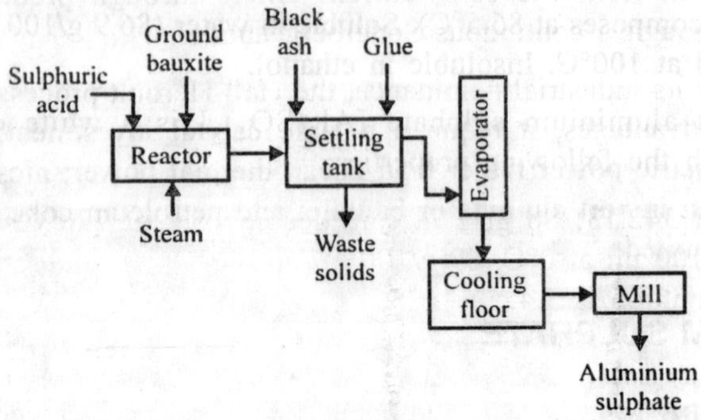

Fig. 10.6. Flow diagram for manufacturing of aluminium sulphate from bauxite.

The process is generally operated in a continuous manner by using a battery of combined reaction and settling tanks. A common variation of this process is the Dorr procedure, which utilises reaction agitators in series. The reactants are thoroughly mixed and heated, using mechanical agitators, air, and live steam. Blach ash (barium sulphide) is added to the last reactor to reduce the ferric sulphate. The reaction mixture is sent through a series of thickeners, operating counter-currently, which remove the undissolved material. At the same time the waste is washed thoroughly, so that it contains practically no aluminium sulphate when discarded. Glue is generally added to the first thickener as a coagulant.

The clarified aluminium sulphate solution, from the counter-current decantation system, is concentrated in open steam-coil-heated, lead-lined evaporators. Here, the specific gravity is increased from about 1.3 to about 1.7. The concentrated solution is run into flat iron pans or onto a cooling table. The liquid quickly and completely solidifies, and when cool is broken up and ground to a uniform powder for shipment.

Commercial aluminium sulphate generally contains only about 13 or 14 moles of water instead of the theoretical 18 moles. Also, it is usually in the basic form containing excess alumina. Anhydrous aluminium sulphate may be obtained by dehydration. The yield of aluminium sulphate based on the amount of aluminium oxide in both the finished product and raw material is 90 to 95%.

Uses

Pulp and paper; Water purification; etc.

Properties

Colourless or white monoclinic crystals. Formula wt. 666.40; Sp gr 1.69 (17°C); mp decomposes at 86.5°C). Soluble in water (86.9 g/100 ml at 0°C, 1104 g/100 ml at 100°C. Insoluble in ethanol.

Anhydrous aluminium sulphate [$Al_2(SO_4)_3$] is a white crystalline compound with the following properties:

Formula wt 342.13; Sp gr 2.71; mp decomposes at 770°C. Soluble in water (31.3 g/100 ml at 0°C, 89 g/100 ml at 100°C). Insoluble in ethanol. pH of a 1% solution is 3.4.

ALUMS

This series of double sulphates has the general formula :

$$M_2SO_4.N_2(SO_4)_3.24H_2O,$$

where M is univalent ion such as Na^+, K^+, or NH_4^+, and N is a trivalent ion such as Fe_3^+, Al_3^+, or Cr_3^+.

e.g.	$K_2SO_4.Al_2(SO_4)_3.24H_2O$	potash alum
	$(NH_4)_2SO_4.Al_2(SO_4)_3.24H_2O$	ammonium alum
	$K_2SO_4.Cr_2(SO_4)_3.24H_2O$	chrome alum
	$(NH_4)_2SO_4.Fe_2(SO_4)_3.24H_2O$	ferric alum.

They are formed by mixing hot concentrated solutions of their component sulphates and cooling, when the alum crystals are deposited because of their low solubility in cold water. They are easily purified by recrystallisation because their solubility increases markedly with rising temperature.

The following special methods of preparation are sometimes used: *Potash Alum:* Enough aluminium foil is added to some caustic potash solution to leave some undissolved at the end of the reaction. The solution of potassium aluminate is filtered, and dilute sulphuric acid is added until the solution is acidic (four gram-equivalents of acid are required for every gram-equivalent of alkali taken). The solution is then concentrated until it deposits crystals of the alum on cooling. These are filtered off, washed with a little cold water and dried; they can be recrystallised if necessary. Equations for the reactions are:

$$2Al + 2KOH + 2H_2O = 2KAlO_2 + 3H_2\uparrow$$

$$2KAlO_2 + 4H_2SO_4 = \underbrace{K_2SO_4 + Al_2(SO_4)_3}_{\text{alum}} + 4H_2O$$

Chrome Alum

A steady stream of sulphur dioxide is passed through a solution of potassium dichromate, acidified with sulphuric acid, until the solution turns green. Deep purple crystals of chrome alum separate from the concentrated solution on cooling. The equation for the reduction is :

$$K_2Cr_2O_7 + 3SO_2 + H_2SO_4 = K_2SO_4 + \underset{\text{alum}}{Cr_2(SO_4)_3} + H_2O$$

The crystals are filtered off, washed, and dried. The alums are isomorphous with each other, readily forming overgrowths; their crystals are usually octahedral in shape. It is important to realise that the alums are double salts and not complex salts. In solution they behave simply as a mixture of their component sulphates and give the reactions of their individual cations.

ALUMINIUM CHLORIDE

From Aluminium Metal and Chlorine

Process

Scrap aluminium, or a mixture of scrap and pig aluminium, is charged to a refractory crucible furnace in which it is melted (660°C). Dry chlorine is passed into the molten charge and forms aluminium chloride which vapourises and leaves the furnace through a vapour duct in the top. The vapours are then passed into air-cooled iron condensers, where aluminium chloride sublimes below 178°C. The aluminium chloride is removed periodically (Fig. 10.7).

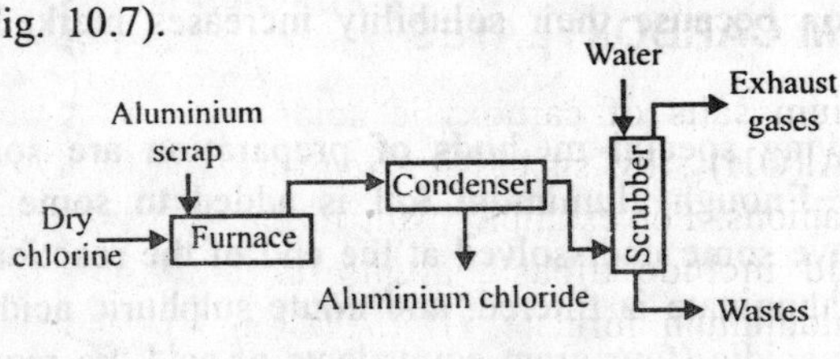

Fig. 10.7. Flow diagram for manufacture of aluminium chloride from aluminium metal and chlorine.

Reaction

$$2Al + 3Cl_2 \rightarrow 2AlCl_3$$

By proper regulation of the chlorine rate, no free chlorine remains in the exit gases from the furnace, but a protective absorber is usually placed beyond the aluminium chloride condenser to absorb unreacted chlorine and any aluminium chloride that may pass through the condenser.

Purity of the aluminium chloride depends on the quality of the aluminium used, as well as on the tightness of the system; moisture in the chlorine introduces corrosion products of the materials of construction. Usually high-purity aluminium chloride is obtained.

Special high-purity grades of crystalline aluminium chloride ($AlCl_3.6H_2O$) and solutions, are made by reaction of purified aluminium hydrate with hydrochloric acid.

Uses

Catalyst in: Ethyl benzene; Dyestuff intermediates; Detergent alkylate; Hydrocarbon resins; Ethyl chloride; Pharmaceutical and cosmetic; Pigments; Roofing granules; Special papers; etc.

Properties

Anhydrous—white deliquescent hexagonal crystals, which sublime readily. Formula wt 133.34; Sp gr 2.44 25°C/4; mp 194°C (520 kPa); bp 182.7°C (100 kPa); sublimes at 178°C. Soluble in water (69.87 g/100 ml at 15°C, decomposes in hot water), ether, chloroform, and carbon tetrachloride. Insoluble in benzene.

Aluminium chloride crystallises from hydrochloric acid as the hexahydrate ($AlCl_3.6H_2O$) in the form of colourless deliquescent trigonal crystals.

ALUMINIUM CARBOXYLATES

The aluminium salts of carboxylic acids are derived from aluminium hydroxide, $Al(OH)_3$, by successive replacement of hydroxyl groups by carboxylate anions. For example, with formic acid the aluminium formate family would include dibasic aluminium formate, $(HO)_2Al(OOCH)$; monobasic aluminium formate, $(HO)Al(OOCH)_2$; and normal aluminium formate, $Al(OOCH)_3$.

General methods of preparation for the mono- and dibasic aluminium carboxylates include direct reaction of aluminium as the catalytic electrode with an aqueous solution of the acid in an electrolytic process; reaction of the metal with the acid in the presence of catalytic amounts of mercury or mercuric chloride; and reaction by double replacement between aluminium alkoxide and the sodium salt of the organic acid. Normal aluminium carboxylates, e.g., tris(formato)aluminium, are prepared by the direct action of the acid with aluminium chloride in an organic solvent or by a double displacement reaction using a soluble aluminium salt and sodium salt of the organic acid. (Aluminium distearate is the largest selling aluminium carboxylate).

Uses

About 15–20 of the aluminium carboxylates are industrially important compounds with commercial applications. Commercial applications fall into three general areas: pharmaceutical preparations, particularly aluminium acetate and aluminium formate, based on the very low toxicity of the salts and upon either their basic or astringent properties; gelling agents, particularly aluminium mono-, di-, and tristearates, aluminium palmitate, and aluminium octanoate (2-ethylhexanoate); the manufacture of cosmetics, coatings, and rocket fuels; and aluminium carboxylates, which are used in the textile industry as finishing agents for water-proofing fabric and as mordants in the dyeing process.

ALUMINIUM CHLORIDE HEXAHYDRATE

Aluminium chloride hexahydrate is usually made by dissolving aluminium hydroxide, $Al(OH)_3$, in concentrated hydrochloric acid. It is used primarily in deodorant, antiperspirant, and fungicidal preparations; for water treatment; in the manufacture of alumina and alumina–silica refractories; and in textile finishing. It may be toxic on ingestion.

ALUMINIUM BROMIDE AND ALUMINIUM IODIDE

Aluminium bromide and aluminium iodide are produced commercially only in small quantities. Anhydrous aluminium bromide, AlB_3, (density (25°C) 3.01 g/cm^3; melting point 97°C; boiling point 256°C) is used as a catalyst in Friedel-Crafts reactions. Anhydrous aluminium iodide, AlI_3, (density (25°C) 3.98 g/cm^3; melting point = 191°C) has limited use as a catalyst in the laboratory.

ALUMINIUM NITRATE

Aluminium nitrate, $Al(NO_3)_3.9H_2O$, is a white crystalline material, mp 73.5°C, soluble in cold water, alcohols, and acetone; decomposes to nitric acid and basic aluminium nitrates, $Al(OH)x(NO_3)y$ ($x + y = 3$); and dissociates to aluminium oxide and oxides of nitrogen above 500°C. It is commercially produced from aluminous materials such as bauxite and nitric acid.

Hydrated aluminium nitrate is used primarily as a salting-out agent in the extraction of actinides.

ALUMINIUM HYDROXIDES

Alumina monohydrates and hydrated alumina do not contain water of hydration but only hydroxide and oxide groups. Preparations of aluminium hydroxide obtained from solution of aluminium salts, alkaline aluminates, or aluminium alcoholates that are mainly amorphous are called gels. They are amphoteric, with low solubility in aqueous solutions of intermediate pH. Colloidal aluminium hydroxide is hydrophilic. Depending on the method of preparation, not all gelatinous aluminas are amorphous.

α-Alumina trihydrate is obtained by the Bayer process from crystallised alumina trihydroxide (known as gibbsite or hydrargillite, the principal constituent of tropical bauxites). Grains of α-trihydrate precipitated by the Bayer process are aggregates of relatively spherical shape, 50–100 μm in size. Solubility of α-trihydrate and heat of solution in various bases and acids is important in connection with the manufacture of aluminium chemicals. It contains about 35% chemically bound water. Typical Bayer hydrate contains 0.4 wt% of sodium ions in the crystals lattice; along with other alumina hydrates, it chemisorbs various anions effectively.

β-Alumina trihydrate is not a product of the Bayer process but is prepared by aging of gels, gassing of sodium aluminate liquor with carbon dioxide, and calcination of aluminium oxychloride. Nordstandite, a new trihydrate.

α-Alumina oxide hydroxide, a well-crystallised modification of AlOOH, also known as α-alumina monohydrate or boehmite, is a principal constituent of many bauxites of the Mediterranean type. Crystallised boehmite is prepared by hydrothermal digestion of aqueous slurries of trihydroxides or by solid-state reaction when gibbsite is heated in air at 383–573 K. Boehmite usually contains water in excess of the theoretical 15%. Another well-crystallised modification of AlOOH, β-alumina oxide hydroxide or diaspore, has been produced in the laboratory.

ACTIVATED ALUMINAS

These chemicals are obtained from various hydrated aluminas by controlled heating which eliminates most of the water of constitution. Their crystal structure is that of the transition aluminas: γ, η, χ or ρ alumina. Structural properties of the transition aluminas are given in Table 10.1. These transition aluminas are known as activated alumina or active alumina.

Table 10.1. Structural properties of transition aluminas.

Form	*Crystal system*	*Molecules per unit cell*	*Unit axis length, nm* *a*	*b*	*c*	*Density, g/cm³*
γ	Tetragonal		0.562	0.780		3.2[a]
δ	Orthorhombic	12	0.425	1.275	1.021	3.2[a]
	tetragonal		0.796		2.34	
η	Cubic (spinel)[b]	10	0.790			2.5–3.6
θ	Monoclinic[c]	4	1.124	0.572	1.174	3.56
χ	Cubic	10	0.795			3.0[a]
	hexagonal		0.556		1.344	
κ	Hexagonal	28	0.971		1.786	3.1–3.3
	hexagonal		1.678		1.786	
ε	Orthohombic	4	0.773	0.778	0.292	3.71[a]
	orthorhombic[d]	3	0.759	0.767	0.287	3.0

[a] Estimated.

[b] Space group O_h^7

[c] Space group C_{2h}^3; angle of crystal, 103°20'

[d] Space group D_{2h}^9 or C_{2v}^8.

Preparation

Type 1, the oldest commercial form still widely used, is made from Bayer α-trihydrate. Processes produce various activated aluminas in cylindrical or spherical forms, but the pore distribution is not broad and few pores are greater than several tens of nanometers in diameter. Surface areas of these products are 300–500 m²/g. Type 2, activated bauxites with properties similar to Type 1, is obtained by thermal activation of bauxite containing alumina in the form of gibbsite. Type 3, another activated alumina, is obtained by very rapid activation of Bayer hydrate at 673–1073 K and has a pore surface area of 300–350 m²/g. Type 4 is derived from alumina gels, principally those prepared from solutions of $Al_2(SO_4)_3$ and NH_3, from $NaAlO_2$ and an acid, or from $NaAlO_2$ and $Al_2(SO_4)_3$, and produce corresponding by-product salts. Surface areas of pores are 300–600 m²/g.

CALCINED ALUMINA

These are generally obtained from Bayer hydroxide. Commercial calcined aluminas are heated to obtain α-Al_2O_3, the stable form of anhydrous alumina known also as the naturally occurring substance corundum. α-Al_2O_3 also results from solidification of molten alumina or by sintering processes. Although α-Al_2O_3 has a mp of 2326 K, the relatively fine hexagonal crystal plates obtained during calcination permit sintering to occur at much lower temperatures for ceramic-product manufacture. Higher sintering temperatures improve thermal stability for refractory applications, and sintered and fused α-Al_2O_3 products made from calcined aluminas exhibit extreme hardness, resistance to wear and abrasion, chemical inertness, outstanding electrical and electronic properties, good thermal shock resistance and dimensional stability, and high mechanical strength at elevated temperatures.

Preparation

Calcination is performed in rotary kilns and/or fluid calciners similar to those used for preparing alumina for the manufacture of aluminium by electrolysis. The particle size of the unground calcined Bayer aluminas is controlled, and the resulting porous agglomerates of alumina crystals (< 1–$10\ \mu m$) are nominally 44–149 μm 100–325 mesh. Physical properties begin to exhibit the effect of crystal size above ca 2–3 μm. Bulk density decreases, angle of repose increases, attrition rate and dustiness increase, and bulk handling characteristics worsen with increasing crystal size. Special Bayer calcined aluminas can be broadly categorised according to sodium content and crystal size.

Dense sintered alumina containing large (typically 50–300 μm) tablet like crystals of α-Al_2O_3 i.e. commonly known as tabular alumina. Fused alumina is available as larger tonnage brown alumina and higher quality white alumina. Calcined sodium β-Al_2O_3, in the forms of $Na_2O.11Al_2O_3$ and $Na_2O.5Al_2O_3$, is also available.

Uses

About 90% of alumina is used in the production of aluminium metal. The remainder, including Bayer-source alumina, chemical-grade bauxite, and calcined abrasive and refractory-grade bauxite, is consumed is other applications such as flame-retardant fillers, preparation of aluminium compounds, pigments, adsorbents, catalysts, ceramics, refractories, and abrasives.

COPPER SULPHATE

From Copper and Sulphuric Acid

Process

Copper sulphate is made by the reaction of scrap or shot copper with sulphuric acid and oxygen (air). The copper is fed to an oxidising tower; and a sulphuric acid-water mixture (20% sulphuric acid) is circulated through the bed while air is blown through the bottom of the tower. Copper sulphate pentahydrate is recovered from the resulting solution by crystallisation; the liquor is recycled to the tower along with makeup sulphuric acid (Fig. 10.8).

Reaction

$$2Cu + 2H_2SO_4 + O_2 \text{ (air)} + 8H_2O \rightarrow 2\, CuSO_4{\cdot}5H_2O$$

Much of the copper sulphate on the market is a by-product in copper refining. Copper is often leached from its ores with sulphuric acid; the resulting solution is treated in an electrolytic cell using an insoluble anode. Copper is plated at the cathode, regenerating sulphuric acid. The spent electrolyte may then be used on a fresh batch of ore. The cell solution after reuse may be pumped to lead-lined tanks for evaporation and crystallisation of copper sulphate pentahydrate.

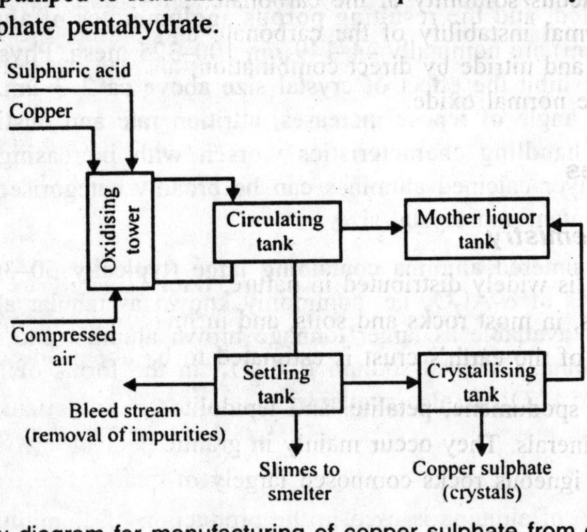

Fig. 10.8. Flow diagram for manufacturing of copper sulphate from copper and sulphuric acid.

Properties

$CuSO_4.5H_2O$ crystallises from water as the pentahydrate in the form of blue triclinic crystals which slowly effloresce in air. Formula wt 249.67; Sp gr

2.286 15.6°C/4; mp Loses $4H_2O$ at 110°C; Loses $5H_2O$ at 250°C. Soluble in water (24.3 g/100 g at 0°C, 205 g/100 g at 100°C). The anhydrous salt ($CuSO_4$) is obtained by dehydration as greenish-white rhombic crystals. Formula wt 159.60; Sp gr 3.606 (15°C); mp 200°C; bp forms CuO at 650°C.

Uses

Fungicides; Algaecides; Feed supplement; Soil nutrient; Intermediates; Flotation; Electroplating; Petroleum etc.

LITHIUM AND LITHIUM COMPOUNDS

The applications of lithium metal are mainly in metallurgy and batteries and in the manufacture of lithium compounds, such as the amide, hydride, and nitride, and organolithium compounds.

Many of the properties of lithium are similar to those of magnesium and of the alkaline-earth metals. The resemblance to magnesium includes the high solubility of the halides (except the fluoride) in both water and polar organic solvents and the high solubility of the alkyls in hydrocarbons; the low aqueous solubility of the carbonate, phosphate, fluoride, and oxalate; the thermal instability of the carbonate and nitrate; the formation of the carbide and nitride by direct combination; and the reaction with oxygen to form the normal oxide.

Sources

Geochemistry

Lithium is widely distributed in nature; trace amounts are present in many minerals, in most rocks and soils, and in many natural waters. The lithium content of the earth's crust is estimated to be about 20 ppm.

One spodumene, petalite, and lepidolite are important lithium sources from minerals. They occur mainly in granite pegmatites, which are coarse-grained igneous rocks composed largely of quartz, feldspar, and mica.

Production

Recovery from ores and brines

Spodumene is the chief lithium source for the two domestic producers of lithium products. The preferred method of extraction of lithium from spodumene is the sulphuric acid process. Methods suitable for extraction from spodumene also can be used for petalite, since the latter mineral converts to a β-spodumene-SiO_2 solid solution on heating to a high temperature.

Other recovery processes from silicate ores

Most of the numerous processes that have been described for the extraction of lithium can be classified as either alkaline methods or ion-exchange methods.

Recovery from brines

Natural predominately chloride brines vary widely in composition, and the economical recovery of lithium from such sources depends not only on the lithium content but on the concentration of interfering ions, especially calcium and magnesium. The location and the availability of solar evaporation capability also are important factors.

Lithium Metal

Properties

Lithium is an alkali metal with a silvery lustre and an atomic weight of 6.941. It is the first member of Group IA in the periodic system. Two stable isotopes are present in natural lithium (^7Li has an abundance of 92.6 wt% and ^6Li 7.4 wt%). Lithium has a density of 0.531 g/cm^3 at 20°C, and is the lightest of all solid elements. In general, the properties of lithium are similar to those of the other alkali metals. Physical properties of lithium are listed in Table 10.2.

The reaction of hydrogen and lithium readily gives a hydride, LiH, which is stable at temperatures from the melting point up to 800°C. Lithium reacts with nitrogen, even at ordinary temperatures, to form the reddish-brown nitride, Li_3N. Lithium burns when heated in oxygen to form the white oxide, Li_2O.

Manufacture

An electrolytic process consists of molten-salt electrolysis from a lithium chloride-potassium chloride mixture using a graphite rod as an anode. Modern industrial installations employ a 55 wt% LiCl–45 wt% KCl electrolyte at about 460°C using high purity lithium chloride as a feed. The current efficiency is about 80% and the lithium recovery is better than 98%, based on the charged chloride. The purity of the metal is 99.8% or better and the metallic impurities are less than 0.1%.

Table 10.2. Physical properties of lithium.

Property	*Value*
At wt	6.941
At vol, cm[a]	13.0
mp, °C	180.5
bp, °C	1336
Electronic configuration	$1s^2\ 2s^1$
First ionisation potential, kJ/mol[a]	519
Electron affinity, kJ/mol[a]	52.3
Crystal structure	bcc
Lattice constant, pm	350
Metallic radius, pm	122.5
Ionic radius, pm	60
d_{20}, g/cm^3	0.531
Specific heat at 25°C, J/g[a]	3.55
Specific heat of liquid at mp, J/g[a]	4.39
Heat of fusion, J/g[a]	431.8
Heat of vapourisation, kJ/g[a]	ca 21.3
Electrical resistivity at 20°C, μΩ.cm	9.446
Characteristic spectrum lines, nm	
red	670.8
orange	610.4
Vapour pressure at 702°C, kPa[b]	0.065

[a] To convert J to cal, divide by 4.184.

[b] To convert kPa to mm Hg. multiply by 7.5.

Inorganic Lithium Compounds

Lithium acetate

Lithium acetate, $CH_3COOLi.2H_2O$, is used as an alcoholysis catalyst for alkyd resin manufacture.

Lithium amide

Lithium amide, $LiNH_2$, is used in the pharmaceutical industry in the synthesis of antihistamines and analgesics.

Preparation

It can be prepared by the reaction of lithium hydride with ammonia. It is flammable.

Properties

White crystalline solid; ammonia-like odour; sp. gr. 1.18; melts 380–400°C; decomposes in water, to form ammonia.

Uses

Organic synthesis, including antihistamines and other pharmaceuticals.

Lithium borates

Lithium metaborate, $LiBO_2.2H_2O$, mp 849°C, is used in special glass and enamel formulations. Lithium tetraborate, $Li_2B_4O_7$, mp 917°C, is used in the ceramic industry and as a flux in emission x-ray spectroscopy.

Lithium carbonate

Lithium carbonate is one of the most important lithium salts because it may be prepared from most of its water-soluble compounds.

Preparation

(1) Finely ground ore is roasted with sulphuric acid at 250°C. Lithium sulphate is leached from the mass and converted to the carbonate by precipitation with soda ash.

(2) Reaction of lithium oxide with carbon dioxide or ammonium carbonate solution. Water solution is strong irritant.

Properties

It is in the form of white powder; sp. gr. 2.111; m.p. 735°C; b.p., decomposes at 1200°C. Slightly soluble in water; insoluble in alcohol; soluble in dilute acid.

Uses

Ceramics and porcelain glazes; pharmaceuticals; catalyst; other lithium compounds; coating of arc-welding electrodes; nucleonics; luminescent paints, varnishes and dyes; glass ceramics; aluminium production.

MOLYBDENUM COMPOUNDS

Halides and Oxyhalides

Molybdenum forms halogen compounds of widely different degrees of stability. The hexafluoride and pentachloride deposit molybdenum metal in the vapour phase as result of reaction with hydrogen.

The highest member of each series (MoF_6, $MoCl_5$, $MoBr_4$, and MoI_3) can be made by direct halogenation of molybdenum metal. The hexafluoride, pentafluoride, pentachloride, oxytetrafluoride, oxytetrachloride, and dioxydichloride are volatile at moderate temperatures. The entire series exhibits a wide range of stability, colour, and volatility. Few of the compounds are monomeric in their normal state. Generally, the lower halides are prepared by reduction of the highest member of the series with molybdenum metal, hydrogen, or a hydrocarbon.

Oxides and Hydroxy Compounds

Molybdenum trioxide

MoO_3, is a white crystalline powder from which most molybdenum compounds are prepared directly or indirectly. Technical grade MoO_3 is prepared commercially by oxidising molybdenite, MoS_2, in a multiple-hearth roaster. Air flow and temperature are carefully controlled to provide a low sulphur product with minimum MoO_2 content. High purity chemical grades are prepared either by sublimation of the technical grade or calcination of crystallised ammonium dimolybdate (ADM). Halogens and hydrohalides react with MoO_3 to form oxyhalides. Reaction with thionyl chloride, $SOCl_2$, gives molybdenum oxytetrachloride or pentachloride, depending on the reaction time.

Hydroxides

The structures of molybdenum blues and the series of hydrated molybdenum oxides are intimately related to those of the oxides and are characterised as genotypic. Hydrated oxides are prepared by the carefully controlled reduction of MoO_3 with either atomic hydrogen, zinc and hydrochloric acid, molybdenum metal powder and water, or lithium aluminium hydride.

Molybdates

Isopolymolybdates, a large class of inorganic compounds, consist of a cation and a condensed molybdate anion. The naming of isopolymolybdates is straightforward: they are called simply di, tri, etc, molybdates depending upon the degree of condensation.

Normal or orthomolybdates

$M_2O.MoO_3.xH_2O$ or $M_2MoO_4.xH_2O$, are prepared by direct combination of the oxides, neutralisation of slurries of MoO_3 with MOH or M_2CO_3, or by precipitation from the molybdate solution by salts of the desired metals. Acidification of molybdate ions in alkaline solution in which they exist as $(MoO_4)^{2-}$ results in the formation of polynuclear species.

Heteropolymolybdates

These form a large family of salts and free acids with each member containing a complex and high molecular weight anion.

Sulphides, Selenides, and Tellurides

Molybdenum forms a series of homologous compounds with sulphur, selenium, and tellurium. The disulphide, MoS_2, diselenide, $MoSe_2$, and ditelluride, $MoTe_2$, are isomorphous. These chalcogenides occur as shiny gray plates and may be prepared by the direct combination of the elements or by heating MoO_3 and the appropriate element in potassium carbonate to a high temperature.

Cyanide, Thiocyanates, and Carbonyls

Cyanides and thiocyanates form complex anions with molybdenum in a lower valence state. In most of these, molybdenum atoms have a coordination number of eight. These compounds are highly coloured. The cyanides and the thiocyanates are light sensitive, but are generally more stable in air than the complex halides.

The red thiocyanate complex is formed by adding a solution of a soluble thiocyanate to a reduced acid solution of molybdenum. It is utilised in the colourimetric determination of molybdenum. The compound contains a thiocyanate to molybdenum ratio of three; the Mo is pentavalent.

Organomolybdenum Compounds

In addition to the substituted molybdenum hexacarbonyl compounds, organomolybdenum sulphur compounds containing phosphorus or nitrogen have been prepared. Many of these compounds are oil-soluble and, thus, the low friction benefits of the molybdenum-sulphur interaction can be imparted to engine and gear oils.

Health and safety

Non-toxic in humans. In some areas where soil molybdenum is high, and consequently the molybdenum content of forage crops is high, animals develop a toxic reaction called molybdenosis, which can be overcome by injection or feed implementations of copper salts.

Uses

The chemical applications of molybdenum have increased substantially over the past twenty years. New applications have been developed in catalysis, lubrication, corrosion inhibition, protective coatings, inhibitive pigments, and flame and smoke retardants. Molybdenum compounds are used in the transportation and petroleum and petrochemical industries; molybdenum catalysts are prime candidates in the conversion of coal to liquid fuels.

STRONTIUM AND STRONTIUM COMPOUNDS

Strontium is in Group IIA of the periodic table between calcium and barium. Strontium is present at 0.02–0.03% concentration in the earth's crust and is the fifth most abundant metallic ion in sea water. Strontium rarely forms independent minerals in igneous rocks but usually occurs as a minor constituent of rock-forming minerals.

Strontium is a hard, white metal with a melting range of 768–791°C, a boiling range of 1350–1387°C, and a density of 2.6 g/cm^3, crystallising in the fcc system. Strontium-90 is a radioisotope produced by nuclear fission; four stable isotopes of strontium are known.

Strontium is produced by reduction of the oxide with aluminium *in vacuo*. There are no commercial uses of strontium metal. It has been used to remove traces of gas from vacuum tubes.

Strontium Compounds

Although many strontium compounds are known, only a few have commercial importance; the carbonate and nitrate are made in large quantities.

The principal strontium mineral is celestite, $SrSO_4$. The carbonate is found as the mineral strontianite; no economically workable deposits are known. Almost all celestite is used to make strontium carbonate.

Strontium carbonate

Strontium carbonate is white impalable powder. Soluble in acids, carbonated water and solutions to ammonium salts, slightly soluble in water. Sp. gr. 3.62; loses CO_2 at 1340°C. Low toxocity. It is prepared when celestite is boiled with a solution of ammonium carbonate or is fused with sodium carbonate. It is used as a catalyst, in radiation-resistant for colour television tubes; ceramic ferrites; pyrotechnics.

Strontium chlorate

It is white crystalline powder, soluble in water, slightly soluble in alcohol. Sp. gr. 3.152; m.p., decomposes at 120°C.

It is prepared when strontium hydroxide solution warmed and chlorine is passed in with subsequent crystallisation. It is used in the manufacture of red-fire and other pyrotechnics; tracer bullets.

Strontium chloride

It is white in the form of crystalline needles, odourless, sharp, bitter taste, soluble in water and alcohol. Sp. gr. 3.054; m.p. 872°C; b.p. 1250°C.

It is prepared when strontium carbonate is fused with calcium chloride, the melt extracted with water, the solution is then concentrated and crystallised. It is used in strontium salts pyrotechnics; electron tubes.

BORON

Boron, the fifth element in the periodic table, is composed of two stable isotopes with mass numbers of 10 and 11. Although widespread in nature, it has been estimated to constitute only 0.0001% of the earth's crust, usually occurring as alkali or alkaline-earth borates, or as boric acid. Pure forms of the element are difficult to prepare; the most common technique involves vapour deposition from a boron halide, usually in admixtures with hydrogen. Research continues for methods of obtaining commercial quantities of the pure element, particularly as filamentary reinforcement for advanced composites. In nuclear technology, thin films of boron are used for neutron counters, and dispersions of powdered boron in poly(vinyl chloride) or polyethylene castings are effective for shielding against thermal neutrons. Table 10.3 lists physical properties of boron.

Table 10.3. Physical properties of boron.

Property	*Value*
Atomic weight	10.811 ± 0.003
Melting point, °C	2190 ± 20
Boiling point, °C	3660
Coefficient of thermal expansion per °C	
from 25 to 1050°C	$(5–7) \times 10^{-6}$
Hardness	
Knoop, HK	2110–2580
Mohs, modified scale[a]	11
Vickers, HV	5000
Density	
Liquid[b]	2.08
α-rhombohedral crystals	2.46
Filamentary boron	
Tensile strength, MPa (psi)	3450–4830 (500,000–700,000)
Young's modulus, MPa (psi)	3040–3330 (440,000–480,000)
Structural modifications[c]	
Preparation temp, °C	
amorphous	800
α-rhombohedral	800–1100
α-tetragonal	1100–1300
β-rhombohedral	1300
Density, g/cm^3	
amorphous	2.3
α-rhombohedral	2.46
α-tetragonal	2.31
β-rhombohedral	2.35

[a] Diamond = 15.

[b] Just above melting point.

[c] The crystalline forms not listed here are designated tetragonal II (β-tetragonal), tetragonal III, and hexagonal.

BORON COMPOUNDS

At present, borax (tincal), colemanite, probertite, ulexite, and szaibelyite are the only borate minerals of commercial importance. Borax and colemanite are the most important.

Boron Oxides

Boric oxide, B_2O_3, is the only commercially important oxide; however, one high oxide and several suboxides have been reported. Boric oxide, also known as diboron trioxide, boric anhydride, or anhydrous boric acid, normally is encountered in the vitreous state. This colourless, glassy solid usually is prepared by dehydration of boric acid at elevated temperatures. It is quite hygroscopic at room temperature, and the commercially available (Table 10.4).

Table 10.4. Physical properties of vitreous boric oxide.

Property	*Value*
Vapour pressure[a], 1331–1808 K	$\log P_{kPa} = 5.849 - \frac{16960}{T}$
Heat of vapourisation[b], ΔH, kJ/mol, 1500 K	390.4
298 K	431.4
Boiling point, °C extrapolated	2316
Viscosity, log η , mPa.s (=cP)	
1000°C	4.00
Density, g/cm^3, 0°C	1.8766
18–25°C,	1.844
well-annealed	
1000°C	1.528
Index of refraction, 14.4°C	1.463
Heat capacity (specific)[b], J/(kg.K) 1000 K	131.38
Heat of formation, ΔH_f, kJ, 298.15 K[b]	
for $2B(s) + 3/2\ O_2(g) = B_2O_3$ (glass)	–1252.2 ± 1.7

[a] To convert kPa to torr, multiply by 7.5.

[b] To convert J to cal, divide by 4.184.

A high purity grade of vitreous boric oxide (99% B_2O_3) is produced by fusing refined, granular boric acid in a glass furnace fired by oil or gas. Principal uses of boric oxide relate to its behaviour as a flux, an acid catalyst, or a chemical intermediate.

Boric Acid

The name boric acid usually is associated with orthoboric acid, which is the only commercially important compound. Three forms of metaboric acid also exist. Orthoboric acid, H_3BO_3, crystallises from aqueous solutions as white, waxy plates, mp 170.9°C. When heated slowly, it loses water to form metaboric acid, HBO_2.

Most boric acid is produced by the reaction of inorganic borates with sulphuric acid in an aqueous medium. Sodium borates are the principal raw material and Boric acid serves as a source of B_2O_3 in many fused products such as textile fibreglass and other borosilicate glasses. An important new use of boric acid is as a fire retardant in cellulosic materials. It also serves as a component of fluxes for welding and brazing. A number of boron chemicals are prepared directly from boric acid. It also catalyses the air oxidation of hydrocarbons and increases the yield of alcohol by forming esters that prevent further oxidation of hydroxyl groups to ketones and carboxylic acids. The bacteriostatic and fungicidal properties of boric acid, although weak, have led to its use as a preservative in natural products. NF-grade boric acid serves as a mild, non-irritating antiseptic. Boric acid also is quite poisonous to many insects and has been used to control cockroaches and to protect wood against insect damage. A special quality grade of boric acid is added to nuclear-reactor cooling water.

Solutions of Boric Acid and Borates

Boric acid is essentially monomeric in dilute aqueous solutions, but polymeric species may form at concentrations > 0.1 *M*. The conjugate base of boric acid in water is the tetrahydroxyborate anion, $B(OH)_4^-$. This species also is the principal anion in solutions of alkali-metal (1:1) borates such as $Na_2O.B_2O_3.4H_2O$. Mixtures of $B(OH)_3$ and $B(OH)_4^-$ would appear to form classical buffer systems where the solution pH is governed primarily by the acid-salt ratio. This relationship is nearly correct for solutions of sodium or potassium (1:2) borates, where the mole ratio $B(OH)_3$: $B(OH)_4^- \equiv 1$, and the pH remains near 9.0 over a wide range of concentrations. However, for solutions that have pH values much greater or less than 9.0, the pH changes greatly on dilution. This anomalous pH behaviour is due to the presence of polyborates which dissociate into $B(OH)_3$ and $B(OH)_4^-$ as the solutions are diluted. Formation of polyborates also greatly enhances the mutual solubilities of boric acid and alkali borates.

From a series of very accurate pH studies, Ingri calculated equilibrium constants involving the species $B(OH)_3$ and $B(OH)_4^-$, and the polyions $B_3O_3(OH)_5^{2-}$, $B_3O_3(OH)_4^-$, $B_5O_6(OH)_4^-$ and $B_4O_5(OH)_4^{2-}$. The ratio between the total anionic charge and the number of borons per ion increases with increasing pH.

Sodium Borates

Disodium tetraborate decahydrate (borax decahydrate), $Na_2O.2B_2O_3.10H_2O$ [formula wt 381.43, monoclinic, sp gr 1.71, specific heat 1.611 kJ/(kg.K) (0.385 cal/g.°C)) at 25–50°C, heat of formation –6.2643 MJ/mol (–1497.2 kcal/mol)] exists in nature as the mineral borax. Disodium tetraborate pentahydrate (borax pentahydrate), $Na_2O.\ 2B_2O_3.5H_2O$ [formula wt 291.35, trigonal, sp gr 1.88, specific heat 1.32 kJ/(kg.K) (0.316 cal/(g.°C)), heat of formation –4.788 MJ/mol (–1143.5 kcal/mol)] is found in nature as a fine-grained deposit formed by dehydration of borax. Disodium tetraborate tetrahydrate, $Na_2O.5B_2O_3.4H_2O$ [formula wt 273.34, monoclinic, sp gr 1.91, specific heat ca 1.2 kJ/(kg.K) (0.287 cal/(g.°C)), heat of formation –4.4890 MJ/mol (–1072.9 kcal/mol)] exists in nature as the mineral kernite. Disodium tetraborate (anhydrous borax), $Na_2O{:}2B_2O_3$ [formula wt 201.27, sp gr (glass) 2.367, heat of formation (glass) –3.2566 MJ/mol (–778.4 kcal/mol)] exists in several crystalline forms as well as a glassy form. Sodium pentaborate decahydrate, $Na_2O{:}2B_2O_3.10H_2O$ [formula wt 590.34, monoclinic, sp gr 1.71] exists in nature as the mineral sborgite. Sodium metaborate octahydrate, $Na_2O.B_2O_3.8H_2O$ [formula wt 295.76, triclinic, sp gr 1.74] and sodium metaborate tetrahydrate, $Na_2O.B_2O_3.4H_2O$ [formula wt 203.68, triclinic, sp gr 191] can also be prepared.

Borax decahydrate and pentahydrate are produced from sodium borate ores, from dry lake brines, from colemanite, and from magnesium borate ores. In general, the production of fused materials is much more energy intensive than that of hydrated products, and this difference is reflected in their prices.

The bulk borate products, borax decahydrate and pentahydrate, anhydrous borax, boric acid and oxide, and upgraded colemanite and ulexite, account in both tonnage and monetary terms for > 99% of sales of the boron primary products industry. A large increase in demand has been related to the use of borates in energy-conserving products (i.e., insulation).

Poisonings by boric acid have been reported following its use over larger areas of burned or denuded skin. The handling of borax or boric acid is generally not considered dangerous, however.

Borates are used mainly for glass manufacture. They also are used as fluxing agents for porcelain enamels and ceramic glazes, in soap and cleaning compositions, in agriculture as a fertiliser and herbicide, as a catalyst in the air oxidation of hydrocarbons, in the manufacture of alloys and refractories, as flux in metallurgy, as fire retardants, and as neutron absorbers in nuclear reactors.

Alkali-Metal and Ammonium Borates

These include dipotassium tetraborate tetrahydrate, $K_2O.2B_2O_3.4H_2O$ [formula wt 305.51 orthorhombic, sp gr 1.92]; potassium pentaborate octahydrate, $K_2O.5B_2O_3.8\ H_2O$ [formula wt 586.42, orthorhombic, sp gr 1.74, heat capacity 329.0 J/(mol.K) (78.6 cal/(mol.K)) at 296.6 K); and diammonium tetraborate tetrahydrate, $(NH_4)_2O.2\ B_2O_3.4H_2O$ [formula wt 263.38, tetragonal, sp gr 1.58]. Ammonium pentaborate octahydrate, $(NH_4)_2.5B_2O_3.8\ H_2O$ [formula wt 544.4, sp gr 1.58, heat capacity 359.4 J/(mol.K) (85.9 cal/(mol.K)) at 301.2 K] exists in two crystalline forms, orthorhombic (α) and monoclinic (β).

Potassium tetraborate tetrahydrate may be prepared from an aqueous solution of KOH and boric acid with a B_2O_3:K_2O mole ratio of ca 2.0, or by separation from KCl–borax solution. Potassium pentaborate is prepared in a manner analogous to that used for the tetraborate, but the strong liquor has a B_2O_3:K_2O mole ratio near 5. Ammonium tetraborate tetrahydrate is prepared with a B_2O_3:$(NH_4)_2O$ mole ratio of 1.8–2.1. Ammonium pentaborate is similarly produced from an aqueous solution of boric acid and ammonia with a B_2O_3:$(NH_4)_2O$ mole ratio of 5.

The potassium and ammonium borates are low volume products with production figures of hundreds of metric tons per year for the tetra- and pentaborates. Dipotassium tetraborate tetrahydrate is used to replace borax in applications where an alkali-metal borate is needed but sodium salts cannot be used, or where a more soluble form is required. The potassium compound is used as a component in lubricants, as a solvent for casein, as a constituent in welding fluxes, and as a component in diazo-type developer solutions. Potassium pentaborate octahydrate is used in fluxes .or welding and brazing of stainless steels and non-ferrous metals. Diammonium tetraborate tetrahydrate is used when a highly soluble borate is desired, but alkali metals cannot be tolerated. Ammonium pentaborate octahydrate is used as a component in electrolytes for electrolytic capacitors, as an ingredient in flame-proofing formulations, and in paper coating.

Calcium-Containing Borates

Dicalcium hexaborate pentahydrate, $2CaO.3B_2O_3.5H_2O$ [formula wt 411.16, monoclinic, sp gr 2.42, heat of formulation – 349 kJ/mol (–0.38 kcal/mol)] exists in nature as the mineral colemanite. Sodium calcium pentaborate hexadecahydrate, $Na_2O.2CaO.5B_2O_3.16H_2O$ [formula wt 810.60, triclinic, sp gr 1.95] exists in nature as the mineral ulexite. Sodium calcium pentaborate decahydrate, $Na_2O.2CaO.5B_2O_3.10H_2O$ [formula wt 702.50, monoclinic, sp gr 2.15] exists in nature as the mineral probertite.

Colemanite is used in the production of boric acid and borax, as well as for the manufacture of the "E" glass used in textile glass fibres and plastic reinforcements (where sodium cannot be tolerated). It also has limited application as a slagging material in steel manufacture and as a precursor to some boron alloys. Ulexite and proberite are used in the production of insulation fibre-glass and borosilicate glass as well as in the manufacture of other borates.

Borate Melts and Glasses

Most of the interest in metal borate glasses has centred on reports that indicated the existence of maxima and minima in some of the physical properties of the glasses with increasing metal oxide content. This phenomenon has been called the boron oxide anomaly. Modern theory on borate glass structure, however, indicates that the changes in the properties with alkali content are not anomalous, but are the result of well-defined structural changes in the glass at the molecular level. The borate glass compounds of commercial importance are boric oxide and disodium tetraborate pentahydrate.

Other Metal Borates

Borate salts or complexes of virtually every metal have been prepared. For most metals, a series of hydrated and anhydrous compounds may be obtained by varying the starting materials and/or reactions conditions. In general, hydrated borates of heavy metals are prepared by mixing aqueous solution or suspensions of the metal oxides, sulphates, or halides with boric acid or alkali-metal borates (e.g., borax). Anhydrous metal borates may be prepared by heating the hydrated salts to 300–500°C of by direct fusion of the metal oxide with boric acid or B_2O_3.

Barium metaborate, $BaO.B_2O_3$, is used as an additive to impart fire-retardant and mildew-resistant properties to latex paints, plastics, textiles, and paper products, and as a preservative in protein-based glues described as technical grade cobalt tetraborate, $CaO.2B_2O_3.4H_2O$, is actually a mixture of metaborates and hexaborates. Hydrated copper metaborate, $CuO.B_2O_3.2H_2O$, has been used as a fungicide for treatment of lumber and other cellulose materials. The anhydrous salt, $CuO.B_2O_3$, is used as an oil pigment.

Zinc borates

A series of hydrated zinc borates has been developed for use as fire-retardant additives in coatings and polymers. Worldwide consumption of these zinc salts is currently several thousand metric tons per year. A substantial portion of this total is used in vinyl plastics where zinc borates are added alone or in combination with other fire retardants such as antimony oxide. The most commonly encountered zinc borate is $2ZnO.3B_2O_3.3.5H_2O$, which is formed when boric acid is added to solutions of soluble zinc salts.

Boron Phosphate

BPO_4 is a white, infusible solid that vapourises slowly above 1450°C, without apparent decomposition. It is normally prepared by dehydrating mixtures of boric acid and phosphoric acid at temperatures up to 1200°C. The principal application of boron phosphate is as a heterogeneous acid catalyst, and a high degree of chemical purity is often unnecessary.

Boric Acid Esters

The term boric acid esters refers to compounds with the general formula $B(OR)_3$. Much of the chemistry of these compounds is related to the electrophilic nature of boron. A series of related compounds can be formed when electron-deficient boron accepts a fourth nucleophilic substituent leading to tetrahedral boron structures such as $NaB(OR)_4$.

Properties

Trialkoxy (or aryloxy) borones range from colourless low boiling liquids such as trimethoxyborane (commonly referred to as methyl borate) to high melting solids. Trialkoxy (or aryloxy) boranes are typically monomeric, soluble in most organic solvents, and dissolve in water with accompanying hydrolysis occurs rapidly except in cases where the boron atom is protected by bulky alkyl or aryl substituent groups. The boron atom in trialkoxy (or aryloxy) boranes is in a trigonal coplanar state with sp^2 bond hybridisation, and has a vacant p orbital along the threefold symmetry axis perpendicular to the BO_3 plane. This vacant orbital is readily available for acceptance of nucleophiles, such as water or alcohols. Structural studies of trialkoxy- and triaryloxyboranes have confirmed this structure having the angle between C—O bonds at 120°C except in those cases where the angles are slightly distorted by bulky substituent groups. The susceptibility of the boron atom to attack by nucleophiles is similar to that of the carbonyl carbon in carboxylic acid esters and leads to the analogy of referring to trialkoxy (or aryloxy) boranes as boric acid esters.

Trialkoxyboranes from straight-chain alcohols and triaryloxyboranes are stable to relatively high temperatures. Attempts to use this potentially useful property in borate ester-based high temperature lubricants and heat-transfer media have met with limited success because of the reactivity of these compounds to water and oxygen.

Preparation

The most common preparative method for trialkoxy- and triaryloxyboranes is the reaction of the appropriate alcohol or phenol with boric acid, an inexpensive and readily available boron source. The boric acid ester prepared in the largest quantities is methyl borate, and most of it is used captively in the production of sodium borohydride. Other borate esters that have been offered commercially include ethyl borate, *n*-propyl borate, isopropyl borate, *n*-butyl borate, cresyl borate (from a mixture of *meta* and *para* cresols), and tri(hexylene glycol) diborate.

Uses

The principal application of methyl borate is as an intermediate in the commercial production of sodium borohydride. Another commercial outlet is the use of methyl borate azeotrope as gaseous flux for welding and brazing. Smaller quantities of borate esters are finding use as epoxy resin curing agents, and specific types of borate esters are used as gasoline additives to reduce engine knocking and engine deposits. Borate esters are used in hydraulic fluids and lubricants and as a variety of polymer additives. Glycol borate esters also constitute the active ingredients in biocides used in jet and diesel fuels.

Boron Halides

The important physical and thermochemical properties of the boron trihalides are given in Table 10.5. The thermal conductivity of BBr_3 at 20°C is 0.112 W/(m.K). The boron trihalides are planar (sp^2) molecules with X–B–X angles of 120°C. Orbital energy assignments have been made for the trihalides based on their photoelectron spectra.

Boron trihalides are highly toxic and contact should be avoided. They react vigorously (sometimes explosively) with water to yield hydrogen halides. At high temperatures they decompose to yield toxic halogen-containing fumes.

Table 10.5. Physical properties of the boron trihalides.

Property	BCl_3	BBr_3	BI_3
Melting point, °C	−107	−46	−49.9
Boiling point , °C	12.5	91.3	210
Density ρ^a (liq), g/mL	1.434_4^0 1.349_4^{11}	2.643_4^{18}	3.35
Critical temperature °C	178.8	300	
Critical pressure, kPa[b]	3901.0		
ΔH_f^0, kJ/mol gas[c]	−403	−206	−18
ΔH_{vap}, kJ/mol[c]	23.8	34.3	
C_p, J/(mol. °C), for gas at 25°C[c]	62.8	67.78	
C_p, J/(mol. °C), for liquid at 25°C[c]	121	128	
ΔH_{hydrol} kJ/mol liquid at 25°C[c]	−289	−351	
ΔH_{fusion}, J/g at mp[c]	18		
B—X bond energy, kJ/mol[c]	443.9	368.2	266.5
B—X distance, nm	0.173	0.187	0.210

[a] For BCl_3: ρ (g/mL) = 1.3730–2.159 × 10^{-3}°C; 8.377 × 10^{-7}°C; −44 to 5°C.
For BBr_3: ρ (g/mL) = 2.698–2.996 × 10^{-3}°C; −20 to 90°C.

[b] To convert kPa to mm Hg, multiply by 7.50.

[c] To convert J to cal, divide by 4.184.

RARE-EARTH ELEMENTS

The rare earths comprise a group of 17 elements in the periodic table and have similar properties in aqueous solutions. All are metals in the elemental state and all form salts that a[illegible] ,trong electrolytes when dissolved in water. They ionise in this medium to give triply charged ions and, because of the high charge on these ions, react strongly with water dipoles to form a tight sheath of water molecules about them. Other ions in aqueous solutions only contact this sheath, giving rise to the similar properties of rare-earth cations in water.

The group consists of the following elements : scandium (^{21}Sc), yttrium (^{39}Y), lanthanum (^{57}La), all of which appear in the group IIIB of the periodic table, and cerium (^{58}Ce), praseodymium (^{59}Pr), neodymium (^{60}Nd), promethium (^{61}Pm), samarium (^{62}Sm), europium (^{63}Eu), gadolinium (^{64}Gd), terbium (^{65}Tb), dysprosium (^{66}Dy), holmium (^{6}Ho), erbium (^{68}Er), thulium (^{69}Tm), ytterbium (^{70}Yb), and lutetium (^{71}Lu).

The rare earths are widely distributed in low concentrations throughout the earth's crust. They occur as mixtures in many massive rock formations, e.g., basalts, granites, gneisses, shales, and silicate rocks, in which they are present in amounts of 10–300 ppm. They also occur in ca. 160 discrete minerals, most of which are rare, but in which the rare-earth content, expressed as R_2O_3 or Ln_2O_3 (REO), can be as high as 60% (REO is an abbreviation used in industry for rare-earth oxide content). Ln_2O_3 is used when Y_2O_3 is not present in the ore or mineral. Approximately ten of these occur in sufficient quantities that they may furnish some REO to commerce, but more than 95% of the REO occurs in three minerals: monazite and bastnasite for the light rare earths, and xenotime for yttrium and the heavy rare earths. Xenotime occurs mixed with monazite in alluvial deposits.

Properties

Physical

The rare earth metals alloy with most metals to form intermetallic compounds and occasionally solid solution. In rare-earth-rich alloys, other elements can change the properties of the pure metal by drastically lowering (or, more rarely, raising) the melting point by 200–300°C in some cases. Alloying with other elements can make the rare earth either pyrophoric or corrosion-resistant. Some properties of the rare earths are listed in Table 10.6.

Chemical

The chlorides, bromides, nitrates, bromates, and perchlorate salts are soluble in water and, when their aqueous solutions evaporate, they precipitate as hydrated crystalline salts. The acetates, iodates, and iodides are somewhat less soluble. The sulphates are sparingly soluble and are unique in that they become less soluble with increasing temperature. The oxides, sulphides, fluorides, carbonates, oxalates, and phosphates are insoluble in water. The oxalate, which is important in the recovery of highly pure Ln, can be calcined directly to the oxide.

Anhydrous rare-earth salts usually cannot be prepared by evaporating the water of crystallisation.

The lanthanides can form hydrides of any composition up to LnH_3. Small amounts of hydrogen dissolve interstitially but, with increasing amounts of hydrogen, a second phase (LnH_2) appears. These alloys are metallic in their properties and lose hydrogen at relatively high temperatures.

Table 10.6. Some properties of the rare earths.

	Scandium	*Yttrium*	*Lanthanum*	*Cerium*	*Praseodymium*	*Neodymium*
Atomic number	21	39	57	58	59	60
Atomic weight	44.9559	88.9059	138.9055	140.12	14.9077	144.24
Colour	silvery	silvery	silvery	silvery	silvery	silvery
Melting point, °C	1541	1522	918	798	931	1021
Boiling point, °C	2836	3338	3464	3433	3520	3074
Density, g/cm^3	2.989	4.469	6.146	6.770	6.773	7.008
Heat of fusion, kJ/mol[a]	14.10	11.40	6.20	5.46	6.89	7.14

	Promethium	*Samarium*	*Europium*	*Gadolinium*	*Terbium*
Atomic number	61	62	63	64	65
Atomic weight	145	150.36	151.96	157.25	158.9254
Colour	silvery	silvery	silvery	silvery	silvery
Melting point, °C	1042	1074	822	1313	1356
Boiling point, °C	3000 (estd)	1794	1527	3274	3230
Density, g/cm^3	7.264	7.520	5.244	7.901	8.230
Heat of fusion, kJ/mol[a]	7.6 (estd)	8.62	9.21	10.05	10.79

	Dysprosium	*Holmium*	*Erbium*	*Thulium*	*Ytterbium*	*Lutertium*
Atomic number	66	67	68	69	70	71
Atomic weight	162.50	164.9304	167.26	168.9342	173.04	174.967
Colour	silvery	silvery	silvery	silvery	silvery	silvery
Melting point, °C	1412	1474	1529	1545	819	1663
Boiling point, °C	2567	2700	2868	1950	1196	3402
Density, g/cm^3	8.551	8.795	9.066	9.321	6.966	9.841
Heat of fusion, kJ/mol[a]	11.06	16.87	19.90	16.84	7.66	18.65

[a] To convert J to cal, divide by 4.184.

The rare earths form many compounds with organic ligands; some of which are water-soluble. Chelating agents, which are organic molecules that engulf rare-earth ions and displace some of the water of hydration, form complexes with rare-earth ions that have a much wider range in their formation constants than do those of ordinary mineral-acid salts.

Production and Processing

Monazite and bastnasite are the minerals used commercially to supply most of the rare-earth chemicals. Monazite is a brown, dense phosphate mineral. There are extensive deposits of enriched monazite sands on beaches in many parts of the world, eg, along the southwest coast of India and the east coast of Brazil. Such deposits are dredged, pulverised if necessary, and further enriched by flotation methods. Sometimes they are also subjected to cross-belt magnetic separation since they are weakly magnetic.

Liquid-liquid extraction

As the industrial demand for individual rare earths increased, some of the rare-earth companies developed liquid-liquid extraction systems. An organic stream immiscible with water is flowed counter-current to the aqueous stream containing the rare-earth mixture. The organic phase may absorb neutral rare-earth molecules by complexing with them, eg, as does tributyl phosphate, or it may contain a complexing molecule or anion added to it for the same purpose. Also, complexing ligands may be added to the aqueous phase so as to effect a synergistic equilibrium between the rare-earth ions and the complexant. If the various equilibria between the organic and aqueous phases have overall equilibrium constants sufficiently different for various rare earths, separation occurs.

Health and Safety

The rare earths are considered only slightly toxic and can be handled safely with ordinary care.

Uses

The rare earths are used in gasoline-cracking catalysts; in carbon arcs; as additives to steel and cast irons; as polishing compounds; to made optical glass; as both colourants and decolourants for glass, depending upon the rare earth element used; and in magnets. In the electronic area, the most important industrial rare-earth compounds are garnet-based materials, eg, yttrium iron garnet and gadolium gallium garnet, used in microwave devices, memory storage, and oxygen sensors (Y_2O_3-stabilised ZrO_2). In nuclear reactors, the rare earths are used in the form of oxides as absorbers of neutrons. They are also used as additions to superalloys, as hydrogen-storing materials, and as synthetic gems.

SODIUM CHROMATE

From Chromite Ore

Process

Sodium dichromate is produced by the reaction of sulphuric acid on sodium chromate, which is turn is obtained by calcining a mixture of chromite ore, limestone, and soda ash (Fig. 10.9).

Reaction

$$4FeCr_2O_4 + 8Na_2CO_3 + 7O_2 \rightarrow 8Na_2CrO_4 + 2Fe_2O_3 + 8CO_2$$

$$2Na_2CrO_4 + H_2SO_4 \rightarrow Na_2Cr_2O_7 + H_2O + Na_2SO_4$$

Chromite, or chrome ironstone, is a chromium iron oxide ore containing principally ferrous chromite ($FeCr_2O_4$ or $FeO.Cr_2O_3$) plus small amounts of alumina, silica, and magnesia. High-grade chromite contains about 50% chromium (III) oxide (Cr_2O_3).

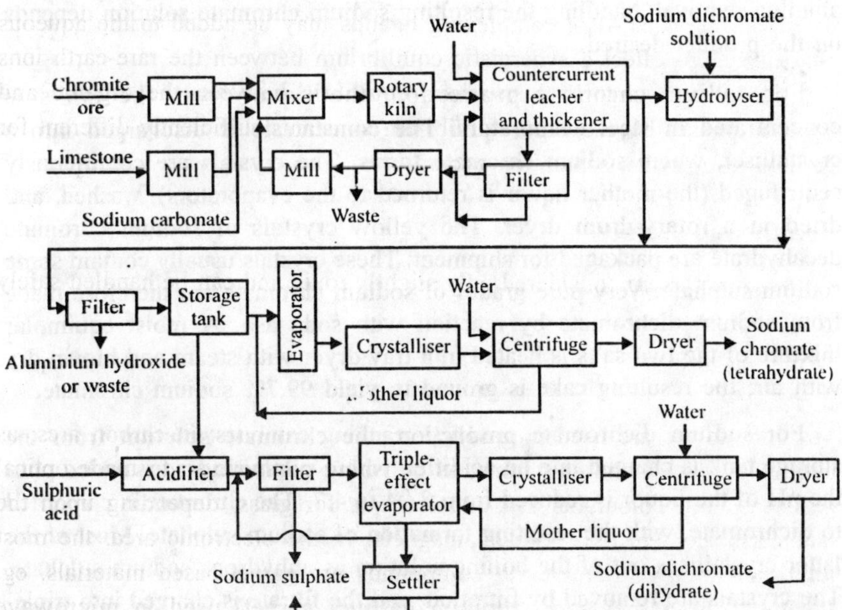

Fig. 10.9. Flow diagram for manufacture of sodium chromate from chromite ore.

Chromite is crushed and ground to a fine powder (90 to 98% through 200 mesh) in ball mills. It is then mixed with at least 90% of the theoretical soda ash, a varying amount of lime, and sufficient filter residue (from farther along in the process) to adjust the chromic oxide content to 15 to 20%. The mixture is then blended in a rotary mixer and calcined in a rotary kiln, similar to those used in the cement industry, at 1100 to 1150°C. After a 4-h retention time the roast from the kiln is sent directly to leaching tanks or thickeners, where the soluble chromate is dissolved in hot water and concentrated. The underflow from the last tray in the thickener is filtered. The leached residue is dried in rotary driers. A portion is recycled to the kiln feed, where it serves to keep the reacting mass in a relatively dry condition. The mother liquor from the filter is returned to the leaching tank.

The overflow from the leacher is also filtered and sent to hydrolysing tanks. Here any sodium aluminate is hydrolysed to insoluble aluminium hydrate by the slow addition of sodium dichromate solution, according to the reaction :

$$2NaAlO_2 + Na_2Cr_2O_7 + 3H_2O \rightarrow Al_2O_3.3H_2O + 2Na_2CrO_4$$

The addition of sodium dichromate solution must be closely controlled to obtain a filterable alumina hydrate instead of a gelatinous mass. After alumina removal, handling the resulting sodium chromate solution depends on the product desired.

For sodium chromate production, the liquor from the tanks is concentrated in steel evaporators. The concentrate is discharged into a crystalliser, where sodium chromate forms. The crystals are continuously centrifuged (the mother liquor is returned to the evaporators), washed, and dried in a rotary-drum dryer. The yellow crystals of sodium chromate decahydrate are packaged for shipment. These crystals usually contain some sodium sulphate. Very pure grades of sodium chromate are therefore made from sodium dichromate by reaction with soda ash. A moist equimolar mixture of the two salts is heated in a tray dryer with steam and blown dry with air; the resulting cake is ground to yield 99.7% sodium chromate.

For sodium dichromate production, the chromate solution from the storage tank is charged into an acidifier, where sulphuric acid is added until the pH of the liquor is reduced from 8.04 to 4.7. The chromate is converted to dichromate, with the resulting formation of sodium sulphate. Most of the latter crystallises out of the boiling solution as anhydrous sodium sulphate. The crystals are removed by filtration, and the filtrate is charged into triple-effect evaporators, where the remainder of the sulphate crystallises. The clear, hot, saturated dichromate solution (sp gr 1.7) is fed into a water-cooled crystalliser. Here, at about 20°C, approximately 70% of the sodium dichromate crystallises and is removed by continuous centrifuging. The mother liquor is returned to the evaporators. After drying in a rotary-drum dryer, the red crystals of sodium dichromate dihydrate are packaged. Both sodium chromate and dichromate may be purified by recrystallisation from water.

Properties

Sodium chromate

(Na_2CrO_4)Yellow, rhombic crystals. Formula wt 162.00; Sp gr 2.723; mp 792°C. Soluble in cold water and hot water (126 g/100 g at 100°C).

Crystallises from water as the decahydrate ($Na_2CrO_4.10H_2O$) in the form of yellow deliquescent monoclinic crystals. Formula wt 342.16; Sp gr 1.483; mp 19.9°C. Very soluble in cold water. Soluble in all proportions in hot water. Slightly soluble in ethanol.

Uses

Pigments; chromic acid; leather tanning; metal treatment; textile and dyes.

Sodium dichromate

$Na_2Cr_2O_7.2H_2O$ Bright-orange, deliquescent, monoclinic crystals. Formula wt 298.05; Sp gr 2.348 (25°C); mp Loses $2H_2O$ at 84.6°C; anhydrous at 356.7°C bp Decomposes at 400°C. Soluble in water (70.6% at 0°C, 88.4% at 80°C).

HYDROGEN PEROXIDE

Process

Hydrogen peroxide is manufactured by the hydrolysis of a solution containing the persulphate ($S_2O_8^{2-}$) ion. The particular solution used (persulphuric acid, ammonium persulphate, or potassium persulphate) is prepared in an electrolytic cell in which the corresponding sulphate is oxidised anodically. Hydrolysis of the persulphate (outside the cell) is carried out under conditions that regenerate the sulphate and vapourise a mixture of water and hydrogen peroxide. The latter are separated and the peroxide concentrated by vacuum distillation.

In the embodiment of the process as shown in the flowsheet, a regenerated solution of ammonium sulphate and sulphuric acid (equivalent to ammonium bisulphate) is adjusted to proper concentration and fed to a series of electrolytic cells. In one plant the porcelain cells are 70 cm wide and 95 cm deep. The platinum anodes and graphite cathodes are so arranged that water in cooling tubes may keep the electrolyte temperature below 35°C. The potential drop across each cell, operated at 5400 A, is 5.7 V. Ammonium thiocyanate (0.1 g/litre) is usually added to the electrolyte to increase oxygen overvoltage. Anode current density is 4000 to 11000 A/m^2. A current of air is swept continuously through the cell to keep the hydrogen concentration below 5%, so as to avoid explosion.

The persulphate solution leaving the cells goes to an evaporator, where it is heated by both live steam and by means of steam coils. Hydrogen peroxide and water vapours go overhead. The ammonium bisulphate solution is cooled, filtered to remove any sediment, adjusted in concentration, and returned to the cells. The cathode liquor is similarly cooled, filtered, and reused. The hydrogen peroxide-water vapour mixture is sent to stoneware distillation towers which are operated under vacuum (40 mm Hg or 5.3 kPa). A 30 to 40% hydrogen peroxide solution is removed from the bottom of the tower. This solution may be concentrated to 70% or 90 to 98% H_2O_2 in two stages.

The 30 to 40% hydrogen peroxide solution recovered from the cell liquors is adjusted with sulphuric acid (about 0.5 g/litre) and fed to a stoneware or porcelain evaporator, where the solution is heated to 65°C at 40 to 50 mm Hg (5.3 to 6.6 kPa).

The vapours evolved are scrubbed in a packed tower with distilled or demineralised water in such ratio to the peroxide vapours that a 70% hydrogen peroxide solution leaves the bottom of the tower. This product may be fed to a second evaporator (at 75°C and the same pressure) and a second packed tower to produce an 80 to 85% solution of hydrogen peroxide. The product is cooled, stabilised with acids or an organic oxidation inhibitor, and sent to storage.

This concentration process may be run continuously by fractionating and recycling the vapours from the first evaporator. The resulting 70% hydrogen peroxide solution may be concentrated to a 90 to 98% solution in a similar second stage. From time to time, liquid bottoms from the first evaporator must be drawn off to remove impurities.

The ammonium persulphate process described differs only in details from the persulphuric acid process and the potassium persulphate process. All three have been used commercially. The pure-acid process has a lower current efficiency (70 to 75%) but has the advantage that crystallisation problems within the system are eliminated.

In the potassium persulphate process, an ammonium bisulphate solution is electrolysed as before. After electrolysis potassium acid sulphate is added to the electrolyte :

$$(NH_4)_2S_2O_8 + 2KHSO_4 \rightarrow K_2S_2O_8 + 2(NH_4)HSO_4$$

The potassium salt is crystallised and separated by filtration. The solid potassium salt is then added to a sulphuric acid solution to produce a slurry. On heating with live steam, the persulphate hydrolyses to produce hydrogen peroxide vapours. As in the other processes, the potassium salt may be recovered and recycled.

By Oxidation of Alkylhydroanthraquinones

Process

The modern chemical method for manufacturing hydrogen peroxide is an organic autoxidation process in which a suitable organic compound (usually a hydroquinone type) in solution is oxidised to the quinone form with the concurrent formation of hydrogen peroxide. The peroxide is extracted and concentrated; the oxidised organic compound is reduced catalytically and reused (Fig. 10.10).

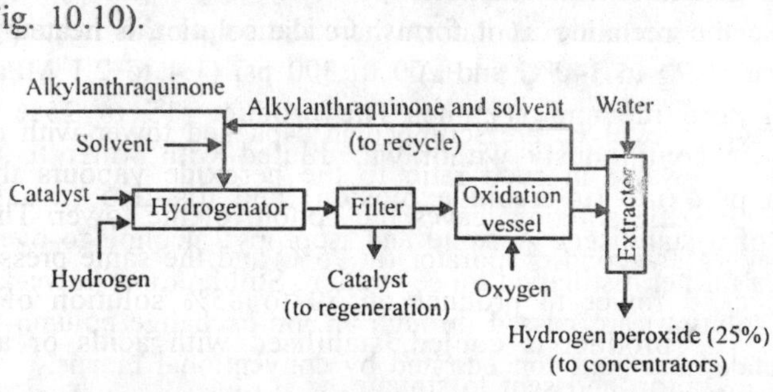

Fig. 10.10. Flow diagram for manufacturing of hydrogen peroxide from oxidation of alkylhydroanthraquinones.

In one adaptation of this process, the raw material is a 10% solution of mixed alkylated anthraquinones and their tetrahydro derivatives. The solvent is a mixture of primary and secondary nonyl alcohols with methyl or dimethylnaphthalene. This mixture is reduced with hydrogen in the presence of a catalyst of activated alumina coated with 0.7% palladium. The reduction may be carried out at 40°C and under hydrogen pressures of 1 to 3 atm (100 to 300 kPa). After removal of the catalyst granules by filtration, the resulting hydroanthraquinone still in solution is blown with oxygen or an oxygen-nitrogen mixture at 30 to 60°C to produce hydrogen peroxide

(5 to 6 g/litre) and regenerate the original anthraquinones. The hydrogen peroxide is extracted with a sufficient quantity of water to yield a 20 to 25% hydrogen peroxide solution. The anthraquinones are recycled. The hydrogen peroxide may be concentrated as described in the previous process.

This process is similar to a process developed in Germany and used there during World War II. The German process made use of ethylanthraquinone in a benzene-cyclohexanol solvent and used Raney nickel as a hydrogenation catalyst. High losses of Raney nickel and the partial solubility of the solvent in water led to the economic modifications of the process described.

By Oxidation of Isopropyl Alcohol

Process

Hydrogen peroxide and acetone are jointly produced by the liquid-phase oxidation of isopropyl alcohol with molecular oxygen. Isopropyl alcohol and hydrogen peroxide (1 wt. %) (to eliminate the prolonged incubation period) are fed to a reactor along with 80 to 95% oxygen. The reactor is lined with glass, enamel, aluminium, or some other material that will not decompose the peroxide as it forms. Oxidation of the isopropyl alcohol takes place at 90 to 140°C and 200 to 300 psi (1.4 to 2.1 MPa) to form hydrogen peroxide and acetone. The product (15 to 25% hydrogen peroxide) is continuously withdrawn, diluted with water to 6 to 10% hydrogen peroxide (to avoid explosion) and distilled (usually in the presence of a stabiliser). Acetone and isopropyl alcohol go overhead and thence to a further distillation for separation. Still bottoms (a weak hydrogen peroxide solution) are passed through an ion-exchange column to remove heavy metals, and then concentrated by conventional means.

Reaction

$$CH_3CH(OH)CH_3 + O_2 \rightarrow \underset{\text{87\% yield}}{CH_3COCH_3 + H_2O_2}$$

At a conversion of 15% per pass, the yield of hydrogen peroxide is 87% based on isopropyl alcohol, and the yield acetone is 93% (Fig. 10.11).

Properties

100% hydrogen peroxide is a water-white sirupy liquid with oxidising properties. Formula wt 34.02; Sp gr 1.438 20°C/4; mp –0.461°C; bp 155.5°C.

Uses

Textiles; pulp and paper; chemical synthesis.

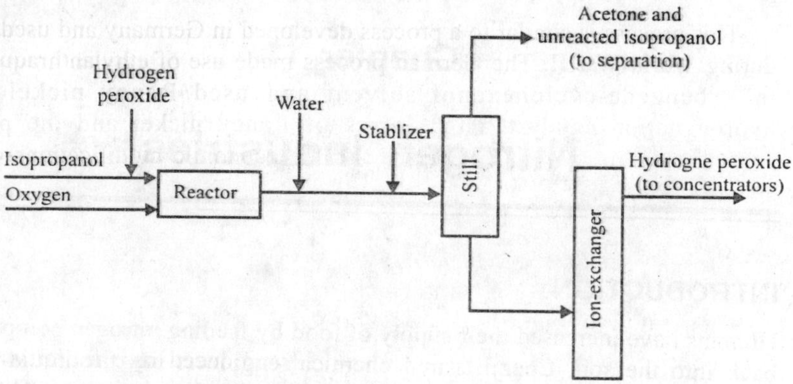

Fig. 10.11. Flow diagram for manufacture of hydrogen peroxide from oxidation of isopropyl alcohol.

Miscible with water in all proportions. Soluble in ether. Insoluble in hydrocarbons. In pure form, either as is or in solution, hydrogen peroxide is quite stable. It decomposes, sometimes violently, in the presence of even traces of metallic impurities. In contact with readily oxidisable organic substances it may cause spontaneous combustion.

Chapter 11

Nitrogen Industries

INTRODUCTION

Humans have increased their supply of food by feeding nitrogen compounds back into the soil. Chemists and chemical engineers have found ways of making nitrogen derivatives out of air in an economical way. The first successful process, the arc process, required much cheap electric energy. However, the present solution to the nitrogen-fixation problem was obtained by reacting nitrogen with low-cost hydrogen to make ammonia under conditions requiring less power and low conversion expense. This made the arc process obsolete. Ammonia has now become one of our heavy chemicals, produced in enormous tonnage around the globe, and at such low prices as to dominate the world supply of nitrogen fertilisers and most nitrogen compounds. This is probably one of the most important chemical engineering achievements in history.

The development of a practical synthetic ammonia process was carried out through the chemical engineering efforts of Haber and Bosch and their co-workers, and the growth of this industry has continued. Cyanamide- and the ammonia-manufacturing procedures are currently those of technical importance for supplying the world with fixed nitrogen.

The chemical nitrogen industries include not only nitrogen fixed by humans, but also by-product ammonia from coke ovens and such natural nitrogen deposits both of which are subjected to manufacturing processes. Ammonia is used in heat-treating and paper pulping, nitric acid for nitro compounds, high explosives, and propellants. Hydrogen cyanide, made largely from ammonia, is used mainly for acrylates and methacrylates, although a little is used for fumigation. Acrylonitrile uses substantial amounts of nitrogen compounds. Hydroxylamine and hydrazine also are smaller-volume nitrogen consumers. Amines of fatty acids enjoy wide use as surface-active agents and in ore flotation.

CYANAMIDE

Cyanamide, (carbamic acid nitrile; carbamodiimide), $H_2NC{\equiv}N$, is a weak acid with a very high solubility in water. The major use of cyanamide was at one time agricultural. Its employment as a chemical raw material has become increasingly important, however, the largest chemical use being for the preparation of the dimer dicyandiamide. This dimer is polymerised to produce the trimer melamine, which has applications in plastic resins. Other commercial derivatives are produced by way of a crude calcium cyanide (48 to 50% expressed as NaCN) from a high temperature melt of cyanamide with excess carbon and NaCl. This is used directly for cyanidation of ores and to manufacture ferrocyanides. The use of cyanamide fertilisers is enjoying a minor resurgence, because these chemicals act as pesticides to snails and larvae and also enrich the soil with nitrogen. Cyanamide compounds are used to fire-proof shingles and cotton fabrics.

Reactions and Energy Changes

The essential reactions for the production of calcium cyanamide are :

$$CaCO_3(s) \longrightarrow CaO(s) + CO_2(g) \qquad \Delta H = +\ 43.5 \text{ kcal} \quad ...(11.1)$$

$$CaO(s) + 3C(amorph) \longrightarrow CaC_2(s) + CO(g) \qquad \Delta H = +\ 103.0 \text{ kcal} \quad ...(11.2)$$

$$CaC_2(s) + N_2(g) \longrightarrow CaCN_2(s) + (C(amorph) \qquad \Delta H = -68.0 \text{ kcal} \quad ...(11.3)$$

Various catalysts or fluxes are used to increase the rate of reaction or cause it to proceed at lower temperatures. Generally uses of calcium fluoride, and reaction-rate studies have shown that it reduces the temperature of optimum reactivity and increases the velocity of the reaction by 4.5 times at 1000°C. Reaction (11.3) takes place at 900 to 1000°C. Reaction (11.2) is carried out in two 20,000- and two 10,000-kW furnaces at about 2000 to 2200°C. Under normal conditions reaction (11.2) yields up to 90% calcium carbide. Considerable overall energy is needed, principally to secure the high temperature for reaction (11.3) to start when it is self-sustaining and for making the calcium carbide in reaction (11.2). The source of carbon is usually coke. Coal is required to burn the limestone and to dry the raw materials (Fig. 11.1).

The following *physical operations* and *chemical conversions* are needed to commercialise the reactions on which Fig. 11.1 is based on :

(1) Limestone, coal and coke are pulverised, separately.

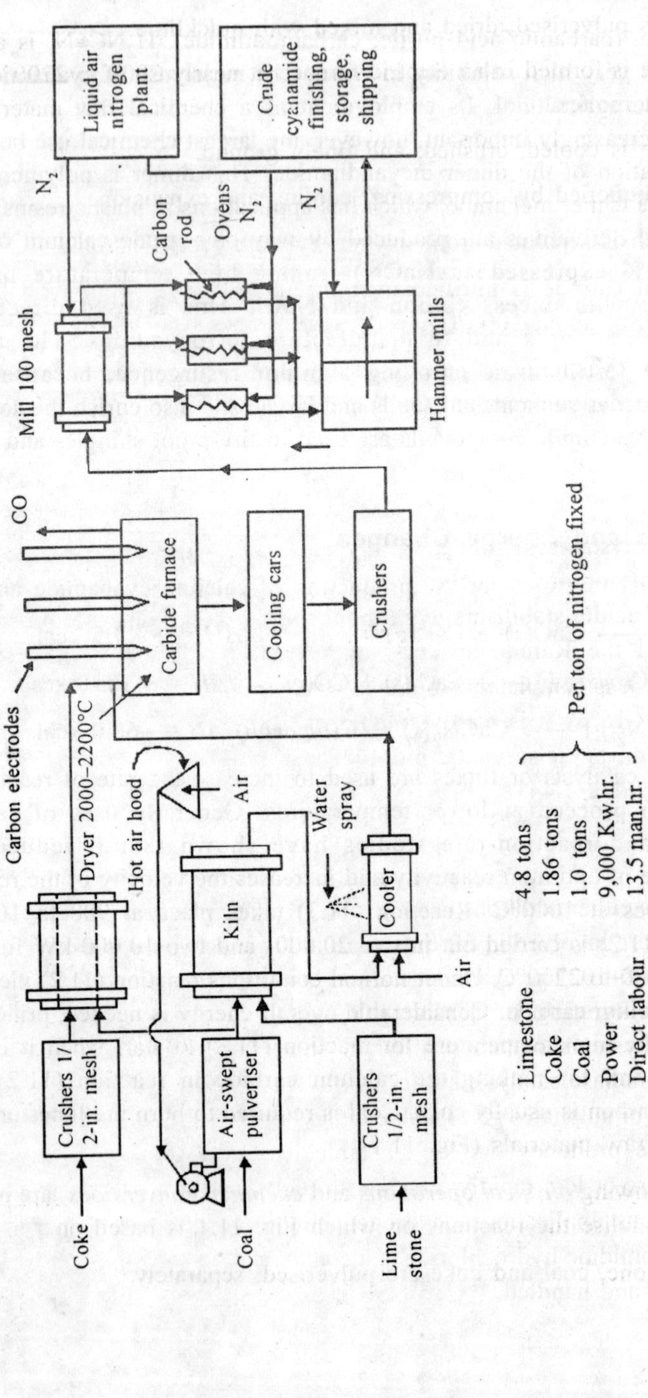

Fig. 11.1 Flowchart for the manufacture of calcium carbide and calcium cyanamide.

(2) Limestone is calcined to quicklime.

(3) Coke is pulverised, dried and mixed with quicklime.

(4) Carbide is formed in an electric furnace at nearly 2000 to 2200°C and run out molten.

(5) Carbide is cooled, crushed, and finely ground.

(6) Air is liquefied by compressing, cooling and expansion.

(7) Nitrogen is separated from oxygen by liquid rectification.

(8) Calcium carbide is nitrified over the course of 40 hour with 99.9% nitrogen at about 1000°C.

(9) Calcium cyanamide is pulverised and treated with a small amount of water to hydrate residual CaO and CaC_2. It may be oiled to reduce dust.

Properties

Cyanamide crystallises from a variety of solvents as somewhat unstable, colourless, orthorhombic, deliquescent crystals. Dimerisation is prevented by traces of acidic stabilisers, e.g., monosodium phosphate. Studies of the infrared and the Raman spectra support the *N*-cyanoamine structure, NH_2—C≡N. It is completely soluble at 43°C, and has a minimum solubility (eutectic) at –15°C. It is highly soluble in polar organic solvents, and less soluble in non-polar solvents, mp 46°C; bp 140°C (101.3 kPa); d 1.282 g/cm^3; ref. index 1.4418 at 48°C.

Reactions

Reactions are either additions to the nitrile group or substitution at the amino group. Both are involved in the dimerisation to dicyandiamide, which occurs readily at pH 8–10. The cyanamide anion is strongly nucleophilic and reacts with most alkylating or acylating reagents; addition to a variety of unsaturated systems occurs readily. In some cases, a cyanamide salt is used; in others base catalysis suffices. Alkylation with a variety of common alkyl halides or sulphates gives stable dialkyl cyanamides. The reaction with formaldehyde produces first an unstable hydroxymethyl derivative that resinifies more or less rapidly, passing through a water-soluble stage which permits various applications. Hydrogen chloride gives a dihydrochloride. Chloroform-amidine hydrochloride is a convenient anhydrous form which is easily stored and handled.

Manufacture

The basic process for manufacture comprises four stages: (i) lime is made from high grade limestone; (ii) calcium carbide is manufactured from lime and coal or coke; (iii) calcium cyanamide is produced by passing gaseous nitrogen through a bed of calcium carbide, which is heated to 1000-1100°C in order to start the reaction (the heat source is then removed as the reaction continues because of its strong exothermic character); and (iv) cyanamide is manufactured from calcium cyanamide by continuous carbonation in aqueous medium. For production of commercial 50% solution, and for recovery of crystalline cyanamide, this process is modified to improve purity and concentration, e.g., calcium and iron may be removed by ion-exchange treatment.

Other processes include the reaction of lime with hydrogen cyanide; the reaction of limestone with ammonia; and the reaction of lime and urea to form calcium cyanate which is then converted to calcium cyanurate which gives calcium cyanamide at a higher temperature.

Dicyandiamide

Dicyandiamide (1) cyanoguanidine is the dimer of cyanamide and crystallises in colourless monoclinic prisms. It is amphoteric, and generally soluble in polar solvents and insoluble in nonpolar solvents. Mp 208°C; bp dec.

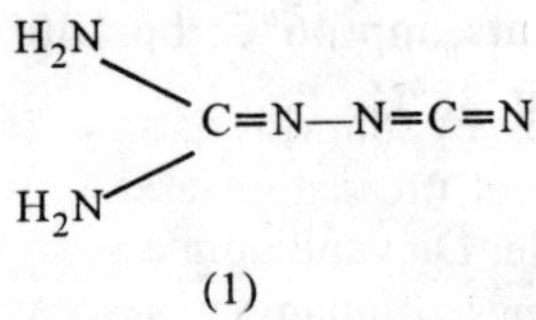

(1)

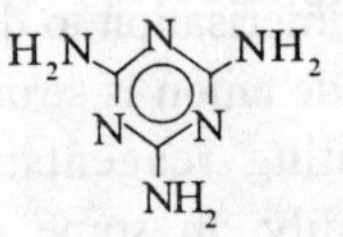

(2)

Reaction

The reactions resemble those of cyanamide. However, cyclisations take place easily and the nitrile group is less reactive. For example, under pressure and in the presence of ammonia, it cyclises to melamine. Guanamines are obtained when dicyandiamide is heated with alkyl or aryl nitriles in the presence of small amounts of alkali. Reaction with ammonium salts gives biguanide salts which react further with the ammonium salt forming guanidine salts; guanidine nitrate is manufactured by this route. Hydrolysis occurs easily at elevated temperatures. It can be treated with formaldehyde to produce resinous compositions of varying propertions.

Dicyandiamide is manufactured by the dimerisation of cyanamide in aqueous solution, with pH adjusted to 8–9 and held at ca 80°C for 2 hour to give complete conversion.

Melamine

Melamine (2) (cyanurotriamide, cyanuramide, 2,4,6-triamino-1,3,5-triazine) is a white crystalline material that is insoluble in most organic solvents. It is appreciably soluble in water, mp 3550°C; d 1.573 g/cm^{-1}. Although moderately basic, it is better considered as the triamide of cyanuric acid than as an aromatic amine. Its reactivity is poor in nearly all reactions considered typical for amines. Melamine was formerly manufactured by the conversion of dicyandiamide under heat. The dehydration condensation of urea has displaced the dicyandiamide process.

Health and Safety

Manufacture of cyanamide does not present any serious health hazard. Contact or ingestation must be avoided. Dicyandiamide is essentially nontoxic, but may cause dermatitis. Melamine crystal may be handled in ordinary industrial use without special hygienic precautions.

Uses

Cyanamide and calcium cyanamide are used as fertilisers. Calcium cyanamide is used for steel nitridation and desulphurisation; it is the raw material for dicyandiamide. Dicyandiamide is used as a raw material for the manufacture of guanamines, biguanide, and guanidine salts, and various resins. Most of the melamine produced is used in the form of melamine-formaldehyde resins.

SYNTHETIC AMMONIA

Ammonia is one of the truly fundamental raw materials utilised in modern civilisation. The largest consumer of anhydrous ammonia is the fertiliser industry, which utilises it in the production of calcium and sodium nitrate, ammonium sulphate, nitrate and phosphate, ammoniated superphosphates, urea, and aqueous ammonia. Fig. 11.2. highlights the raw materials used for manufacture ammonia and nitrogen fertilisers. Applied by special machines, anhydrous ammonia has become a very important fertiliser for direct application to soils. In combination with ammonium nitrate, an increasing amount is used each year. Ammonia is the starting point of nearly all military explosives. Scarcely an industry is untouched, since ammonia is required for the making of soda ash, nitric acid, nylon, plastics, lacquers, dyes, rubber, and other products. The product is handled and shipped in two forms, ammonia liquor and anhydrous ammonia. Commercial grades of the liquor usually contain 28% ammonia.

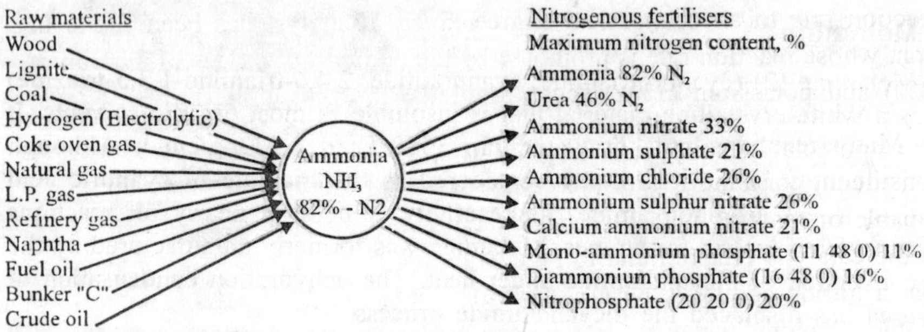

Fig. 11.2. Raw materials used throughout the world to manufacture ammonia and nitrogenous fertilisers produced from ammonia.

Reactions and Equilibriums

For the reaction

$$\tfrac{1}{2}N_2(g) + \tfrac{3}{2}H_2(g) \leftrightharpoons NH_3(g)$$

an equilibrium constant can be expressed as

$$K_p = \frac{p_{NH_3}}{p_{N_2}^{1/2} \times p_{H_2}^{3/2}}$$

Since the volume of ammonia obtained is less than the combined volume of nitrogen and hydrogen, a pressure increase, according to the *principle of Le Chatelier*, gives a higher percentage of ammonia at equilibrium. The conversion percentage increased several fold as the pressure increases from 100 to 1,000 atm, but the percentage of ammonia in equilibrium with the reacting gases decreases continually with a temperature rise up to 1100°C and reaches a minimum at this temperature. When operating below 1100°C the lower the temperature for a given pressure at which an ammonia converter can be run, the larger the possible percentage of ammonia in the ammonia-nitrogen-hydrogen mixture, but the longer the time required to attain equilibrium. Accordingl;, temperatures in converters are kept as low as is consistent with a sufficiently rapid reaction rate to be economical with the catalyst used and with the investment required.

Rate and Catalysis of the Reaction

To be economical, the rate of this reaction must be increased, because hydrogen and nitrogen alone react very slowly. The basis of the commercial synthesis of ammonia rests upon an efficient catalyst to speed up the reaction rate to an economical degree. Such a catalyst has been found in iron, whose reaction rate is promoted by the addition of oxides of aluminum (3%) and potassium (1%). These promoters prevent sintering.

Many reactions are known which occur as a result of one reacting constituent combining with the catalyst to form an intermediate compound capable of reacting with a second reacting constituent to form the product of the catalytic reaction and to regenerate the catalyst. Thus it is postulated that a nitride adduct is possible with the active iron atoms on the surface of the ammonia catalyst. Iron seems to be by far the most satisfactory *catalyst* material, though it seems to lose its activity rapidly if heated to temperatures above 520°C.

The catalyst is promoted by metallic oxides, the activity being increased most by adding both an amphoteric oxide of metal such as aluminum, zirconium, or silicon and an alkaline oxide such as potassium oxide. It can be summarised: The percentage ammonia is a gaseous mixture of pure 3:1 hydrogen-nitrogen gas passed over such a doubly promoted catalyst at 100 atm pressure, 5,000 space velocity, and at 450°C, is 13 to 14 for the doubly promoted catalyst in contrast to 8 or 9 for the singly promoted one and to 3 to 5% for the pure iron catalyst. Modern catalysts are made from magnetite containing K_2O, CaO, MgO, Al_2O_3, SiO_2, and traces of TiO_2, ZrO_2, and V_2O_5.

The promoters make the reduced catalyst more porous. Catalysts more active at a lower temperature would be most valuable. These catalysts are ruined by contact with many substances such as copper, phosphorus, arsenic, and sulphur that affect the electronic structure of the iron; carbon monoxide greatly reduces their activity. Hence much money must be spent in purifying the hydrogen and nitrogen for ammonia synthesis.

A detailed mechanism for the catalysis leading to ammonia is as follows, and is itself only speculation, but to a major degree meets most observed facts. The controlling aspect is nitrogen activation.

$$N_2 + 2Fe = 2Fe - N_{ads}$$
$$H_2 + 2Fe = 2Fe - H_{ads}$$
$$N_{ads} + H_{ads} = NH_{ads}$$
$$NH_{ads} + H_{ads} + NH_{2ads}$$
$$NH_{2ads} + H_{ads} = NH_{3ads}$$
$$NH_{3ads} = NH_{3desorb}$$

Space velocity is the number of cubic feet of gases, corrected to standard conditions (0°C and 760 mm), that pass over 1 ft^3 of catalyst space per hour. The space velocity used in commercial operations differs considerably in different processes and in different plants using the same process. Although the percentage of ammonia in the gas stream issuing from a converter goes down as the space velocity goes up, the amount of ammonia per cubic foot of catalyst space per hour increases. Too high a space velocity, however, disturbs the thermal balance of a converter, involves increased cost of ammonia removal because of the smaller percentage present in the exit gases, and makes necessary the recirculation of large volumes of gas.

AMMONIA

Ammonia is a colourless, alkaline gas, lighter than air, and possessing a unique, penetrating odour.

Physical Properties

Bp, – 33.35°C; fp, –77.7°C; critical temperature, 133.0°C; specific heat (at 0°C), 2097.2; soluble to the extent of 33.1 wt% in H_2O at 20°C; specific gravity (anhydrous ammonia) at –40°C, 0.690; and density of aqueous ammonia is 0.618 g/cm^3 (15°C). The flammable limits of ammonia in air are 16 and 25 vol %; in oxygen the range is 15–79 vol%. Such mixtures can explode, although ammonia–air mixtures are quite difficult to ignite (ignition temperature ca 650°C). Ammonia is readily adsorbed in water to make ammonia liquor. Approximately 2180 kJ (521 kcal) of heat is evolved when 1 kg of ammonia gas dissolves in water.

The alkali and alkaline-earth metals (except beryllium) are readily soluble in ammonia, metallic magnesium only slightly. Iodine, sulphur, and phosphorus dissolve in ammonia; ammonia readily attacks copper in the presence of oxygen. Most fluorides are insoluble in liquid ammonia. Potassium, silver, and uranium are only slightly soluble. Both ammonium and beryllium chloride are very soluble, whereas most other metallic chlorides are slightly soluble or insoluble. Bromides are in general more soluble in ammonia than chlorides, and most of the iodides are soluble. Oxides, hydroxides, sulphates, sulphites, and carbonates are insoluble. Nitrates, e.g., ammonium nitrate, and urea are soluble in anhydrous and aqueous ammonia (this property makes the production of fertiliser nitrogen solutions possible). Many organic compounds, e.g., amines, nitro compounds, and aromatic sulphonic acids, also dissolve in liquid ammonia.

Chemical Properties

Ammonia is comparatively stable at ordinary temperatures but decomposes into hydrogen and nitrogen at elevated temperature. The rate of decomposition is greatly affected by the nature of the surfaces with which the gas comes into contact. Whereas glass is very inactive, porcelain and pumice have a distinctly accelerating effect, and metals, such as iron, nickel, osmium, zinc, and uranium, even more.

At atmospheric pressure, dissociation of ammonia begins at about 450-500°C, whereas in the presence of catalysts, it begins as low as 300°C and is nearly complete at 500-600°C. However, at 1000°C a trace of ammonia remains. Ammonia reacts readily with a large variety of substances.

One of the more important reactions is oxidation giving nitrogen and water. Gaseous ammonia is oxidised by many oxides of the less positive metals (e.g., cupric oxide) when heated to a relatively high temperature. Powerful oxidising agents react similarly at ordinary temperatures, e.g., potassium permanganate :

$$2\ NH_3 + 2\ KMnO_4 \rightarrow 2\ KOH + 2\ MnO_2 + 2\ H_2O + N_2$$

The action of chlorine on ammonia can also be regarded as an oxidation reaction :

$$8\ NH_3 + 3\ Cl_2 \rightarrow N_2 + 6\ NH_4Cl$$

With a platinum-rhodium catalyst, ammonia can be oxidised to nitric oxide and water which is a means of producing nitric acid. Base metal catalysts are also used to promote ammonia oxidation. Reducing agents normally do not react. The neutralisation of acids is of commercial importance, e.g., three major fertilisers, ammonium nitrate, ammonium sulphate, and ammonium phosphate, are made from ammonia.

The reaction between ammonia and water is reversible :

$$NH_3 + H_2O \rightleftarrows [NH_4OH] \rightleftarrows NH_4^+ + OH^-$$

With increasing temperature, the solubility of ammonia decreases rapidly. Ammonia is a comparatively weak base and ionises in water much less than sodium hydroxide. Aqueous ammonia also acts as a base in precipitating metallic hydroxides from solutions of their salts and forming complex ions in excess ammonia solution. For example, with copper sulphate solution the cupric hydroxide, which is at first precipitated, redissolves in excess ammonia solution owing to the formation of the complex cuprammonium, tetramminecopper(II) ion, $[Cu(NH_3)_4]^{2+}$. Both potassium and sodium metals dissolve to give the amides, e.g., $NaNH_2$. Other metals, such as magnesium, calcium, strontium, and barium, give the nitrides. Reactions with halogens give unstable trihalides.

Ammonia reacts with chlorine in dilute solution to give chloramine, an important reaction in water purification. Ammonia reacts with phosphorus vapour at red heat to give nitrogen and phosphine. Sulphur also reacts with liquid anhydrous ammonia to produce nitrogen sulphide, N_2S_4. Ammonia and carbon at red heat give ammonium cyanide.

Ammonia forms a great variety of addition or coordination compounds, also called ammoniates, in analogy with hydrates. Thus $CaCl_2.6NH_3$ and $CuSO_4.4NH_3$ are comparable to $CaCl_2.6H_2O$ and $CuSO_4.4H_2O$, respectively. Such compounds when regarded as coordination compounds are called amines and are written as complexes, e.g., $[Cu(NH_3)_4]SO_4$. The solubility in the water of such compounds is often quite different from the solubility of the salts themselves, e.g., silver chloride, AgCl, is almost insoluble in water, whereas $[Ag(NH_3)_2]Cl$ is readily soluble. Thus silver chloride dissolves in aqueous ammonia. Similar reactions take place with other insoluble salts such as silver phosphate and cuprous chloride.

The reaction of ammonia and carbon dioxide via ammonium carbamate to urea is of considerable industrial importance.

Manufacture

Haber and Bosch in Germany developed the first commercial process for the direct synthesis of ammonia (Fig. 11.3). In the mid 1960s, the ammonia industry grew very rapidly owing to improvements in the energy cycle, including recovery of waste heat to generate high pressure steam, and technical advances in compressors, reformers, and converters. Ammonia is synthesized by the reversible reaction of hydrogen and nitrogen:

$$N_2 + 3H_2 \rightleftarrows 2\ NH_3$$

In the design of synthesis facilities, the rate at which the ammonia is formed has to be considered.

The reaction of hydrogen and nitrogen to form ammonia on an industrial scale is always carried out on a catalytic surface based on metallic iron (mostly magnetite) that has been promoted with other oxides. Oxygen and oxygen-containing compounds have a temporary poisoning effect on the catalyst and sulphur, arsenic, phosphorus, chlorine, and heavy hydrocarbons cause permanent poisoning.

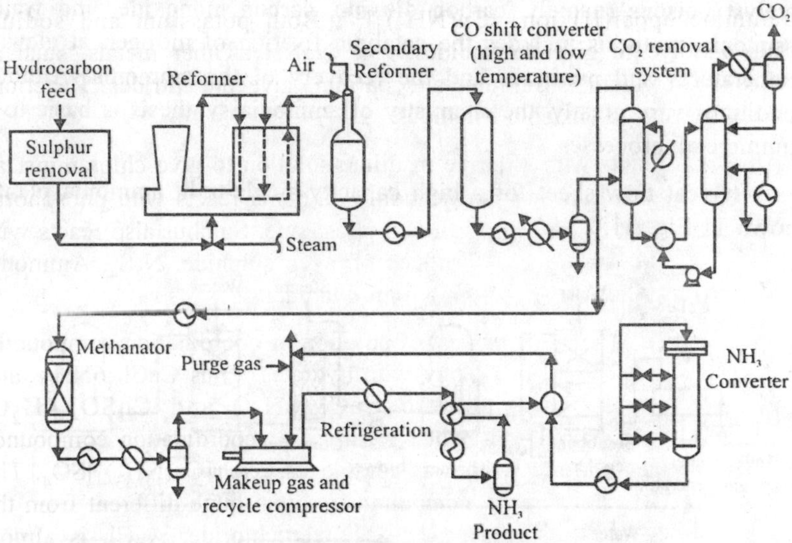

Fig. 11.3. Haber-Bosch process.

The following variables affect ammonia synthesis:

(1) Synthesis pressure, an increase in pressure increases the equilibrium per cent rate of ammonia and the reaction rate.

(2) Synthesis temperature, higher temperatures increases reaction rates but decreases the equilibrium amount of ammonia and increases the thermal degradation of the catalyst.

(3) Space velocity (the ratio of the volumetric rate of gas at standard conditions to the volume of catalysts), the percentage of ammonia in the existing gas decreases with increasing space velocity. However, the same volume of catalyst at increased space velocities can produce more ammonia (normal space velocities for commercial operations are 8,000–60,000 $hour^{-1}$).

(4) Inlet gas composition, effects vary depending on the concentration of inert materials, ammonia concentration, and hydrogen to nitrogen ratio. Most commercial facilities operate at a hydrogen to nitrogen ratio of 3:1; however, the optimum may be something less than that.

(5) Catalyst particle size, catalyst activity increases with smaller particles (most common size 6–10 mm).

The manufacture of synthetic anhydrous ammonia consists of three basic steps: synthesis-gas preparation, purification, and ammonia synthesis. The first two involves the generation of hydrogen, the introduction of nitrogen in the stoichiometric synthesis proportion, and the removal of catalyst poisons (namely, carbon dioxide, carbon monoxide, and water). Ammonia synthesis includes the catalytic fixation of nitrogen at elevated temperatures and pressures and the recovery of the ammonia. Although conditions vary greatly, the chemistry of ammonia synthesis is basic to all commercial processes.

A typical flow sheet for a high capacity single-train ammonia plant is shown in Fig. 11.4.

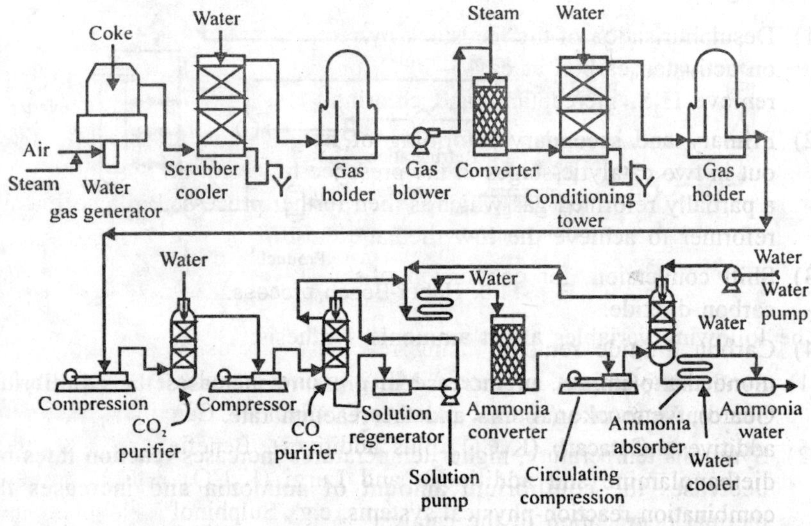

Fig. 11.4. Flow sheet for a high capacity single-train ammonia plant.

By and large, the raw-material sources in modern ammonia plants are coal, petroleum fractions, and natural gas. In general, the most economical feedstock has the highest hydrogen to carbon ratio.

The two basic generation techniques for processing these raw materials are partial oxidation and steam reforming. Steam reforming is most often used in commercial plants, but partial oxidation processes are employed where steam-reformable feeds are not available.

The heart of a coal-based partial oxidation process is the gasification step. The two important commercial processes are the Lurgi, employing a fixed-bed reactor using oxygen and steam at ca 2000-3000 kPa (20–30 atm) and temperatures ranging from 550 to 620°C, and the Koppers-Totzek, in which gasification takes place at low pressures and much higher temperatures (ca 1480°C). In recent years, much attention has been given to production of synthetic gas via coal-gasification processes.

Partial oxidation of hydrocarbon feedstocks for the production of synthetic gas uses the Shell and Texaco processes with operating conditions in the gas generator varying from 1200 to 1370°C and 3200-8370 kPa gauge (465-1215 psig) using heavy oils as hydrocarbon feed.

The bulk (75–80%) of worldwide ammonia production uses hydrocarbon steam-reforming operations, approximately 60-65% use; natural gas feed. The process involves the following:

(1) Desulphurisation of the feedstock by removal of sulphur via adsorption on activated carbon at ca 15–50°C, or by reaction with zinc oxide to remove H_2S, mercaptans, and chlorine.

(2) Primary and secondary reforming of hydrocarbon feedstock, carried out in two catalytic stages in the presence of steam: the first producing a partially reformed gas which is then further processed in a secondary reformer to achieve the low methane content desired.

(3) Shift conversion, for conversion of carbon dioxide to hydrogen and carbon dioxide.

(4) Carbon dioxide removal using reaction systems, e.g., MEA (20% monoethanolamine), promoted MEA (monoethanolamine plus Amine Guard), Vetrocoke (K_2CO_3 plus As_2O_3-glycine), Carsol (K_2CO_3 plus additives), Catacarb (K_2CO_3 plus additives), Benefield (K_2CO_3 plus diethanolamine and additives), and Lurgi (K_2CO_3 plus additives); combination reaction-physical systems, e.g., Sulphinol (sulpholane and 1, 1'-iminobis-2-propanol) and TEA-MEA (triethanolamine and mono ethanolamine); and physical absorption systems that use various solvents, e.g., Purisol (NMP) (*N*-methyl-2-pyrrolidinone), Rectisol (methanol), Fluor solvent (propylene carbonate), and Selexol (dimethyl ether of polyethylene glycol).

(5) Synthesis gas purification, to remove or convert to inert species residual carbon monoxide and carbon dioxide, either by methanation or cryogenic purification.

(6) Ammonia synthesis and recovery, the actual step in which the ammonia is made. Currently, the operating pressure for most modern synthesis

loops fall in the range of 13,785 to 34,475 kPa (2000-5000 psi.), whereas new plants favour the lower range. The economics of the more recent large tonnage ammonia plants is more favourable at these pressures.

The question of what is the right synthesis facility varies from project to project with the synthesis pressure the most important consideration, followed by loop configuration, makeup gas-recycle compression, and the converter. The design of modern ammonia plants reflects a high degree of integration between the process and energy systems. Waste process heat is used to provide energy for boiler-feedwater heating and steam generation. Energy recovery in the convection section of the primary reformer includes satisfying similar utility demands of boiler-feedwater heating, and steam generation and superheating.

Health and Safety

Effects of ammonia can range from a mild irritation to severe corrosion of sensitive membranes of the eyes, nose, throat, and lungs, depending on the concentration. Because of its great affinity for water, it is particularly irritating to moist skin surfaces.

Uses

The largest market for ammonia is the fertiliser industry. Other important markets include commercial explosives (nitric acid, initially derived from ammonia, is used in manufacture of explosives); and fibres and plastics.

AMMONIUM COMPOUNDS

The ammonium compounds consist of a large number of salts, many of which are of considerable industrial importance, e.g., ammonium sulphate, ammonium chloride, and ammonium nitrate. The ammonium salts are much like salts of the alkali metals in such properties as solubility, but they differ in being considerably less stable. Except for complexes containing metallic radicals, they are completely volatile on heating or ashing. They are more like the salts of potassium than those of sodium in some ways, e.g., in the lesser solubility of the ammonium and potassium perchlorates and chloroplatinates.

Ammonium Chloride

Ammonium chloride, known as sal ammoniac, occurs naturally in crevices in the vicinity of volcanoes as a sublimation product. It has specific heat of 1.55 J/g (0.371 cal/g) in the temperature range 1–55°C. It is soluble to the extent of 29.4 g/100 g H_2O at 0°C. Vapour pressure at 250°C 6.5 kPa; at 320°C 60.9 kPa; and at 338°C 101.1 kPa. The notable high vapour pressure of ammonium chloride at elevated temperatures causes it to sublime readily into a vapour consisting of equal volumes of ammonia and hydrogen chloride. Like other ammonium salts of strong acids, it has an acid reaction in aqueous solution and the solid tends to lose ammonia and become more acid on exposure and in storage. Its aqueous solutions have a notable tendency to attack ferrous metal and other metal surfaces.

Ammonium chloride manufacture is based on metathesis, or double decomposition reactions of sodium chloride; the ammonium salt is used directly or formed *in situ*. Ammonium chloride is obtained as a by-product of the classic ammonia-soda or Solvay process. Double decomposition using the ammonium sulphate-sodium chloride process is another method used. Preparation by direct neutralisation of hydrochloric acid with ammonia is simple but not as attractive economically.

Ammonium Nitrate

Ammonium nitrate does not occur in nature and was first made in 1659 by Glauber, who called it *nitrum flammans* because of the difference of its yellow flame from that of potassium nitrate. It is the most important of the ammonium compounds from the standpoints of volume of production and major uses.

Ammonium nitrate is a white crystalline solid, d_4^{20} 1.725 g/cm^3. It is extremely soluble in water, and therefore very hygroscopic (Table 11.1). In the solid state, it occurs in five different crystallographic modifications.

Table. 11.1 Solubility of ammonium nitrate.

Temperature °C	*Solubility of NH_4NO_3, g in 100 g Water*	*in 100 g soln*
0	118	54.2
20	187	65.2
40	297	74.8
60	410	80.4
80	576	85.2
100	843	89.4

The change from one crystal form to another at various transition points can be detected by time-temperature cooling curves.

The solid salt has a specific heat of 1.70 J/g (0.406 cal/g) between 0 and 31°C. It has a negative heat of solution in water and can be effectively used to prepare freezing mixtures.

Ammonium nitrate decomposes under extreme shock or at elevated temperatures in two widely divergent manners. (Fig.11.5.)

$$NH_4NO_3 \rightarrow N_2O + 2\,H_2O$$

$$2\,NH_4NO_3 \rightarrow 2\,N_2 + 4\,H_2O + O_2$$

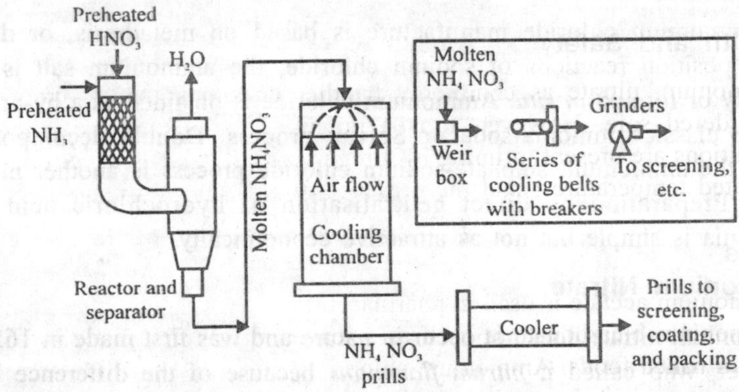

Fig. 11.5. Simplified flowchart for Stengel process for ammonium nitrate manufacture.

The first reaction, which occurs when ammonium nitrate is heated to temperatures from 200 to 260°C, can be carried out safely and can be controlled even when rapid. The second reaction takes place with great rapidity and violence when ammonium nitrate detonates.

Ammonium nitrate acts as an oxidising agent in many reactions at ordinary temperatures and in aqueous solutions it is reduced by various metals. Ammonium nitrate manufacturing processes depend almost entirely on the neutralisation of nitric acid with ammonia in liquid or gaseous form and involve three essential steps: neutralisation, evaporation of the neutralised solution, and control of the particle size and characteristics of the dry product.

Ammonium Sulphate

Ammonium sulphate is a white crystalline solid, whose crystals are of rhombic structure, d_4^{20} 1.769 g/cm^3. The melting point as determined in a closed system is 513 ± 2°C. Upon heating in an open system, the sulphate begins to decompose at 100°C and yields ammonium bisulphate, NH_4HSO_4, which has a melting point of 146.9°C. The solubility of ammonium sulphate in 100 g water is 70.6 g at 0°C and 103.8 g at 100°C. It is insoluble in alcohol and acetone.

Ammonium sulphate is produced by reaction of by-product ammonia from coke ovens with sulphuric acid. Additional amounts have increasingly been produced by the reaction of synthetic ammonia and sulphuric acid; from the process for production of caprolactam; and via a method using gypsum, $CaSO_4.2H_2O$.

Health and Safety

Ammonium nitrate as commonly handled in small amounts in properly considered safe. However, it is a potential high explosive when three conditions are present: priming by a high velocity explosive; confinement at elevated temperatures; and the presence of an oxidizable substance.

Uses

Ammonium acetate is used in pharmaceuticals as a diaphoretic and diuretic. Ammonium bicarbonate is used as a leavening agent in the production of certain baked foods. Ammonium chloride is used in the manufacture of dry cells; as a metal cleaner in soldering; as a flux in tinning and galvanizing; and in pharmaceuticals. Ammonium nitrate is used as a fertiliser and as an industrial explosive, as well as in the manufacture of nitrous oxide. Ammonium sulphate is used as a fertiliser, particularly for rice; and as an additive to supply nitrogen in fermentation processes.

UREA

Urea can be considered the amide of carbamic acid, NH_2COOH, or the diamide of carbonic acid, $CO(OH)_2$. Properties are shown in Table 11.2. At atmospheric pressure and at its melting point, urea decomposes to ammonia, biuret (i), cyanuric acid (ii), ammelide (iii), and triuret (iv). Biuret is the principal and least desirable by-product present in commercial urea.

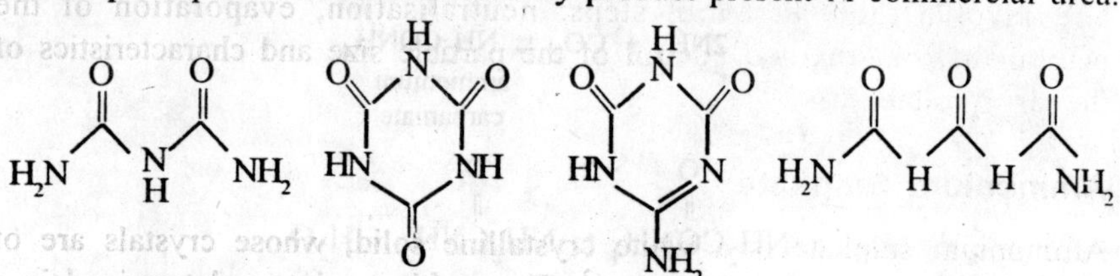

(i) biuret (ii) cyanuric acid (iii) ammelide (iv) triuret

Urea acts as a monobasic substance and form salts with acids. With alcohol urethanes. Urea and malonic acid give barbituric acid, an important compound in medicinal chemistry.

Table 11.2. Properties of urea.

Property	*Value*
Melting point, °C	135
Index of refraction, *n*	1.484, 1.602
Density, d_4^{20}, g/cm³	1.3230
Crystalline form and habit	tetragonal, needles or prisms
Free energy of formation, at 25°C, J/mol[a]	—197.150
Heat of fusion, J/g[a]	251[b]
Heat of solution in water, J/g[a]	243[b]
Heat of crystallisation, 70% aqueous urea solution, J/g[a]	460[c]
Bulk density, g/cm³	0.74
Specific heat, J/(kg·K)[a]	
at 0°C	1.439
at 100°C	1.887

[a] To convert, J to cal, divide by 4.184.

[b] Endothermic.

[c] Exothermic.

Manufacture

Urea is produced from liquid ammonia and gaseous CO_2 at high pressure and temperature. The formation of ammonium carbamate and the dehydration to urea take place simultaneously, for all practical purposes.

$$2NH_3 + CO_2 \leftrightarrows \underset{\text{ammonium carbamate}}{NH_2\overset{\overset{O}{\|}}{C}ONH_4}$$

$$NH_2CONH_4 \leftrightarrows \underset{\text{urea}}{NH_2\overset{\overset{O}{\|}}{C}NH_2} + H_2O$$

Ammonium carbamate is a white crystalline solid that forms by passing ammonia gas over dry ice. The conversion to urea begins below 100°C. The maximum conversion is attainable at 185°C, ca 53%. The ammonia and CO_2 gases recovered from the effluent mixture is several pressure-staged decomposition sections are absorbed in water and recycled.

In the Mitsui-Toatsu total recycle process, the reactor is operated at ca 25 MPA (246 atm) and 195°C at an NH_3-to-CO_2 overall mol ratio (fresh feed plus recycle) of ca 4:1. A relatively high conversion of carbamate to urea per pass is reported (67-70%). Many urea plants with capacities of up to 1800 metric tons per day are using this process.

In the Montedison urea process, the reactor is operated at ca 20-22 MPa (197-217 atm) and an overall NH_3-to-CO_2 mol ratio of ca 3.5:1 (fresh feed plus recycle). A 62-63% conversion of carbamate is reported. In an improvement, based on the new isobaric double-recycle (IDR) technology, the reactor effluent is first stripped with NH_3 gas and then with CO_2 gas. Considerable reduction in steam consumption is reported.

The UTI heat-recycle process offers the following new concepts: An isothermal reactor is provided with an internal open-ended coil for countercurrent heat transfer from the strongly exothermic process of carbamate formation to the endothermic formation of urea; ca 40-50% of the makeup CO_2 feed is injected into the medium-pressure carbamate decomposition section, and more than 70% of the exothermic heat of carbamate formation in the medium-pressure absorption system is exchanged with relatively colder streams within the process.

High pressure gas stripping, is based on high pressure CO_2 gas stripping at reactor pressure and relatively high temperature. The unconverted carbamate is decomposed to NH_3 and CO_2 by the stream of gaseous CO_2 passed through the reactor effluent solution at reactor pressure and condensed and recycled. Because of its energy efficiency, the stripping process accounts for almost half of the world's urea production.

In the Stamicarbon CO_2 stripping process, reactor, carbamate decomposer (stripper), and carbamate condenser each operate at ca 14 MPa (140 atm) at an NH_3:CO_2 mol ratio of ca 2.8:1.

In the Snamprogetti NH_3 stripping process, a synthesis loop is operated at ca 15 MPa (150 atm) and an NH_3: CO_2 overall mol ratio of 3.8:1. Carbamate conversion to urea per pass is reported to be ca 65–67%.

Waste-water Treatment

Under the pressure of government regulations with regard to residual NH_3 and urea in waste-waters, the fertiliser industry made an effort to improve waste-water treatment. It is, however, difficult to reduce the NH_3 and urea is hydrolysed to ammonium carbamate, and CO_2 and NH_3 are stripped off. The Stamicarbon system uses a hydrolyser for urea and a separate dual-desorption system to strip the ammonia. The residual content of NH_3 and urea is reported to be reduced to about 70 and 80 ppm, respectively.

The UTI hydrolyser stripper (Fig. 11.6) consists of a single stainless-steel tower in which urea is hydrolysed and NH_3 stripped simultaneously by means of steam and CO_2 stripping.

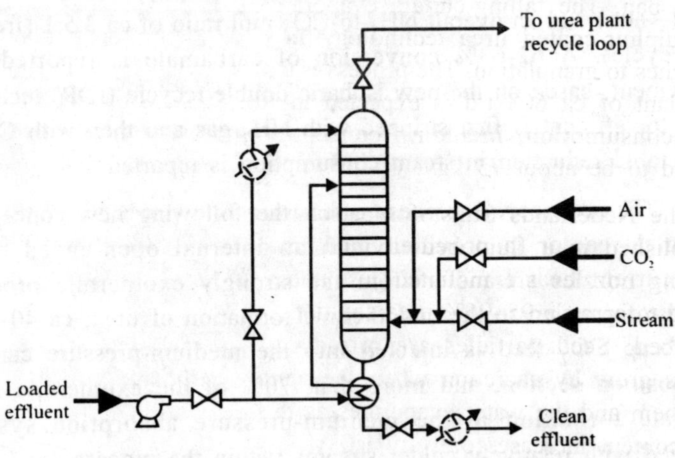

Fig. 11.6. Hydrolyser-stripper clean-up process.

Finishing Processes

Urea processes provide an aqueous solution containing 70–80% urea. This solution can be used directly for nitrogen-fertiliser suspensions or solutions. The water is evaporated from the steam-heated solutions under reduced pressure with or without the addition of hot air as drying agent.

A combination of crystallisation and remelting removes biuret, which is detrimental in the technical-grade urea used for the manufacture of plastics. The solid urea thus obtained is relatively pure and well-suited for such applications. Until recently, most solid urea was produced by prilling. Molten urea obtained either by evaporation or by crystal melting is sprayed as droplets from the top of a tall cylindrical tower (50-60 m high) and allowed to fall counter-currently through a stream of cold air emanating from a fluid-bed cooler at the base of the tower. Prill towers require large quantities of air, and pollution abatement is very expensive. In addition, granulation offers far more flexibility, and therefore the industries all over the world are shifting to this technique.

In granulation, urea is concentrated to ca 99.5% and sprayed through nozzles onto a combination of falling granules and a cascading bed of recycled fines in a rotating cylindrical granulation drum. The physical characteristics of the granules are good, with usually < 0.15% H_2O; sizes range between 6.7–4 mm or 3.4–1.7 mm, and the granules are hard and spherical. Biuret content is 1–2%. In pan granulation, a concentrated urea solution is sprayed onto a cascading bed of recycled fines in an inclined rotating pan. The falling-curtain evaporative cooling process is based on TVA sulphur-coated urea technology in combination with some novel approaches to granulation. The process is extremely energy efficient. A full-scale plant of ca 608 t/d is expected to use ca 22.1 kW.h/t of product. Steam consumption, including concentration of scrubbing solution, is expected to be about 75 kg/t.

In the Nederlands Stikstof Maatschappij N.V process, granulation is accomplished in a fluid bed divided into several chambers. Pneumatic atomising nozzles are mounted just above the air-distribution plate and oriented to spray a 95–96% urea solution at 130–135%°C upward into the active bed. Seed particles are introduced into the first chamber. The particles grow by accretion when the atomised droplets of urea solution strike them and the water evaporates as the concentrated urea crystallises; biuret content increases only by 0.03%.

In the Mitsui-Toatsu and Toyo engineering process, concentrated urea solution is sprayed onto the surface of particles in a fluidising granulation process; the heat released by solidification of the urea feed is removed by a combination of cooled recycle, fluidising air, and some water evaporating. Hardness may be increased with additives. Granules with a 3-mm diameter are reported to have a crushing hardness of 20 N (2 kgf). A production capacity of 100 t/d per granulator is estimated to be the optimum size.

Uses

Solid urea containing 0.8–2.0% biuret is primarily used for direct application to the soil as nitrogen-release fertiliser. Urea is also used as feed supplement for ruminants, where it assists in the utilisation of protein. Urea is a raw material for urea-formaldehyde resins. Reagent-grade urea is needed in some pharmaceutical preparations.

Clathrates

Urea has the remarkable property of forming crystalline complexes or adducts with straight-chain organic compounds. These complexes consist of a hollow channel, formed by the crystallised urea molecules, in which the hydrocarbon is completely occluded. Urea clathrates are used in petroleum refining for the production of jet fuels and for the dewaxing of lubricating oils.

NITRIC ACID

Raw Materials

The essential raw materials for the modern manufacture of nitric acid are anhydrous ammonia, air, water, and platinum-rhodium gauze as a catalyst. Because of its low molecular weight, the ammonia can be shipped economically from large primary nitrogen-fixation plants to various oxidation plants at consuming centers. Furthermore, anhydrous ammonia can be shipped in steel cars, in comparison with aqueous nitric acid, which requires stainless-steel tank cars weighing much more.

Reactions and Energy Changes

The essential reactions for the production of nitric acid by the oxidation of ammonia may be represented as follows :

$$4NH_3(g) + 5O_2(g) \longrightarrow 4NO(g) + 6H_2O(g) \quad \Delta H_{298°C} = -216.6 \text{ kcal} \quad ...(11.4)$$

$$2NO(g) + O_2(g) \longrightarrow 2NO_2(g) \quad \Delta H_{298°C} = -27.1 \text{ kcal} \quad ...(11.5)$$

$$3NO_2(g) + H_2O(l) \longrightarrow 2HNO_3(aq) + NO(g) \quad \Delta H_{298°C} = -32.2 \text{ kcal} \quad ...(11.6)$$

$$NH_3(g) + O_2(g) \longrightarrow \frac{1}{2} N_2O(g) + \frac{3}{2} H_2O(g) \quad \Delta H = -65.9 \text{ kcal} \quad ...(11.7)$$

Several side reactions reduce somewhat the yield of reaction (11.1):

$$4NH_3(g) + 3O_2(g) \longrightarrow 2N_2(g) + 6H_2O(g) \quad \Delta H_{298°C} = -302.7 \text{ kcal} \quad ...(11.8)$$

$$4NH_3(g) + 6NO(g) \longrightarrow 5N_2(g) + 6H_2O(g) \quad \Delta H_{298°C} = -431.9 \text{ kcal} \quad ...(11.9)$$

$$2NO_2(g) \longrightarrow N_2O_4 \quad \Delta H = -13.9 \text{ kcal} \quad ...(11.10)$$

Reaction (11.4) is essentially a very rapid catalytic one, carried out by passing about 10% ammonia by volume, mixed with preheated air, through the multilayered, silk-fine platinum-10% rhodium gauze at a temperature of approximately 920°C or lower (Fig. 11.7) here, once ignited, the stream of ammonia continuous to burn. The yield is 94 to 95%. The equilibrium constant for reactions (11.4) and (11.5) is 9.94×10^{14} for NO_2 production at 627°C. This reaction is usually carried out at a pressure of 100 psi. In Europe the process pressure is frequently 40lb and the temperature is about 815°C, with consequent smaller platinum losses and a lower production rate. Much has been published on the commercialisation of these reactions. As can be seen from reaction (11.4), there is only a small increase in volume, so that the *principle of Le Chatelier* does not affect the equilibrium very substantially. However, the increase in pressure, by compressing the reactants, permits greater space velocity to be maintained, resulting in a plant saving until such a pressure is reached that the cost of the increased thickness of the stainless steel required more than counterbalances the saving in volume of equipment per pound produced. Pressure oxidation also furnishes an acid containing 60 to 70% HNO_3 in comparison with the 50 to 55% HNO_3 obtained from atmospheric oxidation. The speed of ammonia oxidation is extraordinarily high, giving an excellent conversion in the short contact time of 3×10^{-4} s at 750°C with a fine platinum–10% rhodium gauze catalyst. Hence, in the industrial procedure, it has been found economical to mix initially with the ammonia all the air needed for reactions (11.4) and (11.5). The oxidation of the NO to NO_2 is the slowest reaction, but the equilibrium is more favourable at lower temperatures. Hence this reaction is carried out in absorbers of considerable capacity, which are provided with cooling in all the upper trays. Because of a decrease in volume, this reaction is favoured under pressure, according to the principle of Le Chatelier. Although these factors raise the cost of the equipment for carrying out the oxidation of nitric oxide, they increase the conversion. It is necessary in the design of a plant to know how long this reaction will take, in order to calculate the volume necessary for the equipment.

$$3NO_2 + H_2O \rightleftharpoons 2HNO_3 + NO$$

Equation (11.6) is really an absorption phenomenon. This reaction, in the opinion of Taylor, Chilton, and Handforth, is the controlling one in making nitric acid, and its rate was increased by employing an absorption tower under pressure and cooling and with countercurrent graded strengths of acid for the absorption. Warm air is introduced into a short raschig-ring-packed section between the tower and acid trap. This provides for the reoxidation of the NO formed and also desorbs (bleaches) the dissolved nitrous oxides which would colour the acid.

From Ammonia

Process

Air is compressed to approximately 100 psi (0.7 MPa), filtered, and preheated to about 300°C by passing through a heat exchanger. It is mixed with anhydrous ammonia, previously vapourised in a continuous-steam evaporator. The resulting mixture, containing about 10% ammonia by volume, is passed through the reactor, which contains a platinum-rhodium (2 to 10% rhodium) wire-gauze catalyst. The platinum-rhodium alloy is used in the form of a very fine gauze, 80-mesh and 75-μm-diameter wire, packed in layers of 10 to 30 sheets so that the gas travels downward through the gauze sheets. The catalyst temperature is about 750°C, the pressure 100 psi (0.7 MPa), and the contact time about 3×10^{-4} sec. (Fig. 11.7).

$$4NH_3 + 5O_2 \rightarrow 4NO + 6H_2O$$
$$2NO + O_2 \rightarrow 2NO_2$$
$$3NO_2 + H_2O \rightarrow 2HNO_3 + NO$$

93–95% yield

The hot nitric oxide and excess air mixture (about 10% nitric oxide) from the reactor is partially cooled in a heat exchanger and further cooled in a water cooler. The cooled gas is introduced into a stainless-steel absorption tower with more air being added for the further oxidation of nitric oxide to nitrogen dioxide. Since the equilibrium of this reaction is more favourable at lower temperatures, most of the reaction occurs in the absorption tower which is continuously cooled externally with water. Small quantities of water are added to hydrate the nitrogen dioxide and partially to scrub the gases. The gas from the top of the tower generally is heated in a heat exchanger and then expanded through a power-recovery compressor for part of the air supply before being released to the atmosphere (Fig. 11.7).

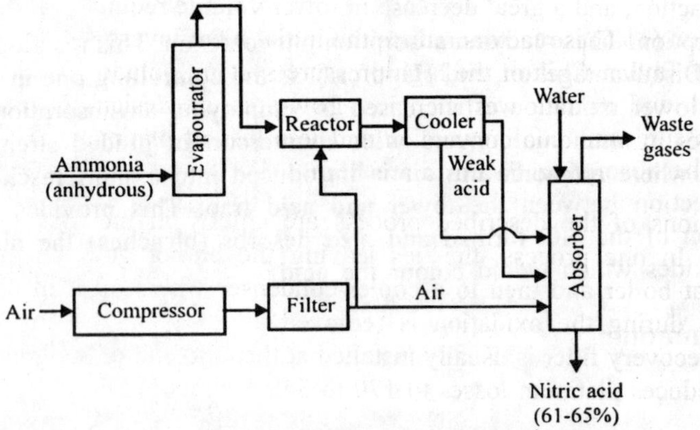

Fig. 11.7. Flow diagram for manufacturing of nitrogen and its compounds from ammonia.

The bottom of the tower yields nitric acid of 61 to 65% strength, which may be concentrated if so desired. This may be done in silicon-iron or stoneware towers by feeding concentrated sulphuric acid, sp gr 1.84, to the top, and 61 to 65% nitric acid below the top. The vapour leaving the tower is 90% nitric acid or more, and the bottoms consist of dilute (70 to 75%) sulphuric acid. The latter may be reconcentrated, or used as is, for example, for making superphosphate fertilisers.

Other processes are available for concentrating the nitric acid to nearly 100%. In one such process, magnesium nitrate is used as the drying agent. Magnesium carbonate is treated with nitric acid to yield magnesium nitrate; this is fed, with weak nitric acid, to a tower where the magnesium nitrate combines with the water; the magnesium nitrate is dehydrated for recycle in a separate concentrator.

In another technique, nitrogen oxides from the ammonia-oxidation unit are processed directly. Some of these gases are dried and further oxidised to nitrogen tetroxide; a mixture of the tetroxide, weak acid, and air is heated to 70°C at 50 atm (5 MPa) pressure for about 4 h to yield 98 to 99% nitric acid.

These techniques for concentration nitric acid are necessary because simple distillation is not applicable; nitric acid forms an azeotrope with water at about 68% acid.

The process described is termed a high-pressure process, operating at 8 to 10 atm (0.8 to 1 MPa). Low- or atmospheric-pressure and medium-pressure (3 to 5 atm, 0.3 to 0.5 MPa) plants are also used, but to a lesser extent. High-pressure plants have the advantages of lower capital costs, because of the increase in acid strength obtained, a 50-fold increase in the rate of reaction, and a great decrease in tower volume required for oxidation and absorption. These factors reduce the initial plant investment by as much as 50%. Disadvantages of the high-pressure system include greater catalyst loss and lower oxidation efficiencies. Low-pressure designs have the best efficiencies in ammonia conversion and less loss of platinum, and may be preferred where ammonia costs are high.

Variations of the described process are chiefly in methods of power recovery. In one process, the gas leaving the burner goes directly to a waste-heat boiler and then to a cooler-condenser where most of the water produced during the oxidation is removed as 40 to 45% nitric acid. A catalyst-recovery filter is usually installed at the cold end of the boiler. Such a filter reduces platinum losses to 170 to 340 mg/metric ton of 100% nitric acid.

In another common variation, tail gas from the absorber, which contains unused oxygen, is mixed with methane and passed over an oxidation catalyst. The resulting hot gases are sent to a power-recovery turbine, which recovers sufficient energy to make the nitric acid plant self-sufficient in power.

Uses

Ammonium nitrate, including military, Adipic acid, Miscellaneous fertilisers, Military other than ammonium nitrate, Isocyanates, Nitrobenzene, Potassium nitrate, Steel pickling, Other.

Properties

Pure 100% nitric acid is a colourless, highly corrosive liquid with powerful oxidising properties. The commercial acid varies from colourless to a yellowish-red colour, owing to the presence of dissolved nitrogen oxides.

Mol. wt 63.01, mp –42°C, Sp gr 1.502, bp 86°C (partial decomposition). Threshold limit value, nirtrogen oxides, 5 ppm, 9 mg/m^3.

Soluble in all proportions in water and forms a constant-boiling mixture at 760 mm (100 kPa) pressure of 68% acid and 32% water with a boiling point of 110°C. Solutions containing more than 86% nitric acid are known as fuming nitric acids.

The largest application for nitric acid is for conversion to ammonium nitrate; most of this conversion is accomplished on site. Thus nitric acid manufacture is tied very closely to the fertiliser industry, which is notoriously subject to wild swings in demand, although with worldwide food needs the trend must be upward. However, it is not certain that demand for ammonium nitrate will necessarily follow the trend, since alternative nitrogen sources such as urea and anhydrous ammonia may become more economical. However, nitric acid may still be needed for such newer fertilisers as nitric phosphates, ammonium phosphate nitrates, sulphate nitrates, which do not require the separate production and isolation of ammonium nitrate.

Practically all nitric acid production comes from the catalytic oxidation of ammonia. The relative economics of the three types of process—high-, medium-, and low-pressure—have been discussed. Although the platinum-rhodium alloy catalyst remains standard, other catalysts have been developed, some using less precious metal and at least one none at all.

SODIUM AND POTASSIUM NITRATES

Sodium Nitrate

Sodium nitrate, $NaNO_3$, occurs in nature associated with sodium and potassium chlorides, potassium nitrate, sodium sulphate, magnesium chloride, and other salts. Chilean nitrate is the only inorganic nitrogen fertiliser produced today are given in Table 11.3.

Table. 11.3. Selected properties of sodium nitrate.

Property	*Value*
Crystal system	trigonal, rhombohedral
Mp, °C	308
Refractive index, n_D^{20}	
trigonal	1.587
rhombohedral	1.336
Density (solid), g/cm³	2.257
Solubility in H_2O, molality (± 2%)	
at 0°C	8.62
at 80°C	17.42
Specific conductivity at (300°C), S/cm	0.95
Viscosity[a], mPa·s (=cP) at 590 K	2.85
Heat of fusion, J/g[b]	189.5
Heat capacity, J/g[b]	
Solid at 0°C	1.035 ± 0.005
Liquid at 350°C	1.80 ± 0.02

[a] Measurement method: capillary.

[b] To convert J to cal, divide by 4.184.

Manufacture and processing

Guggenheim process

The Guggenheim nitrate process demonstrated that caliche ores as low as 7 wt% in nitrate content could be economically exploited. The mined ore is transferred to crushing units where it is crushed in three stages. The crushed ore is leached in 8–10 vats, of which four are used at any given time at an average temperature of 40°C. The leaching cycle lasts 40 hour, the total vat cycle ca 168 hour.

The fines produced during crushing are mixed in the filter plant with liquor containing 150–200 g/L of nitrate, at a mixing ratio of two tons of solids per cubic metre of liquor; the leaching is carried out at 40–50°C. The rich liquor from the leaching vats containing 450 g/L of sodium nitrate is sent to crystallisers where it is chilled to 0–5°C. At the exit of the last crystallisers, the mother liquor and the crystals are pumped out as a slurry to continuous centrifuges.

The sodium nitrate crystals are transferred to the graining plant, which consists of large, oil-fired, reverberatory-type furnaces, where the crystals are heated to 315–325°C. The molten sodium nitrate is pumped to large spray chambers where the spray solidifies as prills. The Guggenheim process recovers 80–85 wt% of the $NaNO_3$ contained in coarse fractions at a total fuel consumption of 0.1 metric ton per metric ton of sodium nitrate produced.

In the modified Guggenheim process, the ore is leached countercurrently with water in vats. From the weak sodium nitrate brines, sodium sulphate is extracted by a chilling process that yields Glauber's salt. Rich sodium nitrate solution is passed to the crystallisers and mother liquor to the solar-pond system. Continuous evaporation in the ponds results in concentration increases of potassium nitrate and iodine. Potassium nitrate is removed through crystallisation. Synthetic sodium nitrate is produced by neutralisation of nitric acid with soda ash or caustic soda. The sodium nitrate is melted and prilled.

Health and safety factors

The acceptable daily intake by adults for nitrates (suggested by WHO) is 5 mg/kg in addition to naturally occurring nitrates; large doses are lethal. Accidental ingestion of ca 8–15 g or more causes severe abdominal pain. Poisonings from the ingestion of meat containing sodium nitrate and nitrite have occurred.

Uses

Sodium nitrate is used primarily in agriculture as a fertiliser. The sodium in sodium nitrate corrects soil acidity and liberates phosphates, and is a partial substitute for potassium. The nitrogen from sodium nitrate is available faster than nitrogen from ammonia.

The main industrial use is in the manufacture of explosives, e.g., dynamite. Sodium nitrate is also used in the manufacture of glass, fibre glass, enamels, and porcelain as an oxidising and fluxing agent. It is an ingredient in the production of charcoal briquettes. Sodium nitrate has application in the production of certain antibiotics and pharmaceuticals. In curing beef, bacon, and other meats, it is used as a preservative.

Chapter 12

Phosphorus Industries

INTRODUCTION

The use of artificial fertilisers, phosphoric acid, and phosphate salts and derivatives has increased greatly, chiefly because of aggressive and intelligent consumption promotion on the part of various manufacturers. However, before full consumption of these products could be achieved, more efficient and less expensive methods of production had to be developed. During recent decades, the various phosphate industries have made rapid strides in cutting the costs both of production and distribution and have thus enabled phosphorus, phosphoric acid, and its salts to be employed in wider fields and newer derivatives to be introduced. Supplementing the development of more efficient phosphorus industries have been the epoch-making pure chemical studies of phosphorus, in its old and in its new compounds.

These phosphates are not simple inorganic chemicals, as was assumed several decades ago, and their study has become a unique and complicated branch of chemistry that may some day be compared with the carbon (organic) or silicon branches of today. The properties of phosphorus chemicals are unique because of the important role of phosphorus in many biochemical processes, the ability of polyphosphates to complex or sequester many metal cations, and versatility in forming various types of organic and inorganic polymers.

Table 12.1 is a compilation of phosphate-rock treatment processes. Tricalcium phosphate in raw and/or steamed and degreased bones and in basic slag is also used after grinding as a direct phosphate fertiliser. A small percentage of the former chemicals. Large tonnages of phosphate rock are converted to phosphorus or phosphoric acid and their derivatives.

Table 12.1. Phosphate-rock processing, products, and by-products.

Process	*Raw materials and reagents*	*Main products and derivatives*	*By-products*
Acidulation	Phosphate rock, sulphuric, acid, nitric acid, phosphoric acid, hydrochloric acid ammonia, potassium chloride	Superphosphate, phosphoric acid (wet process), triple superphosphate, monoammonium phosphate, diammonium phosphate, monopotassium phosphate	Fluorine compounds, vanadium, uranium (limited)
Electric-furnace reduction	Phosphate rock, siliceous flux, coke (for reduction), electrical energy, condensing water	Phosphorus, phosphoric acid, triple superphosphate, various Na, K, NH_4, Ca salts; phosphorus pentoxide and halides	Fluorine compounds, carbon monoxide, slag (for RR ballast aggregate, fillers, etc.), ferrophophorus, vanadium
Calcium metaphosphate	Phosphate rock, phosphorus, air or oxygen, fuel	Calcium metaphosphate	Fluorine compounds
Calcination or defluorination	Phosphate rock silica, water or steam, fuel	Defluorinated phosphate	Fluorine compounds

Phosphorus, the twelfth most abundant element, is fairly widely distributed in igneous and sedimentary rocks. The only important source is apatite, $Ca_5X(PO_4)_3$ (X = F, OH, or 0.5 carbonate). The coordination number of phosphorus (at. wt 30.98) is 3 or 4, but 1, 2, 5, and 6 are also known. There is one stable nuclide. ^{31}P, and four radioactive isotopes.

PHOSPHORUS AND PHOSPHIDES

Properties

Phosphorus is a colourless or white waxy solid, d 1.28 g/cm^3, which melts at 44.1°C to a clear colourless liquid (bp 280.5°C). Commercial white phosphorus (99.9%) is slightly yellowish. Red phosphorus forms from molten white phosphorus at ca 580°C. Amorphous or crystalline black phosphorus (d 2.70 g/cm^3) is made from the white form under high pressure (> 980 MPa or 9670 atm) and at high temperature. The black modifications resemble graphite and are good conductors of electricity.

White phosphorus is more reactive than the red modification, which is more reactive than the black. White or liquid phosphorus ignites spontaneously in air and is usually protected from oxidation by a layer of water. Commercial red phosphorus is more stable and reacts only slowly with oxygen and water vapour. Spontaneous combustion, catalysed by traces of iron, sometimes takes place in storage piles. Oxygen at sub-atmospheric pressures or moist air causes white phosphorus to emit a greenish glow.

Halogens, sulphur, and oxidising acids give halides, sulphides, and oxo-acids, respectively, of phosphorus. Phosphine, PH_3, is formed when phosphorus is heated with aqueous solutions of strong alkalies.

Manufacture

Elemental phosphorus is manufactured either in the electric or the blast furnace; both depend on silica as a flux for the calcium present in the phosphate rock. (Fig. 12.1).

$$\underset{\text{fluoroapatite}}{4Ca_5F(PO_4)_3} + \underset{\text{silica}}{18SiO_2} + \underset{\text{coke}}{30C} \rightarrow \underset{\text{slag}}{18CaO.SiO_2.\frac{1}{9}CaF_2} + 30CO\uparrow + 3P_4\uparrow$$

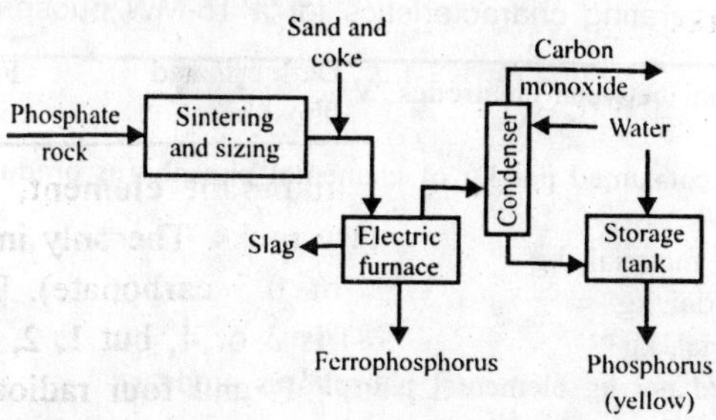

Fig. 12.1. Flow diagram for phosphorus from phosphate rock (Electric Furnace).

Modern furnaces are three-phase units with potentials of 200-300 V. The outer shell is of welded steel sheets and is often cooled by a water spray. The floor is of monolithic carbon which extends a few meters up the side and above the surface of the molten slag. The upper walls are of firebrick or cast refractory cement and the roof is a cast monolithic structure. Operating data are given in Table 12.2; capacities of 50 MW are common.

Most furnaces have two tap holes, a lower one for ferrophosphorus, and an upper one for slag. From furnaces with only one tap hole, the molten stream runs into a catch basin from which the lighter slag overflows. The practice of selling granulated slag as a liming agent has been discontinued because of emission, albeit low, of radioactivity.

Hot gases are emitted at ca 100 m^3/min; ca 93% is carbon monoxide and the remainder is primarily phosphorus with some silicon tetrafluoride and dust. The phosphorus is condensed in towers equipped with water sprays of 45–55°C temperatures. The carbon monoxide is usually burned as fuel.

Red phosphorus is manufactured by a batch process, although continuous methods have been developed. White phosphorus is heated gradually in a steel or cast-iron vessel equipped with a condenser. The temperature is kept at 400°C for several hours. After cooling, the product is wet-ground and boiled with sodium carbonate solution to remove remnants of the white modification. The red phosphorus is sieved, washed, and vacuum dried. It can be stabilised by suspension in a 1 wt% solution of sodium aluminate, the slurry being aerated for several hours.

Phosphorus is poisonous and can cause bone necrosis and fatty degeneration of the viscera. It must be handled with special safety measures. The vapour is highly poisonous, and the established exposure limit is 0.1 mg/m^3 air. Phosphine is much more lethal than elemental phosphorus, but it does not exhibit long-term effects.

Table 12.2. Operating characteristics for a 15-MW phosphorus furnace.

Average potential between electrodes, V	300.00
Power factor	0.97
Raw materials consumed per kg of elemental phosphorus produced	
power, kWh	14.30
phosphatic material, kg	10.00
silica material, kg	1.50
coke material, kg	1.50
Products formed per kg elemental phosphorus produced	
slag, kg	4.00
ferrophosphorus, kg	0.30
carbon monoxide, kg	2.80
recovery, based on the element, of the charged phosphorus, %	87.00
Temperature of off-gases, °C	370.00
Temperature of slag at tapping, °C	1500.00

Uses

Nearly all of the phosphorus production is converted to phosphoric acid or other phosphorus compounds. White phosphorus is rodent poisons. The military uses phosphorus to produce smoke clouds or to ignite gasoline bombs. Red phosphorus is used in safety matches and for fireworks.

Metallic Phosphides and Alloys

Phosphorus forms phosphides with most metals. Sodium phosphides (Na_2P, Na_3P, Na_3P_{11}) are reddish-brown to black materials thermally stable to 650°C; they react instantly with moisture to give phosphine. Magnesium phosphide, aluminium phosphide, or the mixed alloys are shiny gray-to-yellow crystalline materials which are stable in dry air but which decompose upon contact with water. They are used with an igniting agent in sea flares. Addition of ferrophosphorus to high strength, low alloy steel eliminates quenching or tempering. Phosphor copper is used for the deoxidation of copper and its alloys, and phosphor tin for deoxidation of bronzes and German silver. Phosphor bronzes are copper and tin alloys containing 1.25–11 wt% tin; 0.03–0.35 wt% phosphorus deoxidises the alloy. Zinc phosphide, Zn_3P_2, is used as a rodenticide. Silver solder contains small amounts of silver phosphide.

PHOSPHORUS COMPOUNDS

Phosphorus exhibits oxidation states from –3 to +5. The largest group of phosphorus compounds contains oxygen, and the oxo acids form the basis for the most systematic nomenclature. However, the term phosphorous acid is often used to describe phosphonic acid, $H(H_2PO_3)$, which is commonly written as H_3PO_3 or $(HO)_3$ P. The least ambiguous method for interpreting the nomenclature of phosphorus compounds is by referring to the structural formula.

Many phosphorus compounds are named as salts with phosphorus being the metallic or electropositive element, i.e., phosphorus trichloride, PCl_3; phosphorus oxybromide (phosphoryl bromide), $POBr_3$; phosphorus triamide, $P(NH_2)_3$; and tetraphosphorus decasulphide (phosphorus(V) sulphide), P_4S_{10}. Some compounds are named as derivatives of the simple phosphorus hydrides, phosphine, PH_3, diphosphine, P_2H_4, and their oxides.

An unshared electron pair, characteristic of many trivalent phosphorus compounds, can be described in terms of the general formula

$$\left.\begin{array}{l} R \\ R' \\ R'' \end{array}\right\rangle P$$

where R, R′, and R″ can be hydrogen or alkyl, aryl, alkoxy, amino, halo, mercapto, or other groups.

The strength of a complex formed between a phosphorus donor and an electron acceptor depends on the nature of the acceptor. Strong acids provide stronger complexes than weak acids. Of phosphonium halides, only the iodide has a low dissociation pressure, which imparts stability to the compound under ambient conditions. The unshared electron pair on phosphorus reacts with oxidising agents, e.g., hydrogen peroxide, sulphur, or halogens :

$$(RO)_3P: + H_2O_2 \rightarrow H_2O + (RO)_3P{=}O$$

$$R_3P: + S \rightarrow R_3P{=}S$$

$$:PCl_3 + Cl_2 \rightarrow PCl_5$$

Another reaction of the unshared electron pair is quaternisation with alkyl halides :

$$R_3P: + R'X \rightarrow [R_3PR']^+ X^-$$

The tetrahedral structure is the most common coordination pattern for phosphorus. Frequently, there are distortions in a pure tetrahedral environment that result from steric and electronic effects. Pure tetrahedral coordination probably occurs only in species where there are four identical groups and no packing distortions.

Phosphorus Sulphides

Phosphorus and sulphur form the binary compounds tetraphosphorus trisulphide (phosphorus sesquisulphide), P_4S_3, mp 171–172.5°C, the most stable; tetraphosphorus pentasulphide, P_4S_5; tetraphosphorus heptasulphide, P_4S_7; and phosphorus(V) sulphide (tetraphosphorus decasulphide), P_4S_{10}. A stable oxysulphide, $P_4O_6S_4$, exists as a colourless, deliquescent crystalline solid, mp 102°C (Table 12.3).

Table 12.3. Physical properties of the phosphorus sulphides.

Property	P_4S_3	P_4S_5	P_4S_7	P_4S_{10}
Mp, °C	171–172.5	170–220	305–310	286–290
Bp, °C	407–408		523	513–515
Sp gr at 17° C	2.03	2.17	2.19	2.09
Colour				
solid	yellow	sulphur yellow	almost white	yellow
liquid	brownish yellow		light yellow	reddish brown
Solubility at 17°C, g/100 g CS_2	100	ca 10	0.029	0.222
Action with cold water	scarcely attacked		fairly readily decomposed	slowly decomposed
Stability of solid on standing in air	slowly oxidised		decomposes	slowly decomposed

O,O'-Dialkyl or diaryl thiophosphoric acids, e.g., $(C_6H_5O)_2$ P(O)SH, are obtained readily by alcoholysis of phosphorus(V) sulphide. They provide the basis for high pressure lubricants, oil additives, and flotation agents. The insecticides parathion and methylparathion are made from dialkyl thiophosphates.

The phosphorus sulphides are manufactured by direct union of the elements. The pentasulphide, P_4S_{10}, is purified by distillation. Exposed to moisture, it hydrolyses, giving off H_2S; ventilation in storage avoids excessive H_2S build-up and fire and explosion hazards.

Phosphorus Halides

Phosphorus forms well-defined halogen compounds of the type PX_3, PX_5, POX_3, and PSX_3. All, except the pentaiodide and the oxy- and sulphoiodides are known. Physical properties of some halides are given in Table 12.4.

Table 12.4. Physical properties of some phosphorus halides.

Compound	*Mp, °C*	*Bp, °C*	*Specific gravity*	*Heat of vapourisation, kJ/mol*[a]	*Critical temperature, °C*	*Colour*[b]	*State at STP*
PCl_3	—111.8 (-91)	75.5	1.57	30.5	285.5		liquid
PCl_5	167 (under pressure[c]	sublimes	1.6	64.9	372	white to pale yellow	solid
$POCl_3$	–1.2	106.5	1.68	34.7	331.8		liquid
PBr_3	–40	175.3	2.89	38.9			liquid
PBr_5	< 100 (dec)	dec				yellow	solid
PI_3	61	dec 120[d]				dark red	solid
PF_3	–151.5	–101.1	3.91	14.6	–2.05		gas
PF_5	93.7	–84.5	5.84				gas
POF_3	–39.4	–39.8	4.65	21.1	73.3		gas

[a] To convert J to cal, divide by 4.184.

[b] Colourless unless otherwise noted.

[c] Sublimes at 101.3 kPa (1 atm) at 159°C; log P_{kPa} = 10.159 - 16,100 / 4.57 K; (log P_{mmHg} = 11.034 – 16,100/4.57 K).

[d] At 2.0 kPa (15 mm Hg).

The trihalides, obtained by direct halogenation under controlled conditions, are Lewis bases. Examination by electron diffraction has confirmed pyramidal structures for the gaseous molecules.

Phosphorus trichloride is a clear volatile liquid with a pungent, irritating odour. Depending on the mole ratio of H_2O : PCl_3, hydrolysis produces phosphonic acid, a mixture of phosphonic and pyrophosphonic acids, or an unstable polymer of indefinite composition. Although it is nearly insoluble in water, PCl_3 hydrolyses rapidly, and a polymer forms at low H_2O : HCl ratios. It also forms upon storage in contact with moist air. Phosphorus trichloride is made by combining the chlorine and phosphorus in the presence of a precharge of phosphorus trichloride which is refluxed continuously (Fig. 12.2).

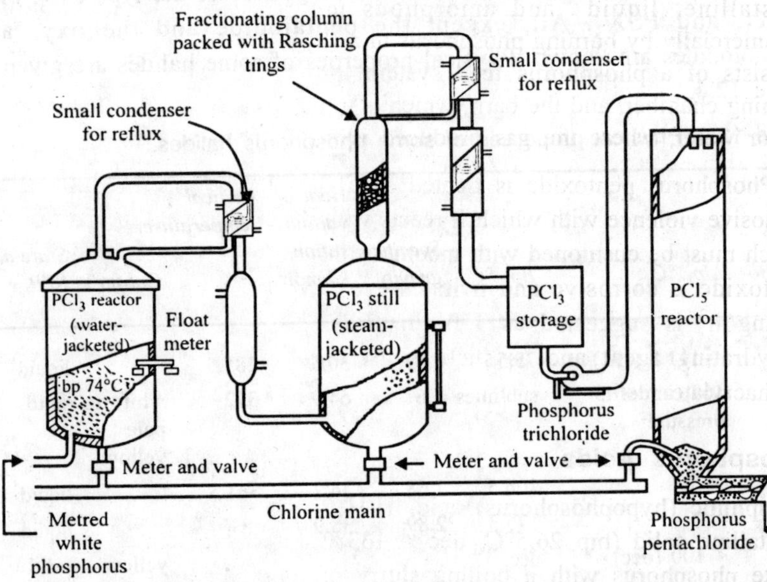

Fig. 12.2. Plant for the continuous production of phosphorus trichloride and phosphorus pentachloride.

Phosphorus oxychloride (phosphoryl chloride), $POCl_3$, is a colourless fuming liquid with a pungent disagreeable odour. It is made by heating PCl_5 with oxalic or boric acid. It is stable above 300°C, yields phosphoric acid upon hydrolysis, and is used extensively for the manufacture of alkyl and aryl orthophosphate esters.

Phosphorus pentachloride is a pale greenish solid with a pungent odour. It is made from PCl_3 and chlorine or by burning phosphorus in excess chlorine; reaction with water is violent. It serves as chlorinating agent and catalyst in organic synthesis.

Phosphorus halides are shipped in glass containers and must be handled with care. They are all irritating to skin, eyes, and mucous membranes; contaminated clothing must be removed immediately.

Phosphorus Oxides

Phosphorus forms five well-defined oxides: phosphorus(III) oxide, P_4O_6; phosphorus(V) oxide (phosphorus pentoxide), P_4O_{10}; phosphorus tetroxide, P_2O_4; phosphorus heptoxide, P_4O_7; and phosphorus nonaoxide, P_4O_9. They are obtained by direct oxidation of phosphorus.

The pentoxide is a very stable, white solid which exists in several crystalline, liquid and amorphous modifications. It is produced commercially by burning phosphorus in a stream of dry air. The usual plant consists of a phosphorus feed system, provisions for drying the air, a burning chamber, and the barn, which is a large room, cooled externally by air or water, where the gas produced is condensed.

Phosphorus pentoxide is treated as a flammable solid because of the explosive violence with which it reacts with water. It is sold in glass bottles which must be cushioned with incombustible packing material. Phosphorus pentoxide is corrosive and irritates eyes, skin, and mucous membranes (1 mg/m^3 is suggested as TLV in air). It is an important drying and dehydrating agent and as such is used in the manufacture of methyl methacrylate resins.

Phosphorus Acids

Phosphinic (hypophosphorus) acid, $H(H_2PO_2)$ or H_3PO_2, a deliquescent crystalline solid (mp 26.5°C, dec > 133°C), is manufactured by treating white phosphorus with a boiling slurry of lime. Calcium hypophosphite remains in solution, whereas insoluble calcium phosphite precipitates. The reducing capacity of hypophosphorus acid and its salts is utilised in electroless-plating processes, in which nickel salts are chemically reduced to form a smooth adherent surface plate. Phosphonic (phosphorus) acid, $H_2(HPO_3)$ or H_3PO_3, is a white deliquescent crystalline compound, mp 73.6°C, manufactured by the hydrolysis of phosphorus trichloride; the reaction can be violent.

Phosphonic acid and hydrogen phosphonates are used as strong but slow-acting reducing agents, e.g., to precipitate heavy metals from solutions. Aminoalkylphosphonic acids are used in the production of herbicides and water-treatment systems.

Phosphazenes and other Phosphorus-Nitrogen Compounds

The phosphazenes $(NPX_2)_n$, are linear or cyclic compounds with alternating phosphorus and nitrogen atoms; they are toxic and irritating. The lower linear chlorophosphazenes are oils, the cyclic members are crystalline white solids. The latter can be polymerised to chains of highly controlled molecular weight. Chlorophosphazenes are prepared from phosphorus(V) chloride and ammonium chloride. Of the binary phosphorus-nitrogen compounds, only triphosphorus pentanitride has been obtained in a pure state by treating P_2S_5 with ammonia.

Phosphine and its Derivatives

Phosphine, PH_3, is produced in a number of ways; however, the inadvertant evolution of phosphine in an otherwise safe reaction is an element of hazard in many procedures involving phosphorus chemicals. Phosphine is conveniently produced by the hydrolysis of an active metal phosphide, e.g., Ca_3P_2 or AlP, or from the acid- or base-catalysed reaction of elemental phosphorus with water. The addition of P—H bonds in phosphines across on olefinic double bond is an economical method to produce alkyl phosphines. Aryl groups are conveniently introduced by Friedel-Crafts reactions of phosphorus halides. Aliphatic phosphines may be gases, volatile liquids, or oils. Aromatic phosphines are frequently crystalline, although many are oils. Because of the bonding ability of phosphines to acceptors, e.g., metals, they are widely used as ligands in catalysts. A phosphine can block a specific site on a central metal under all conditions, or a site to some species but not to others, or simply make a metal more soluble and thus dispersible in a liquid medium. Phosphine, a central nervous system and liver toxin, is highly toxic and lethal to adults in a 0.5-1-h exposure of 0.05 mg/L. Transport on both passenger and cargo aircraft is forbidden.

PHOSPHORIC ACIDS AND PHOSPHATES

Phosphates may be defined as compounds containing four phosphorus-oxygen (P-O) linkages. Compounds containing discrete, i.e., monomeric PO_4^{3-} ions are known as orthophosphates or simply phosphates, linear condensed P-O-P chains as polyphosphates, cyclic rings as metaphosphates, branched polymeric materials and cage anions as ultraphosphates. Stoichiometrically, phosphates have been represented as combinations of oxides, e.g., H_3PO_4, as $P_2O_5.3H_2O$ and Na_2HPO_4 as $P_2O_5.2Na_2O.H_2O$ (Table 12.5).

Table. 12.5. Traditional classification of sodium phosphates.

Oxide ratio, R $(Na_2O + H_2O_{comp})$: P_2O_5	*Designation*	*General formula*	*Structure*
> 3	phosphate + metal oxide (includes double salts and solid solutions)		mixtures
3	phosphate or orthophosphate, $n \leq 3$	$Na_nH_{3-n}PO_4$	one phosphorus atom
3 > R < 2	mixture of ortho and pyrophosphates		
2	pyrophosphate, $n \leq 4$	$Na_nH_{4-n}P_2O_7$	two phosphorus atoms
2 > R > 1	polyphosphates, $n = 2, 3, 4 \ldots$	$Na_{n+2}P_nO_{3n+1}$	linear chains
1	metaphosphates, $n = 3, 4, 5 \ldots$	$Na_n(PO_3)_n$	cyclic or extremely long chains
1 > R > 0	ultraphosphates, $0 < x < 1$	$(xNa_2O)P_2O_5$	cross linked chains and/or rings
0	phosphorus pentoxide	$(P_2O_5)_n$	P_4O_{10} or continuous structures

Table 12.6. Physical properties of aqueous solutions of phosphoric acid.

Concentration, wt % H_3PO_4	P_2O_5	*Density, 25°C, g/cm³*	*Bp, °C*	*Fp, °C*	*Viscosity, mPa·s (=cP) at* 20°C	60°C	100°C
0	0	0.997	100.0	0	1.0	0.48	0.30
5	3.62	1.025	100.1	-0.8	1.1	0.54	0.33
10	7.24	1.053	100.2	-2.1	1.2	0.61	0.38
20	14.49	1.113	100.8	-6.0	1.6	0.78	0.48
30	21.73	1.182	101.8	-11.8	2.2	1.0	0.62
50	36.22	1.333	108	-44.0	4.3	1.8	1.1
75	54.32	1.573	135	-17.5	15	4.8	2.4
85	61.57	1.685	158	21.1	28	8.1	3.8
100	72.43	1.864	261	42.35	140	25	9.2
105	76.10	1.925	>300	16.0	600	70	19
115	83.29	2.044	>500			1500	250

The double-neutralisation process is used to purify phosphoric acid during the production of large-volume detergent-builder phosphates. In the furnace process, elemental white (yellow) phosphorus is burned in excess air; the resulting phosphorus pentoxide is hydrated, the heats of combustion and hydration are removed, and the phosphoric acid mist is collected. The processes are called wetted-wall, water-cooled, or air-cooled, depending on the protection of the combustion-chamber wall.

A potential environmental problem is created by the burning of phosphorus, which produces a persistent white cloud of phosphorus pentoxide and phosphoric acid droplets of such high obscuring power that it is used as a screening smoke by the military. Neither precipitators nor Venturi scrubbers reduce the pentoxide content to a level acceptable in states with low plume-opacity regulations; therefore, high efficiency mist eliminators are now standard practice.

Elemental phosphorus produced by the electrothermal process is a distilled product of high purity, which yields phosphoric acid pure enough for most industrial uses. The high heat of combustion of phosphorus 3.05 MJ/mol (730 kcal/mol) can be recovered for process use such as evaporation of water from dilute phosphate solutions. Furnace-grade phosphoric acid is used in metal treatment, refractories, catalysts, and food applications.

PHOSPHORIC ACID

From Phosphate Rock by Electric Furnace Process

In the electric-arc furnace process, phosphate rock is reduced to elemental phosphorus by the action of coke and heat in the presence of sand. Subsequent oxidation by air to phosphorus pentoxide followed by hydration yields phosphoric acid (Fig.12.3).

Phosphate rock is charged into a sintering oven, where it is nodulised to facilitate escape of phosphorus vapours in the electric furnace and to prevent the entrainment of dust or fines (in the vapours). The raw material is sized, and the fines are returned to the sintering oven. Coke (generally in the form of breeze) and sand are added in carefully controlled ratios, determined by rock analysis, to the sintered rock, and the mixture is charged into the shaft of an electric furnace. In the shaft hang three carbon electrodes, which are connected to a three-phase alternating current.

The charge on reaching the level of the arc, is fused at approximately 1315°C, resulting in reduction of the phosphate rock with liberation of elemental phosphorus vapours. Since phosphate rock usually contains fluorides as impurities, (or as a constituent of the rock itself, which is often written as fluoroapatite, $Ca_5F(PO_4)_3$), calcium fluoride, fluosilicates, and silicon tetrafluoride are also formed. The slag (mostly calcium silicate) from the furnace is usually tapped periodically and subsequently crushed for use as aggregate for road construction. Ferrophosphorus (resulting from the iron impurities) runs out with the slag. The amount of this material produced may be increased by adding iron slugs to the furnace charge. The ferrophosphorus is separated from the slag and marketed.

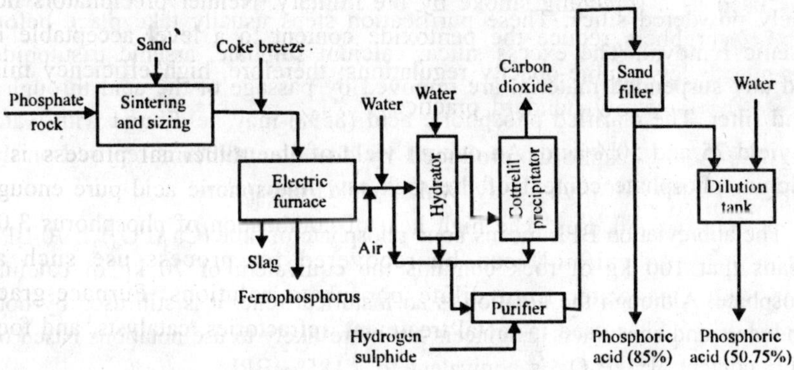

Fig. 12.3. Flow diagram for manufacture of phosphoric acid from phosphate rock by electric furnace.

Reactions

$$2Ca_3(PO_4)_2 + 6SiO_2 + 10C \rightarrow P_4 + 10CO + 6CaSiO_3$$

$$P_4 + 10C0 + 10O_2 \rightarrow 2P_2O_5 + 10CO_2$$

$$2P_2O_5 + 6H_2O \rightarrow 4H_3PO_4$$

$$\text{Overall: } Ca_3(PO_4)_2 + 3SiO_2 + 5C + 5O_2 + 3H_2O \rightarrow 3CaSiO_3 + 5CO_2 + 2H_3PO_4$$

87 to 92% yield

The gases from the furnace, phosphorus and carbon monoxide, are withdrawn from the furnace by means of a fan. In the one-step system, a current of air is drawn down through the charge by the suction induced by the fan. The two-step method produces phosphorus, which is stored for subsequent processing. The two currents (reaction gases and air) mix in the flue at a temperature sufficient to burn the phosphorus to phosphorus pentoxide (P_2O_5), and the carbon monoxide to the dioxide. The gases pass into a tall, packed tower, where they are sprayed with water (forming a mist of phosphoric acid), and then through a Cottrell electrostatic precipitator made of graphite (to resist the action of hydrofluoric acid) to remove any remaining phosphoric acid.

The crude phosphoric acid (85%) is generally purified with respect to arsenic by the action of hydrogen sulphide. Depending on conditions, the acid may be purified further by adding sulphuric acid to remove calcium salts. Sufficient sulphuric acid is used to precipitate calcium sulphate and also to leave a slight excess to inhibit the corrosive action of the phosphoric acid. This slight excess of sulphuric acid permits the use of lead-lined equipment. Residual hydrofluoric acid may be removed by the addition of finely powdered silica. These purification steps usually take place before arsenic removal. The excess silica, calcium sulphate, arsenic trisulphide, and any suspended material are removed by passage of the acid through a sand filter. The clarified phosphoric acid (85%) may be diluted with water to yield 75 and 50% acid. An overall yield of about 90% is realised on the calcium phosphate content of the rock raw material.

The abbreviation BPL means bone phosphate of lime ($Ca_3(PO_4)_2$); 70 BPL means that 100 kg of rock contains the equivalent of 70 kg of calcium phosphate. Although the notation is an historical relic, it is still used by both producers and consumers. Engineers are more likely to use notations based on P_2O_5 content; 1% P_2O_5 is equivalent to 2.185% BPL.

Reaction

$$P_4 + 5O_2 \rightarrow 2P_2O_5$$

$$P_2O_5 + 3H_2O \rightarrow 2H_3PO_4$$

$$\text{Overall: } P_4 + 5O_2 + 6H_2O \rightarrow 4H_3PO_4$$

94 to 97% yield

Process

Elemental phosphorus is often converted to phosphoric acid at locations other than the original point of production. The conversion involves oxidation of molten phosphorus to phosphorus pentoxide and subsequent hydration of the oxide to phosphoric acid. (Fig. 12.4)

Molten phosphorus is sprayed into a combustion chamber along with air and steam. Flame temperature approaches 1980°C. Chamber design varies from one plant to another, but construction materials are usually limited to acid-proof brick, structural carbon, and stainless steel. Both vertical and horizontal chambers are used. In either case some glassy metaphosphoric acid is formed and drops to the bottom of the chamber.

The effluent gases leaving the chamber (P_2O_5, steam, nitrogen, and some oxygen) are mixed with a spray of dilute phosphoric acid in the hydration tower. If the tower is externally cooled, it too may be built of stainless steel. By proper adjustment of reacting streams, various strengths of phosphoric acid (as high as 116%—equivalent to 84% P_2O_5) may be produced. The usual product is in the range of 75 to 85% phosphoric acid. Some acid mist leaves the hydrator and must be recovered by appropriate means, such as a packed tower or electrostatic precipitator. In the process shown in the flow diagram a combination cyclone scrubber and glass-wood filter is used.

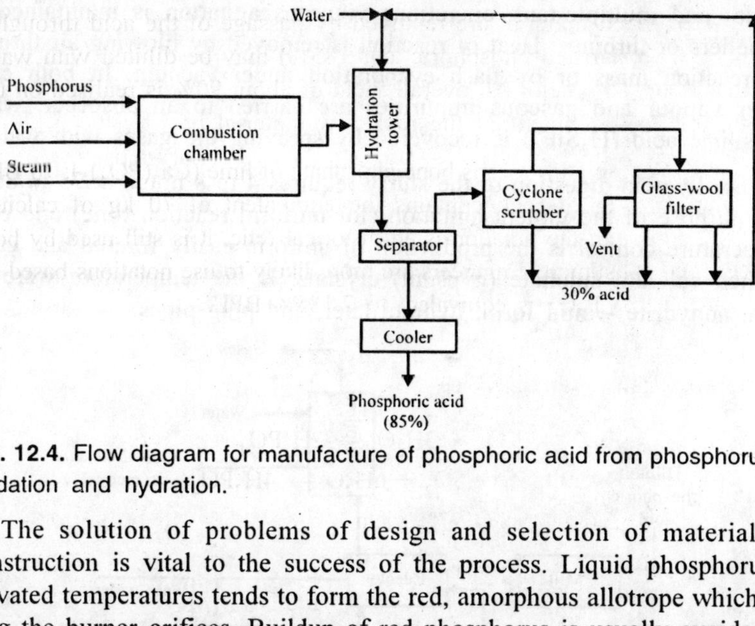

Fig. 12.4. Flow diagram for manufacture of phosphoric acid from phosphorus by oxidation and hydration.

The solution of problems of design and selection of materials of construction is vital to the success of the process. Liquid phosphorus at elevated temperatures tends to form the red, amorphous allotrope which can clog the burner orifices. Buildup of red phosphorus is usually avoided by a design in which rapid atomisation of the phosphorus occurs. Corrosion problems are also severe. Hot P_2O_5 vapours are very reactive, and the phosphoric acid formed from the moisture in the air tends to coat the walls with a viscous, highly reactive layer which attacks most materials. Graphite is used in many installations, unless the units are cooled by sprays, in which case stainless steel is satisfactory.

Wet Process

The sulphuric acid or wet process produces phosphoric acid by the action of sulphuric acid on phosphate rock, accompanied by the precipitation of calcium sulphate. The resulting phosphoric acid solution is separated from the precipitated gypsum by filtration, and then concentrated by evaporation (Fig. 12.5).

$$\underset{\text{Fluoroapatite}}{Ca_5F(PO_4)_3} + 5H_2SO_4 + 10H_2O \xrightarrow{\text{95 \% yield}} \underset{\text{Gypsum}}{5CaSO_4 2H_2O} + HF + 3H_3PO_4$$

Phosphate rock is charged into a special ball mill, where it is ground in a dilute solution of phosphoric acid obtained as a filtrate from an ensuing operation. The slurry is passed into single or multiple tanks, where it is allowed to react with sulphuric acid diluted with sufficient phosphoric acid to yield a 55% sulphuric acid solution. In multiple-tank operations, acid is recycled to maintain uniform temperatures in the several reactors. In both single- and multiple-tank operations vigorous agitation is maintained by propellers or turbines. Heat of reaction is removed by blowing air through the reaction mass or by flash evaporation under vaccum. In both cases water vapour and gaseous impurities are carried to an absorber, where fluosilicic acid (H_2SiF_6) is recovered by spraying the gases with water.

Usually acid digestion of the slurry requires 4 to 8 hours at 75 to 80°C. The purpose of the violent agitation (for uniform reaction time) and close temperature control is the production of uniform easily filtered and easily washed calcium sulphate (gypsum) crystals. If the temperature were too high, anhydrite would form, hydrate later, and plug pipes.

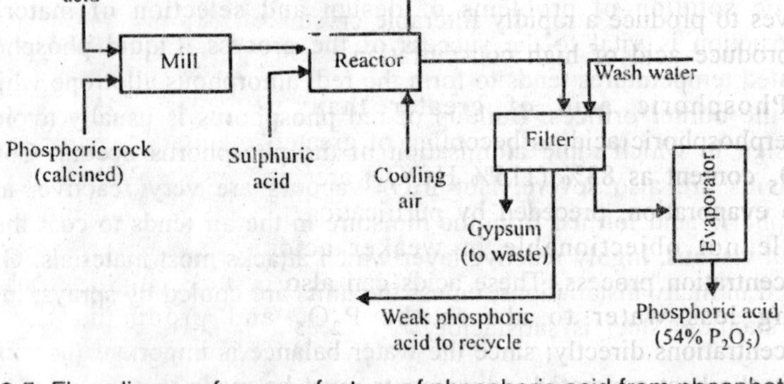

Fig. 12.5. Flow diagram for manufacture of phosphoric acid from phosphate rock and sulphuric acid (Wet Process).

The slurry from the digester passes to a horizontal, rotating, tilting-pan-type vacuum filter, where phosphoric acid (30 to 35% P_2O_5) is removed from the cake. Most plants use tilting-pan filters because liquors from the various washing stages can be kept separate, thus increasing recovery and minimising dilution. Polypropylene cloth makes an excellent filter medium. The washed gypsum cake is slurried with waste-water to a settling pond from which water is ordinarily pumped back to the plant.

The acid filtrate is then evaporated to the desired concentration, usually 54% P_2O_5. Forced-circulation evaporators with outside heat exchangers are most commonly used. Fluorine-containing compounds may be recovered from the flash chamber condensate. Submerged-combustion evaporators are also used, but high-efficiency scrubbers are required to recover P_2O_5 fumes.

Throughout the plant, corrosion-resistant materials of construction must be used. The most common ones are structural carbon or nickel alloy for evaporator heat exchangers; rubber or carbon-brick for reactor linings; and polyester-fibre glass in pipes, ducts, and small vessels. The yield of phosphoric acid based on the phosphorus content of raw material is 95%.

Various modifications of the process described, known as the dihydrate process, are in use. In one, conditions are modified to obtain a hemihydrate of calcium sulphate ($CaSO_4$. ½ H_2O) instead of gypsum; this is known as the hemihydrate or HDH process. Advantages claimed for its most recent modification include improved extraction efficiency (99% versus 95 to 97%) of the phosphorus values, higher purity and more concentrated product, 40 to 54% P_2O_5, without the need for evaporation. The calcium sulphate is said to entrain less phosphate and fluoride ions, thus rendering it more suitable as building material. Whatever the modifications used, each strives to produce a rapidly filterable crystal, to minimise P_2O_5 losses, and to produce acid of high concentration.

Phosphoric acid of greater than 100% strength, known as superphosphoric acid, is becoming of greater importance. Acids as high in P_2O_5 content as 83% (115% H_3PO_4) are produced, essentially by a two-step evaporation, preceded by purification to remove contaminants which while not objectionable in weaker acids would interfere with the concentration process. These acids can also be made from phosphorus by using less water to absorb the P_2O_5 and producing the higher concentrations directly; since the water balance is important for conducting away the heat generated, adjustments must be made to compensate for this factor. The superacids contain varying amounts of polyphosphoric acids. Among the advantages claimed for them are lowered corrosivity and reduced shipping costs.

Other phosphorus-containing acids

Phosphoric acid (H_2PO_4) may be converted into pyrophosphoric acid ($H_4P_2O_7$) and metaphosphoric acid (HPO_3) by heating. Pyrophosphoric acid is formed at 250 to 260°C. On heating either the ortho- or pyro- to a red heat, metaphosphoric acid is formed. The latter may also be prepared by treating phosphorus pentoxide with the calculated amount of cold water.

Raw Material

Phosphate rock, also known as rock phosphate, phosphorite, and raw phosphate, is the primary source of practically all phosphatic chemicals. These deposits are classified roughly as residual, replacement, and sedimentary. Residual phosphate (e.g., Tennessee brown rock) is derived from phosphatic limestone. Replacement phosphate is phosphatised limestone formed by the reaction of limestone and phosphoric acid of organic origin. Virtually all commercial deposits of phosphate rock are amorphous, impure varieties of the mineral fluorapatite [$Ca_5F(PO_4)_3$]. The workable deposits contain from 18 to 90% available tricalcium phosphate [$Ca_3(PO_4)_2$]. The chief impurities in domestic phosphate rock are iron, aluminium, and silicon oxides, as well as calcite, magnesite, dolomite, sand, clay, and organic matter. Most of the undesirable impurities are removed in the washing and sintering operations prior to phosphoric acid manufacture.

Uses

Fertilisers, Sodium phosphates, Dicalcium phosphate, Tetrapotassium pyrophosphate.

Properties

Colourless, crystalline solid. Mol. wt 98.00, mp 42.35°C, Sp gr 1.834 (18.2°C), bp loses 1/2 H_2O at 213°C. Soluble in water (2340 g/100 ml at 26°C) and ethanol. Commercial phosphoric acid is a clear, colourless, sparkling, mobile liquid which reaches sirupy consistency at higher concentrations (85%). At 88% concentration, a crystalline hydrate, $H_3PO_4.1/2\ H_2O$, separates at room temperature. Threshold limit value (mg/m^3) 1.

Both the wet process and the electric furnace process are plagued with by-products of low value which require disposal. Calcium sulphate waste from the wet process is not utilised at present, and the slag, carbon dioxide, and ferrosilicon from the electric furnace are only partially utilised. Fluorine is a particularly troublesome by-product of the wet-process. Antipollution laws greatly restrict the amounts that can be wasted into the air or into streams. Thus gas streams must be scrubbed and liquid streams neutralised.

Some fluorine remains in wet-process acid, but the trend toward higher P_2O_5 concentrations means greater release during evaporation. The calcium sulphate (gypsum) and calcium silicate (slag) by-products are often not marketable, and these sludges also pose disposal problems.

Wet-process acid is mostly (93%) used for fertiliser production; of furnace acid, only about 20% goes to fertilisers, the remainder being used for applications where its higher purity is demanded. Since the principal growth area for phosphoric acid will be in fertiliser production for an increasingly hungry world, expansion in capacity will no doubt be in wet-process installations unless new technology is developed. Phosphate rock itself is too insoluble to be a useful source of phosphorus for agriculture.

Polyphosphoric Acids

The only clearly defined crystalline compositions of the H_2O–P_2O_5 system are three forms of phosphoric acid and their hemihydrates, pyrophosphoric acid, and crystalline P_4O_{10}. Amorphous condensed phosphoric acids are hygroscopic; they may be viscous, oily, gummy, and may consist of a mixture of glassy and crystalline materials.

Linear-polyphosphoric acids are strongly hygroscopic and undergo viscosity changes and hydrolysis to less complex forms when exposed to moist air. Upon dissolution in excess water, hydrolytic degradation to phosphoric acid occurs.

Pyrophosphoric (diphosphoric) acid, $H_4P_2O_7$, crystallises in two forms (mp = 54.3 and 71.5°C); acid salts are known. Tripolyphosphoric (triphosphoric) acid, $H_5P_3O_{10}$, occurs in varying amounts as a component of condensed phosphoric acids containing more than ca 72% P_2O_5.

The largest use of polyphosphoric (superphosphoric) acids is as an intermediate in the production of liquid fertilisers. Condensed acids of 82–84% P_2O_5 are employed as catalysts in the petroleum industry. Polyphosphoric acid is also used as a dehydrating agent and in the production of phosphoric esters and agricultural chemicals.

Polyphosphates

The condensed phosphates are derived from phosphates by the loss of water. These materials range from simple diphosphates (pyrophosphates) to long-chain polymeric structures with molecular weights in the millions (10^6).

Polyphosphates are resistant to chemical attack, but are susceptible to hydrolysis. Short-chain polyphosphates hydrolyse without the concurrent formation of cyclic metaphosphates; in this respect they differ from the long-chain acids. Many polyphosphates form water-soluble complex ions, a phenomenon called sequestration. This property forms the basis for the water-treatment and detergent applications.

Pyrophosphates

The simplest linear condensed phosphates ($M_{n+2}P_nO_{3n+1}$, where $n = 2, 3, 4...$) are pyrophosphates ($M_4P_2O_7$). A water molecule is eliminated from condensed orthophosphates:

$$2MH_2PO_4 \xrightarrow{\Delta} M_2H_2P_2O_7 + H_2O$$

$$2M_2HPO_4 \xrightarrow{\Delta} M_4P_2O_7 + H_2O$$

Insoluble pyrophosphates are obtained by treating a soluble salt of the desired cation with a sodium pyrophosphate solution.

Tetra- and disodium pyrophosphates (sodium acid phosphate) are prepared by thermal dehydration of di- and monosodium orthophosphate, respectively. Tetrasodium pyrophosphate is used as a builder in detergent and cleaning formulations, in food applications, and as a deflocculant in drilling muds, dyes and inks. Sodium acid pyrophosphate is used as a leavening and chelating agent. (Fig. 12.6)

Calcium pyrophosphate exists in three polymorphic modifications; they form progressively upon dehydration of calcium hydrogen phosphate dihydrate. Calcium pyrophosphates are used primarily as abrasives in fluoridated toothpaste.

Sodium tripolyphosphate (STP, pentasodium triphosphate), $Na_5P_3O_{10}$, occurs as the anhydrous forms I (STP-I) (thermodynamically stable) and II (STP-II), the low temperature form; the transition temperature is 417 ± 8°C. They are differentiated by x-ray diffraction, IR, Raman spectra, and the temperature-rise test (ASTM D 501, 30). A hexahydrate forms by the addition of either anhydrous form to water or by the hydrolysis of sodium trimetaphosphate, $(NaPO_3)_3$. STP is produced commercially by calcination of a mixture of mono- and disodium phosphates.

The solubility and hydration behavior of STP are of particular importance in its industrial applications. As a builder for synthetic detergents, STP is the largest-volume phosphate salt for purposes other than fertilisers. Thermal dehydration of monosodium phosphate gives rise to numerous condensed polyphosphates, e.g., Graham's, Madrell's, and Kurrol's salts. In the manufacture of phosphate salts, phosphoric acid is treated with a base, e.g., carbonate, hydroxide, or ammonia, to form a solution or slurry. Most phosphates crystallise readily from solution and are separated by conventional techniques; the solutions or slurries are often evaporated to dryness.

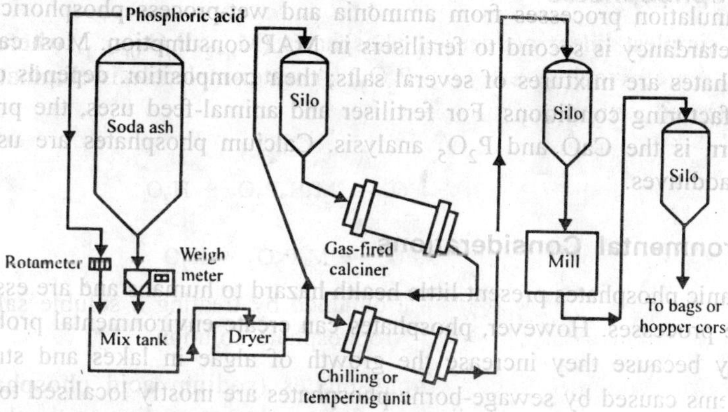

Fig. 12.6. Diagram of a plant for the manufacture of crystalline sodium polyphosphate.

The phosphate glasses are manufactured in refractory-type furnaces, where they are heated to 1000°C, then quenched rapidly to a solid glass; trade names are used for glasses of only slightly different composition. Sodium tripolyphosphate (90-95% pure) is manufactured by drying and subsequent calcination of a solution or slurry with a Na_2O/P_2O_5 mole ratio of 1.67, corresponding to two mol disodium phosphate and one mole monosodium phosphate. Calcining conditions are less critical than drying conditions; combinations of dryer/calcining units are available.

Disodium phosphate, Na_2HPO_4, is marketed as the dihydrate and the anhydrous salt. Tetrasodium pyrophosphate, $Na_4P_2O_7$, is obtained by calcination of disodium phosphate or any of its hydrates. Commercial manufacture is similar to that of STP, except that the final Na_2O/P_2O_5 ratio is adjusted with NaOH; the same equipment can be used.

Chlorinated TSP, the second largest volume sodium phosphate salt, is a complex mixture approximating $(Na_3PO_4.11H_2O)_4$. NaOCl. It is made by the addition of sodium hypochlorite solution to a hot concentrated sodium phosphate solution, followed by cooling, crystallisation, and granulation.

Chlorinated TSP is unstable above 40°C and up to 20% of the available chlorine may be lost on heating to elevated temperatures during processing.

In the manufacture of ammonium phosphates, the high partial pressure of ammonia has to be considered, because it rises rapidly with the temperature or the NH_3: P_2O_5 mole ratio. Phosphoric acid reacts quickly with ammonia vapour and is used as a scrubber fluid to prevent ammonia emissions and recover ammonia values. MAP and DAP fertilisers are made in granulation processes from ammonia and wet-process phosphoric acid. Fire retardancy is second to fertilisers in MAP consumption. Most calcium phosphates are mixtures of several salts; their composition depends on the manufacturing conditions. For fertiliser and animal-feed uses, the primary concern is the CaO and P_2O_5 analysis. Calcium phosphates are used as food additives.

Environmental Considerations

Inorganic phosphates present little health hazard to humans and are essential to life processes. However, phosphates can create environmental problems, mainly because they increase the growth of algae in lakes and streams. Problems caused by sewage-borne phosphates are mostly localised to areas that employ lakes as receiving waters for sewage effluents. Average concentrations vary according to industrial input, storm waters, seasonal fluctuations, and population density.

Pharmaceutical and food grades

Pharmaceutical and Food Grades (purer) of dicalcium phosphate dihydrate for use as a polishing agent in dentifrices, ammonium phosphate for use as yeast nutrient, calcium pyrophosphates, and monocalcium phosphate are manufactured from furnace-grade phosphoric acid.

Monocalcium phosphate $[Ca(H_2PO_4)_2.H_2O]$ is manufactured by the crystallising, after evaporation and some cooling, of a hot solution of lime and strong furnace phosphoric acid. The crystals are centrifuged, and the highly acidic mother liquor returned for reuse. This acid salt is also made by spray-drying a slurry of the reaction product of lime and phosphoric acid. This product is used for baking powder.

SODIUM PHOSPHATES

Monobasic (NaH_2PO_4), Dibasic (Na_2HPO_4), Tribasic (Na_3PO_4), Tetrabasic ($Na_4P_2O_7$), Acid ($Na_2H_2P_2O_7$), Meta ($(NaPO_3)_6$), Tripoly ($Na_5P_3O_{10}$).

Process

Sodium phosphates are produced by processing the reaction products of phosphoric acid and soda alkalies. Since phosphoric acid (orthophosphoric acid) is a tribasic acid, it forms three sodium salts. Sodium meta- and pyrophosphates are obtained by heating the corresponding sodium orthophosphate. The very important sodium tripolyphosphate is made by calcining a mixture of monobasic and dibasic sodium orthophosphates (Fig. 12.7).

Reaction

$$Na_2CO_3 + H_3PO_4 \longrightarrow Na_2HPO_4 + CO_2 + H_2O$$

$$Na_2HPO_4 + NaOH \longrightarrow Na_3PO_4 + H_2O$$

90—95% yield

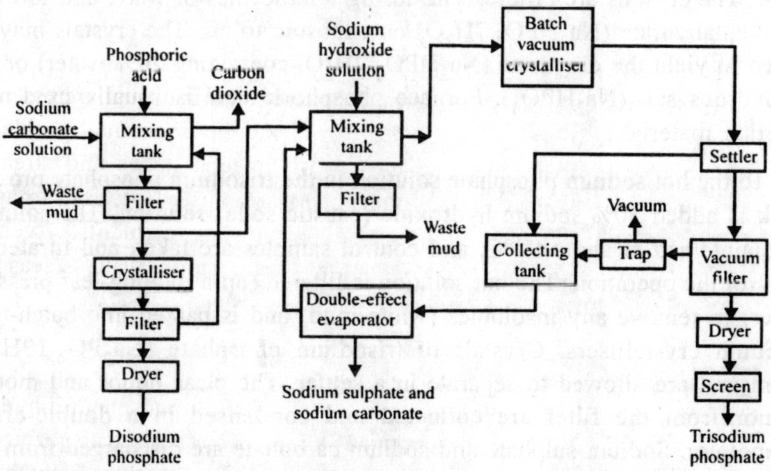

Fig. 12.7. Flow diagram for manufacturing phosphoric acid from phosphoric acid, sodium carbonate, and sodium hydroxide.

Disodium phosphate (Na_2HPO_4) is the ordinary sodium phosphate of commerce and may be considered the base of all the other phosphates. It is the intermediate in the production of trisodium phosphate (Na_3PO_4), sodium pyrophosphate ($Na_4P_2O_7$), and sodium tripolyphosphate ($Na_5P_3O_{10}$). A description of the manufacture of disodium phosphate and trisodium phosphate follows.

Sodium carbonate (soda ash) is introduced into a mixing tank, either as an aqueous solution or as a solution made with hot liquor from the evaporators or mother liquor from the filters. Phosphoric acid (60 to 65% H_3PO_4) is added at the surface of the tank, so that carbon dioxide may be liberated easily. A slight excess of sodium carbonate over the theoretical ratio is added, and the solution is boiled with steam until all the carbon dioxide has been driven off. The resulting disodium phosphate solution is filtered hot (85 to 100°C) and is split into two portions. A small amount of white mud, consisting of silica and iron and aluminum phosphates, remains on the filter and is discharged to waste. If trisodium phosphate is to be made, part of the clear solution of disodium phosphate, containing about 14.5% phosphorus pentoxide and 13% sodium oxide, is pumped to a trisodium phosphate process tank.

The remainder of the solution is cooled in a crystalliser to yield crystals of disodium phosphate, containing 12 molecules of water of crystallisation (60% water). The crystals ($Na_2HPO_4.12H_2O$) are centrifuged and packaged in moisture-proof containers. The mother liquor is returned to the mixing tank. The crystals are efflorescent, losing 5 molecules of water and forming the heptahydrate ($Na_2HPO_4.7H_2O$) on exposure to air. The crystals may be dried to yield the dihydrate ($Na_2HPO_4.2H_2O$, containing 20% water) or the anhydrous salt (Na_2HPO_4). Furnace phosphoric acid is usually used as a starting material.

To the hot sodium phosphate solution in the trisodium phosphate process tank is added 50% sodium hydroxide (caustic soda) solution. The solution is maintained at about 90°C, and control samples are taken and titrated to control the operation. The hot solution is filtered (on a rotating-leaf pressure filter) to remove any insolubles (white mud) and is passed into batch-type vacuum crystallisers. Crystals of trisodium phosphate ($Na_3PO_4.12H_2O$) form and are allowed to separate in a settler. The clear liquor and mother liquor from the filter are collected and condensed in a double-effect evaporator. Sodium sulphate and sodium carbonate are discharged from the bottom of the evaporator, whereas the concentrated liquor is returned to one of the mixing tanks for reprocessing.

The settled crystals of trisodium phosphate are separated from the remaining mother liquor on rotary vacuum filters. The crystals are further dried in rotary dryers (below 70°C), screened, and packaged. By drying above 100°C, the hydrate loses 11 molecules of water to yield trisodium phosphate monohydrate ($Na_3PO_4.H_2O$). Trisodium phosphate ($Na_3PO_4.12H_2O$) is obtained in a 90 to 95% yield based on the weight of phosphoric acid charged.

Monosodium phosphate is produced by adding the requisite amount of phosphoric acid to a concentrated solution of disodium phosphate. The resulting solution, an evaporation, yields crystals of monosodium phosphate monohydrate ($NaH_2PO_4.H_2O$). The anhydrous salt (NaH_2PO_4) may be produced by desiccating the hydrate at normal temperatures.

By heating the phosphate between 225 and 250°C for 6 to 12 hour, the reaction $2NaH_2PO_4 \longrightarrow Na_2H_2P_2O_7 + H_2O$ takes place, yielding the non-hygroscopic salt, sodium-acid pyrophosphate. This salt is more stable than the monosodium phosphate and is used generally for the same purposes (baking-powder formulations).

On heating monosodium phosphate to 350 to 400°C, an insoluble metaphosphate is formed. If the monosodium salt is fused to approximately 760°C and quickly cooled [$6NaH_2PO_4 \longrightarrow (NaPO_3)_6 + 6H_2O$], sodium hexametaphosphate results. Tetrasodium pyrophosphate ($Na_4P_2O_7$) is manufactured by dehydrating disodium phosphate in a rotary kiln according to the reaction: $2Na_2HPO_4 \rightarrow Na_4P_2O_7 + H_2O$. Both the anhydrous salt ($Na_4P_2O_7$) and the crystalline decahydrate ($Na_4P_2O_7 \cdot 10H_2O$) are produced, depending on the operating conditions.

Sodium tripolyphosphate (STPP) is made from a mixture of mono- and disodium phosphates. The mixture is made directly in solution in a manner similar to that described for disodium phosphate. The ratio of phosphoric acid to soda ash is so adjusted as to produce a Na_2O/P_2O_5 ratio of 1.67, which corresponds to a mixture of 1 mole of mono- to 2 moles of disodium phosphate. After filtration the solution is sprayed into a rotary continuous kiln, where drying, molecular dehydration, conversion to STPP, annealing, and cooling take place in sequence. After grinding, the product is packaged for shipment.

Various other polyphosphates may be produced from mixtures of mono- and disodium phosphates. These materials are generally of rather indefinite molecular composition. Only STPP is crystalline. It is usually sold as the anhydrous salt but also exists as the hexahydrate.

Sodium metaphosphates are also of somewhat indefinite composition. The formula of the compound sodium hexametaphosphate, usually written $(NaPO_3)_6$, is only approximate; generalised formulae of this class range from $(NaPO_3)_x$ to $Na_{x+2}P_xO_{3x+1}$; many are amorphous and insoluble glasses.

Uses

Detergents and water treatment, foods, textiles, dentifrices, feeds, clay processing, drilling muds, etc.

Properties

Monosodium phosphate

Sodium phosphate monobasic, sodium acid phosphate, sodium biphosphate, sodium dihydrogen phosphate, monosodium orthophosphate.

Properties

Anhydrous, white, crystalline powder; soluble in water. Formula wt 119.98, mp decomposes at 225 to 250°C to acid pyrophosphate. Monohydrate, large, transparent crystals, acid reaction, very soluble in water, insoluble in ethanol. Formula wt 138.0, mp loses water at 100°C, Sp gr 2.040.

Disodium phosphate

DSP, sodium phosphate dibasic, secondary sodium phosphate, sodium hydrogen phosphate, phosphate of soda, sodium phosphate, or disodium orthophosphate.

Properties

Anhydrous (Na_2HPO_4), colourless, translucent crystals or white powder; soluble in water. Formula wt 141.965, mp converts to pyrophosphate 240°C. Dihydrate, $Na_2HPO_4{\cdot}2H_2O$–Colourless, translucent crystals. Formula wt. 178.0. Sp gr 2.066 (15°C), mp loses H_2O at 92.5°C.

Trisodium phosphate

TSP, sodium phosphate tribasic, normal sodium phosphate, tertiary sodium phosphate, or trisodium orthophosphate.

Properties

Formula wt 380.12; Sp gr. 1.62 (20°C); mp 73.4°C, with decomposition; bp loses $11H_2O$ at 100°C; pH of 1% solution 12.0.

Formula wt 380.12, mp 73.4°C, with decomposition, Sp gr 1.62 (20°C), bp loses $11H_2O$ at 100°C; pH of 1% solution 12.0. Soluble in water (28.3 g/100 g at 15°C, in all proportions in hot water). Insoluble in carbon disulphide.

Tetrasodium pyrophosphate

TSPP, sodium phosphate tetrabasic, sodium tetraphosphate, sodium pyrophosphate, pyrophosphate of soda, normal pyrosodium phosphate, or alkaline sodium pyrophosphate.

Properties

Crystalline ($Na_4P_2O_7 \cdot 10H_2O$), colourless monoclinic crystals. Formula wt 446.1, mp loses water 94°C, Sp gr 1.82. Soluble in water (5.4 g/100 g at 0°C, 93 g/100 g at 100°C). Insoluble in ethanol and ammonia.

Anhydrous ($Na_4P_2O_7$), white crystals. Formula wt 265.9, mp 988°C, Sp gr 2.45. pH of 1% solution 10.2. Soluble in water (3.2 g/100 g at 0°C, 40.3 g/100 g at 100°C). Decomposes in ethanol.

Sodium acid pyrophosphate

Disodium dihydrogen pyrophosphate, disodium pyrophosphate, acid sodium pyrophosphate.

Properties

Anhydrous ($Na_2H_2P_2O_7$), white powder. Formula wt 221.95, mp decomposes 220°C, Sp gr 1.862. Soluble in water.

Uses

Electroplating metal cleaning and phosphatising, drilling muds, baking powders etc.

Sodium metaphosphate

Sodium hexametaphosphate, Graham's salt.

Properties

[$(NaPO_3)_6$], colourless or white flakes. Formula wt 612.10, mp 640°C, Sp gr 2.181, bp sublimes > 1,000°C. Soluble in hot and cold water.

Uses

Dental polishing agent, detergent binders, water softening, sequestrants emulsifier, food additive, textile processing, laundering.

Sodium tripolyphosphate

Properties

Anhydrous ($Na_5P_3O_{10}$), white powder or granules. Formula wt 367.88, pH of 1% solution 9.7. Soluble in water (14.5 g/100 g at 25°C, 32.5 g/100 at 100°C).

Uses

Water softening, sequestering, peptising, food additive, texturiser.

Chapter 13

Salt and Miscellaneous Sodium Compounds

INTRODUCTION

The salt industry is as old as mankind. Salt has long been an essential part of the human diet. It has served as an object of worship and as a medium of exchange, lumps of salt being used in Tibet and Mongolia for money. Its distribution was employed as a political weapon by ancient governments, and in Oriental countries high taxes were placed on salt. Salt is a vital basic commodity for life but is also a source of many of the chemicals which are now the mainstay of our complex industrial civilisation.

Many sodium salts are of definite industrial necessity. Most of them are derived directly or indirectly from ordinary salt so far as their sodium content is concerned. In one sense, the sodium may be viewed simply as a carrier for the more active anion to which the compound owes its industrial importance. For instance, in sodium sulphide, it is the sulphide part that is the more active. Similarly, this is the case with sodium thiosulphate and sodium silicate. The corresponding potassium salt could be used in most cases; however, sodium salts can be manufactured more cheaply and in sufficient purity to meet industrial demands.

Sodium chloride is the basic raw material of a great many chemical compounds, such as sodium hydroxide, sodium carbonate, sodium sulphate, hydrochloric acid, sodium phosphates, sodium chlorate and chlorite, and it is the source of many other compounds through its derivatives. Practically all the chlorine produced in the world is manufactured by the electrolysis of sodium chloride. Salt is used in the regeneration of sodium zeolite water softeners and has many applications in the manufacture of organic chemicals.

Salt is obtained in three different ways, namely, solar evaporation of sea water on the Pacific Coast or from western salt-lake brines, from mining of rock salt, and from well brines. The purity of the salt obtained from the evaporation of salt water is usually about 98 to 99%. Mined salt varies very widely in composition, depending on the locality. Some rock salt, however, runs as high as 99.5% pure. The solution obtained from wells is usually about 98% pure and depends to a great extent on the purity of the water forced down into the well to dissolve the salt from the rock bed. For many purposes, the salt obtained from mines and by direct evaporation of salt solutions is pure enough for use; however, a large portion must be purified to remove such materials as calcium and magnesium chlorides. Solar evaporation is used extensively in dry climates, where the rate of evaporation depends on the humidity of the air, wind velocity, and amount of solar energy absorbed. The mining of rock salt uses methods similar to coal mining. As the salt is removed, large rooms are formed, supported by pillars of salt.

SODIUM CHLORIDE (COMMON SALT)

Manufacture

From saturated brine by multiple-effect evaporation process

Process

Saturated brine for the production of "evaporated" salt is usually obtained by pumping water into an underground salt deposit and removing the saturated solution from an adjacent interconnected well, or from the same well by means of an annular pipe. Besides sodium chloride, the brine will contain more or less calcium sulphate, calcium chloride, and magnesium chloride, plus traces of hydrogen sulphide and ferrous ions (Fig. 13.1). Such a solution may have the following approximate composition :

Ingredients	*Per cent*
Water	73.5
Sodium chloride	26.3
Calcium sulphate	0.12
Calcium chloride	0.003
Magnesium chloride	0.007
	99.93

Other minor impurities account for the remainder. Dissolved hydrogen sulphide (H_2S) may amount to about 0.015%.

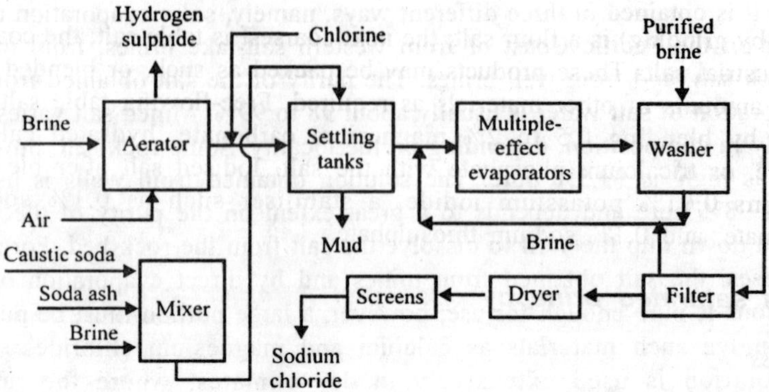

Fig. 13.1. Flow diagram for manufacture of sodium chloride from saturated brine by multiple-effect evaporation.

The chemical treatment the brine receives varies from one plant to another, and in a few cases, where very pure rock salt deposits are available, it may require none at all. Typically, however, the brine is first aerated to remove most of the hydrogen sulphide. Addition of a small amount of chlorine will complete hydrogen sulphide removal by oxidation, and also oxidise ferrous ions to ferric. The brine is then pumped to settling tanks, where it may be treated with a dilute solution of caustic soda and soda ash to remove most of the calcium, magnesium, and ferric ions. After clarification the treated brine is pumped to multiple-effect evaporators. As water is removed, salt crystals formed and are removed as a slurry. After screening to remove lumps, the slurry is sent to a conical washer, where the salt crystals are washed counter-currently with fresh brine. By this washing, actually hydraulic classification, fine, light crystals of calcium sulphate are removed from the mother liquor of the slurry and returned to the evaporator. Eventually, the calcium sulphate concentration in the evaporation system builds up to the point where it must be removed by "boiling-out" the evaporators.

The washed slurry is filtered, the mother liquor is returned to the evaporators, and the salt crystals from the filter are dried and screened. Salt thus produced from the typical brine described is of 99.8% purity or greater. Many salt manufacturers do not treat the raw brine, but control calcium and magnesium impurities by watching the concentrations in the evaporators and bleeding off sufficient brine to maintain a predetermined level. By such methods, salt of better than 99.5% purity can be made consistently. In either case, the final screening of the dried salt yields various grades, depending on particle size. The finest grade (sometimes made by grinding) is a flour salt; the next coarsest is table salt; the coarsest is industrial salt. These products may be packed as such, or blended with small amounts of other materials as required. Free-flowing table salts are made by blending 0.5 to 2% magnesium carbonate, hydrated calcium silicate, or tricalcium phosphate with the salt. Iodised salt after blending contains 0.01% potassium iodide, a stabiliser such as 0.1% sodium carbonate, and 0.1% sodium thiosulphate.

From saturated brine by open pan (Grainer) process

Process

Salt in the form of a hopper-like crystal (grainer salt) is made by causing the salt crystal to form on the surface of hot, quiescent brine held in an open pan or grainer. The grainer is a flat, open pan, 4.5 to 6 m wide, 45 to 60 m long, and about 60 cm deep. Beneath the submerged steam coils is a system of reciprocating rakes for salt removal (Fig. 13.2).

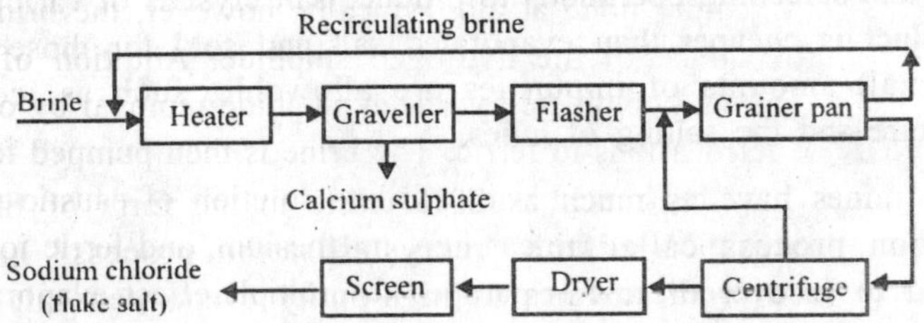

Fig. 13.2. Flow diagram for manufacture of sodium chloride from saturated brine by open pan ("Grainer") process.

As in the case of vacuum-pan salt, the brine may or may not be chemically treated. Also, different methods may be used to reduce calcium sulphate contamination of the finished salt. In the process shown in the flow diagram, saturated brine mixed with recirculating brine from the grainer is heated to 110°C, at which temperature calcium sulphate is much less soluble than at lower temperatures. The precipitated calcium sulphate is then removed in a graveller or bed of stones. The purified brine is then flash-cooled to such a temperature that sodium chloride is still retained in solution, but any remaining calcium sulphate crystals re-dissolve. The slurry is then pumped to the grainer, where evaporation takes place slowly at 96°C. Flat, hopper-shaped crystals formed on the surface and then fall to the bottom of the grainer, where they grow further before being removed by the rake system. The wet crystals are centrifuged, dried, and screened. When the incoming brine has been treated, salt of 99.98% sodium chloride can be obtained. Even without chemical treatment, salt of 99.8% purity is not unusual.

Because of its high surface-volume relationship, grainer salt dissolves rapidly even though it is coarse. It is preferred in the butter and cheese industries. The grainer process is very wasteful of heat, and there are only a few plants operating to satisfy the demand for this special type of salt.

From Rock Salt by Mining

About 35% of the salt produced comes from mines; mines in eight states are operated to produce rock salt. The salt deposits vary in colour from light reddish-brown to a sort of off-gray. Purity is usually about 98.5%, but one mine has deposits as high as 99.4% pure. After chunks of salt are blasted loose, they are preliminarily crushed in the mine and then crushed again at the surface. The remaining processing consists of a series of grinding and screening operations to produce salt crystals of various sizes. The product is cheaper than evaporated salt and sold for those uses in which small amounts of impurities are allowable, such as ice cream manufacture and the salting of hides.

Some mines have as much as 20% waste in the form of fines. An evaporation process called the "recrystallisation process" has been developed to recover these fines as high-purity evaporated salt without resorting to chemical treatment. A slurry of rock salt fines is heated and sent to a saturator, where all the salt dissolves and part of the calcium sulphate separates. Final separation of salt and the remaining calcium sulphate is accomplished in a double-effect crystallising evaporator. Salt of 99.99% purity is produced.

From Sea Water by Solar Evaporation

At points along the West Coast of the Great Salt Lake, salt is produced from sea water by solar evaporation. In these areas, there are no rock salt deposits, and annual evaporation exceeds precipitation. For example, during the 7-month period from April to October, there is less than 125 mm of rain, but evaporation is equivalent to about 840 mm. Here sea water (containing about 3.7% solids of which about 78% is sodium chloride) is caught in ponds at high tide. After the water evaporates to about sp. gr. 1.21, at which concentration most of the calcium sulphate precipitates, it is pumped to another pond from which it flows to the rest of the system by gravity. When salt crystals start to form, the brine is run into crystallising ponds. Here about 75% of the salt separates as the evaporation continues. When the specific gravity reaches 1.25 to 1.29, the mother liquor (bittern) contains 300 to 400 g/litre of solids and is run off. It may either be discarded or worked up for other soluble salts.

At the end of the season, the ponds are drained, and the salt is harvested mechanically. It is crushed, washed with salt brine, and dried to produce an industrial grade (about 95% pure). If table salt or dairy salt is desired, the solar salt is re-dissolved and evaporated.

Other Processes

A process has been proposed whereby salt will be produced as a by-product of the production of fresh water by salt-water desalination. Multiple-effect evaporation of sea water would convert 75% of the water to fresh, potable water. The remaining brine would be sent to solar evaporation ponds, and the harvested salt would be purified by the recrystallisation process previously described for upgrading rock salt. A similar effect would occur if other means for desalination, such as reverse osmosis or dialysis, are employed.

Sylvinite is a mixed crystal of sylvite (KCl) and halite (NaCl). It is usually worked for the potassium chloride content with the halite going to waste, but in a few cases the sodium chloride is recovered.

Uses

Chlor-alkali industry, soda ash, food processing, other chemicals, feed dealers and mixers.

Properties

Colourless or white hygroscopic cubic crystals. Flake salt has flat, hopper-shaped crystals. Formula wt 58.44, mp 801°C, sp. gr. 2.165, bp 1413°C (also reported as 1473°C). Soluble in water (35.7 g/100 g at 0°C, 39.8 g at 100°C). Slightly soluble in ethanol. pH of aqueous solution 6.7 to 7.3.

SODIUM SULPHATE (SALT CAKE)

From Natural Brines

Process

Natural brines are now the principal sources of sodium sulphate. From the Searles Lake brines, sodium sulphate may be considered either a co-product or a by-product of the production of sodium carbonate, and the process is considered under that heading (Fig. 13.3).

Reaction

$$Na_2SO_4 + 10H_2O \rightarrow Na_2SO_4 \cdot 10H_2O$$
$$Na_2SO_4 \cdot 10H_2O \rightarrow Na_2SO_4 + 10H_2O$$

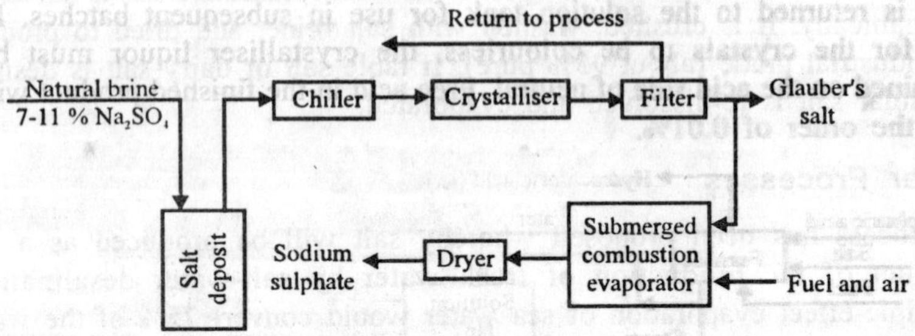

Fig. 13.3. Flow diagram for manufacture of sodium sulphate from natural brines.

The specific process used for recovery of natural sodium sulphate depends of course on the composition of the brine, and to some extent on its location. The natural brines available there run from 7 to 11% sodium sulphate, plus some sodium chloride and magnesium sulphate. To lower the sodium sulphate solubility of the brine it is saturated with sodium chloride by pumping it into a salt deposit. The salt-enriched brine leaving the well is chilled to –10 to –6°C in ammonia-cooled coils and sent to a crystalliser. The resulting Glauber's salt crystals are separated from the mother liquor by filtration. The mother liquor is returned to the process. The Glauber's salt crystals are charged to a submerged combustion evaporator where they are melted, and most of the resulting water is removed by evaporation. The wet salt product is then dried in a rotary kiln.

From Salt and Sulphuric Acid

Process

Still a major source of salt cake, but of diminishing importance, is the salt–sulphuric acid process for the production of hydrochloric acid. Salt cake is a by-product. Salt and sulphuric acid, sp. gr. 1.72 (slight excess) are charged to a furnace equipped with a rake agitator (Mannheim furnace), where the reacting mass is slowly heated to a temperature just below fusion (843°C). Hydrogen chloride is evolved and led through a cooling and condensing system to the absorbers. Salt c. (crude sodium sulphate) is continuously discharged from the periphery of the furnace (Fig. 13.4).

When Glauber's salt is desired, the salt cake is dissolved in hot water to form a solution of sp. gr. 1.29. Either soda ash or lime is then added to neutralise excess sulphuric acid and to precipitate iron and alumina. The precipitate is allowed to settle, and the clear supernatant liquor is pumped to the crystalliser. The muddy bottom layer is filtered and also sent to the crystalliser. The filtered mud cake is discarded. After crystallisation the Glauber's salt is stored in closed bins to prevent desiccation. The mother liquor is returned to the solution tank for use in subsequent batches. In order for the crystals to be colourless, the crystalliser liquor must be maintained on the acid side of neutral. Free acid in the finished product will be of the order of 0.01%.

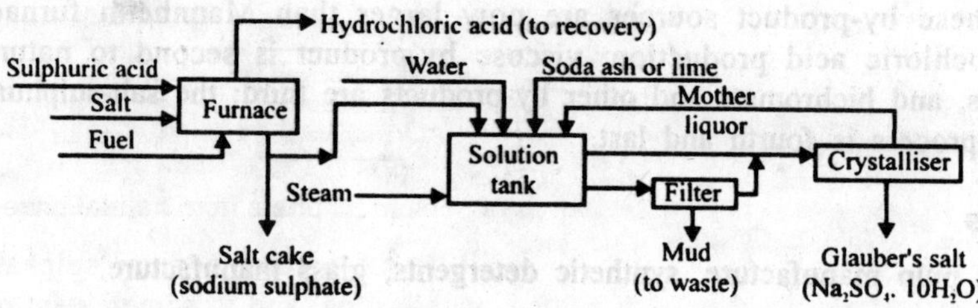

Fig. 13.4. Flow diagram for manufacture of sodium sulphate from salt and sulphuric acid.

Reaction

$$2NaCl + H_2SO_4 \rightarrow 2HCl + Na_2SO_4$$

98% yield

$$Na_2SO_4 + 10H_2O \rightarrow Na_2SO_4 \cdot 10H_2O$$

95% yield

Niter cake (sodium bisulphate) may be used in the Mannheim process in place of sulphuric acid; it is mixed with the salt and fed to the furnace.

From Rayon Spin Bath

In the manufacture of viscose rayon the coagulating or spin bath is a solution containing 9 to 11% sulphuric acid and 20% sodium sulphate, plus minor amounts of other materials. During the spinning process, 1.1 lb sodium sulphate is produced for each pound of rayon spun. The spent spin bath is recovered by further concentration of the liquor, followed by cooling in a vacuum crystalliser. Glauber's salt crystallises and may be sold as such or converted to anhydrous sodium sulphate. The mother liquor from the crystalliser is fortified with sulphuric acid and returned to the spin bath.

Other Processes

A process of some importance is Hargreaves-Robinson process. In this process sulphur dioxide, air, and steam are passed over specially prepared porous salt granules. The equation expressing the reaction is:

$$2NaCl + SO_2 + \frac{1}{2}O_2 + H_2O \rightarrow Na_2SO_4 + 2HCl$$

The yield is 93 to 98%.

Two other forms of by-product salt cake of commercial importance are *chrome cake* and *phenol cake*. The former is a green form of salt cake containing a small amount of chromium. It is formed in the manufacture of chrome salts. Yellow *phenol cake* is a by-product of the manufacture of phenol by the sulphonation process.

These by-product sources are now larger than Mannheim furnace hydrochloric acid production; viscose by-product is second to natural brines, and bichromate and other by-products are third; the salt-sulphuric acid process is fourth and last.

Uses

Kraft pulp manufacture, synthetic detergents, glass manufacture.

Properties

White solid or crystals. Crystalline form is thenardite. At 100°C the rhombic crystal form changes to monoclinic, and at 500°C to hexagonal. Formula wt 142.04, mp 884°C, sp. gr. 2.70. Soluble in water (5 g/100 g water at 0°C; 40.8 g/100 g at 30°C).

SODIUM THIOSULPHATE (HYPO)

From Soda Ash and Sulphur Dioxide

Process

Soda ash and sulphur dioxide are simultaneously added to water to form a slurry of sodium sulphite. The sulphur dioxide is sparged well below the surface to insure complete absorption. The batch is transferred to the digestion vessel, where sulphur is added and the mixture is heated, forming sodium thiosulphate. The liquor is decanted from unreacted sulphur and concentrated to sp. gr. 1.6, causing precipitation of sodium sulphate which is filtered off and discarded. The sodium thiosulphate pentahydrate formed in a vacuum crystalliser is centrifuged, dried, and packaged in conventional equipment. To ensure product quality, most equipment in the process is of stainless steel (Fig. 13.5).

Reaction

$$Na_2CO_3 + SO_2 \rightarrow Na_2SO_3 + CO_2$$
$$Na_2SO_3 + S \rightarrow Na_2S_2O_3$$

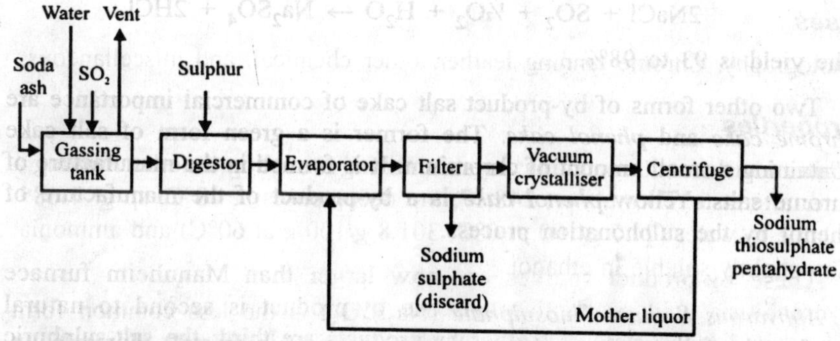

Fig. 13.5. Flow diagram for manufacture of sodium thiosulphate from soda ash and sulphur dioxide.

From Sodium Sulphite and Sulphur

In many plants the starting material for sodium thiosulphate manufacture is sodium sulphite. In such cases the plants are essentially the same as that described beginning with the cast-iron reaction kettle.

From By-Product Sodium Sulphide

In the manufacture of sodium sulphide, a by-product sulphide-carbonate liquor (8% Na_2S, 6% Na_2CO_3) is obtained, which may be converted to sodium thiosulphate by reaction with sulphur dioxide. The reaction taking place is essentially :

$$2Na_2S + Na_2CO_3 + 4SO_2 \rightarrow 3Na_2S_2O_3 + CO_2$$

Any excess sodium sulphide reacts with sulphur dioxide to yield sodium sulphite :

$$Na_2S + SO_2 + H_2O \rightarrow Na_2SO_3 + H_2S$$

In such a case the sulphite may be converted to the thiosulphate by treatment with sulphur. Accordingly, the plant used to produce sodium thiosulphate from by-product sodium sulphide is almost identical with that described using soda ash, sulphur dioxide, and sulphur.

From By-Product of Sulphur-Dye Manufacture

In the manufacture of sulphur dyes by fusion of sulphur, caustic soda, and certain organic compounds, the fusion product is leached with water, and the leach liquor is filtered to remove the insoluble dye. The filtrate resulting from this process contains most of the original sulphur and caustic soda in the form of sodium thiosulphate. This filtrate may be concentrated and crystallised to yield sodium thiosulphate of good quality.

Uses

Photography, Chrome tanning leather, other chemical, and miscellaneous.

Properties

White, translucent, monoclinic crystals having a cooling taste but leaving a bitter aftertaste. Formula wt 248.17, mp decomposes at 48°C, sp gr 1.685. Soluble in water (74.7 g/100 at 0°C, 301.8 g/100 g at 60°C) and ammonia. Very slightly soluble in ethanol.

Anhydrous Sodium Thiosulphate ($Na_2S_2O_3$) is the less common form and has the following properties: White, monoclinic crystals. Formula wt 158.10, Sp gr 1.667. Soluble in water (50 g/100 g at 0°C, 231 g/100 g at 80°C).

SODIUM SILICATE

Ingredints	*Formula*	*Ratio*
Sodium tetrasilicate (water glass)	$Na_2Si_4O_9$	4
Sodium metasilicate	Na_2SiO_3	1
Sodium sesquisilicate	$Na_3HSiO_4{\cdot}5H_2O$	0.67
Sodium orthosilicate	Na_4SiO_4	0.5

From Sodium Carbonate and Silica (Sand)

Process

A variety of compounds, ranging in chemical composition from $Na_2O{\cdot}4SiO_2$ to $2Na_2O{\cdot}SiO_2$, is produced by the fusion of silica (sand) and sodium carbonate (soda ash). By properly proportioning the reactants, the ratio of the constituent parts (Na_2O and SiO_2) may be varied to obtain a number of desired properties. Sodium silicates, varying in ratio from $Na_2O{\cdot}1.6SiO_2$ to $Na_2O{\cdot}4SiO_2$, are known as colloidal silicates. These are generally sold as 20 to 50% aqueous solutions called water glass. They are so named because they solidify to a glass that is water-soluble (Fig. 13.6). The expression "ratio" signifies the molar ratio of SiO_2 to Na_2O when the formulas are written in the mixed oxide form; sodium metasilicate has a ratio of 1, and water glass has ratios between 2.4 and 3.25.

Sodium metasilicate (Na_2SiO_3) is a definite crystalline compound which forms various hydrates. Substances having higher sodium oxide content are sodium sesquisilicate ($1\frac{1}{2}\ Na_2O{\cdot}SiO_2$) and sodium orthosilicate (Na_4SiO_4) ($2Na_2O{\cdot}SiO_2$).

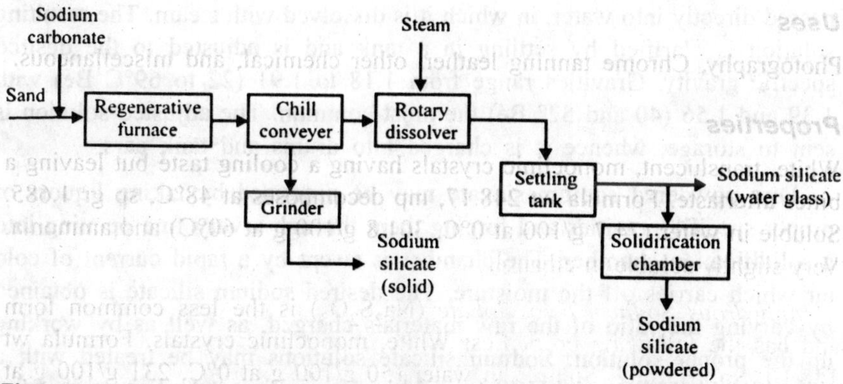

Fig. 13.6. Flow diagram for manufacture of sodium silicate from sodium carbonate and silica (Sand).

Reaction

$$Na_2CO_3 + SiO_2 \rightarrow \underset{\text{Sodium metasillicate}}{\underset{(Na_2SiO_3)}{Na_2O{\cdot}SiO_2}} + CO_2$$

The more siliceous sodium silicates are glasses, typical non-crystalline solid solutions, which are important mostly for their adhesive and binding properties. The more alkaline silicates (including sodium metasilicate) are crystalline materials with definite structures and characteristic properties. These are used chiefly as cleaners and detergents.

Sand and sodium carbonate, in selected proportions, are charged batchwise into a regenerative tank furnace resembling that used for the manufacture of glass. Fuel gas (producer, coke-oven, or natural) and air are mixed, preheated, and burned to maintain oven temperatures of 1200 to 1425°C in the hot zone. The melted materials gradually flow through the furnace evolving carbon dioxide. There is a normal shrinkage (approximately 10%) in the weight of the charge, due to the loss of gases and volatilisation of alkali oxides.

The fused melt is drawn from the furnace continuously or periodically as a thin stream. This is solidified by passage onto a moving chilled conveyer of steel moulds, in which the melt cools to a semi-transparent solid. If the hot melt is sprayed with a stream of cold water, it is shattered into fragments. The fragments are either charged into grinding and screening equipment to yield solid sodium silicates (granular) or are passed into a rotary dissolver. Here the solid material is dissolved by superheated (80-psi, 560-kPa) steam. In some plants the hot melt from the furnace is passed directly into water, in which it is dissolved with steam. The resulting solution is clarified by settling in a tank and is adjusted to the desired specific gravity. Gravities range from 1.18 to 1.91 (22 to 69°C Bé) with 1.39 and 1.56 (40 and 52° Bé) the most common. The adjusted solution is sent to storage, whence it is charged into drums and tank cars.

Dry, powdered sodium silicate may be produced by taking liquors of proper specific gravities and forcing them through a very fine opening into a solidification chamber. The chamber is swept by a rapid current of cold air which carries off the moisture. The desired sodium silicate is obtained by varying the ratio of the raw materials charged, as well as by working up the proper solution. Sodium silicate solutions may be treated with a mineral acid (H_2SO_4 and HCl) to yield hydrous silica and silicic acid. After the product is washed with water and dehydrated, silica gel results.

Sodium sulphate and carbon may be used in place of soda ash; and caustic soda solutions are used to dissolve silica in autoclaves. Potassium silicates are made by procedures analogous to that described; in much smaller quantities, lithium silicates and quaternary ammonium silicates are also produced.

Use

Silica gel and catalysts, Soaps and detergents, pigments, Boxboard adhesive, Water, paper and or treatment, Roofing granules, Other.

Properties

Sodium tetrasilicate

Water glass, Amorphous, white, deliquescent powder and lumps or clear to cloudy, colourless, aqueous solutions of varying density and viscosity. Formula wt 302.32, mp 1100°C (liquid temperature).

Soluble in hot and cold water. Insoluble in ethanol. The colloidal silicates or water glasses vary in chemical composition from $Na_2O{\cdot}1.6SiO_2$ to $Na_2O{\cdot}4SiO_2$. A solution containing 32% $Na_2O{\cdot}1.69SiO_2$ has a specific gravity of 1.365 20°C/4 (38.8°Bé) and a viscosity of 29 centipoises (20°C°); a 32% $Na_2O.3.9SiO_2$ solution has a specific gravity of 1.298 20°C/4 (33.3° Bé) and a viscosity of 180 centipoises (20°C).

Sodium disilicate

Rhombic, pearly lustre, crystals. Formula wt 185.15, mp 874°C, sp. gr. 2.5. Soluble in hot and cold water.

Sodium metasilicate

Monoclinic, colourless crystals or white powder containing approximately 51.7% Na_2O and 46.2% SiO_2. Formula wt 122.06, mp 1.088°C, Sp gr 2.4. Soluble in cold and hot water, decomposing in the latter. Insoluble in ethanol. Crystallises from water as the pentahydrate and nonahydrate.

Sodium metasilicate pentahydrate

White, granular or crystalline material containing about 29.2% Na_2O and 28.7% SiO_2. Formula wt 212.14, mp 71.8°C. Very soluble in cold and hot water.

Sodium metasilicate nonahydrate

White, efflorescent rhombic crystals. Formula wt 284.20, mp 40–48°C, bp loses $6H_2O$, 100°C. Very soluble in cold and hot water; soluble (29%) in *N/2* sodium hydroxide at 18°C. Insoluble in ethanol and acids.

Sodium sesquisilicate

White, granular powder containing approximately 36% Na_2O and 24% SiO_2. mp 90–95°C. pH—11.6 (concentration of 0.1% by weight). Soluble in water.

Sodium orthosilicate

White, hexagonal crystals or powder. Formula wt 184.04, mp 1,018°C. Soluble in water.

SODIUM SULPHATE DECAHYDRATE

Properties

Large transparent crystals, small needles, or granular powder; sp. gr. 1.464 (crystals); m.p. 33°C (liquefies); loses water of hydration at 100°C. Energy storage capacity is over 7 times that of water. Soluble in water and glycerol; insoluble in alcohol; solutions neutral to litmus. Non-toxic; non-flammable.

Preparation

It is prepared by crystallisation of sodium sulphate from water solutions. (Glauber's salt); also occurs in nature as mirabilite.

Uses

It is used in solar heat storage; air conditioning.

SODIUM BISULPHATE

Properties

It is in the form of colourless crystals or white, fused lumps; aqueous solutions is strongly acid. Soluble in water. Sp. gr. 2.103 (13°C); m.p. 58.5°C. Non-combustible.

Preparation

As a by-product in the manufacture of hydrochloric and nitric acids. Strong irritant to tissue. It can be purified by recrystallisation.

Uses

Flux for decomposing minerals; substitute for sulphuric acid in dyeing; disinfectant; manufacture of sodium hydrosulphide, sodium sulphate and soda alum; liberating CO_2 in carbonic acid baths; in thermophores; carbonising wool; manufacture of magnesia cements, paper, soap, perfumes; foods; industrial cleaners; metal pickling compounds; laboratory reagent.

SODIUM SULPHITE

Properties

It is in the form of white crystals or powder; saline, sulphurous taste. Soluble in water; sparingly soluble in alcohol. sp. gr.: (a) 2.633; (b) M.p.:1.539. (a) decomposes; (b) loses $7H_2O$ at 150°C.

Preparation

(1) Sulphur dioxide is reacted with soda ash and water, and a solution of the resulting sodium bisulphite is treated with additional soda ash.

(2) By-product of the caustic fusion process for phenol.

Its use prohibited in meats and other sources of Vitamin B_1.

Uses

It is used in paper industry (semichemical pulp); reducing agent (dyes); water treatment; photographic developer; food preservative and antioxidant; textile bleaching (antichlor).

SODIUM SULPHIDE

Properties

It is in the form of yellow or brick red lumps or flakes or deliquescent crystals; (a) sp. gr. 1.856 (14°C); m.p. 1180°C; (b) sp. gr. 1.427 (16°C); decomposes at 920°C. Soluble in water; slightly soluble in alcohol; insoluble in ether; largely hydrolysed to sodium acid sulphide and sodium hydroxide.

Preparation

It can be prepared by heating sodium acid sulphate with salt and coal to above 950°C, extraction with water, and crystallisation. It is flammable, dangerous fire and explosion risk. Strong irritant to skin and tissue. Liberates toxic hydrogen sulphide on contact with acids.

Uses

Organic chemicals; dyes (sulphur); intermediates; viscose rayon (sulphur removal); leather (depilatory); paper pulp; hydrometallurgy of gold ores; sulphiding oxidised lead and copper ores preparatory to flotation; sheep dips; photographic reagent; engraving and lithography; analytical reagent.

SODIUM HYDROSULPHIDE

Properties

It is in the form of colourless needles to lemon-coloured flakes. Soluble in water, alcohol, and ether. 70–72% NaSH; m.p. 55°C; water of crystallisation 26–28%.

Preparation

It can be prepared from calcium sulphide by treating it in the cold with sodium bisulphate. Contact with acids causes evolution of toxic gases.

Uses

It is used in paper pulping; dyestuffs processing; rayon and cellophane desulphurising; unhairing hides; bleaching reagent.

SODIUM NITRITE

Properties

Ii is slightly yellowish or in the form of white crystals, pellets, sticks or powder. Oxidises on exposure to air. Soluble in water; slightly soluble in alcohol and ether. Sp. gr. 2.157; m.p. 271°C; explodes at 1000°F (537°C); decompses at 320°C.

Dangerous fire and explosion risk when heated to 537°C (1000°F) or in contact with reducing materials. A strong oxidising agent. Sodium nitrite has been found to cause cancer in test animals. Its use in curing fish and meat products is restricted to 100 ppm.

Uses

It is used as diazotisation (by reaction with HCl to form nitrous acid); rubber accelerators; colour fixative and preservative in cured meats, meat products, fish; pharmaceuticals; photographic and analytical reagent; dye manufacture; antidote for cyanide poisoning.

SODIUM PEROXIDE

Properties

It is in the form of yellowish white powder, turning yellow when heated. Absorbs water and carbon dioxide from air. Active oxygen content approximately 20% by weight; sp. gr. 2.805; m.p. 460°C; b.p. 657°C. Soluble in cold water with evolution of heat.

Preparation

It is prepared when metallic sodium is heated at 300°C in aluminium trays in a retort in a current of dry air, from which the carbon dioxide has been removed. It is dangerous fire and explosion risk in contact with water, alcohols, acids, powdered metals, and organic materials. Keep dry. Toxic and irritant. Strong oxidising agent.

Uses

It can be used as an oxidising agent; bleaching of miscellaneous materials including paper and textiles; deodorant; antiseptic; organic chemicals; water purification; pharmaceuticals; oxygen generation for diving bells, submarines, etc.; textile dyeing and printing; ore processing; analytical reagent; calorimetry; germicidal soaps.

SODIUM PERBORATE

Properties

It is in the form of white odourless crystals or powder; salty taste. M.p. 63°C; loses H_2O at 130–150°C Stable in cool, dry air but decomposes with evolution of oxygen in warm or moist air. Moderately soluble in water (with decomposition) and glycerol. pH of aqueous solutions 10.0 to 10.3. Active oxygen content 10% min.

Preparation

(1) Electrolysis of a solution of borax and soda ash.

(2) Crystallisation from solution of borax or boric acid, sodium peroxide, and hydrogen peroxide.

It is moderately toxic by ingestion. Fire risk in contact with organic materials; Strong oxidising agent.

Uses

It can be used in developing vat dyes; textile bleaching; synthetic detergents; neutralising cold wave preparations; dental compositions; clectroplating; laboratory reagent; germicide; deodorant; mouthwash.

SODIUM AMIDE

Properties

It is in the form of white crystalline powder with ammonia odour. Decomposes in water and hot alcohol. M.p. 210°C; b.p. 400°C.

Preparation

It can be prepared when dry ammonia gas is passed over metallic sodium at 350°C. It is flammable, dangerous fire risk.

Uses

It is used in the manufacture of sodium cyanide; organic synthesis; laboratory reagent; dehydrating agent.

SODIUM CYANIDE

Properties

It is in the form of white deliquescent, crystalline powder. Soluble in water; slightly soluble in alcohol; m.p. 563°C; b.p. 149°C. The aqueous solution is strongly akaline and decomposes rapidly on standing.

Preparation

It can be prepared by absorption of hydrocyanic acid in a solution of sodium hydroxide, with subsequent vacuum evaporation. It is toxic by ingestion and inhalation. Tolerance (as CN), 5 mg per cubic metre of air.

Use

It is used in the extraction of gold and silver from ores; electroplating; heat treatment of metals (case-hardening); making hydrocyanic acid; insecticide; cleaning metals; fumigation; manufacture of dyes and pigments; nylon intermediates; chelating compounds; ore flotation.

SODIUM FERROCYANIDE

Properties

It is in the form of yellow, semi-transparent crystals. Partially soluble in water; insoluble in organic solvents. Sp. gr. 1.458. Low toxicity.

Uses

It is used in the manufacture of sodium ferricyanide; blue pigments; blueprint paper; anticaking agent for salt; ore flotation; pickling metals; polymerisation catalyst; photographic fixing agent.

Chapter 14

Alkali and Chlorine Products

INTRODUCTION

Soda ash, chemically known as sodium carbonate (Na_2CO_3), is a versatile inorganic chemical with a wide variety of applications in the industry and domestic sector. Its major application is in the soaps and detergents industry. Glass industry is also one of the large users of soda ash and ranks second. Apart from these two industries, it is used in a number of other industries such as sheets/hollow-wares/sleeves/tubes, paper, textile, dyes and dyestuffs, sodium silicate, sodium bicarbonate, sodium dichromate, etc. Large quantities of soda ash is used in the country for washing of clothes by dhobis and laundries, purification of industrial water and as an auxiliary chemical in other industries like resins, vanaspati, petroleum etc.

Caustic soda and chlorine are two basic products widely used in the chemical industry in India, either as raw material or as auxiliary chemical. Caustic soda is mainly used in the manufacture of pulp and paper, newsprint, viscose yarn, staple fibre, aluminium, cotton, textiles, toilet and laundry soaps, detergents, dyestuffs, drugs and pharmaceuticals, vanaspati, petroleum refining etc. Chlorine is used in the manufacture of PVC, pulp and paper, bleaching powder, textiles and a host of other inorganic and organic chlorinated compounds like metallic chlorides, refrigerants, chlorinated solvents etc. Large quantities of chlorine are also used for water purification.

Sodium carbonate Na_2CO_3, better known as soda ash, is the neutral salt of carbonic acid. It is a white crystalline, hygroscopic solid and is one of the most important raw materials used in the chemical industry. Presently it ranks eleventh in the list of commodity chemicals by volume. The normal article of commerce is a highly purified material of over 99% purity, and differences in bulk density are the only major distinction between the various grades. Soda ash competes with sodium hydroxide (caustic soda) in the production of some sodium compounds, and in certain applications some switching between the two is seen depending on economic conditions, importantly availability and price.

Production of the alkaline materials (sodium carbonate and potassium carbonate) was carried out from ancient times until the 1800s by the combustion of marine and land vegetation, followed by calcination at red heat and leaching of the ash. The term 'soda ash' can be traced to this process. The soda ash obtained in this way was a low-purity material, was very expensive and the processes too primitive for mass production, especially as they consumed vast quantities of vegetation.

Significant development have taken place since, and the technologies for manufacture based on readily available raw materials have seen significant improvements, which have made soda ash a readily available, high volume, low value inorganic commodity chemical of significant commercial importance to several industries. This chapter highlights the technical, as well as commercial aspects of soda ash, and includes the international and Indian scenario.

SODA ASH (SODIUM CARBONATE)

Properties

Physical properties

Sodium carbonate i.e. soda ash exists as various hydrates – anhydrous, monohydrate, heptahydrate and decahydrate. These have differing number of water molecules associated, as indicated by their names, and show differing physical properties (Table 14.1).

Most of the commercial product is anhydrous sodium carbonate. Minor quantities of sodium carbonate monohydrate and sodium carbonate decahydrate are sold and used in speciality applications.

Anhydrous sodium carbonate is an odourless, hygroscopic powder of alkaline taste. On exposure to air it gradually absorbs one mol. of water–about 15%. In air at 96% relative humidity its weight can increase 1.5% within 30 minutes. One part of soda ash dissolves in 3.5 parts water at room temperature, in 2.2 parts water at 35°C. It combines with water with evolution of heat. Its aqueous solution is strongly alkaline with a pH of 11.6. Soda ash containers should be kept well closed. Soda ash is soluble in glycerol but insoluble in alcohol.

Table. 14.1. Physical properties of sodium carbonate and its hydrates.

	Anhydrous	*Monohydrate*	*Heptahydrate*	*Decahydrate*
Formula	Na_2CO_3	$Na_2CO_3H_2O$	Na_2CO_37 H_2O	Na_2CO_3 $10H_2O$
M1	105.99	124.00	232.10	286.14
Density at 20°C, g/cm^3	2.533	2.25	1.51	1.469
Melting point, °C	851	105^a	35.37^b	32.0
Heat of fusion, J/g	316			
Specific heat capacity at 25°C, Jg^{-1} K^{-1}	1.043	1.265	1.864	1.877
Heat of formation, Jg	10.676			
Heat of hydration, J/g		133.14	646.02	858.3
Crystal structure	monoclinic	rhombic bipyramidal	rhombic pseudo-hexagonal	monoclinic
Refractive indices, $\eta_\alpha, \eta_\beta, \eta_\gamma$	1.410, 1.537, 1.544	1.420, 1.506, 1.524		1.405, 1.425 1.440
Heat of solution^c, J/g	–222	–79.6	197	243

^{a} in its own water of crystallisation.

^{b} Incongruent.

c integrated enthalpy of solution at infinite dilution.

Sodium carbonate monohydrate is an odourless, crystalline powder with alkaline taste. It is stable at ordinary temperature, but dries out somewhat in warm, dry air or above 50°C and becomes anhydrous at 100°C. One part is soluble in 3 parts of water, 1.8 parts of boiling water, and 7 parts of glycerol, but is insoluble in alcohol.

Sodium carbonate decahydrate is also known as Nevite or Soda. The technical product is called as Sal soda or washing soda. It has transparent crystals and it efforesces on exposure to air. One part is soluble in 2 parts of cold water, in 0.25 parts boiling water, and in glycerol. It is also insoluble in alcohol.

Chemical properties

Even though the melting point of anhydrous sodium carbonate is 851°C, water vapour reacts with sodium carbonate about 400°C to form sodium hydroxide and carbon dioxide. If sodium carbonate is stored under moist conditions, its alkalinity decreases due to absorption of moisture and carbon dioxide from the atmosphere. Soda ash is decomposed by acids with effervescene.

The thermal decomposition of sodium carbonate to sodium oxide and carbon dioxide in a vacuum in the absence of chemically activating substances such as water vapour begins at about 1000°C. The dissociation pressure is about 200 Pa and increases to 10.3 kPa at 1450°C.

Sodium carbonate reacts exothermically with chlorine above about 150°C to form NaCl, CO_2, O_2 and $NaClO_4$. The elements platinum, gold, vanadium, titanium, zirconium, aluminium, molybdenum, tungsten and iron (1200°C) are attacked by fused sodium carbonate, with liberation of carbon dioxide and formation of complex metal oxide—sodium oxide compounds. At higher temperature, the Na_2O formed reacts with excess metal to form sodium metal vapour. Sodium carbonate reacts very slowly with copper and nickel at high temperature (more than 1500°C).

HEALTH AND SAFETY

Due to its alkaline reaction, sodium carbonate has irritating effect on the skin and mucous membranes. However, exposure to soda ash is ordinarily not hazardous but soda ash dust may produce temporary irritation of the nose and throat. Although some become accustomed to working in the dust and suffer relatively little discomfort, others are allergic to alkaline materials and develop a condition of dermatitis. Dermal effects due to soda ash range from a transient reddening and inflammation to mild burns to abraded skin areas. The skin irritations experienced by workmen exposed to soda ash dust in hot weather are usually more severe because soda ash is likely to dissolve in perspiration.

Soda ash is corrosive to the eyes. It produces several corneal, irital an conjunctival effects (tissue destruction). If skin is attacked, the affected part should be thoroughly washed with large quantities of water, and a non-irritant dressing applied, if necessary. If eyes are splashed with sodium carbonate, the eyelids should be held open while the eye is treated for several minutes with running water or saline solution.

Soda ash is harmful if ingested and may be corrosive to the linings of the stomach. The lowest known lethal dose of sodium carbonate is 4000 mg/kg (rat, oral). For humans, oral ingestion of more than 15 gm is potentially lethal. In case of ingestion, vomiting should not be induced, but large amounts of water and dilute lemon juice or vinegar (two tablespoons per glass of water) should be drunk. The stomach can be carefully irrigated for 15 minutes at most with the usual precautions, but use of this treatment for longer periods is strongly contraindicated (danger of perforation). Medical help should be sought.

Simultaneous exposure to the dusts of lime and soda ash should be avoided. The two react to form caustic soda in the presence of moisture or perspiration.

MANUFACTURE

Some of the processes used for manufacturing sodium carbonate are discussed as under :

Leblanc Process

In the Leblanc process, common salt was treated with sulphuric acid to make sodium sulphate and hydrochloric acid. The sodium sulphate was treated with limestone and coal to produce black ash, which contained sodium carbonate, calcium sulphide, and some unreacted coal. The sodium carbonate was leached with water from the black ash. The chemical reactions of the Leblanc process are:

$$2NaCl + H_2SO_4 \rightarrow Na_2SO_4 + 2HCl$$

$$Na_2SO_4 + 2C \rightarrow Na_2S + 2CO_2$$

$$Na_2S + CaCO_3 \rightarrow Na_2CO_3 + CaS$$

A major disadvantage of the Leblanc process compared to the Solvay process is that it involves mainly solid-phase reactions and consumes large amounts of energy (at red heat or just below). The waste products (calcium sulphide and hydrochloric acid) are another disadvantage. Calcium sulphide causes both atmospheric and water pollution.

Modern Processes

The modern processes for the manufacture of soda ash may be classified into two categories: Natural Processes and Synthetic Processes. The natural processes utilise the natural deposits of sodium carbonate for extracting soda ash; while the synthetic or chemical processes use various chemicals to produce soda ash. In the last two decades the importance of natural processes has significantly increased.

Natural Soda Ash Processes

From ores

Natural evaporate deposits containing sodium carbonate in some form are found throughout the world. Some brines also contain sodium carbonate. The proportion of sodium carbonate in the salt mixture in these natural deposits seldom exceeds 40%. The accompanying compounds include sodium hydrogen carbonate, borax, sodium sulphate, sodium chloride and potassium chloride. The composition of sodium carbonate containing minerals is given below:

Trona:	$Na_2CO_3.NaHCO_3.2H_2O$
Natron:	$Na_2CO_3.10H_2O$
Thermonantrite:	$Na_2CO.H_2O$
Hanksite:	$2Na_2CO_3.9Na_2SO_4.KCl$
Pirssonite:	$Na_2CO_3.CaCO_3.2H_2O$
Gaylussite:	$Na_2CO_3.CaCO_3.5H_2O$

Only trona is of commercial importance. Each deposit throughout the world has its own distinctive characteristics and each requires different processing techniques.

A sodium carbonate deposit is normally, a massive, brown matrix of trona containing other alkaline minerals, as well as shales. The bed material which is mined varies slightly in quality but is usually 90% sodium sesquicarbonate. The insoluble materials are primarily dolomite shales and clays. The typical trona analysis on a water-free basis is given in Table 14.2.

Table 14.2. Typical trona analysis.

Major components	*Weight %*
Na_2CO_3, $NaHCO_3$	90.1
NaCl	0.08
Na_2SO_4	0.03
SiO_2	0.002
Organic	Trace
Water insoluble	9.8

Mining

Operations for each mine are generally concentrated in one thick bed of trona. The seams above and below the major bed are not recovered. Access is gained by circular, vertical concrete-lined shafts. The mines are worked by mining techniques that are similar to coal mining: room and pillar, long wall, and short wall. However, the equipment is modified to withstand the heavier and more abrasive trona ore. The extraction rates of in-place ore range from 50 to 62%. Longwall and shortwall mining techniques which involve hydraulically supporting the overburden while undercutting the trona seam have recently been introduced. The raw techniques allow over 70% recovery of the ore compared to 50% recovery for conventional room and pillar methods.

Solution mining of trona is made difficult by its low solubility in cold water and the 'blinding' of dissolving face with insoluble impurities. This method avoids expensive mine development costs and allows ore recovery in beds that are too deep for conventional mining. Pairs of wells are drilled into the ore bodies and frac-conneted using high-pressure pumps. The solvent is then circulated through the wells and sent to the processing plant.

Another mining process involves the recovery of sodium carbonate decahydrate from alkaline ponds. FMC mines this material from its solar evaporation pond using a bucket wheel dredge. The decahydrate slurry is dewatered, melted, and processed to soda ash.

Purification of trona

Several processes are used to refine trona ore and arc simpler than the Solvay process. Generally, two processes, named the monohydrate and sesquicarbonate according to the crystalline intermediates, are used to produce refined soda ash from trona. Both the processes involve the same unit operations but in different sequences.

In the monohydrate process, the trona ore is calcined to impure soda ash, which is then purified. The sesquicarbonate process produces soda ash by calcination of purified sodium sesquicarbonate obtained from the trona.

The simplified flow diagrams of the monohydrate process and the sesquicarbonate process are shown in Fig. 14.1. Most of the soda ash is made using the monohydrate process. However, earlier, the sesquicarbonate process was used.

Waste disposal

The new natural soda ash plants operate without discharge of liquid wastes to local waterways. Solar evaporation ponds are used for liquid waste disposal. Solid wastes, which for trona based processes are predominantly the insoluble material in the ore, are stockpiled. Although, environmental regulations increase construction and operating costs, their impact on natural soda ash plants is less severe than on ammonia-soda ash plants.

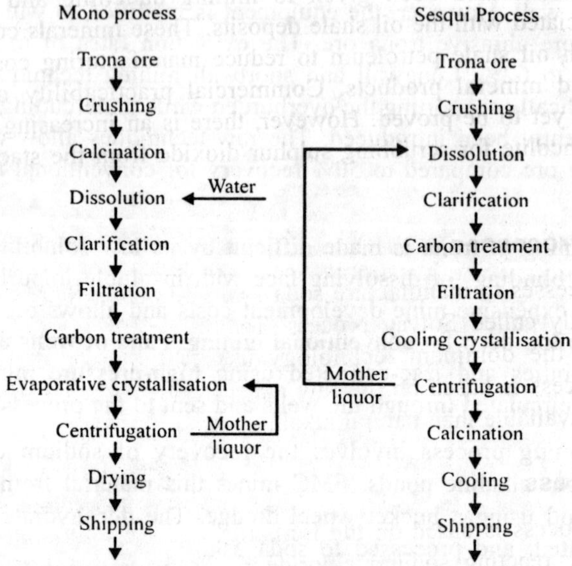

Fig. 14.1. Simplified flow diagrams for soda ash from trona.

Production from brine

Soda ash is also produced from natural brines available from lakes. More complicated processes are required than for trona because of the complex nature of brines. The soda ash is obtained by carbonating and cooling the brine to crystallise sodium bicarbonate, the bicarbonate is filtered and calcined to light soda ash which is densified by conversion to the monohydrate followed by calcining. The procedure results in a dense ash with properties equivalent to Wyoming trona derived ash. Other salts, most notably borax, are also recovered from the lake.

In another process, brine extracted from brine lenses in the lake is evaporated to give sodium carbonate in the form of the crystalline double salt, burkeite ($Na_2CO_3.2Na_2SO_4$). Lithium compounds may also precipitate and are eventually separated from the sodium sulphate. The brine is further processed for other by-products. The burkeite is redissolved and cooled to precipitate the sodium sulphate, which is removed. The remaining solution is further treated to recover sodium carbonate. By-products include sodium sulphate, potassium chloride and borax.

Oil shale deposits

Consideration has also been given to mining nahcolite and dawsonite deposits associated with the oil shale deposits. These minerals could be co-produced with oil shale petroleum to reduce manufacturing costs of both petroleum and mineral products. Commercial practicability of such an operation has yet to be proved. However, there is an increasing interest in the use of nahcolite for scrubbing sulphur dioxide from the stack gases of power plants.

Synthetic Processes

Synthetic processes to manufacture soda ash are mainly the ammonia-soda process, usually called Solvay process. Today, the Solvay process, or its variations, is the dominant technology used throughout the world. Its continued success is based on the raw materials (salt and limestone) being more readily available than natural alkali.

Solvay Process

The Solvay process is based on the formation of sparingly soluble sodium bicarbonate by reacting sodium chloride with ammonium bicarbonate in aqueous solution.

$$NaCl + NH_4HCO_3 \longrightarrow NaHCO_3 + NH_4Cl$$

All the soda ash processes are based on the manipulation of saline phase chemistry, an understanding of which is important both to improving current processes and to the economic development of new alkali resources.

In the Solvay process, the raw materials are common salt and limestone. Ammonia enters into the process but is not consumed, and only a very small amount is lost. Consequently, it is not a raw material in the usual sense. The overall chemical reaction for the entire process is:

$$CaCO_3 + 2NaCl \longrightarrow Na_2CO_3 + CaCl_2$$

However, this does not take place directly, but in a number of steps. These reactions and the required apparatus for the process are shown in Fig. 14.2.

The Solvay process includes the following stages:

(1) Burning of limestone or chalk to produce carbon dioxide and lime, and production of milk of lime, $Ca(OH)_2$.

(2) Production of a saturated salt solution (Brine).

(3) Saturation of the salt solution with ammonia (Ammonia absorption).
(4) Precipitation of bicarbonate by the introduction of carbon dioxide (from stages 1 and 6).
(5) Filtering and washing of precipitated bicarbonate.
(6) Thermal decomposition of bicarbonate to sodium carbonate (Calcination).
(7) Recovery of ammonia by distillation of the mother liquor from stage 4 (bicarbonate precipitation) with milk of lime (the ammonia liberated is recycled to stage 3).

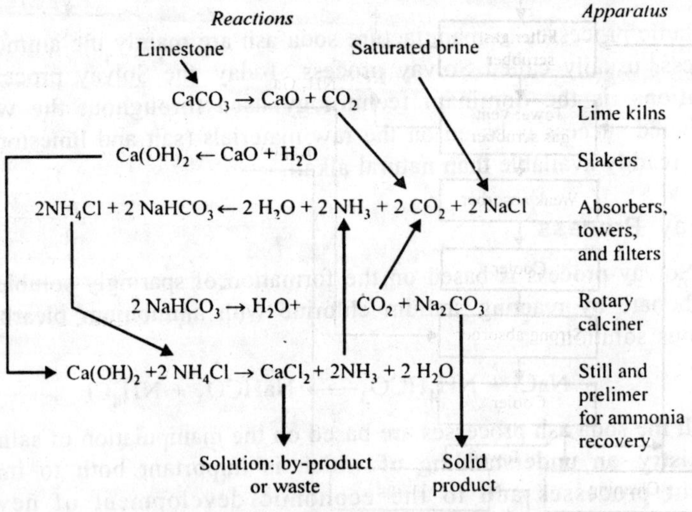

Fig. 14.2. Reaction scheme of the Solvay ammonia-soda process.

It is convenient to consider the calcination of limestone with fuel in a kiln to produce carbon dioxide and lime (CaO), as the first step. The lime is discharged from the kiln and may be slaked with excess water to form a thick milk of lime. Common salt, in the form of near saturated purified brine, is treated with ammonia and carbon dioxide producing ammonium chloride and precipitating sodium bicarbonate, which is filtered and thermally decomposed, at about 200°C, to sodium carbonate. The filtrate contains ammonium chloride, some unreacted sodium chloride and the excess of both ammonia and carbon dioxide; the latter probably exists as bicarbonate ions, although the ammonia is not fully bicarbonated. All of the comparatively expensive ammonia is recovered in two steps. Heating alone suffices to drive off the 'free' ammonia corresponding to the bicarbonate and hydroxide ions. The hot solution, containing only ammonium chloride and unreacted salt is then treated with the milk of lime to recover 'fixed' ammonia. The liquid remaining after distillation is usually discarded in its entirely, because only a small fraction can be used for calcium chloride production, depending on demand. The reactions in Fig. 14.2 are reversible. The bicarbonation of the sodium salt is never complete and the reaction is carried only as far as economically feasible. The Solvay ammonia-soda process for the manufacture of soda ash is shown schematically in Fig. 14.3. The process employs rugged large-scale equipment, the design of which has evolved over the years.

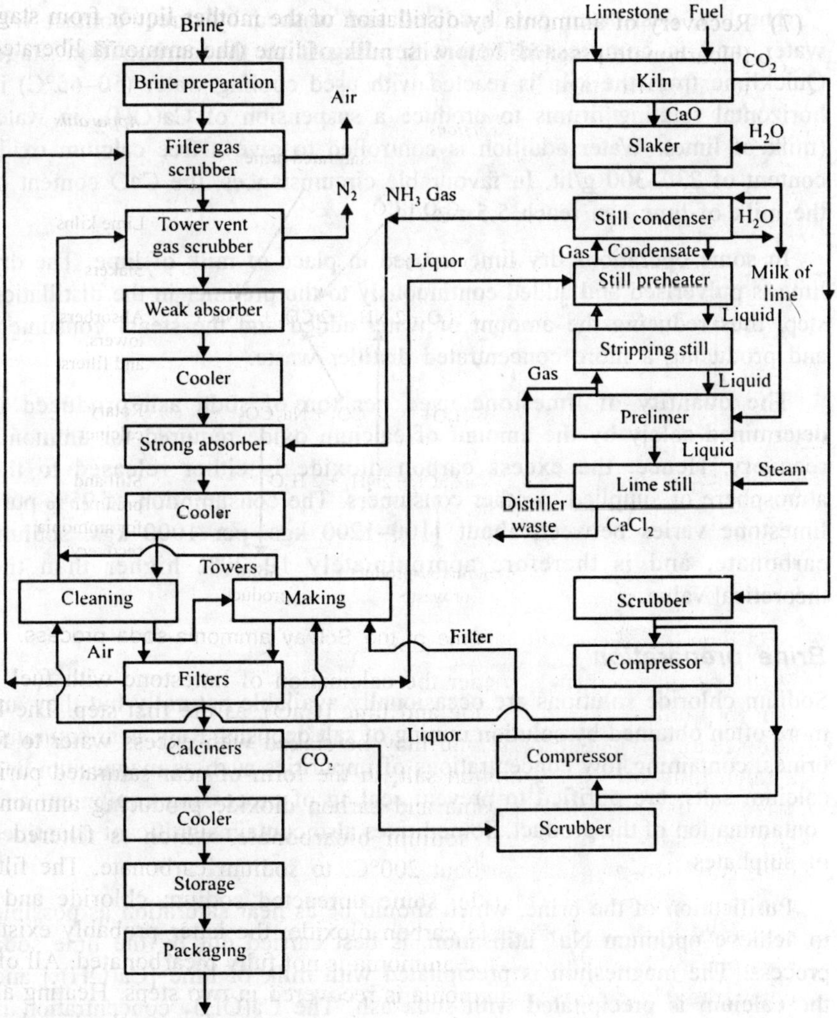

Fig. 14.3. Flow diagram for the production of ammonia-soda by the Solvay process.

Limestone burning and lime slaking

The limestone (of high purity, SiO_2 less than 3%, Fe_2O_3–Al_2O_3 less than 1.5%) is graded to a reasonably uniform coarse size. Although other fuels may be used, the limestone is usually mixed with about 7% metallurgical-grade coke or anthracite and then burned in vertical shaft kilns. Furnace gas CO_2 contents of 37–42 vol. % can be obtained. If liquid or gaseous fuels are used, the CO_2 contents of the gas is lower.

The hot, dust-laden gas is cooled and washed by direct contact with water, and is compressed before sending it to the carbonation stage. Quicklime from the kiln is reacted with used cooling water (50–65°C) in horizontal rotating drums to produce a suspension of $Ca(OH)_2$ in water (milk of lime). Water addition is controlled to give a free calcium oxide content of 230–300 g/lit. In favourable circumstances, the CaO content of the milk of lime can reach 5.5 mol/lit.

In some operations dry lime is used in place of milk of lime. The dry lime is pulverised and added continuously to the prelimer in the distillation step, thus reducing the amount of water added and the steam consumed, and producing a more concentrated distiller waste.

The quantity of limestone used per ton of soda ash produced is determined solely by the amount of calcium oxide required for ammonia recovery. Hence, the excess carbon dioxide is either released to the atmosphere or supplied to other consumers. The consumption of 95% pure limestone varies between about 1100–1200 kgs. per 1000 kgs. sodium carbonate, and is therefore approximately 10–20% higher than the theoretical value.

Brine preparation

Sodium chloride solutions are occasionally available naturally but they are more often obtained by solution mining of salt deposits. Raw, near-saturated brines, containing low concentrations of impurities such as magnesium and calcium salts, are purified to prevent sealing of processing equipment and contamination of the product. Some brines also contain significant quantities of sulphates.

Purification of the brine, which should be as near saturation as possible to achieve optimum Na^+ utilisation, is best carried out by the lime soda process. The magnesium is precipitated with milk of lime [$Ca(OH)_2$] and the calcium is precipitated with soda ash. The $Ca(OH)_2$ concentration is 170–185 kg/m^3 and the Na_2CO_3 concentration is 65–68 kg/m^3. Reaction times, crystallisation rates and settling rates are improved if the brine with the added reagent solutions is mixed with a recycled seed slurry. Under these conditions, precipitates of calcium carbonate and magnesium hydroxide can be produced with settling rates upto 1 m/h and, in some cases, even 3 m/h. After separation from the separated impurities, the brine is sent to the ammonia absorbers.

Ammonia absorption

The strong brine is saturated with ammonia gas containing water vapour and carbon dioxide in an absorption tower equipped with effective internal or external coolers to remove the heat of reaction. The mixture requires cooling of about 1650 mJ/T (=394 Kcal/kg.) of product soda ash. A typical absorption plant consists of a tower containing eight bubble trays and an external intermediate cooler. The brine descends through the main part of the absorber counter-current to the rising ammoniacal gases. This operation is generally carried out at slightly less than atmospheric pressure. Most of the ammonia is obtained from various process steps. Gas from distillation stage contains ammonia and carbon dioxide and is saturated with water vapour. However, small amounts are added to make up for losses. In the absorption stage, the total amount of ammonia required in the process and about 20% of the required carbon dioxide are added. The ammoniacal brine obtained contains about 85–90 kg/m^3 NH_3 and 40–50 kg/m^3 CO_2. The NaCl content is reduced from about 300–260 g/lit due to the increase in the specific volume of the liquor caused by the addition of ammonia and by dilution with water. For favourable yields, more then the theoretical equivalent of NH_3 must be present and in practice about half a ton of NH_3 is recycled per ton of Na_2CO_3 made.

Precipitation of bicarbonate (making)

The ammoniated brine from the absorber coolers is pumped to the top of one column in a block of columns used to precipitate bicarbonate. It is an exothermic process in which the ammoniacal brine is reacted with carbon dioxide from the lime kiln or from the calcination stage. Lime kiln gas, compressed to about 414 Kpa (60 psi), enters the bottom of the cleaning column and bubbles up through the solution absorbing most of the carbon dioxide. This process is accompanied by the evolution of considerable heat, 1450 MJ/T (347 Kcal/kg), that must be removed to maintain yield. In most ammonia-soda plants, cast iron columns are used. These bubble-tray columns are equipped with tubular coolers in the lower part. Bicarbonate crystals gradually foul the heat exchange surfaces and crystallising columns must be alternately used as the cleaning column (every 3–4 days). Gases, mostly nitrogen but some carbon dioxide and ammonia, are vented from the crystallising tower columns and collected before being recycled to the absorber.

A few soda plants still use the Honigmann process in which carbonation is carried out in three to five vessels in series. The temp. profile in both column and vessel processes exhibits a maximum of 50–60°C. The suspension of bicarbonate in mother liquor leaves the carbonation stage at 30°C.

Filtration of bicarbonate

Sodium bicarbonate is generally separated from the mother liquor in continuous rotary vacuum filters or band filters, or in centrifuges. Air drawn through the vacuum filter (or the vent gas from the centrifuge operation) is returned to the ammonia absorber. Mother liquor adhering to the bicarbonate crystals is washed off with condensate produced during the production process or with softened water. The quantity of washing water required depends on the particle size of the precipitated product and varies between 0.3–1 m^3/T of bicarbonate. The filter cake is carefully washed to control residual chloride and to meet customer specifications. The filter cake often called crude bicarbonate or ammonia soda contains 9–10% moisture as residual diluted filter liquor. The cake is made up of a sodium bicarbonate and small amounts (5 mol% on a dry basis) of ammonia primarily as ammonium bicarbonate. Crude bicarbonate has the following approximate composition:

$$NaHCO_3\ 75.6\%,\ Na_2CO_3\ 6.9\%,$$

$$NH_4HCO_3\ 3.4\%,\ NaCl + NH_4Cl\ 0.4\%,\ H_2O\ 13.7\%$$

Centrifuged products contain only 7–9% water. Calcination of 1 MT of crude bicarbonate yields about 520–560 kgs. of soda ash.

Calcination of Bicarbonate

Crude filtered bicarbonate is calcined continuously by indirect heating. Various techniques are used to heat [2430 KJ/(581kcal/kg)] this material to 175–225°C in the calciners. The decomposition temperature of $NaHCO_3$ is 87.7°C at a total CO_2–H_2O pressure of 0.1 MPa. Thermal decomposition of crude bicarbonate to carbonate liberates carbon dioxide, ammonia and water vapour. The product is technical grade soda ash, which contains sodium chloride. The hot soda ash discharged from the calciner is cooled, screened, and packaged, or shipped in bulk. This product called light ash has a bulk density of around 590 kg/m^3. A certain amount is sold in this form; the majority, however is converted to dense ash.

Dense soda ash is manufactured by hydrating light ash to produce the larger sodium carbonate monohydrate crystals and then dehydrating them. Hydration can be carried out in two ways: by feeding light ash and water to mixers or blenders, or by adding light ash to a saturated soda ash solution containing a slurry of monohydrate crystals. The monohydrate crystals are then fed to a continuous dryer and the dehydrated product is screened before packing and shipping. Bulk densities typically run between 960–1040 kg/m^3. Wet calcination can also be used for this purpose. This is carried out in a tower with steam in counter-current flow. In this thermal decomposition, a sodium carbonate-sodium bicarbonate equilibrium is established, which limits the yield to 85–88%.

Today, calcination is almost always carried out in rotary calciners. These are heated either externally with oil, gas or coal, or internally with steam, which passes through heating tubes in the calciner counter-current to the flow of solid material. Condensing steam gives much better heat transfer than combustion gas, so that the output from steam-heated rotary calciners is considerably higher than that of calciners heated by combustion processes. The advantages of steam heated calciners compared with those heated by combustion include the absence of thermal damage to the calciner, smaller space requirements and hence lower equipment costs, and about 15% less fuel combustion.

Gases extracted from the calciners contain CO_2, NH_3, H_2O and some leakage air, together with large amounts of dust (sodium carbonate and bicarbonate), which is removed in cyclones or electrostatic filters. The gas is then cooled and washed with brine and water until it is free of ammonia. Carbon dioxide, produced at 95% or higher purity, is compressed and recycled to the carbonating tower in order to enrich the make up kiln-gas feed. The condensate produced by gas-washing is fed to the ammonia recovery stage.

Recovery of ammonia

The filter liquor contains unreacted sodium chloride and substantially all the ammonia with which the brine was originally saturated. The ammonia may be fixed or free. Fixed ammonia (ammonia chloride) corresponds stoichiometrically to the precipitated sodium bicarbonate. Free ammonia includes salts such as ammonium hydroxide, bicarbonate, and carbonate and the several possible carbon-ammonia compounds that decompose at moderate temperatures. A sulphide solution may be added to the filter liquor or corrosion protection. The sulphide is distilled for eventual absorption by the brine in the absorber. As the filter liquor enters the distiller, it is preheated by indirect contact with departing gases. The warmed liquor enters the main cake, tile or bubble-cap-filled sections of the distiller where heat decomposes the free ammonium compounds and steam strips the ammonia and carbon dioxide from the solution.

The carbon dioxide-free solution is usually treated in an external, well-agitated lining tank called a 'prelimer'. Then the ammonium chloride reacts with milk of lime and the resultant ammonia gas is vented back to the distiller. Hot calcium chloride solution, containing residual ammonia in the form of ammonium hydroxide, flows back to a lower section of the distiller. Low pressure steam sweeps practically all of the ammonia out of the lined solution. The final solution known as 'distiller waste' contains calcium chloride, unreacted sodium chloride, and excess lime. It is diluted by the condensed steam and the water in which the lime was conveyed to the reaction. Distiller waste also contains inert solids brought in with the lime.

The waste liquors are usually pumped to settling basins where the suspended solids are deposited. In some plants, calcium chloride, $CaCl_2$, is recovered from part of this solution, close control of the distillation process is required in order to thoroughly strip carbon dioxide, avoid waste of lime, and achieve nearly complete ammonia recovery. The hot (56°C) mixture of wet ammonia and carbon dioxide leaving the top of the distiller is cooled to remove water vapour before being sent back to the ammonia absorber.

Waste disposal

Large volumes of liquid waste containing suspended and dissolved solids are produced in an ammonia–soda plant. The largest quantity comes from the distiller operations where, for every ton of product soda ash, nearly 10 cubic metres of liquid wastes are produced. This waste contains about one ton of calcium chloride, one-half ton of sodium chloride, and other soluble and suspended impurities. Traditionally, after settling the suspended solids in large basins, the waste was discharged into local water-ways, a practice no longer acceptable in many countries. In other countries, waste is sent to the ocean rivers or underground wells. The cost to comply with environmental regulations and increasing operating costs relative to natural ash production have forced shutdown of all synthetic soda ash plants in USA. However, several processes have been proposed to reduce or eliminate the waste streams.

Modified Ammonia-soda Processes

Many variants of the ammonia-soda process exist, and are used depending on regional requirements.

One such process is the ammonium chloride process (known as the dual process as it combines soda ash production with ammonium chloride production). The importance of the process is due to the high cost of imported rock salt and use of the ammonium chloride as a fertiliser. The dual process requires the use of solid sodium chloride to form NH_4Cl, but gives higher sodium conversion (+ 90%) than that obtained in the Solvay process.

Important differences exist between equipment used for the conventional Solvay process and that used for the dual process. For example, in the dual process, ammonia is not recovered, hence no distillation equipment is required. Also, lime kilns are not required if other sources of CO_2 are available.

This process has two variations named the co-production process and the mono-production process. In the co-production process, solid ammonium chloride is produced as a by-product by direct contact cooling crystallisation. In the mono-production process, the crystalline ammonium chloride reacts with a lime slurry to release the ammonia for recycle, sending the resulting calcium chloride solution to waste. Most Japanese plants have adopted some aspects of the new Asahi process which it is claimed requires less energy than the basic Solvay process and, by using direct contact cooling instead of conventional heat exchangers, reduces fouling.

OTHER SYNTHETIC ROUTES

Production from Nepheline

In the former Soviet Union (USSR), pottery clay is produced from nepheline. The process also yields some sodium carbonate. Finely ground nepheline on sinterine with very pure limestone and after grinding the clinker with water gives the aluminate solution, which is treated with carbon dioxide to precipitate aluminium oxide hydrate and give a solution of alkali-metal carbonates. This solution is evaporated and fractionated to give sodium and potassium carbonate. In the process for converting nepheline to aluminium oxide, 4 MT of nepheline gives about 1 MT of sodium carbonate.

Huls Process

Chemische Werke Huls AG has developed a process to produce soda ash and hydrochloric acid from salt via an amine-solvent system. A potential advantage of the Huls process is that, under some market conditions hydrochloric acid may be more easily sold than either ammonium or calcium chloride.

Akzo Process

Akzo Zout Chemie has developed a process to produce vinyl chloride and soda ash from salt using an amine-solvent system catalysed by a copper-iodide mixture. This procedure theoretically requires half the energy of the conventional Solvay process.

NATURAL vs SYNTHETIC

The dense soda ash produced from the ammonia-soda process is chemically and physically equivalent to that produced from trona (natural). It contains slightly more impurities (Table 14.3) than the trona-based soda ash, but these slightly higher impurities cause no serious problems for the consumers.

There is wide pricing disparity between natural and synthetic soda ash. The basic reason for this is the difference in manufacturing cost between natural and synthetic ash. Local cost differentials, e.g., in fuel, may also be important (extremely site specific).

The natural plant requires approximately one-third less energy than synthetic plants. The operating labour requirements for the natural soda ash plant are less than half the synthetic. In general terms, the synthetic ash plant is a sophisticated chemical process made up of many different steps, all of which require careful control reflected in labour requirements. Maintenance labour and supplies for the natural ash plant are about one-half those of the synthetic, showing that much less equipment is required to manufacture natural soda ash compared with synthetic soda ash.

Also, the waste disposal problems in the natural process is far less serious than that in the synthetic process. The natural soda ash processes produce no large volumes of associated wastes, as against large volumes from Solvay processes.

ANALYSES AND QUALITY SPECIFICATIONS

Anhydrous, technical grade soda ash is produced either by ammonia-soda or natural process. The synthetic ash is commercially available in two forms: light ash and dense ash; while the natural ash is available as medium and dense ash. Typical analysis of these forms and their granulometry are given in the Table 14.3.

Table 14.3. Soda ash analyses, % and granulometry.

Constituent	Synthetic ash Light	Synthetic ash Dense	Natural ash Medium	Natural ash Dense
Na_2O	58.3	58.3	58.4	58.4
Na_2CO_3	99.7	99.7	99.8	99.8
Na_2SO_4	0.015	0.015	0.04	0.04
NaCl	0.18	0.18	0.03	0.03
Fe	0.0010	0.0010	0.0006	0.0006
H_2O	0.10	0.10	0.12	0.12

Light soda ash has a bulk density of 0.5–0.6 kg/lit, while the dense soda ash has a bulk density of 1.05–1.15 kg/lit.

In addition to particle size and bulk density, soda ash quality is also assessed by several chemical tests. The different physical properties of the various grades are dependent both on the crystalline product calcined (bicarbonate, monohydrate or sesquicarbonate) and on calcination conditions.

Typical properties of commercial soda ash are listed in the Table 14.4. Dense ash is preferred by the glass industry because its size and bulk density minimise segregation when mixed with other glass batch materials. Light and intermediate (medium) grades of soda ash are preferred for some detergent applications where surfactant carrying capacity and dissolution are important.

Table 14.4. Typical properties of commercial soda ash.

Property		Synthetic ash Light	Synthetic ash Dense	Natural ash (U.S.) Intermediate	Natural ash (U.S.) Dense
Bulk density, g/ml.		0.59	1.04	0.80	1.04
NaCl, ppm		8000-1000	5000-300	300	300
Na_2SO_4, ppm		200	200	300	300
FeO_3, ppm		80-20	80-20	5	3
Size distribution, cumulative % on					
Sieve no.	Size, μm				
+40	420	3	11	90	11
+100	149	17	90	89	90
+200	74	63	99	99	99
+325	44	86	99	99	99
Pore volume, mL/g		0.4	0.1	0.3	0.1
Shape[a]		A	M	N	M

[a] Particle shapes are classified as acicular. A, Monoclinic, and blocky, M, or monoclinic and needle, N.

The above grades of soda ash are suitable for most applications and usually adequate for the food industry, other, special grades include dense granulated soda and briquette soda. So-called washing soda crystals, the decahydrate is now of minor importance.

Indian Standards

Indian Standards for technical grade of soda ash is given by IS-251 (1982). The requirements are listed in the Table 14.5.

Table 14.5. Requirements for soda ash, technical.

Characteristic	*Value (on dry basis)*
Total alkalinity as Na_2CO_3, % by mass, min	98.5
Matter insoluble in water % by mass, max	0.15
Sulphates as Na_2SO_4, % by mass, max	0.08
Chlorides as NaCl, % by mass, max	1.00 (0.7 for sodium dichromate industry)
Iron as Fe_2O_3, % by mass max	0.007

STORAGE AND TRANSPORT

Calcined soda ash is hygroscopic, gradually taking up moisture and carbon dioxide during transportation and storage. This converts the surface of sodium carbonate to the bicarbonate and leads to a weight increase of up to 17%. Soda ash is therefore best stored in closed, dry areas.

Soda ash is supplied either bulk in silo wagons that can be discharged pneumatically or in paper sacks. Hopper cars, pneumatic trucks, super-sacks, and multi-wall kraft bags with polyethylene liners are the usual shipping containers.

Soda crystals (washing soda) can effloresce or dissolve in the water of crystallisation, depending on the temperature and relative humidity. They melt at 32–33°C. At lower temperature, they are stable, provided the relative humidity is within the limits listed below:

T,°C	0	10	15	25
R.H., %	59–98	64–96	68–94	76–87

At lower humidity, soda crystals effloresce, and at higher humidity the surface becomes wet.

APPLICATIONS

Sodium carbonate or soda ash has a wide variety of applications in different sectors of the chemical industry. Its sodium content results in fluxing properties that make it important in the glass and silicate industries. The biggest single use of soda ash is in glass manufacture, which has steadily increased in the past. Large quantities are used for producing other chemicals. Pulp & paper and soaps & detergent also consume significant quantities of soda ash.

It is used to neutralise inorganic and organic acids or acidic salts, and to maintain a constant pH in processes where acids are liberated. It is also used in the production of sodium salts, e.g., phosphates, nitrates, chromates, citrates, tartarates, salts of fatty acids. Sodium silicates and sodium phosphates account for the largest share of the use of soda ash for chemicals.

Soda ash can be used in aqueous solution to remove sulphur dioxide from process gas or off-gases, forming sodium sulphite and sodium bicarbonate.

These reactions are also important in the production of paper pulp by the sulphite process. Other applications of sodium carbonate are summarised here:

(1) *Chemical industry:* Production of bleaching agents, borax, chromate's and dichromate's, fertilisers, dyes, fillers, tanning agents, industrial cleaning agents, catalysts, cryolite, adhesives, metal carbonates, sodium nitrate, perborates, phosphates, silicates, ultramarine pigments, soluble silicates, etc.

(2) *Soaps and detergents:* Manufacture of detergents and saponification of fats.

(3) *Petrochemicals:* Neutralisation.

(4) *Pulp and paper:* Cooking wood, neutralisation, cleaning, bleaching, and treatment of recycled paper.

(5) *Artificial silk industry:* Deacidification of artificial silk.

(6) *Textile industry:* Dyeing, bleaching, and finishing of wool and cotton.

(7) *Coke ovens, gas works and hydrogenation plants:* Gas purification (desulphurisation).

(8) *Iron and steel industry:* Removal of sulphur and phosphorus from pig iron, cast iron and steel, ore benefication, flotation agents, and fluxing agents.

(9) *Heavy metal industry:* Digestion and benefication of ores, e.g. of antimony, lead, chrome, cobalt, nickel, bismuth and tin.

(10) *Glass industry:* Raw material for the glass melt and for reacting with sand.

(11) *Aluminium industry:* Reaction with bauxite.

(12) *Ceramics industry:* Production of refractory materials and glazes.

(13) *Enamel industry:* As a flux.

(14) *Food industry:* Hydrolysis of proteins, production of margarine and starch, and softening of sugar beet juice.

(15) *Various branches of industry:* Water treatment, metal degreasing.

(16) *Environmental protection:* Purification of flue gases by injecting sodium carbonate or sodium hydrogen carbonates (dry process); Regeneration of acidic lakes by the introduction of briquettes of soda ash, so that organic sediments exhibit an alkaline reaction over a long period.

(17) *Miscellaneous:* General cleanser, in photography, as reagent in analytical chemistry, pharmaceutics aid (alkaliser). In veterinary treatment, soda ash has been used as an emetic. In solution it is used to cleanse skin, and in eczema, to soften scabs of ringworm.

SODIUM BICARBONATE (BAKING SODA)

Manufacture

From sodium carbonate (soda ash)

Reaction

$$Na_2CO_3 + H_2O + CO_2 \rightarrow 2NaHCO_3$$

Close to 100% yield

Process

Sodium bicarbonate, also known as bicarbonate of soda and baking soda, is produced by treating a saturated solution of sodium carbonate with carbon dioxide (Fig. 14.4).

Sodium carbonate (soda ash) is dissolved in the return liquor (mainly filtrate) in a dissolver to yield an essentially saturated solution; this is filtered and pumped to the top of a carbonating tower, where it comes in intimate contact with carbon dioxide gas. Sodium bicarbonate crystals are precipitated, forming a slurry.

The suspension of sodium bicarbonate thus formed is withdrawn from the bottom of the tower and filtered; the filter cake is washed. Several types of filters and centrifuges may be used. The filtrate is recycled to the dissolver. The washed filter cake, which contains about 8% moisture, is then dried on either a continuous-belt conveyor (enclosed in a chamber) at 70°C or in a vertical-tube dryer, in which the temperature of the air is maintained at 70 to 90°C.

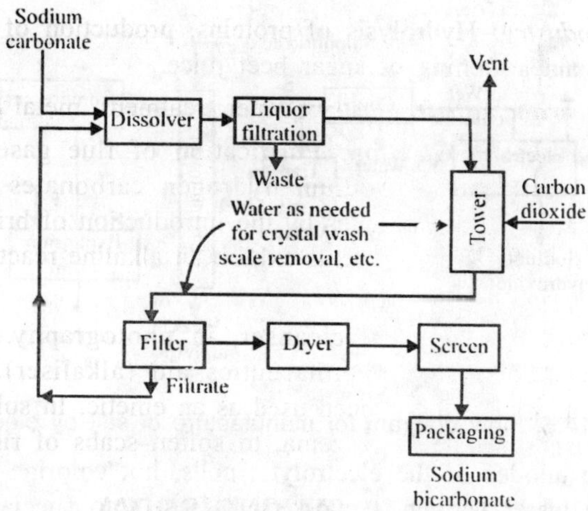

Fig. 14.4. Flow diagram for manufacture of sodium bicarbonate from sodium carbonate (soda ash).

Following the dryer the product goes to a screening plant, where it is screened to various grades, and then to storage bins for packaging. Sodium bicarbonate of about 99.9% purity is obtained.

Uses

Food; industrial and household; pharmaceutical; fire extinguishers; soap and detergents; textile, paper, and leather; rubber and plastics etc.

Properties

White, monoclinic crystals or powder with a slightly alkaline taste. Formula wt; 84.01; Sp gr 2.20; Bulk density 850 to 1000 kg/m^3.

Loses carbon dioxide progressively as the temperature is raised; the loss is significant at 100°C, and the rate increases with temperature. Soluble in water (6.9 g/100 g at 0°C, 16.4 g/100 g at 60°C). Insoluble in ethanol.

CHLORINE

From Salt by Electrolysis

Process

In the electrolysis of salt, chlorine is produced as a co-product with sodium hydroxide. Hydrogen is also obtained as a by-product (Fig.14.5).

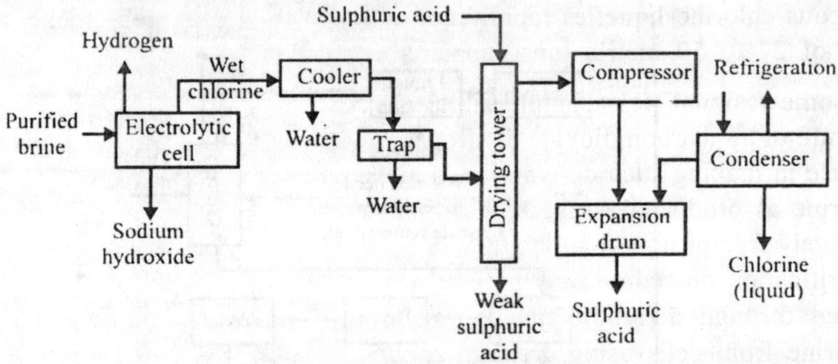

Fig. 14.5. Flow diagram for manufacture of salt by electrolysis.

From the anodes of the electrolytic cells, hot chlorine gas containing considerable water vapour is evolved. It is sent to specially constructed coolers, where the temperature of the gas is reduced to 12 to 14°C, just high enough to avoid the formation of solid chlorine hydrate, thereby circumventing the possible plugging of pipes. Most of the water vapour condenses in these coolers, which are usually of the following types: chemical stoneware chlorine lines submerged in water in trenches, water-cooled glass pipes and rubber-lined steel pipes, polyvinyl chloride (PVC) pipes, or stoneware disk coolers. Refrigeration is sometimes used, depending on the available water temperature.

Reaction

$$2NaCl + 2H_2O \xrightarrow[\text{current}]{\text{Direct}} Cl_2 + 2NaOH + H_2$$

95–97% current efficiency

The partially dried chlorine gas is then passed to drying towers, where the remaining water is removed by scrubbing with sulphuric acid, sp. gr. 1.84. After drying, the chlorine can be handled in iron or steel equipment, whereas the extremely corrosive moist chlorine necessitates chemical stoneware, glass, rubber-lined, or PVC pipes. (In handling bulk chlorine, it is of paramount importance to exclude moisture.) The dried gas goes to compressors, which almost invariably are the means of creating a partial vacuum which withdraws the chlorine from the electrolytic cells. Most plants use carbon-ring unlubricated compressors operating at about 40 psig (275 kPa). Centrifugal compressors are also used successfully. The heat of compression is removed progressively with water and refrigeration. A temperature between –30 and –45°C is attained, using single-stage or multiple-stage refrigeration, depending on the amount of compression. Most commonly employed refrigerants are of the Freon and Genetron types. Gaseous chlorine liquefies rapidly and is piped to steel storage tanks, tank cars of 27 to 50 metric tons capacity, or cylinders.

Some residual (less compressible) gas, called blow gas, an equilibrium mixture of air, carbon dioxide, hydrogen, and chlorine, is formed and vented for use in making such derivatives as bleaching powder. Commercial liquid chlorine as produced assays higher than 99.5%. For chlorine that is to be relatively free of non-volatile material (chlorine content better than 99.9%), a purification procedure may be used. The chlorine to be liquefied may be passed through a bubble-cap column countercurrent to liquid chlorine. Chlorine from cells using graphite anodes may contain small amounts of chlorinated hydrocarbons arising from anode reactions; the increasingly popular metallic anodes—usually platinum-plated or clad-titanium, known as dimensionally stable anodes or DSA do not cause such impurities in the product.

In the manufacture of caustic potash by the electrolysis of potassium chloride, chlorine is produced, as well as hydrogen, by reactions analogous to those already discussed; recovery processes are similar. Chlorine is also obtained commercially as a coproduct in the manufacture of sodium from fused salt. Although chlorine is also a coproduct in the electrolysis of magnesium chloride for magnesium production, this chlorine is recycled in the process and does not enter the market.

Approximately 95% of total domestic chlorine is produced as a coproduct of caustic soda manufacture, as described. Another 4.5% results from other electrolytic processes: production of caustic potash and of sodium. Less than 0.5% is produced by non-electrolytic methods, as described briefly below.

From Potassium Chloride and Nitric Acid

Process

In the nitrosyl chloride process, chlorine and potassium nitrate are produced by treating potassium chloride with nitric acid.

Reaction

$$3KCl + 4HNO_3 \rightarrow Cl_2 + 3KNO_3 + NOCl + 2H_2O$$

$$2NOCl + O_2 \rightleftarrows Cl_2 + 2NO_2$$

$$3NO_2 + H_2O \rightleftarrows 2HNO_3 + NO$$

$$2NO + Cl_2 \rightleftarrows 2NOCl$$

$$2KCl + 4NO_2 \rightarrow 2KNO_3 + 2NOCl$$

Potassium chloride, mixed with 63 to 66% nitric acid in a mole ratio of 1:1.7 to 1.9, is charged into a reaction tower. The reaction takes place at 75 to 125°C and atmospheric pressure to form chlorine gas along with nitrosyl chloride in a ratio of about 2:1. The nitrosyl chloride-chlorine gas mixture goes to a drying column, where it rises counter-current to 63 to 66% nitric acid at –10 to 0°C. The gases leaving the scrubber are dry enough to be handled in corrosion-resistant equipment. The gas is liquefied by cooling with brine at –15 to –20°C in a digester-condenser made of chrome–alloy steel (Fig. 14.6).

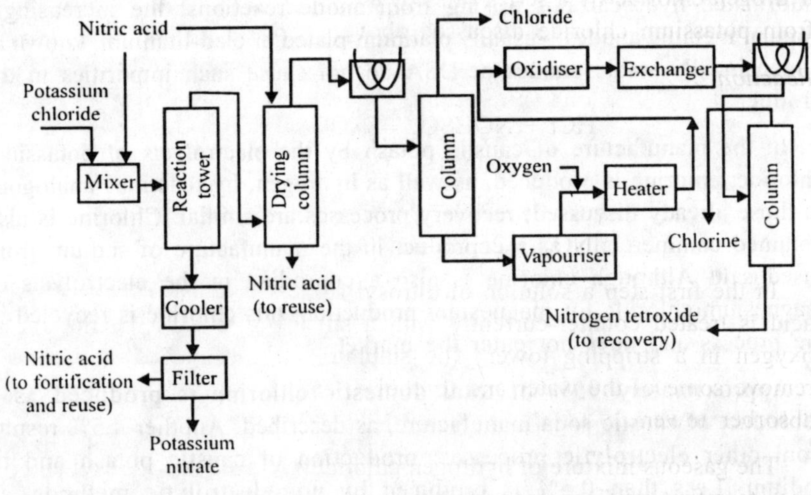

Fig. 14.6. Flow diagram for manufacture of chlorine from potassium chloride and nitric acid.

The condensed liquid is fed to a separation column from which nitrosyl chloride is removed as bottoms and chlorine gas goes overhead. The liquid nitrosyl chloride is then vapourised and heated along with added oxygen to 100 to 200°C, and sent to an oxidiser where it is converted to nitrogen dioxide and chlorine at 200 to 400°C. After cooling the gaseous effluent to 25 to 35°C in an interchanger, the mixture is liquefied with brine at –10 to –20°C. The liquid is fractionated at 15 to 60°C to send chlorine gas overhead and nitrogen tetroxide out the bottom. The nitrogen tetroxide may be converted to nitric acid by conventional methods.

This process was originally operated with salt (sodium chloride) as starting material, the coproduct being sodium nitrate. This proved uneconomical; but the analogous process using potassium salts has the attraction that the coproduct is potassium nitrate which is a superior fertiliser, especially for tobacco and other crops that cannot tolerate the chloride content of potassium chloride as a source of potash.

From Hydrogen Chloride and Air (Kel-Chlor Process)

Process

Many organic chlorinations release large amounts of hydrogen chloride, and much effort has been expended on means of recovering chlorine from what is in effect a waste product. One of the more promising is the oxidation of hydrogen chloride by a series of steps not unlike the production of chlorine from potassium chloride discussed above.

Reaction

$$HCl + NOHSO_4 \rightleftharpoons NOCl + H_2SO_4$$
$$2NOCl \rightleftharpoons 2NO + Cl_2$$
$$2NO + O_2 \rightleftharpoons 2NO_2$$
$$NO_2 + 2HCl \rightarrow NO + Cl_2 + H_2O$$

In the first step a solution of nitrosyl sulphuric acid in dilute sulphuric acid is treated counter-currently with a stream of hydrogen chloride and oxygen in a stripping tower. The sulphuric acid produced is flashed to remove some of the water and to cool it, and is recycled to the top of an absorber tower.

The gaseous mixture of hydrogen chloride, oxygen, and nitrosyl chloride goes to a first oxidiser stage; the temperature is raised enough to dissociate nitrosyl chloride, and the oxygen reacts with the nitric oxide to form nitrogen dioxide; this reaction is carried out under 10 to 15 atm (1 to 1.5 MPa) pressure. The resulting nitrogen dioxide oxidises hydrogen chloride; thus the nitrogenous compounds function as a catalyst (Fig. 14.7).

The gases leaving the first-stage oxidiser are passed to a second-stage oxidiser, where they come in contact with a liquid obtained from the previously mentioned flashing zone. Several reactions occur here, one of which is the reverse of that occurring in the stripper. The hydrogen chloride released is largely oxidised to chlorine in this stage. The gases from the second-stage oxidiser are passed to an absorption zone, and sulphuric acid is introduced, which also removes steam from the reaction products.

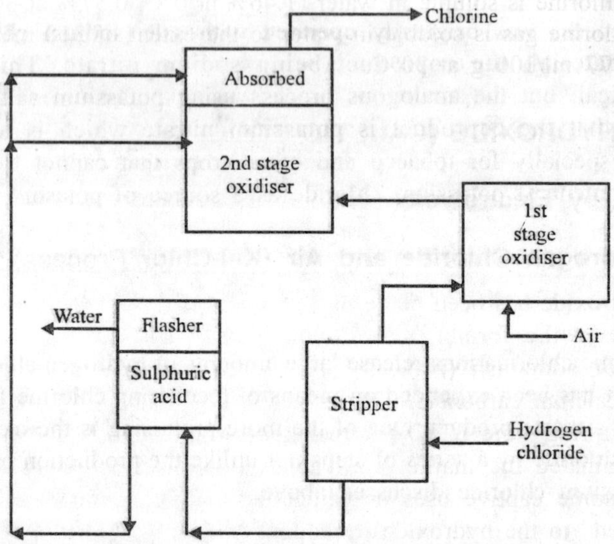

Fig. 14.7. Flow diagram for manufacture of chlorine from hydrogen chloride and Air (Kel-Chlor process).

The chlorine leaving the absorber tower contains a little hydrogen chloride and traces of nitrogen compounds, but is suitable for many uses. It may be further purified if necessary.

Electrolysis of hydrochloric acid, to produce chlorine and hydrogen without coproduct caustic soda, has been extensively researched and even practiced, but so far has not achieved significant acceptance.

Uses

Chlorinated hydrocarbons; pulp and paper; inorganic chemicals; water and sewage; etc.

Properties

Heavy, toxic, greenish-yellow gas with a pungent, irritating odour, which may be compressed to a yellowish liquid at 15°C under a pressure of 5.7 atm or –33.6°C under atmospheric pressure. Mol. wt 70.91; Sp gr. Gas 2.49 (0°C–Air = 1), Liquid 1.56 (–33.6°C); mp –101.6°C; bp –34.6°C.

Liquefaction point 6.8 atm (680 kPa) room temperature or –40°C (1 atm or 0.1 MPa). Critical temperature 146°C; Critical pressure 93.5 atm (9.35 MPa); Weight per cubic metre 3.219 kg (gas), 1468 kg (liquid); Threshold limit value 1 ppm (by volume), 3 mg/m^3.

Liquid chlorine is soluble in water (1.46% at 0°C, 0.57% at 30°C) and alkalies. Chlorine gas is soluble in water to the extent of 310 ml/100 g at 10°C and 177 ml/100 g at 30°C.

SODIUM HYDROXIDE (CAUSTIC SODA)

From Salt by Electrolysis

Process

Sodium hydroxide has been made by two processes: the lime-soda and the electrolytic. In the former, slaked lime [$Ca(OH)_2$] reacts with sodium carbonate or soda ash (Na_2CO_3) to produce sodium hydroxide and precipitate calcium carbonate. This is heated to drive off carbon dioxide, and the resulting lime (CaO) is reslaked and recycled. The lime-soda process dominated the market until about 1940; but it is now obsolete, except for some captive uses in which by-product sodium carbonate is "recausticised" to the hydroxide by adding lime (Fig. 14.8).

Reaction

$$2NaCl + 2H_2O \xrightarrow[\text{current}]{\text{Direct}} 2NaOH + H_2 + Cl_2$$

95 to 97% efficiency

In the electrolytic process, direct current is passed through a sodium chloride (salt) solution in specially designed cells. The salt brine is decomposed by the current to form a 10 to 12% sodium hydroxide solution, with hydrogen gas forming at the cathode and chlorine gas at the anode as coproducts.

Two types of cells, the mercury cathode and the diaphragm, are used; new installations use the latter because of the danger of mercury pollution in the effluent, as well as somewhat more economical operation, although the relative economics of the two types has been a matter of debate.

In the mercury-cell process, solid salt is required for make-up of the electrolyte; the diaphragm cell can use solid salt, natural brines, or sea water concentrated by evaporation or, more recently, by dialysis. In any case, the electrolyte is a saturated (about 25%) solution of salt, which is heated and passed into purification tanks, where it is treated with sodium carbonate and some caustic soda to remove calcium and magnesium compounds. If the brine is high in sulphate, barium chloride may be added. This purification procedure is necessary both to produce a high-grade product and to reduce clogging of the diaphragm, if used, in the electrolytic cell. The purified brine is neutralised with hydrochloric acid, reheated, and fed to the cells for electrolysis.

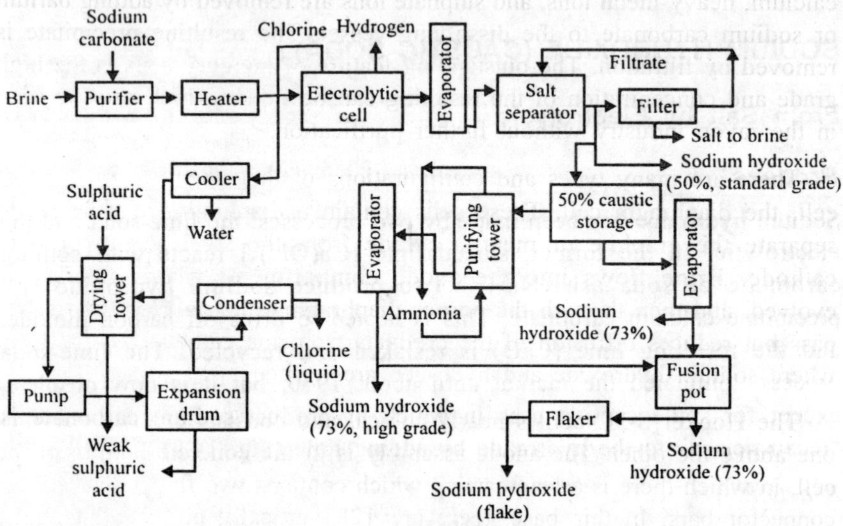

Fig. 14.8. Flow diagram for manufacture of sodium hydroxide from salt by electrolysis.

The mercury cell, of which the Mathieson E-8 cell is typical, is more expensive to operate because of the high initial investment in mercury (ca. 1400 kg/30,000-A cell) and because of the higher voltage of operation. In this cell, mercury is the cathode; metallic sodium, not ordinarily electro-depositable from aqueous solution, deposits into the mercury as an amalgam; this is made possible by the high over-voltage of hydrogen on mercury. The amalgam is decomposed in another compartment of the cell by water to produce hydrogen and a very pure (chloride-free) sodium hydroxide of 50% concentration. Chlorine is formed at the graphite anodes which operate at a current density of 4800 A/m^2 and a voltage drop of 4.5 v/cell. The current efficiency of these cells is about 95%. The cells operate at a temperature of 60 to 70°C.

The brine fed to the mercury cell, 26% sodium chloride, is reduced only to 22% sodium chloride per pass through the cell. Solid salt is used to saturate the recycle brine; this is necessary because, unlike diaphragm-cell operation, there is no water purge in the system. The brine must also be dechlorinated before recycling. The spent brine is acidified with hydrochloric acid to pH 3, and then degassed under vacuum and by blowing air counter-current to it in a spray tower. The last traces of chlorine and chlorate ion are destroyed by adding sodium bisulphite. The brine is then neutralised with caustic soda and resaturated with salt. Magnesium, calcium, heavy-metal ions, and sulphate ions are removed by adding barium or sodium carbonate to the dissolving vessel. The resulting precipitate is removed by filtration. The outstanding feature of mercury cells is the high grade and concentration of the resulting caustic liquor which may be used in the rayon industry without further purification.

There are many types and configurations of the other general type of cell, the diaphragm cell. These cells contain a diaphragm of asbestos to separate the graphite or metallic anode from the steel screen or wire cathode. Brine flows into the anode compartment, where chlorine is evolved, and then through the porous diaphragm which allows the ions to pass but reduces diffusion of the products from the cathode compartment where sodium hydroxide and hydrogen are formed.

The Hooker S-3C cell is nearly cubical and consists of three sections, one above the other. The anode assembly is in the concrete bottom of the cell, in which there is a lead casting which contains two flat copper anode-connector bars. In this base, there are 128 vertically projecting graphite anode blades.

The cathode consists of a steel frame flanged at the top and bottom. A steel screen structure is welded inside the steel frame to give an integral cathode unit. The diaphragm is applied by immersing the cathode in a bath of asbestos slurried in cell liquor and applying a vacuum to allow the asbestos to be deposited on the screen. Then the cathode with the deposited diaphragm is fitted over the graphite anodes; the concentric cell top is put into place and sealed with putty. Brine (322 g/litre and 65 to 75°C) flows into the top of the cell through a tantalum orifice, then into the anode compartment, where chlorine is formed, and then through the porous diaphragm. The cell operates with a 99.5°C catholyte, at 30,000 A or more, and with an original voltage drop of 3.95 V; this increases as the cell ages. Other characteristics of the cell are: 2500 KWh (9050 MJ)/metric ton of chlorine; 914 kg of chlorine and 1030 kg of sodium hydroxide/day per cell; 1.4 kg of NaCl/kg NaOH; anode life 280 days, during which two or three diaphragms have been used.

The caustic soda brine solution from the cathode compartment flows out through a percolation pipe: hydrogen is removed from the back of the cell. Chlorine bubbles through the brine and accumulates at the top of the cell, whence it flows through a pipe to a header.

Piping for the brine, hydrogen, and caustic liquor is usually of ceramic or polyester. Hydrogen is produced at a rate of 26.8 kg/day per cell (300 m^3/cell at standard conditions).

A variety of diaphragm cells is in use, each with slightly different design and operating characteristics but all embodying the same general principles. In particular, new installations tend to dispense with graphite anodes, which slowly disintegrate, thereby changing the geometry of the cell and introducing side products into the chlorine. In their place metallic anodes are increasingly used; these are of platinum-coated titanium, have a much longer life, and are dimensionally stable; in fact, they are known as dimensionally stable anodes (DSA).

The diaphragm cell decomposes about 50% of the brine to give a 10 to 12% caustic soda solution containing undecomposed sodium chloride. The weak caustic is charged into multieffect evaporators lined with nickel or Inconel (to avoid iron contamination), where a 50% sodium hydroxide solution is produced. Most of the salt precipitates out of the concentrated caustic and is recovered in salt separators. After filtering and washing, the salt is returned and used again to make up the charging brine. The 50% caustic soda liquor, which contains about 1% sodium chloride, may be solid as is, as standard-grade caustic liquor, purified, or concentrated further.

Several purification processes may be used and, depending on the procedure, either high-grade 50% liquor or a high-quality concentrated product is obtained. Some of the troublesome impurities in 50% caustic soda solution produced by diaphragm-cell electrolysis are sodium chloride (ca. 2%, dry basis), sodium chlorate (0.1 to 0.5%, dry basis), and colloidal iron. The last may be removed by treating the liquor with 1% by weight of calcium carbonate and filtering the resulting mixture, or by treating with chlorine. The sodium chloride and chlorates may be removed by a countercurrent treatment of the liquor with 70 to 90% ammonia in a liquid-liquid diffusion tower. The resulting 50% high-grade caustic liquor is as pure as that produced in mercury cells, and it is either sold as such or concentrated.

Another method of reducing the salt content consists of cooling the liquor to 20°C and filtering off the crystallised sodium chloride. Other processes involve both purification and concentration. For example, ammonia may be added to the 50% caustic soda solution to precipitate sodium hydroxide hydrates which contain less water than the original liquor.

Alternatively, free-flowing anhydrous crystals of sodium hydroxide may be formed by treating the 50% caustic with anhydrous ammonia in counter-current pressure equipment.

The 50% standard-grade or high-grade caustic liquor may be run into single-effect evaporators, operating at 75 to 100 psi (570 to 690 kPa) steam pressure to produce either 73% standard or high-grade caustic soda liquor, depending on the purity of the dilute liquor used. These 73% solutions are handled in steam-jacketed pipes to prevent solidification, which takes place at 65°C, and are either charged into insulated tank cars equipped with steam coils or further dehydrated to yield solid sodium hydroxide.

The 70 to 75% caustic soda solution from the final evaporator is run to fusion pots constructed of special fine-grained cast iron. Here, at a final temperature of 500 to 600°C, essentially all the water is boiled off (residual H_2O less than 1 %). The molten caustic is generally treated with sulphur to precipitate iron and, after settling, it is discharged into thin steel drums for sale as solid caustic, or the hot anhydrous sodium hydroxide is run to a flaker to produce flaked caustic. Depending on the purity of the 73% caustic liquor charged, governed by purification or manufacturing processes, either standard-grade anhydrous caustic soda, containing 2 to 3% impurities, or high-grade anhydrous sodium hydroxide with less than 1% impurities is obtained. In addition to liquid, solid, and flake forms, caustic soda is produced in ground, powdered, and pelleted grades.

The hot, wet chlorine gas evolved at the anodes of the cells is recovered after cooling. Hydrogen is combined with chlorine to produce a high grade of hydrogen chloride, compressed for sale, used on site for chemical processes requiring it, or burned for fuel.

Uses

Chemicals; pulp and paper; aluminium; petroleum; textiles; rayon; soap and detergents; cellophane etc.

Properties

White; deliquescent pieces; lumps or sticks. Absorbs both carbon dioxide and water from the air. Mol wt 40.00; Sp gr 2.130; mp 318.4°C; bp 1390°C. Soluble in water (42 g/100 at 0°C, 347 g/100 g at 100°C), ethanol, ether, and glycerine. Insoluble in acetone. Threshold limit value (mg/m^3) 2.

BLEACHING POWDER

When chlorine reacts with calcium hydroxide below 35°C, the product is a white, amorphous solid called bleaching powder. Its exact chemical composition is not known, but it behaves chemically as if it contained calcium hypochlorite, $Ca(OCl)_2$, and basic calcium chloride, $CaCl_2.Ca(OH)_2.H_2O$. The manufacture is carried out in towers designed to ensure that the upflowing chlorine mixes intimately with the finely-powdered slaked lime. The reaction is exothermic, and cooling is necessary to prevent the formation of calcium chlorate.

Bleaching powder reacts with carbon dioxide from the atmosphere producing hypochlorous acid, which accounts for its oxidising and bleaching actions. With dilute acids it gives chlorine, even in the cold, and this 'available chlorine', as it is called, is usually expressed as a percentage by weight of the solid, reaching about 36% in freshly made material. It can be estimated by adding a known weight of the bleaching powder to a solution containing excess of acidified potassium iodide, and titrating the liberated iodine with a standard solution of sodium thiosulphate using starch as indicator.

When some bleaching powder is warmed with a dilute solution of a cobalt salt, oxygen is evolved rapidly. This is a catalytic reaction, even traces of the cobalt compound bringing about the decomposition.

Bleaching powder is not completely soluble in water, and an aqueous suspension is normally used for bleaching wood pulp, linen and cotton, and also as a disinfectant ('chloride of lime'). Its main advantage over gaseous chlorine arises from its easier storage and safer transport.

SODIUM HYPOCHLORITE

Sodium hypochlorite is employed as a disinfectant and deodorant in dairies, creameries, water supplies, sewage disposal, and for household purposes. It is also used as a bleach in laundries. As a bleaching agent, it is very useful for cotton, linen, jute, artificial silk, paper pulp, and oranges. Indeed, much of the chlorine bought for bleaching cellulose products is converted to sodium hypochlorite before use. The most common method for making it is the treatment of sodium hydroxide solution with gaseous chlorine.

$$Cl_2 + 2NaOH \longrightarrow NaCl + H_2O + NaOCl$$

The other once widely used method was the electrolysis of a concentrated salt solution whereby the same product was made. These electrolytic cells do not have a diaphragm and are operated at high current density in a nearly neutral solution. The cells are designed to function at a low temperature and to bring the cathode caustic soda solution in contact with the chlorine given off at the anode.

SODIUM CHLORITE

Sodium chlorite ($NaClO_2$), the 80% commercial material has about 125% available chlorine. It is manufactured from chlorine through calcium chlorate to chlorine dioxide, ending with the reaction:

$$4NaOH + Ca(OH)_2 + C + 4ClO_2 \longrightarrow 4NaClO_2 + CaCO_3 + 3H_2O$$

After filtering off the calcium carbonate, the solution of $NaClO_2$ is evaporated and drum-dried. $NaClO_2$ is a powerful but stable oxidising agent. It is capable of bleaching much of the colouration in cellulosic materials without tendering the cellulose. Hence it finds use in the pulp and textile industries, particularly in the final whitening of kraft paper. Besides being employed as an oxidiser. $NaClO_2$ is also the source of another chlorine compound, chlorine dioxide, through the reaction:

$$NaClO_2 + 1/2Cl_2 \longrightarrow NaCl + ClO_2$$

Chlorine dioxide has 2 1/2 times the oxidising power of chlorine and is important in water purification, for odour control, and for pulp bleaching.

Chapter 15

Potassium Industries

INTRODUCTION

Potassium is far too reactive metal, but it is widely distributed in minerals, the most important deposits of which occur in places where island seas have undergone natural evaporation. The main commercial source is carnalite, but kainite and chonite and sylvinite are sometimes used. Potassium is present in most clays and silicate rocks, and also in sea water (about 0.05%) and wood ash (about 30% of which is potassium carbonate).

Potassium salts must be present in the soil in order to have normal plant growth, hence they are essential components of the fertilisers used in food production throughout the world. Various potassium salts are also basically important as raw materials for many commodities such as soaps, detergents, glass, dyes, gunpowder, and pyrotechnics.

The sylvinite is mined and treated to yield high-grade potassium chloride, and langbeinite is processed to make potassium sulphate.

POTASSIUM CHLORIDE

From Sylvinite by Fractional Crystallisation

Process

Sylvinite (42.7% KCl, 56.6% NaCl), is the raw material for the potassium salts. The particular processes used by the various companies vary in their details but are of two general types. In the oldest plant which probably works the highest-grade ores, the sodium and potassium chlorides are separated by fractional crystallisation; the others use flotation processes (Fig. 15.1).

The feed for the fractional crystallisation process is –4 mesh ore. The 8- to 16-mesh material from the grinding operation is separated on tables to yield a granular 50% K_2O product. For fertiliser use, the potassium content of all potassium salts is expressed as the equivalent percentage of K_2O. For instance, 50% K_2O is equivalent to 80% potassium chloride, and 60% K_2O to 97% potassium chloride. The fines (–4 mesh) are fed to a tank where they are partially dissolved in mother liquor returned from further along in the process. This mother liquor, when heated in the dissolver, remains saturated with respect to sodium chloride but has the capacity to dissolve more potassium chloride. Undissolved sodium chloride is removed from the dissolver, dewatered, washed, and sent to waste. The solution, saturated with both salts at 110°C, is clarified in a thickener. The underflow, containing insoluble mud, is concentrated and washed in counter-current decantation units.

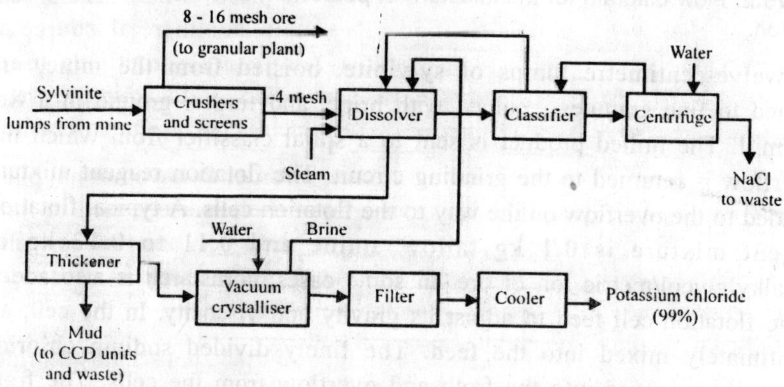

Fig. 15.1. Flow diagram for manufacture of potassium chloride from sylvinite by fractional crystallisation.

The hot clarified solution is then sent to a vacuum crystalliser, where potassium chloride crystallises at 32°C. To prevent sodium chloride from crystallising, water is added to replace that which evaporates. The cooled slurry is sent to a filter from which the mother liquor is recirculated to the dissolving tank. The cake goes to a rotary cooler, and thence to storage.

From Sylvinite by Flotation

Process

Standard 60% (K_2O) muriate of potash can be produced by separating potassium chloride from sodium chloride by froth flotation of finely ground sylvinite ore. (Fig. 15.2). The ore is obtained by shaft mining and is broken from the mine face by drilling and blasting. About 375 g blasting powder is required per metric ton of ore. The ore is crushed underground to 13 cm maximum size, hoisted to the surface, and then ground and screened to smaller size.

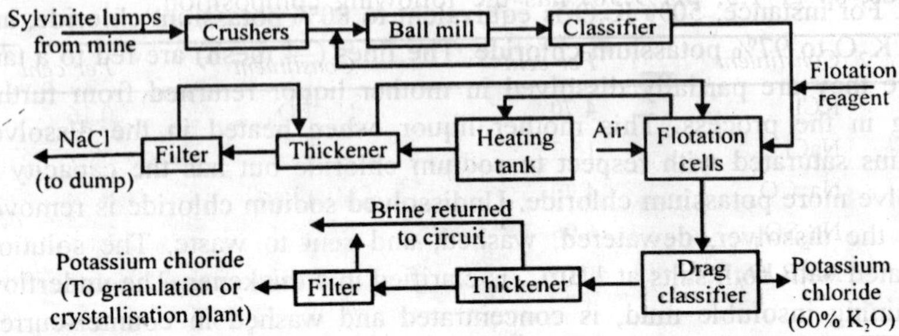

Fig. 15.2. Flow diagram for manufacture of potassium chloride from sylvinite by flotation.

Twelve-centimetre lumps of sylvinite, hoisted from the mine, are crushed to fine granules, pulped with brine, and further ground in a wet ball mill. The milled product is sent to a spiral classifier from which the underflow is returned to the grinding circuit. The flotation reagent mixture is added to the overflow on the way to the flotation cells. A typical flotation reagent mixture is 0.1 kg tallow amine and 0.11 to 0.12 kg of polyalkylglycol/metric ton of ore. In some cases, cornstarch is also added to the flotation-cell feed to adjust its gravity and viscosity. In the cell, air is intimately mixed into the feed. The finely divided sodium chloride particles are carried into the froth and overflow from the cells. The froth breaks and is pumped through a heating tank, a thickener, and a filter in series. Tailings of sodium chloride are sent to the dump; the final filtrate is recycled to the main stream.

The potassium chloride underflow from the flotation cell is pumped to a drag classifier from which a wet solid phase is removed and dewatered by filtration. The product is the standard 60% muriate of potash. The overflow from the classifier is thickened and filtered. The filter cake may be melted in a furnace and flaked on a cold drum to a granular grade 60% K_2O product (97% KCl), or it may be sent to a crystallisation plant for the manufacture of a chemical-grade product (99.9% KCl). In some of the newer plants, sylvinite is floated and sodium salts removed in the underflow. This reversal from the process described is accomplished by proper selection of cell-feed gravity and the flotation agent.

From Searles Lake Brine

Process

Potassium chloride has been recovered from brines. The brine (34.68% solids, sp gr 1.3, pH 9.38) has the following composition.

Constituent	*Per cent*	*Constituent*	*Per cent*
KCl	4.70	Na_3PO_4	0.16
NaCl	16.35	NaF	0.01
Na_2CO_3	4.70	Misc. solids	0.30
Na_2SO_4	6.96	Water	65.32
$Na_2B_4O_7$	1.50		——
		Total	100.00

Raw lake brine is mixed with soda products mother liquor and borax mother liquors, and then pumped through a series of condensers and filters to a triple-effect evaporator to effect a crude separation of the soluble salts into a potassium chloride-borax solution, solid crude sodium chloride, burkeite crystals ($Na_2CO_3 \cdot Na_2SO_4$), and solid dilithium sodium phosphate crystals (Li_2NaPO_4).

The hot, concentrated liquor, saturated with potassium chloride and borax, is clarified and then cooled quickly to 38°C in a three-stage vacuum crystalliser. (Fig. 15.3.) The rate of crystallisation of potassium chloride is so much faster than that of borax that an excellent separation is effected. To prevent crystallisation of sodium chloride, enough additional water is added to replace that which evaporates during the vacuum cooling.

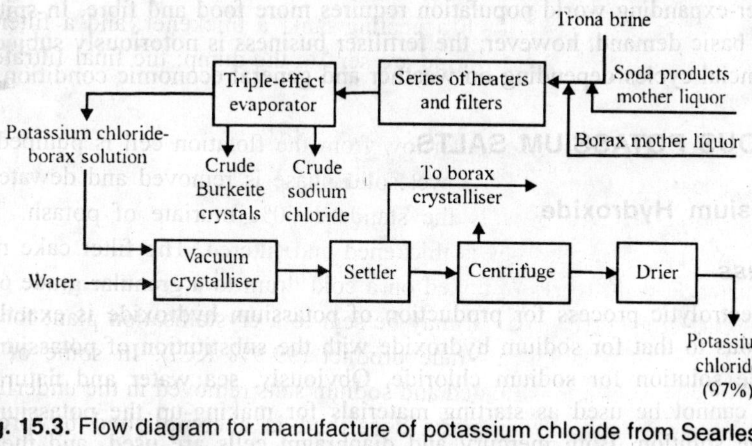

Fig. 15.3. Flow diagram for manufacture of potassium chloride from Searles lake brine.

The suspension of solid potassium chloride crystals in mother liquor goes to a cone settler. The clear overflow liquor is sent to borax crystallisers; the underflow potassium chloride sludge is further concentrated in continuous centrifuges, and then dried in a rotary dryer to yield potassium chloride crystals of 97% purity.

Other Processes

Smaller quantities of potassium chloride are recovered from brines other than the Searles Lake brine. The natural brine, containing about 1% potassium chloride, is concentrated to 30% in ponds. The crystalline product is then brought to about 95% purity by flotation.

Uses

Agricultural and chemical.

Properties

Colourless cubic crystals, granules, or powder. Low toxicity. Formula wt 74.55, mp 772°C, Sp gr 1.988, bp 1411°C. Soluble in water (27.6 g/100 g water at 0°C; 56.7 g/100 g at 100°C). Soluble in ethanol and alkalies.

Muriate of potash (the commercial name for fertiliser-grade potassium chloride) is the principal source of potassium not only for fertiliser but also for the much smaller demand for potassium salts of all kinds, principally potassium hydroxide. It is not, however, the only source; langbeinite, a mixed potassium-magnesium sulphate, is also worked. The greatest demand for potassium salts is for fertiliser; this demand is essentially insatiable as the ever-expanding world population requires more food and fibre. In spite of this basic demand, however, the fertiliser business is notoriously subject to financial cycles depending on weather and general economic conditions.

VARIOUS POTASSIUM SALTS

Potassium Hydroxide

Process

The electrolytic process for production of potassium hydroxide is exactly analogous to that for sodium hydroxide with the substitution of potassium chloride solution for sodium chloride. Obviously, sea water and natural brines cannot be used as starting materials for making-up the potassium chloride solution. Both mercury and diaphragm cells are used, and their relative advantages and disadvantages are the same as for caustic soda.

About 4 to 5% of the electrolytic chlorine supply originates as a co-product of potassium hydroxide manufacture.

Reaction

$$2KCl + 2H_2O \rightarrow 2KOH + H_2 + Cl_2$$

95 to 97% current efficiency

Solubility relationships in the potassium chloride-potassium hydroxide system are different from those in the corresponding sodium system; and potassium salt solutions are more electrically conductive than their sodium counterparts. There are therefore some minor differences in operating parameters. In diaphragm cells, the product liquor contains 10 to 15% potassium hydroxide and about 10% potassium chloride. As with sodium hydroxide, most of the potassium chloride crystallises out during concentrating by evaporation and subsequent cooling, but about 1% potassium chloride is a lower limit to the possibilities of purification by this method. Mercury-cell operation is analogous to the sodium system; the product potassium hydroxide is much purer, and is concentrated enough (45%) to the sold directly.

However, the purification afforded by the mercury cathode does not extend to the sodium impurity in the potassium chloride feed; and sodium ends up as equivalent contamination of the caustic potash with caustic soda. A drawback of mercury cells is that they require solid potassium chloride feed; this accentuates the difficulties of purifying the potassium chloride.

Large amounts of caustic potash are used as 45 or 50% solutions; lesser amounts are concentrated to 61% potassium hydroxide, corresponding to the dihydrate $KOH{\cdot}2H_2O$, and cast into sticks or pellets. The remainder is concentrated to 90 to 92% and sold as solid flakes or lumps. Reducing the water content further would require melting *in vacuo*. Most technical caustic potash contains up to 1.5% potassium carbonate.

Although, like caustic soda, potassium hydroxide may be safely handled in plain steel tanks, at higher temperatures, approaching 90 to 95°C, it is much more corrosive to steel than is sodium hydroxide, so that when heating tanks to avoid freezing, care must be exercised. Nickel-lined tanks are frequently used for this reason.

Uses

Potassium carbonate, soaps, tetrapotassium pyrophosphate, other potassium compounds, liquid fertilisers, dyestuffs etc.

Properties

Soluble in water (112 g/100 g. at 20°C), ethanol, ether, and glycerine. Insoluble in acetone. Since caustic potash is more costly than caustic soda on the bases of both unit weight and alkali equivalent weight, caustic soda will always be used where only the properties of a strong alkali are required; the uses of caustic potash will be restricted to those where its particular properties are necessary.

Potassium soaps are liquid, while sodium soaps are generally solid; thus potassium hydroxide is reauired for liquid soaps and detergents. Tetrapotassium pyrophosphate, for the same reason, is used as the builder in liquid detergents. Potassium is required by plants, whereas sodium may actually be harmful if in excess. Potassium salts in general are more soluble than their sodium analogs, and this advantage often outweighs their higher cost.

Potassium Carbonate

Properties

It is a white, deliquescent, granular translucent powder, soluble in water, giving an alkaline solution by hydrolysis; soluble in alcohol, sp. gr. 2.428 (19°C); m.p. 891 c; b.p, decomposes, Non-combustible. It can exist as tri.-, di.-, and monohydrates, all of which loose their water of crystallisation on heating above 130°C. In its reactions it closely resembles sodium carbonate.

Preparation

(1) Alkyl amines or ion exchange resins can be used with potassium chloride and carbon dioxide to yield potassium bicarbonate, which is calcined to carbonate.

(2) It can be prepared by using magnesium oxide, potassium chloride, and carbon dioxide, by separating Engles salt ($MgCO_3 \cdot KHCO_3 \cdot 4H_2O$). Decomposition leaves potassium bicarbonate in solution, which can be processed to potassium carbonate.

Uses

It is used to make special glasses (optical and colour TV tubes), potassium silicate, pigments; dehydrating agent; printing inks, laboratory reagent, soft soaps, raw wool washing; general purpose food additive.

Potassium Sulphate

Properties

It is colourless or white hard crystals or powder; bitter saline taste. Soluble in water; insoluble in alcohol. Sp gr. 2.66; m. p. 1072°C.

Preparation

(1) By fractional crystallisation from solutions of naturally occurring double salts.

(2) By treatment of potassium chloride either with sulphuric acid or with sulphur dioxide, air, and water.

(3) From salt lake brines.

Uses

Reagent in analytical chemistry; medicine (cathartic); gypsum cements; fertiliser for chloride sensitive crops, such as tobacco and citrus; alum manufacture, glass manufacture; food additive.

Potassium Acid Tartrate

Properties

White crystals or powder; soluble in boiling water; pleasant slightly acid taste; insoluble in alcohol. Sp. gr. 1.984 (18°C). Non-toxic.

Preparation

A domestic tartrate industry has been developed by processing pomace, which is discarded from wine fermentors and contains from 1 to 4% potassium bitartrate (From wine lees by extraction with water and crystalisation).

Uses

Baking powder, preparation of other tartrates; medicine; galvanic tinning of metals; food additive.

Potassium Nitrate

Properties

It is transparent, colourless, or white crystalline powder; slightly hygroscopic; pungent, saline taste, sp. gr. 2.1062; m.p. 337°C; b.p., decomposes at about 400°C. Soluble in water and glycerol; slightly soluble in alcohol. Low toxicity.

Preparation

It is prepared by adding potassium chloride to a list saturated solution of sodium nitrate :

$$NaNO_3 + KCl \rightleftharpoons KNO_3 + NaCl$$

Because sodium chloride is the least soluble of these substances in hot water, much of it is precipitated from solution and filtered off while hot causing position of equilibrium to move to the right. On cooling the mixture, most of potassium nitrate formed crystallises out, because it is much less soluble in cold water than in hot (the solubility in 100g of water is 13 g at 0°C, and 246 g at 100°C).

Uses

Pyrotechnics; explosives; matches; speciality fertiliser; reagent to modify turning properties of tobacco; glass manufacture; tempering steel; curing foods; oxidiser in solid rocket propellants. It is dangerous, fire and explosion risk when shocked or heated, or in contact with organic materials; strong oxidising agent.

Chapter 16

Barium and Its Compounds

INTRODUCTION

Mineral barytes, is the most important mineral of barium. Cudappa district in Andhra Pradesh is a major producer of this mineral in India. There are deposits of this mineral in Salem district of Tamil Nadu, in Alwar district of Rajasthan, and in Chandrapur district of Maharashtra. Good quality barytes is found in large quantities in village Mangampetta of Koduru in Andhra Pradesh. The mineral barytes is generally mined by open cast mining method in India and rarely from underground. Mined mineral is generally hand sorted for grades and colour and accordingly stacked at the mining sites. Total deposits of this mineral in India is estimated around 80 million tonnes, and only 10 to 12% of this mineral is actually converted to barium salts. It is the only economical mineral of barium and is extensively used in industries like rubber, glass, paints, etc. In oil-well drilling, the ground mineral is used in large quantities.

Barytes crystallises in the orthorhombic system. Usually it is white in colour, but often the mineral is seen in grey, brown-pink, yellow, off-white colours. Based on colour, there are three grades, in general, of the mineral viz. snow white, white, and off-white. Chemically mineral barytes is barium sulphate-$BaSO_4$. The percentage of $BaSO_4$ in the mined mineral varies from 80% to 99%. Silica-SiO_4 is generally found associated with barytes from as low as nil percentage to as high as 25%. In addition iron, lead, manganese. etc, are also present in the mineral. The important compounds of barium like carbonate, chloride, nitrate are manufactured from the mineral barytes.

Fig. 16.1 shows the various production methods for barium salts from the mineral barytes. For the manufacture of barium compounds, the barytes

should have $BaSO_4 \geq 94\%$ and as much less silica, alumina, manganese iron, etc. Luckily the Indian barytes have calcium and strontium in traces only. Bleached/unbleached snow-white finely ground to 300 mesh-barytes powder is used in the paint industries as extender and as diluent in coloured pigments. Barytes powder has fire quenching property. Unbleached finely ground–to 300 mesh–mineral powder is used in oil-well drilling.

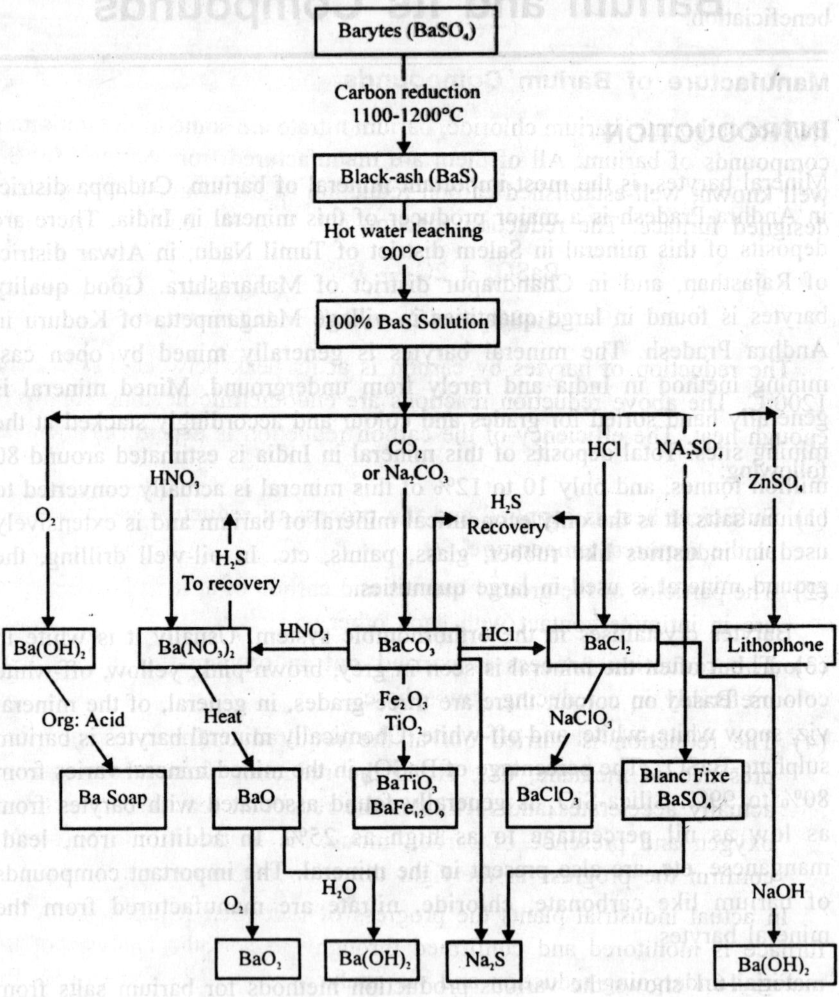

Fig. 16.1. Production pathways to barium salts from mineral barytes.

Benefication of Barytes

In India due to availability in abundance, mined bartyes of assay below 85% $BaSO_4$ is presently wasted at or near the mining sites, though it is possible to beneficiate them to 95% from as low as 65%. Unfortunately no serious efforts and attempts have been made for the beneficiation of low grade barytes. Actually they are at present wasted, often utilised as filling earth. National Mineral Development Corporation has the technology for beneficiation.

Manufacture of Barium Compounds

Barium carbonate, barium chloride, barium nitrate are some of the important compounds of barium. All of them are manufactured from barytes, by the well known, well-established carbon reduction of the mineral in a suitably designed furnace. The reduction reaction can be represented. (Fig. 16.1).

$$BaSO_4 + 2C \rightarrow BaS + 2CO_2$$

$$BaSO_4 + 4C \rightarrow BaS + 4CO$$

The reduction of barytes by carbon is at its peak between 1100°C and 1200°C. The above reduction reactions are endothermic needing supply of enough heat. The efficiency of the carbon reduction is dependent upon the following:

(1) Sufficient heat is supplied and the process of reduction is carried out at the required temperature.

(2) The particles of the ground barytes and carbon of at least 20 mesh size are in intimate contact with each other.

(3) The reduction process is carried out in oxygen-free atmosphere preferably in a reducing atmosphere.

(4) The reduction is carried out at the required optimum temperature, presence of methane, CO, H_2 in the reduction furnace atmosphere, actually accelerates and facilitates the reduction reaction. Absence of oxygen and presence of carbon monoxide in the furnace off-gases confirm the progress in the right direction of reduction reaction.

In actual industrial plants the progress of reduction reaction inside the furnace is monitored and confirmed through the periodic analysis of the material undergoing reduction and via analysis of O_2, CO and CO_2 in the flue gases. Optical pyrometer can be used to measure the temperature inside the temperature. Also one can roughly arrive at the furnace temperature, once the flue gas temperature at the furnace outlet is measured. As the reduction progresses one can see the roasted material inside the furnace undergoing physical changes, the mass starts slowly agglomerating and later trying to stick to furnace wall. At this stage even one can see puffs of gas buffling out from the agglomerated mass. These observations are a must, and they confirm the progress in the right direction of the carbon reduction in the furnace. The CO_2 in the reduction furnace waste gases in the continuous furnaces will be around 30% during the best reduction process. In batch reduction furnaces the CO_2 readings are generally erratic due to secondary air leakage. The roast after the completion of the reduction reaction is discharged from furnace for its further processing. This roast is called black ash, dark grey in colour; and an granular to soft lumpy mass, having a pungent odour similar to SO_2. The impurities like ash and silica in coal or other sources of carbon, and in barytes used for reduction is poison for the reduction since they combine with the sulphide in the black ash, leading to poor efficiency.

Almost all the barium salts manufacturers in India follow the batch wise operated rotary furnace. The reduction process can be carried out continuously via a rotary kiln of length varying from 100 ft. to 200 ft. The manufacture of barium salts is restricted in India to S.S. units, who cannot afford the high investments required for continuous rotary kiln process.

Fluidised kiln reduction process is also possible, but not in operation on a commercial scale. Here mixture of pulverised mineral is granulated with carbon and made as pellets. These pellets are charged to a fluidized bed furnace and the reduction process is carried out. Part of the carbon in the pellets is used for heat generation required for the reduction and balance is utilized for reduction. Mixture of CO and H_2, or coal or CH_4, natural gas or even hydrogen can be deployed for the reduction, in place of carbon. However, no large scale plants of this kind are in operation.

In India with the exception of one manufacturer all are following the batch process using a rotary furnace. The rotary furnace has a cylindrical shell with short cones at either ends, suitably lined inside, having a charging hole in the centre of the cylindrical portion and is provided with girth gear (for rotation of kiln) and two support rings mounted on the cylindrical portion with few typical batch rotary furnaces. Furnace is supported on four cast steel support rollers, and rotated through the girth gear via pinion wheel mounted on a speed reduction gear box. Furnace rotation -r.p.m.- varies from 0.5 to 1.2. At higher r.p.m. loss of material from the furnace by way of carry over through flue gases is more. Furnace is mounted/ erected at such a height from the ground, so that it can easily discharge the roast/black ash conveniently into trolley or conveyor for further processing.

The fuel burning burner with its air supply system is so mounted at the burner end of the furnace, in such a way that the whole assembly can be moved away from furnace as and when required. Furnace is lined with high alumina insulation bricks inside. The flue gas end of the furnace should be flush with the dust catch chamber wall, leaving a gap of 5 to 6 mm between the furnace mouth and opening of the chamber. This clearance all around the furnace mouth with the opening in the D.C. chamber is required: (i) to control and checking of excess air (due to chimney), entry into reduction furnace through the burner end openings; and (ii) to take care of the movement of furnace due to thermal expansion. Waste heat recovery unit is installed after the dust catch chamber, to utilise waste heat for generation of hot water needed for the process and to concentrate dilute sodium sulphide solution. The reduction furnace chimney should have a damper.

Feed preparation for reduction furnace

Raw materials are barytes and source of carbon like coal coke. 94–96% barytes is ground to a size of 60–80 mesh via jaw crusher, Roller mill/Ball mill and sieved to get powdered barytes, 100% passing through a 100 mesh sieve, and stored in bins or on the ground. Similarly good quality coal or coke is also ground to 30/56 mesh and stored. Fine grinding of coal/coke should be avoided. Barytes used for production of barium salts should not be less than 94% $BaSO_4$, and free from silica, iron, lead etc. If a Ball mill is used for the fine grinding of jaw crushed barytes, about 90% of the silica in barytes gets separated in the Ball mill, without undergoing any grinding.

Barytes used for manufacture of barium salts should be as under; for best results :

$BaSO_4$	-	94% mini.
SiO_2	-	1.0% Mx.
Fe_2O_3 & Al_2O_3	-	1.2%Max.

Ca and strontium should be nil or in traces only. Indian barytes from areas in and around Cuddappa in A.P. has been found to have traces of Ca and Sr.

Unfortunately, most of the manufacturers of barium salts do not pay any serious attention to barytes quality. The final yield of saleable barium salts like $BaCO_3$ and chloride is directly related to % $BaSO_4$ in barytes. Sources of carbon are good quality coal (low ash - high carbon) petroleum coke,

coke, coal fines, coke fines, lignite, lecofines, petcoke fines. Unfortunately Indian coal is bad due to its high ash and low carbon content. However, lignite either as lumps or as lecofines, is excellent for the reduction since its ash content is quite low and carbon percentage is high. Lecofines (waste from lignite lump users) of size 0-10 mm has an average carbon of 76/80% with ash around 8-10% only. Balance is water. Lecofines has a tendency to catch fire and burn, on open storage, and hence water is sprayed over the fines to keep it moist/wet. Lecofines, or char fines, as it is also known, is available in plenty from Neyveli Lignite Corporation, a public sector undertaking in the South. There is no need to grind the carbon for reduction when lecofines of site –2mm to +2mm is used. Almost all the barium units in the south India use this as source of carbon.

One can even blend coke fines, coal fines, and lecofines so as to get a source of carbon having atleast 70-75% carbon and maximum ash of 15%. Units located in Maharashtra often follow this blending method for sourcing their carbon. No doubt petroleum coke (or petcoke) is the best source for carbon since it has nil ash, and fixed carbon as high as 92–96%. Unfortunately the cost of this is so high that it is no longer an economic source for carbon. However if petcoke fines (–2mm to 2mm) are available from users of petcoke lumps, it is worth buying it and blend it with lecofines or coal powder/coke breeze/coke fines. Wood charcoal due to its spontaneous combustion characteristics, and low density, is not recommended as source of carbon even though it has fairly high carbon. However some manufacturers are still using it blended with lecofines or coke powder. Coconut shell carbon is very good for reduction, but its availability in large quantities is very limited. However, charcoal powder blended well with 94–96% barytes powder and pelletised, has been found to work fairly well. Data available on this is insufficient to conclude.

As per reduction batch size, measured barytes powder and source of carbon are blended in a mixer/blender very well, before shifting the mixture to the feed mixture storage hopper above reduction furnace. In actual practice 10–15% excess carbon is added to the mixture to take care of excess air leakage into reduction furnace. Mixing of barytes powder and source of carbon is very essential to have good reduction in furnace; intimate contact of particles of $BaSO_4$ and carbon facilitates the reduction process to the right direction.

Reduction process operation

Reduction furnace is heated to the required temperature inside (1100°C), by burning fuel oil, LSHS or coal powder. This normally takes 4 to 5 hours from cold if done properly, otherwise it may take 8–10 hours even. Once the temperature is reached, the furnace is rotated to bring the charging hole up. Same is opened, keeping the burner shut and the feed mixture is charged into the furnace; the chimney damper is kept fully open while charging mixture into R.F. Once the charging is over man hole is fixed back, chimney damper controlled and burner is lighted. Initially large volume of thick black smoke can be seen at top of chimney during the charging of feed to R.F. which after some time reduces to very light grey smoke, which is also an indicator for proper reduction.

Air supply to burner is kept controlled and the chimney damper should be adjusted and kept such, that there is a very slight back pressure at the burner end of the furnace, rather than heavy suction here. As the material undergoes reduction, keep drawing samples from the furnace roast at 15 minutes to half-an-hour intervals and analyse for BaS in it. The roast in the furnace should be watched often through the small opening at burner end to see and monitor the process of reduction. The mass normally start agglomerating after about 1–1/2 hours of reduction, which is a good indication of the reduction process and start sticking to furnace wall after around 2 hours to 2–1/2 hours of reduction. At this stage the burner is reduced considerably, the roast from the furnace is discharged into trolleys. Keep the chimney damper closed fully at the time of roast discharge from furnace. One can even observe puffs of blue flame here and there emanating from the roast discharged down. This is a good indication and good reduction. This is BaS in the roast getting spontaneously oxidised by the atmospheric oxygen at the temperature of the roast. The roast is called black ash and is mainly BaS—the % of BaS should analyze around 70% or more. In well operated controlled and maintained R.F., the batch reduction is completed in 2 1/2 – 3 hours after the charge into R.F. In some units the reduction has been completed in 1-3/4 hours also when operated at 1150 to 1200°C with proper quality and quantity of raw materials mix.

Many units operate the R. Furnace batches for 4 hours to 7 hours time cycle may be: (i) to have closer and better control over the reduction efficiency, (ii) to save on fuel oil used and thereby reduce fuel bill, and (iii) due to use of LSD, HSD or waste oil as fuel.

Such units due to their close proximity to source of raw materials use excess carbon than required for the reduction, and utilize this excess carbon to generate and maintain the required temperature and heat in the furnace. However the roast produced from these longer batches has been found to have more acid soluble barium with 50–52% only of BaS (W.S). Iron and alumina ($Fe_2O_3Al_2O_3$) SiO_2 etc. in the raw materials will reduce the water soluble BaS in the roast, as they combine with the BaS as and when formed and form compounds like silicates, aluminates, which finally leads to generation of more solid waste during the processing of the roast for producing $BaCO_3$, barium nitrate etc. from BaS. If the roast is not discharged after completion of the reduction process: (i) it will start fusing and the fused mass will get stuck to furnace lining, (ii) hard lumps will be formed instead of granular mass, and (iii) roast discharge will not be possible due to fused mass totally sticking inside the furnace.

At this stage removal of the roast is very difficult leading to a major furnace shut down. The efficiency or the extent of the conversion of $BaSO_4$ to BaS is dependent upon, as is explained earlier, on many factors. Theoretically 950 BTU per lb of 100% BaS is required as heat. In actual manufacture more heat supply is required for the reduction.

With quality barytes and sources of quality carbon, and a properly operated reduction process, it is possible to achieve production of black ash having 92-96% BaS or even higher as water solubles and 3-4% of acid soluble BaS, with unreduced barytes in the black ash either nil or around 1% maximum.

Conversion

The percentage of water soluble BaS in the black ash is termed generally as 'conversion'. To have a better and more clear understanding of this 'conversion' term, the total acid solubles in the Roast/Black-Ash should be analysed and estimated for every reduction batch, estimated generally as acid soluble BaS, instead of the water soluble BaS alone. This total acid soluble BaS value of the black-ash is the real or actual conversion because this gives the total saleable products from the roast or barytes used in the batch. One can determine the water soluble BaS, followed by determination of acid soluble BaS in the residue left after estimating the W.S. BaS. The total of the two, i.e. W.S. BaS + A.S. BaS, should be taken as the real conversion or total conversion. In actual industrial production this should never be below 86 per cent taking into consideration, all factors affecting production, and efficiency like R.M., poor supervision, etc.

Acidimetry method is widely used abroad and by some Indian manufacturers for estimation of W.S. BaS in roast, with Std. HC1 and MeO as indicator. No doubt one can use the costly iodine method also. However acidimetry is more quick, simple, less costly, and is very reliable, Higher acid solubles in B. ash indicates: (i) R.M. has more $A1_2O_3$, SiO_2 etc. or (ii) black ash is allowed to get oxidised due to not being discharged at the right time, and allowing excess air into the furnace.

BARIUM COMPOUNDS

Barium Carbonate

This is one of the most important compounds of barium, and finds wide application and use in many fields, like electronics, ceramics, caustic soda, optical and heat resistance glass, colour T.V. picture tubes, C.I. welding electrodes, permanent magnets etc. This is manufactured by the following two methods:

(1) Soda ash process

$$BaS + Na_2CO_3 \rightarrow BaCO_3 + Na_2S$$

Almost all the Indian producers follow this route.

(2) Carbon dioxide process

$$BaS + H_2O + CO_2 \rightarrow BaCO_3 + H_2S$$

Only one Indian manufacturer is following this process route. CO_2, available from the reduction furnace waste gases, is utilised. In the soda ash process calculated quantity of light soda ash is added to BaS solutions in an agitated vessel, and the white, heavy precipitated $BaCO_3$ is allowed to settle down (Fig. 16.2). Top clear layer of sodium sulphide solution is siphoned for storage or further processing. White suspension of $BaCO_3$ is water washed 2-3 times, and further filtered through Nutsche filter or filter press, washed fully free of any sulphide and taken for drying, pulverising and packing. A final rinse of the wet $BaCO_3$ with H_2O_2 and a water-wash after will make it snow white and 99.9% pure. The type of dryer is decided based on capacity; upto 4 tpd, tray driers are O.K. Spray or spin flash dryers are very costly and are suitable for capacities above 10 tpd. For capacities of 4 tpd to 6 tpd, the agitated vacuum dryer, drum dryer or even rotary drier is preferred.

Dilute solutions of Na_2S from wet carbonate washes, of 1.0 to 2.0% Na_2S, are concentrated through waste heat recovery unit of reduction furnace. Those sulphide washings of below 1 % is mixed with make-up process water. Agitated nutsche or washable type P & F filter press are used for $BaCO_3$ washing generally. If the wet $BaCO_3$ has greyish or bluish or bluish-grey colour, then it is not fully washed free of Na_2S. Following points should be strictly followed in the CO_2 process to get quality $BaCO_3$.

(1) BaS solution should be freshly made and free from suspended matter and polysulphides

(2) It should be of concn. around 10 to 12% only, and at 90°C preferably.

(3) CO_2% in compressed off gas should be above 28% with nil SO_2, O_2 tarry matter and suspended particles.

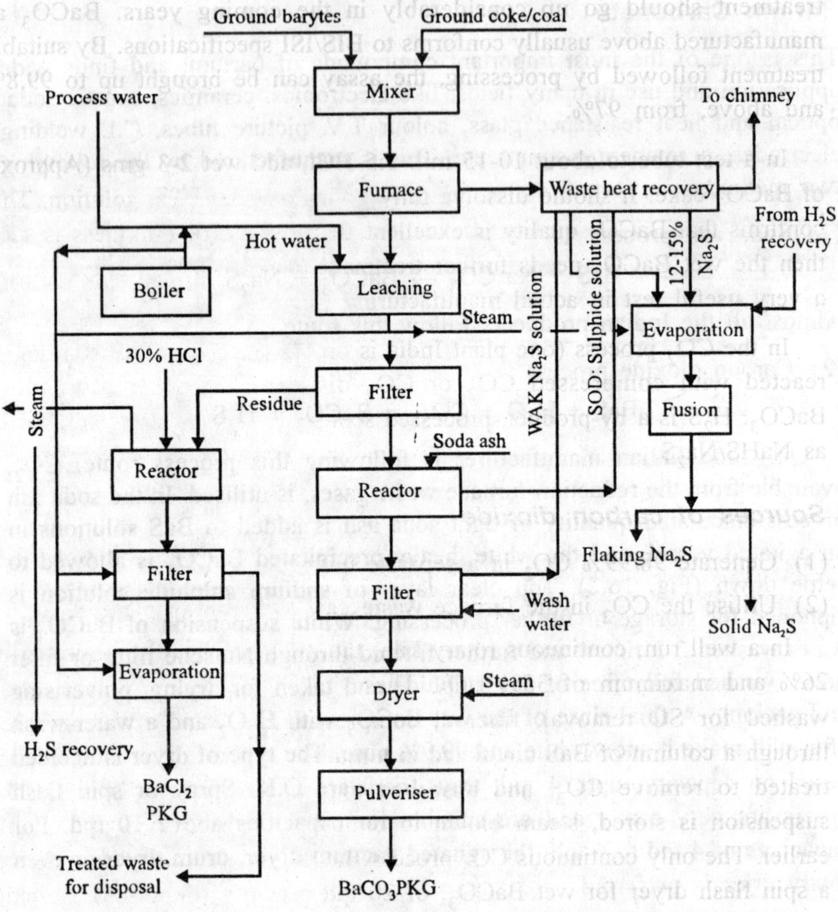

Fig. 16.2. Flow diagram for manufacture of barium carbonate by batch furnace soda ash route.

Barium carbonate is non-combustible, non-explosive, but highly toxic by ingestion. It should not be taken through mouth under any circumstances. Tolerance limit in air is 0.5 mg per cubic metre. Avoid exposing open wounds in the body to $BaCO_3$. Lethan dose for human is 20 mg/kg. Continuous inhalation of fines of $BaCO_3$ for long periods have been found to lead to cramps and bronchial problems. Do not store it with food grains, eatables, fruits. In India, the caustic soda industry is a major consumer of $BaCO_3$ for the removal of sulphates from brine, consuming around 30–40 kg. of $BaCO_3$ per 1000 kgs. of caustic soda. With the increasing switch over to membrane cell processes the consumption of $BaCO_3$ for brine treatment should go up considerably in the coming years. $BaCO_3$ as manufactured above usually conforms to BIS/ISI specifications. By suitable treatment followed by processing, the assay can be brought up to 99.8% and above, from 97%.

In a test tube to about 10-15 mil. 1:5 HCl, add wet 2-3 gms (Approx.) of $BaCO_3$ cake. It should dissolve fully giving a water clear solution. This confirms that $BaCO_3$ quality is excellent. If turbidity, or milkiness is seen then the wet $BaCO_3$ needs further treatment to reach 99% purity. This is a very useful test in actual manufacturing process.

In the CO_2 process (one plant India is on this process), BaS solution is reacted with compressed CO_2 or CO_2 dissolved in water to produce $BaCO_3$. H_2S is a by-product, processed separately to recover as sulphur or as $NaHS/Na_2S$.

Sources of carbon dioxide

(1) Generate 98/99% CO_2 in a separate unit; and

(2) Utilise the CO_2 in the furnace waste gas, available in plenty.

In a well run, continuous rotary kiln, the off gases have a minimum of 26% and maximum of 35% CO_2. These gases are water washed, soda washed for SO removal, filtered, compressed to 70 psig and bubbled through a column of BaS circulated in a reaction tower. By-product H_2S is treated to remove CO_2, and absorbed to produce NaHS. The $BaCO_3$, suspension is stored, steam blown to remove H_2S and dried as detailed earlier. The only continuous CO_2 process $BaCO_3$ manufacturer in India has a spin flash dryer for wet $BaCO_3$, of 20 tpd capacity (imported). Sulphur in $BaCO_3$ can be removed by chemical treatment and by drying at high temperatures.

Solid waste management

During the manufacture of $BaCO_3$ from mineral barytes, as explained in detail earlier, solid waste is generated at the rate of 1000 kgs. per 1000 kgs. of carbonate, at the minimum level. Solid waste generation is high in poorly operated plants with poor quality barytes and source of carbon.

After the extraction of BaS by hot water from the black ash, a residue is generated, often called leached mud or residue, which should not be disposed off as solid waste without proper treatment. A typical analysis of this solid waste is as under :

Total acid solubles	= 20-24% (Avg.)
Unconverted barytes	= 22%
Unused coal/coke	= 2.0%
Others as SiO_2, Al_2O_3 etc.	= 38% (avg.)
Moisture	= Balance

By acid treating the same, the harmful barium can be removed and recovered as saleable barium salt, and the treated solid can be disposed off as filling earth or for making red bricks etc. It is easier to acid treat when the solid waste is fresh. If allowed to accumulate in open ground for long period, the reactivity of the waste to acid treatment is seen to be very poor, needing long hours. Generally, the waste is treated with hydrochloric acid, and the acid extracted barium is recovered as barium chloride. Acid treated solid waste is harmless.

Barium Chloride

It is manufactured from $BaCO_3$, BaS and recovered from solid waste of $BaCO_3$ manufacture. Manufacture from $BaCO_3$ is not economical, and the most economical route is recovery of barium of $BaCl_2$ from solid waste generated during manufacture of $BaCO_3$. But here the capacity is very low, and can be supplemented by manufacture through black ash (Fig. 16.3).

Manufacture from black ash

Black ash when reacted with hydrochloric acid, produces H_2S and $BaCl_2$. The reaction is carried out in acid-proof M.S. reaction vessels and the H_2S is absorbed in caustic producing NaHS or Na_2S. It is preferable to add slurry of black ash to HC1 30% kept in the reaction vessel (preferably closed vessel), thereby generating the H_2S under slight pressure, and utilise this pressure for its easy absorption in 30-40% caustic lye. Reaction control is easy here, and the clear supernatent solution of barium chloride is taken for filtration, if required, or to storage 18-22% solution so obtained is evaporated in suitable evaporator, and crystals centrifuged to get $BaCl_2.2H_2O$. The residue of solids in the reactor is stirred with hot water and dropped to a nutsche filter, to extract the remaining chloride from the residue. This is mixed with centrifuge mother liquor and fed to evaporator.

The residue can now be disposed off safely. pH adjustment of the chloride solution before evaporation is a must to get good quality

$BaCl_2.2H_2O$ and to reduce its corrosion. Steam consumption for evaporation is around 6 kgs/per kg. of chloride for open vessel evaporation, and only 3 kg. of vacuum evaporation.

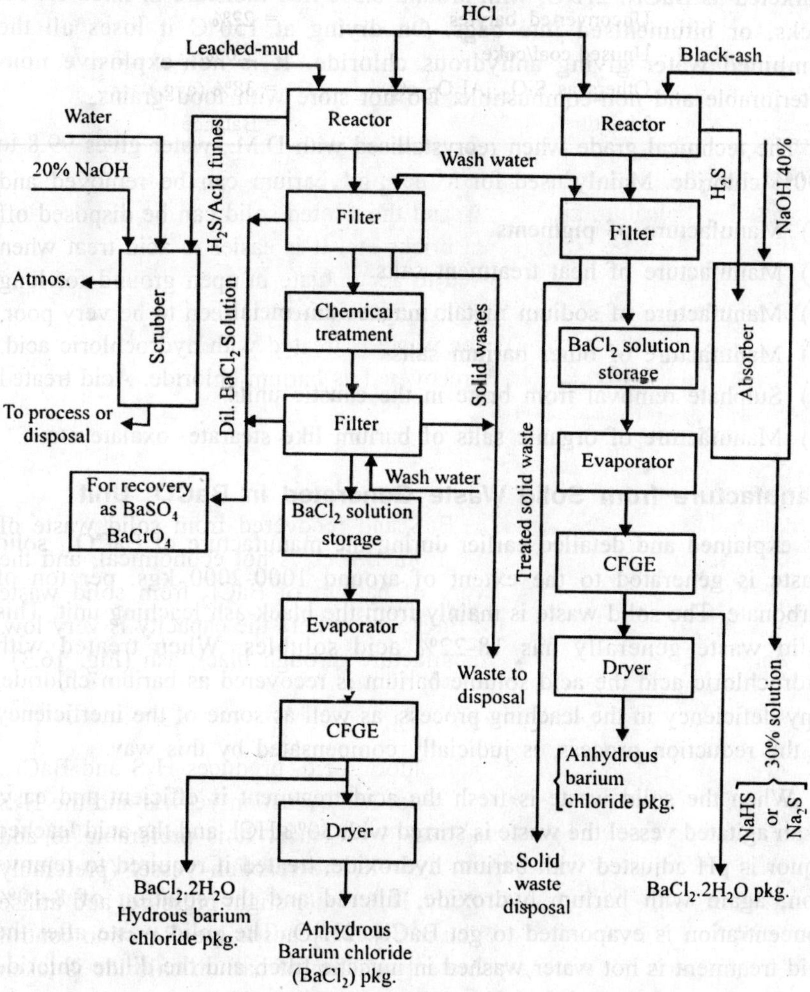

Fig. 16.3. Flow diagram for manufacture of barium chloride.

Safety, quality and packing

Barium chloride is highly toxic by ingestion. A 0.8 gm can prove to be fatal. If taken through mouth, convert it to harmless $BaSO_4$ by the intake of sodium sulphate solution. Avoid exposing open wounds to it. Use hand gloves, face shields while handling it. Wash eyes thoroughly with water in case of accidental fall into eyes. The chloride manufactured as above generally conforms to BIS/ISI specification. The hydrated white material is marketed as $BaCl_2 . 2H_2O$, with around 0.5% free moisture in lined HDPE sacks, or bitumenised jute bags. On drying at 150°C it loses all the combined water giving anhydrous chloride. It is non-explosive non-deteriorable and non-combustible. Do not store with food grains.

The technical grade when recrystallised with D.M., water gives 99.8 to 100% chloride. Mainly used for :

(1) Manufacture of pigments.

(2) Manufacture of heat treatment salts.

(3) Manufacture of sodium metal, magnesium metal.

(4) Manufacture of other barium salts.

(5) Sulphate removal from brine in the caustic units.

(6) Manufacture of organic salts of barium like stearate, oxalate etc.

Manufacture from Solid Waste Generated in $BaCO_3$ Unit

As explained and detailed earlier during the manufacture of $BaCO_3$, solid waste is generated to the extent of around 1000-2000 kgs. per ton of carbonate. The solid waste is mainly from the black-ash leaching unit. This solid waste generally has 18-22% acid solubles. When treated with hydrochloric acid the acid soluble barium is recovered as barium chloride. Any deficiency in the leaching process, as well as some of the inefficiency in the reduction process, is judicially compensated by this way.

When the solid waste is fresh the acid treatment is efficient and easy. In an agitated vessel the waste is stirred with 30% HCl, and the acid leached liquor is pH adjusted with barium hydroxide, treated if required to remove iron, again with barium hydroxide, filtered and the solution of 8-10% concentration is evaporated to get $BaCl_2 . 2H_2O$. The solid waste after the acid treatment is hot water washed in nutsche filter, and the dilute chloride solution is used as under:

(1) Use it in place of water for washing the residue after the acid leaching process.

(2) Produce barium chromate from it or barium sulphate (blanc fixe).

Evaporation of chloride solutions, for low output of chloride like 500 to 700 kgs/day. This can be done in an appropriate size conveniently in open

M.S. trays (of size 3 × 6 × 1 or 4 × 8 × 1 1/2) kept over fire. The pH of the solution should be between 7.0 and 8.0. Manually dig out crystals from the tray and centrifuge it.

By-product recovery

Sodium sulphide is the by-product of $BaCO_3$ manufactured by the soda ash process. Theoretically for every 1000 kgs. of 98% $BaCO_3$ produced 646 kgs. of 60% sodium sulphide Na_2S is generated as by-product. At an achievable 90% yield of recovery efficiency level one should get 580 kgs. of 60% Na_2S. Sodium sulphide is generated as 7 to 12% solution directly and as 3-4% and 1 to 2% solution. For every 1000 kg. $BaCO_3$ output there is a generation of approx. 8,000 lits. of Na_2S solution. 2-4% Na_2S solutions are evaporated to 10% via waste heat and the balance via open type fired vessels. Some time sulphide solutions of upto 2% are recirculated through make-up water required of black ash leaching. Based on observed characteristics of the material and its temperature, 58 to 60% Na_2S is discharged into drums or taken through drum flaker, or as pellets through a conveyor. Generally the solutions, are all handled in M.S. vessels. Though sulphide is very corrosive to M.S. cast iron with about 4% nickel or 6% chromium, has good resistance to Na_2S, even at higher temperatures. Nickel is the best material of construction but is very costly. The material red brick coloured or red orange coloured, due to iron in sulphide, starts frothing up as the concentration in fusion vessel reaches around 56% and above. The temperature at this stage will be around 160-170°C. Sodium sulphide solution should be free from BaS and/or Na_2CO_3 before evaporation/fusion. Otherwise processing of sulphide solution will give many problems during evaporation and fusion. If you have a nearby or next door consumer accepting a solution of sulphide upto 25% concentration you can avoid the fusion operations and its associated headaches. Addition of sodium cyanide, glycerine can reduce corrosion due to sodium sulphide. Sodium sulphide is highly reactive and corrosive, hence packed drums should be sealed. Flakes can also be packed in HDPE bags with inside liner. It should not be stored near oxidising agents like H_2O_2, dichromate, chlorate etc.

Safety precautions

Both solid and solutions are very hazardous, can cause burns which takes very long to heal. Eyes should be well protected by goggles and face by face shield while handling hot or cold sulphide solutions. Nails dissolve in sulphide, and it is a powerful depilatory. Never expose naked eye to solutions or flakes of Na_2S. Those units located far away from sources of barytes and carbon, should be able to compete with those units who enjoy locational advantage to raw materials, by recovering as much chloride as possible from solid waste treatment together with the by-product Na_2S; and also by recovering as $BaSO_4$, or $BaCrO_4$ from the very low concentration solutions of barium chloride available from the process. Iron in sodium sulphide solution can be removed partly by sodium cyanide, and by crystallising the solution to get almost iron free sulphide crystals, $Na_2S.9H_2O$.

Wt. % anhydrous Na_2S	*Density of solution* 30°C	40°C
10%	1.10	1.095
14%	1.138	1.135
16%	1.159	1.155
18%	1.790	1.756
20%	—	1.196
22%	—	1.216

In India the caustic soda industry is a major consumer of carbonate for brine treatment. With the membrane cell process for caustic alkali gaining popularity and importance, due to many factors, the demand for barium carbonate in the caustic alkali industry is bound to go up.

Properties of barium chloride

It is one of the most important inorganic chemical and is colourless in the form of flat crystals. Soluble in water and insoluble in alcohol. Sp. gr. 3.097; m.p. 960°C (anhydrous). Combustible. It is used in artificial barium sulphate, other barium salts. Reagents: lube oil additives, boiler compounds, textile dyeing, pigments, manufacture of white leather. It is toxic by ingestion, 0.8 gm may be fatal.

Barium Nitrate

$Ba(NO_3)_2$ is manufactured as under:

(1) From $BaCO_3$, by reaction with nitric acid as :

$$BaCO_3 + 2HNO_3 \longrightarrow Ba(NO_3)_2 + H_2O + CO_2$$

(2) From 100% BaS by reaction with nitric acid as :

$$BaS + 2HNO_3 \longrightarrow Ba(NO_3)_2 + H_2S$$

(3) From barium chloride by reaction with sodium nitrate :

$$BaCl_2 + 2NaNO_3 \longrightarrow 2NaCl + Ba(NO_3)_2$$

Commercially most of the manufacturers in India (excepting one manufacturer), follow the nitric acid neutralisation with $BaCO_3$ due to its simplicity, easy R.M. availability, low capital deployment etc. (Fig. 16.4). In an agitated reactor in S.S. (304/321/316) $BaCO_3$ is slowly added to 50% nitric acid, acid fumes being properly vented, via a scrubber. The 5 to 6 pH solution is settled or filtered to remove suspended matter and transferred to open crystallising tanks, for slow crystallisation. Crystals are centrifuged through an S.S. centrifuge. Mother liquor is taken for evaporation. White crystals are dried at 100°C in tray drier for capacity up to 3 tpd and pulverised to 300 mesh before packing.

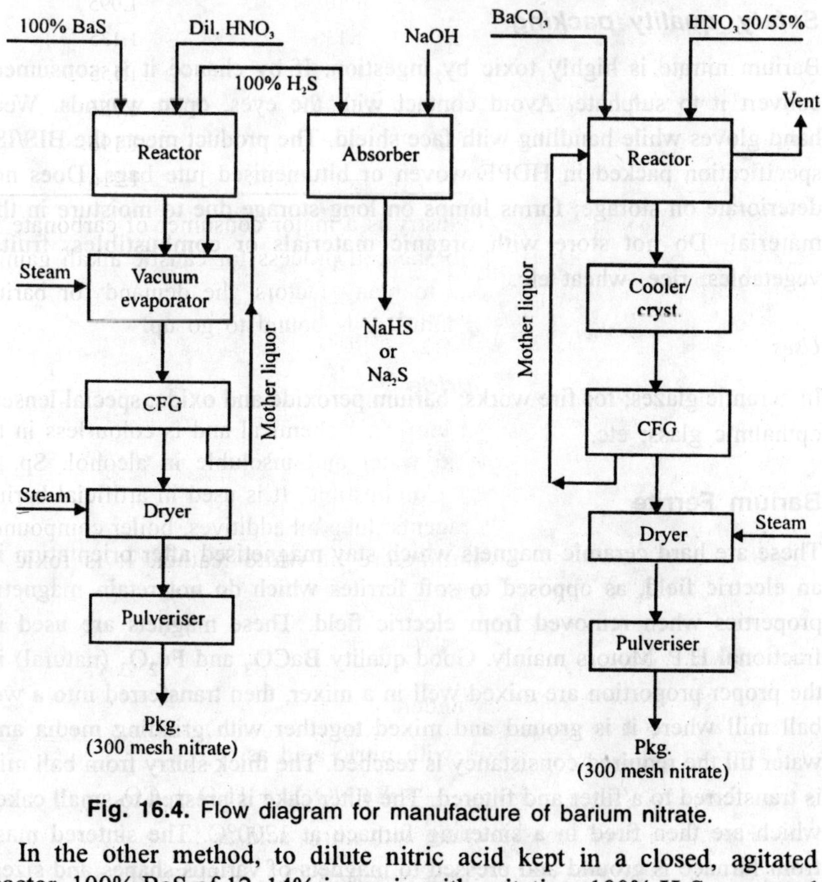

Fig. 16.4. Flow diagram for manufacture of barium nitrate.

In the other method, to dilute nitric acid kept in a closed, agitated reactor, 100% BaS of 12–14% is run in with agitation. 100% H_2S generated is absorbed in 40% alkali. After the reaction is over, as confirmed by its pH, the nitrate solution is transferred to settling tanks for suspended matter to settle down. Its pH is adjusted to around 4 to 5 and is fed to an S.S. evaporator (open or vacuum type) to get crystals of barium nitrate which are centrifuged, dried and pulverised before packing. For capacities above 3 tpd, vacuum evaporator is the most economical one; steam consumption is around 7 to 8 kg. per kg. Of nitrate for open type evaporation and 3.5 to 4.5 for vacuum evaporation. Best quality nitrate is produced by this route. Commercially there is only one manufacturer following this route. The 3rd method of production is economical only if cheap sodium nitrate is available along with cheap steam.

Safety, quality packing

Barium nitrate is highly toxic by ingestion. If by chance it is consúmed, convert it to sulphate. Avoid contact with the eyes, open wounds. Wear hand gloves while handling with face shield. The product meets the BIS/ISI specification packed in HDPE woven or bitumenised jute bags. Does not deteriorate on storage; forms lumps on long storage due to moisture in the material. Do not store with organic materials or combustibles, fruits, vegetables, rice, wheat etc.

Uses

In ceramic glazes; for fire works; barium peroxide and oxide; special lenses, opthalmic glass; etc.

Barium Ferrite

These are hard ceramic magnets which stay magnetised after orientation in an electric field, as opposed to soft ferrites which do not retain magnetic properties when removed from electric field. These magnets are used in fractional H.P. Motors mainly. Good quality $BaCO_3$ and Fe_2O_3 (natural) in the proper proportion are mixed well in a mixer, then transferred into a wet ball mill where it is ground and mixed together with grinding media and water till the required consistancy is reached. The thick slurry from ball mill is transferred to a filter and filtered. The filter cake is pressed to small cakes which are then fired in a sintering furnace at 1200°C. The sintered mass from furnace is ground and pressed to magnets of various shapes and sizes. There are number of manufacturers in India, manufacturing barium ferrites.

Blanc Fixe or Precipitated Barium Sulphate

Precipitated barium sulphate or blanc fixe is another important compound of barium. The only non-toxic compound of barium, it is used in X-ray photography, in opaque medium for gastro intestinal radiography, paper coatings, as filler in plastics, rubber, linoleum, as fire retardant in paint base, for lake colour, lithographic inks, etc. It is manufactured by the following routes:

From fresh BaS solution and iron free good quality sodium sulphate (preferably from rayon process)

$$BaS + Na_2SO_4 \longrightarrow BaSO_4 + Na_2S$$

In agitated tiled or rubber lined M.S. tank, both streams are run simultaneously into water kept at 40–50°C. Alternatively, solid, free flowing Na_2SO_4 is added to BaS solution of concentration 12-16%; precipitated $BaSO_4$ settles down. Na_2S solution from upper layer is removed for concentration, the sulphate is washed twice or thrice with hot water at 50°C. Blanc fixe is now filtered in a washable type F. press and washed free of Na_2S. A wash with 5% H_2SO_4, followed by 5% H_2O_2 followed by thorough water wash till washings are free from Na_2SO_4 and acid is also given. Cake is now dried in tray drier at 120°C on Al or PP trays, pulverised to 300 mesh and packed.

From BaS liquor and 70% sulphuric acid

$$BaS + H_2SO_4 \rightarrow BaSO_4 + H_2S$$

Maintain a pH of 2.5 to 3.0 towards the end. Maintain reaction temperature 50–70°C. After the reaction, drive away the H_2S byline steam. Product is of good quality, free from sulphur and is difficult to filter if pH is not kept at 2.5 to 3.0. For pH adjustments use quality $Ba(OH)_2$.

From $BaCl_2$ solution and sulphuric acid (90-98%)

Both liquids are run into good quality water, in an acid-proof tank at 80–100°C temperature. Course blanc fixe is formed here, washed free of HCl, dried, and pulverised to 300 mesh. Fine grains of blanc fixe is produced by rapid precipitation from concentrated solution at high pH and low temperature. Stirring the precipitated mass for a long period at low pH produces large grain size of blanc fixe. The temperature and pH are important factors to get proper size grains of blanc fixe.

From sulphur free $BaCO_3$ and H_2SO_4 good quality blanc fixe is produced. Commercially this route is not exploited, due to poor economics.

For photographic papers blanc fixe of following specification is used :

Moisture	40/45%
pH of paste	6.5 to 7.5
Dry brightness	98% min.
Fe	6 ppm. max.
S	0.3 ppm. max.
Carbonate	0.2% max.
Total SO_4 over $BaSO_4$	0.1 to 0.40

Blanc fixe can be dried in spray drier also. Blanc fixe is blended with other white pigments, e.g. 30% TiO_2 with 70% of blanc fixe; 29% ZnS with 71% blanc fixe, 15% TiO_2 with 25% ZnS and 60% blanc fixe. Blanc fixe used for, oil paints, should have the following specifications :

Grain size	0.10 to 20/μm
Average grain	0.2 to 2.0/μm
Brightness (ref. MgO 100)	95-99
Oil consumption	12-22 gm. oil per 100 gm.
Tapped bulk dye	0.9 to 2.2 gm/cm^3
Density	4.0 to 4.45 gm/cm^3

Lithophone — a well known white pigment — is a mixture of ZnS and blanc fixe, (29 to 35% ZnS and balance blanc fixe). This is manufactured from freshly prepared 10/12% BaS solution and iron free zinc sulphate solution of 12–15%. Metered quantities of both are run unto an agitated vessel of PP/R. Lined/PVC-FRP, keeping the temperature of mass at 70–80°C. After the reaction is completed, precipitated white solid mixture of ZnS and blanc fixe is filtered, washed and the wet cake is calcined out of contact with air to about 900–1000°C, for a short period. The mass is then quickly quenched in cold, good quality water, and again filtered. Wet cake is then dried and pulverised before packing.

Barium Hydroxide

As monohydrate $Ba(OH)_2.H_2O$ barium hydroxide is a very important compound/salt of barium. It is a white powder, highly alkaline and toxic. It is extensively used for manufacture of oil and grease additives, refining beet sugar, as alkalising agent in water softening, boiler scale softener, catalyst in the manufacture of phenol formaldehyde resins, carbonising agent, for steel etc. This is made from the penta/octahydrate by controlled drying under vacuum at low temperature of 70-75°C. The octahydrate $Ba(OH)_2$-$8H_2O$ is made by the following process routes.

From barium chloride by reaction with caustic alkali

$$BaCl_2 + 2NaOH \longrightarrow Ba(OH)_2 + 2NaCl$$

45% NaOH, is run slowly into 15-18% solution of $BaCl_2$ till all the $BaCl_2$ is converted to hydroxide. Flakes of white $Ba(OH)_2$. $8H_2O$, separated is centrifuged and water washed to reduce sodium chloride. Washed, centrifuged material is dissolved in hot water and re-crystallised, to get 98% $Ba(OH)_2$ $8H_2O$. This should be dried under vacuum below or around 70°C temperature. It melts at 78°C, and loses thereafter all the water gradually. 98–99% $Ba(NO_3)_2$ is heated to 600°C, loses NO_2 to become BaO, which is dissolved in D.M. water and recrystallised. Gives very good quality hydroxide. This is not economical commercially.

From barium silicate, there is a commercial process 50 mesh 96% minimum barytes mixed with same size silica (iron free) in the ratio of 2.5:1, a binder like calcium lignosulphonate is added (0.5 to 1.0%) and mixed thoroughly with water in a pug mill. Wet mass is then pelletised, which are fed to a rotary furnace and calcined at 1200°C and above. Calcined mass is cooled, ground to 150 mesh, and sent for leaching with hot water at 85°C in a series of leaching tanks. From the last leaching tank, the slurry of barium silicate and 30% barium hydroxide concentration of leached liquor, filtered to remove the barium silicate from the barium hydroxide solution at 85°C. The clear hydroxide solution evaporated to 50% concentration, and cooled to get crystals of $Ba(OH)_2$. $8H_2O$. Residue from filter recycled with fresh batch to furnace $Ba(OH)_2$ is centrifuged and dried under vacuum around 70°C to get monohydrate. Melting should be prevented during drying as otherwise all the bound water will be lost leading to anhydrous hydroxide.

From specially formulated barium aluminate also, $Ba(OH)_2$ can be made similar to above.

Barium Stearate

This is a white solid product of barium used extensively in plastic processing to give heat and light stabilising property. It is a water-proofing agent, lubricant in metal working, etc. manufactured from sodium stearate and barium chloride.

Barium Chromate

This is also known as lemon chrome, baryta yellow, ultramarine yellow etc. This is toxic, used in safety matches, as yellow pigment in paints, metallic primers and for ignition control devices. Manufactured from sodium chromate and barium chloride, heavy suspension (yellow) formed is filtered, washed, dried and pulverised to 250/300 mesh before packing. It is widely used in marine paints. In PVC insulation tapes this is generally used, to provide colour and stability to PVC. Clear solutions of chromate of sodium and barium chloride are simultaneously run into an agitated tank PP, R. Lined, wooden—steam heated to about 80-90°C till precipitation is completed. Stop agitation, allow precipitate to settle down and siphon out the supernatant clear solution. Add hot water at 60-80°C to the yellow precipitate stir for some time and filter via. F. press or nutsche. dry wet cake at 100-120°C and pulverise to 300 mesh and packed.

Chapter 17

Fertiliser Industries

INTRODUCTION

Fertilisers are substances which are introduced into the soil to increase the yield of agricultural crops. With each harvest part of the nutrients is withdrawn from the soil, gradually exhausting it, and this results in reduced crops. To obtain stable harvests the losses must be offset by systematic-application of fertilisers. In this way not only can the crop be maintained at a fixed level, but it can be raised as well.

In addition to increasing the amount of agricultural produce, fertilisers also improve their quality. Thus, with the use of fertilisers the starch, sugar, protein and oil content of plants is raised, this in turn increases the content of fats and proteins in livestock products. The use of fertilisers increases the strength of such products as flax, cotton and wool.

Fertilisers increase the resistance of plants to diseases, drought, low temperatures and other conditions which hinder the development of the plant. The use of fertilisers results in a great increase in labour productivity.

To live and develop normally plants require many different substances. Of the 103 elements of the Mendeleyev periodic system about 60 are present in the compounds of which plants are made up. They must all be contained in the soil–in the upper fertile layer of the earth, and in the atmosphere surrounding the plants.

The principal role in plant life is played by carbon, hydrogen, oxygen (these are contained in the cellular tissue), nitrogen, potassium, phosphorus, magnesium, sulphur, calcium, and iron. These 10 elements account for over 99% of the plant weight. The other elements, although contained in plants in very small quantities, are of great importance in their vital processes.

The bulk of the carbon is received by the plants from the air in the carbon dioxide assimilated by photosynthesis, with a small amount of it supplied from the soil through the root system. Oxygen and hydrogen enter into the plant mainly from the soil as ingredients of water; the other elements are contained in water-soluble mineral salts or weak organic acids. Usually the soil quickly becomes impoverished in the water-soluble mineral salts of nitrogen, potassium and phosphorus, because agricultural crops consume relatively large amounts of them.

As for the remaining elements (about 50), their consumption by plants is comparatively small, but if there is a deficiency of any of them in the soil (especially of boron, manganese, cobalt, zinc, copper, etc.) normal development of the plant is disturbed. Therefore, where there is an insufficient amount of these elements in the soil they should also be introduced in small quantities as so-called *minor-element fertilisers.*

To make rational use of fertilisers, one must know the composition of the soil and the nutrient requirements of the individual agricultural crops. This makes it possible to produce the fertilisers into the soil in the amounts essential for normal development and life of the plants.

CLASSIFICATION OF FERTILISERS

Fertilisers are usually classified into two groups :

(1) Organic fertilisers (manure, peat, industrial wastes, green manure etc.)

(2) Mineral, or artificial fertilisers, i.e., specially manufactured products.

In this chapter the production of mineral fertilisers is discussed. With respect to their agricultural effects mineral fertilisers are divided into direct and indirect ones.

Direct fertilisers are substances containing elements essential to plant life in the form of their compounds.

Indirect fertilisers are substances introduced into the soil mainly with the purpose of improving its physical and biological properties. For instance, ground limestone and dolomite reduce the acidity of the soil, gypsum improves the properties of saline soils, etc.

With respect to the nutrients contained in them, mineral fertilisers are divided into *phosphorus, nitrogenous*, and *potassium fertilisers*. The *minor-element fertilisers* are usually treated as a separated group. Fertilisers containing only one nutrient are called *simple,* those with two or more nutrients in them are termed *multiple-element fertilisers.* If such fertilisers are produced by mixing simple (single) fertilisers, they are called *mixed fertilisers*. The most important mineral fertilisers used in agriculture are listed in Table 17.1.

Table 17.1. Mineral fertiliser used in agriculture.

Fertiliser	*Principal components*	*Content of nutrients, %*
Phosphorus fertilisers		*P_2O5*
Superphosphate	$CaH_4(PO_4)_2.H_2O+nCaSO_4.2H_2O$	16-20
Triple superphosphate	$CaH_4(PO_4)_2.H_2O$	38-50
Precipitate (dicalcium phosphate)	$CaHPO_4.2H_2O$	27-42
Phosphate meal	$3Ca_3(PO_4)_2.CaF_2$	19-25
Thomas slag	$4CaO.P_2O_5 + 5CaO.P_2O_5.SiO_2$	14-20
Phosphate meal	$Na_2O.4CaO.P_2O_5.SiO_2$	20–35
Nitrogenous fertilisers		*N*
Ammonium nitrate	NH_4NO_3	33-35
Ammonium sulphate	$(NH_4)_2SO_4$	20-21
Sodium nitrate	$NaNO_3$	16
Calcium nitrate	$Ca(NO_3)_2$	13-15
Calcium-ammonium nitrate	$NH_4NO_3 + Ca(NO_3)_2$	18-20
Carbamide (urea)	$CO(NH_2)_2$	42-46
Potassium fertilisers		*K_2O*
Ground sylvinite	KCl + NaCl	12-15
Potassium chloride	KCl	50-62
Potassium sulphate	K_2SO_4	48-52
Multiple-element fertilisers		
Ammophos	$NH_4H_2PO_4 + (NH_4)_2HPO_4$	11-14 N, 48-55 P_2O_5
Potassium nitrate	KNO_3	13.5 N, 46.5 K_2O
Nitrophosphate	$CaHPO_4.2H_2O + NH_3H_2PO_4 + + NH_4NO_3 + KCl$ (there are three-element fertilisers of other compositions)	12–20 N, 10-14 P_2O_5 10-21 K_2O

Phosphorus Fertilisers

The raw materials for the manufacture of phosphorus fertilisers are naturally occurring minerals—phosphate rock and apatite—whose basic component is tricalcium phosphate, $Ca_3(PO_3)_2$, which is insoluble in water and in weak organic acids. In addition to tricalcium phosphate, phosphate rock and apatite contain calcium fluoride, CaF_2. Apatite containing CaF_2—$3Ca_3(PO_4)_2.CaF_2$, or $Ca_5F(PO_4)_3$—is called fluorapatite.

Phosphate rock occurs either in the form of nodules of various shapes and colours (nodular deposits), or in seams. Most of the phosphate rock deposits are of sedimentary origin. They were formed from animal or plant organisms rich in phosphorus compounds and are deposited in seams. Phosphate rock contains various impurities which reduce the per cent content of the useful ingredient, tricalcium phosphate, in them (it is usually given in terms of per cent of P_2O_5). Phosphate rock is often concentrated to remove the impurities and increase the P_2O_5 content. The P_2O_5 content of concentrated phosphate rock ranges from 20 to 30%.

Apatite ores are predominantly of igneous (volcanic) origin and made up of large crystals. In the deposit of apatite occurs together with nepheline, $(Na \text{ or } K)_2 \cdot Al_2O_3.2SiO_2.2H_2O$, and other valuable minerals. The, apatite-nepheline rock is separated at an ore-beneficiation factory into an apatite concentrate containing 39 to 41% P_2O_5 and a nepheline fraction, used as a raw material from which aluminium, soda products and cement are made.

Plants can only assimilate the phosphorus in compounds which are soluble in water or weak organic acids. Tricalcium phosphate, $Ca_3(PO_4)_2$, contained in phosphate rock and apatite, is insoluble both in water and in weak organic acids. Since plants cannot assimilate the phosphorus in such a compound it must be converted into other compounds available to the plants: such compounds as, for example, acid calcium salts of phosphoric acid—monocalcium phosphate, $CaH_4(PO_4)_2$, which is readily soluble in water, or dicalcium phosphate, $CaHPO_4$, which is insoluble in water but soluble in weak organic acids (for instance, citric acid). The P_2O_5 in dicalcium phosphate, $CaHPO_4$, which is insoluble in water but soluble in weak organic acids (for instance, citric acid). The P_2O_5 in dicalcium phosphate, $CaHPO_4$, is said to be in an available citrate-soluble form. Water-soluble compounds of phosphorus are more rapidly assimilated by plants than citrate-soluble ones. However, in sandy soils, through which rain water easily penetrates, citrate-soluble phosphorus compounds are not quickly washed away and will be more fully, although slowly, utilised by the plants.

For highly acid soils it is advisable to use ground phosphate rock as the fertiliser. In this case the conversion of tricalcium phosphate to compounds available to the plants is by the action of the acids contained in the soil. Simultaneously, the acidity of the soil will be reduced and the physical and chemical conditions for plant growth will improve. Animal bones contain phosphorus as tricalcium phosphate, and therefore, after the fat and glue have been removed from them by extraction, they can be used in the form of bone meal in the same way as phosphate rock and apatite meals.

Naturally occurring phosphate rock and apatites are processed industrially on a large scale into phosphorus fertilisers, i.e., into phosphorus compounds available to plants. Naturally occurring phosphates are also used for producing elemental phosphorus and phosphoric acid—semi-products from which concentrated phosphorus fertilisers are made. In short, the main purpose in the technology of phosphorous fertilisers is to convert insoluble natural compounds of phosphorus into soluble ones by decomposing phosphate rock or apatite with acids, alkalies, or in other ways.

Production of superphosphate

Superphosphate is the product of decomposition of phosphate rock or apatite by sulphuric acid. It may be in the form of a grey powder or of granules consisting practically of the water-soluble phosphorus compound—the monohydrate of monocalcium phosphate, $CaH_4(PO_4)_2.H_2O$, and the low-solubility calcium sulphate dihydrate, $CaSO_4.2H_2O$—gypsum. Superphosphate is a phosphorus fertiliser with a low P_2O_5 content (maximum 20 to 22%). The gypsum ($CaSO_4.2H_2O$) in the superphosphate has no nutrient value for plants.

To make superphosphate, ground phosphate rock or apatite and sulphuric acid are mixed in a definite ratio, and the mixture obtained, in the form of pulp, is poured into a chamber, or den. In this den the mixture gradually solidifies and matures, i.e., the phosphorite and the sulphuric acid react. Then the superphosphate is cut into flakes and sent to the warehouse. The reaction between the phosphate rock and the sulphuric acid is described by the following equation :

$$Ca_3(PO_4)_2+2H_2SO_4+5H_2O \rightarrow CaH_4(PO_4)_2.H_2O+2(CaSO_4.2H_2O)$$

As can be seen, water is also a participant in the reaction of superphosphate formation. Therefore dilute sulphuric acid, of 68% concentration, is used in making superphosphate. One molecule of the water in the acid combines with one molecule of the monocalcium phosphate, four others with calcium sulphate to form two molecules of gypsum, $CaSO_4.\times 2H_2O$; this promotes setting of the pulp and solidification of the superphosphate formed in the den. The reaction of apatite with sulphuric acid to yield superphosphate can be represented in this way :

$$3Ca_3(PO_4)_2.CaF_2+7H_2SO_4+17H_2O\rightarrow$$
$$\rightarrow 3CaH_4(PO_4)_2.H_2O+7(CaSO_4.2H_2O)+2HF$$

This process of superphosphate formation is a complicated one; it involves chemical reactions, diffusion of the acid into the particles of the phosphate raw material and crystallisation of monocalcium phosphate. All these individual processes are affected by the temperature and the concentration of the reacting substances. A temperature of about 110°C is maintained in the den (by the heat released in the decomposition reaction). Curing of the superphosphate does not terminate in the den, but continues during storage.

The oxides Al_2O_3 and Fe_2O_3 in the phosphate raw material are harmful impurities. In the process of treating the phosphate with sulphuric acid the sesquioxides form aluminium sulphate and iron sulphate. During storage of the superphosphate these compounds react with the monocalcium phosphate to convert the phosphorus into compounds which cannot be assimilated by plants. This very undesirable process, which lowers the quality of the superphosphate, is called *retrogradation*. The principal stages in manufacturing superphosphate are: proportioning of ground phosphate rock or apatite and sulphuric acid; obtaining a pulp by agitating the ground phosphate with sulphuric acid in a mixer; setting, solidification and maturing of the mix in the den; final curing of the superphosphate formed in the storage premises.

A schematic diagram of the process of *production of superphosphate* is given in Fig. 17.1. Sulphuric acid from a Glover tower (76% concentration) is continuously supplied to proportioner 1, where it is diluted with water to a concentration of 68%, then it is fed into mixer 3. Ground phosphate rock or apatite from the warehouse is measured out with the aid of automatic scale 2. The ground phosphate is mixed with the sulphuric acid in continuous-action mixer 3. Ordinarily, use is made of screw-type mixers or chamber-type mixers, in which mixing is by means of high-speed agitators. The pulp obtained in the mixer is fed to continuous-action maturing den 4, where it sets (solidifies), matures, and then is cut up and sent to the storage room.

Formerly, setting and maturing of superphosphate were carried out in batch-type dens. After such a den was filled with the mix and when the superphosphate had solidified, the den was opened and the superphosphate was cut up. After then den was freed from the superphosphate it was refilled with the mixture, etc. Batch-type dens have serious disadvantages: their capacity is small, the product obtained is not of uniform composition and properties, the dens are not hermetically sealed and therefore the plant atmosphere may contain noxious gases and dust. Such dens are still in operation, but most newly built and reconstructed plants are equipped with improved, continuous-action, dens.

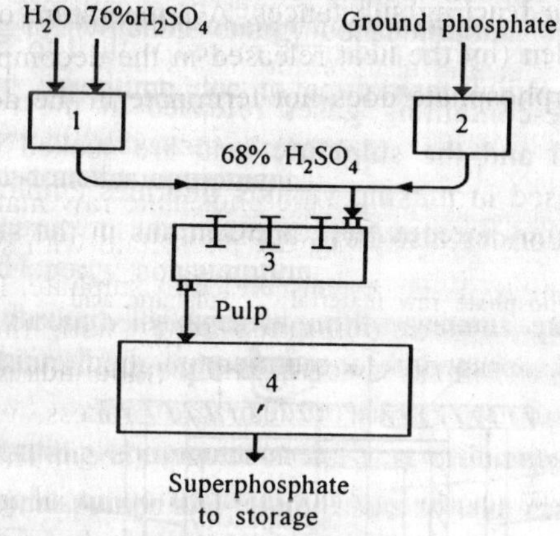

Fig. 17.1. Production of superphosphate.

Fig. 17.2., is a diagram of a continuous-action den; this is a vertical reinforced-concrete cylinder, 1, and a bottom, 2, which rotates slowly (one revolution per 1.5–2.5 hour) on rollers 5. The den has a stationary reinforced concrete lid, 8, resting on supports 6. Inside the den there is a cast-iron tube 4 coaxial with the cylinder and attached to the fixed lid; the tube passes through a packed gland in the bottom of the den; it is used for discharging the superphosphate from the den. Partition 11 is also rigidly fixed to and "suspended" from the den lid. There is a sliding fit between the edges of the partition and the bottom and wall of the den. The partition separates the zone into which the mix is charged from the zone out of which the superphosphate is discharged. From mixer 10 through hatch 9 the mix of ground phosphate and sulphuric acid is continuously fed into the den, where it sets. As the den rotates in the clockwise direction, the solidified mix slowly moves toward the discharge zone, maturing as it goes. The vacated space at the fixed partition is continuously filled with fresh mix.

The discharge zone of the den contains rotary cutter 7 armed with scrapers of knives. The cutter is secured to the den lid and rotates in the direction opposite to the direction of rotation of the den.

As the den rotates the solidified superphosphate slowly moves toward the cutter. The superphosphate is cut up by the knives and falls through tube 4 onto conveyer belt 3. which delivers the superphosphate to the storage premises.

The production capacity of a den, whose inner diameter is 7 m, height 2.5 m, and which rotates at 1.5 rph, is about 50 t/hr of superphosphate.

To speed up the curing process, the superphosphate in the storage premises is mixed using bucket cranes or excavators. It is sometimes also ammoniated or has other additives mixed with it to neutralise free acid.

The fluoride-containing gases released in the reaction between the phosphate meal and the sulphuric acid are sucked out of the chamber, absorbed and used in making various fluorides employed in agriculture as insecticides. Fluorides also have applications in the silicate industry.

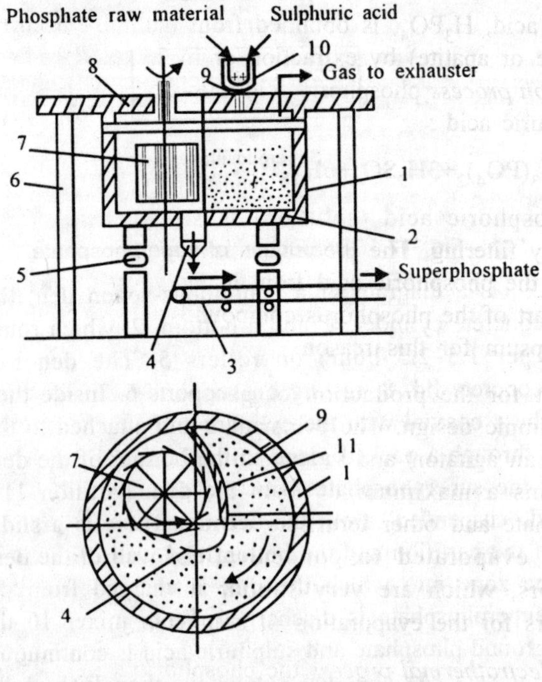

Fig. 17.2. Continuous-action den

To improve their physical and agricultural properties many fertilisers, including superphosphate, are *granulated*, i.e., converted from a powderlike state into granules. Granulated fertilisers are less liable to cake, have a lower moisture content and are convenient to use, since they can be introduced into the soil together with the seeds by the sowing machines.

Superphosphate is granulated in slightly inclined rotating drums. The powdered superphosphate is moistened inside the drum and rolled into granules of different size. The granulated superphosphate is dried with combustion gas in a drum-type dryer. The dried granules are screened into fractions. The larger granules are ground and rescreened. The small fraction is returned to the granulator.

Simple superphosphate is a low P_2O_5-content phosphorus fertiliser (maximum 20%). This leads to unproductive expenses for transportation, storage facilities and containers. Therefore the trend today is toward production of fertilisers with high nutrient content. High P_2O_5-content phosphorus fertilisers can be obtained by using phosphoric acid for their production.

Manufacture of phosphoric acid

Phosphoric acid, H_3PO_4, is obtained from naturally occurring phosphates (phosphorite or apatite) by extraction or by electrothermal procedures. In the *extraction process* phosphoric acid is made from the phosphates mainly using sulphuric acid :

$$Ca_3(PO_4)_2 + 3H_2SO_4 + 6H_2O \rightarrow 2H_3PO_4 + 3(CaSO_4.2H_2O)$$

The phosphoric acid (solution) is separated from the precipitate (gypsum) by filtering. The precipitate is washed with water for complete removal of the phosphoric acid from it. However, even after the washing operation part of the phosphorus compounds remain in the gypsum, called phospho-gypsum for this reason.

The unit for the production of phosphoric acid by extraction is of relatively simple design. The principal items of equipment are extractors (tanks with an agitator) and vacuum filters. Extraction-process phosphoric acid contains a maximum of 36% H_3PO_4. For use in making triple superphosphate and other fertilisers extraction-process phosphoric acid is sometimes evaporated to concentrations of 50–75% in drum-type concentrators, which are very similar in design and operation to the concentrators for the evaporation of sulphuric acid.

In the *electrothermal process* the phosphoric acid is usually obtained in two stages. The first stage yields elemental (yellow) phosphorus. For this purpose phosphorite or apatite, carbon (coke and anthracite), and silicon dioxide (quartzite) are charged into an electric furnace. Upon heating the mixture to a temperature of about 1500°C the following reaction takes place :

$$Ca_3(PO_4)_2 + 5C + nSiO_2 \rightarrow \uparrow P_2 + 3CaO{\cdot}nSiO_2 + \uparrow 5CO$$

Gas containing the phosphorus in the vapour phase is withdrawn from the electric furnace at a temperature of about 600°C and cooled, first by steam and then by water; the phosphorus condenses and accumulates under the water in the yellow phosphorus form. The residual gas, which contains carbon monoxide, can be used as a fuel or in the synthesis of organic compounds.

In the second stage molten yellow phosphorus is pumped from a heated storage tank through an atomiser nozzle into a combustion chamber, where it burns (is oxidised) to P_2O_5. The phosphorus pentoxide combines avidly with the water vapour contained in the flue gases to form phosphoric acid.

Phosphoric acid obtained by the electrothermal process is distinguished by high purity and high concentration (up to 80–95%). It can be used in manufacturing a number of different phosphoric acid salts and for synthesising organic compounds. Where electric power is inexpensive thermal-process phosphoric acid can be used to advantage in the production of concentrated fertilisers (double superphosphate, precipitate, ammophos, etc.).

Manufacture of concentrated phosphorus fertilisers (triple superphosphate, etc.)

Triple superphosphate is made by decomposing the natural phosphates not with sulphuric acid as in the production of simple superphosphate, but with phosphoric acid containing up to 70% H_3PO_4 :

$$Ca_3(PO_4)_2+4H_2PO_4+3H_2O \rightarrow 3Ca(H_2PO_4)_2.H_2O+Q$$

The product, consisting almost entirely of monocalcium phosphate, is soluble in water and is called triple superphosphate. Its P_2O_5 content (40 to 50%) is 2 to 2.5 times that of simple superphosphate. Extraction-process phosphoric acid, of about 36% concentration, is now widely used in manufacturing triple superphosphate.

Precipitated phosphate, dicalciumphosphate, is obtained by neutralisation of phosphoric acid with limestone and then with milk of lime :

$$H_3PO_4+CaCO_3+H_2O \rightarrow CaHPO_4.2H_2O+CO_2$$

$$H_3PO_4+ Ca(OH)_2 \rightarrow CaHPO_4.2H_2O$$

The precipitated phosphate is filtered, dried and packed. The product contains up to 42% P_2O_5 in an available (citrate-soluble) form.

Ammophos, $(NH_4)H_2PO_4$ and $(NH_4)_2HPO_4$, is obtained by the neutralisation of phosphoric acid with gaseous ammonia. Phosphate rock and apatite can also be decomposed with nitric acid; the resulting complex fertiliser contains both nitrogen and phosphorus. By adding potassium salts to the product of nitric acid decomposition of phosphates, fertilisers containing nitrogen can be obtained. Natural phosphates can be converted into products available to plants by sintering them with soda ash at high temperatures (1000 to 1200°C). These fertilisers are called thermal phosphates. The fusing of phosphates and silicon dioxide, magnesium silicates or alkaline aluminosilicates results in a fertiliser called *fused phosphate*. Thermal phosphates and fused phosphates contain 20 to 35% P_2O_5 in an available (citrate-soluble) form. They are only slightly hygroscopic, comparatively inexpensive and are used to advantage is phosphorus fertilisers, especially on acid soils.

When steel is smelted from pig iron with a high phosphorus content slags are produced, which are close in composition and properties to fused phosphates. This waste product of the iron and steel industry is used as a phosphorus fertiliser under the name of *slag*. It contains up to 20% P_2O_5 in a form available to plants and 40 to 50% CaO, which makes it especially good for use on acid soils.

Nitrogen Fertilisers

The nitrogen in nitrogenous fertilisers can be either in the form of the cation NH_4^+, i.e., in the ammonia form, or in the form of the anion NO_3^-, i.e. in the nitrate form; some fertilisers contain both ammonia and nitrate nitrogen. Another nitrogenous fertiliser is carbamide (urea), $CO(NH_2)_2$. All the nitrogenous fertilisers are soluble in water and are readily assimilated by plants.

The most commonly used nitrogenous fertilisers are ammonium nitrate, NH_4NO_3, which contains up to 35% nitrogen, and carbamide (urea), $CO(NH_2)_2$, which has the highest nitrogen concentration (up to 46%) of all the nitrogenous fertilisers. Ammonium sulphate, $(NH_4)_2SO_4$, containing 20 to 21% nitrogen is also used as a nitrogenous fertiliser.

Manufacture of ammonium nitrate

Ammonium nitrate, NH_4NO_3, contains nitrogen in both the ammonia and nitrate forms and is successfully used for all types of soils and agricultural crops. It is an extremely valuable nitrogenous fertiliser. However, there are some physical properties of ammonium nitrate which impede its application. In a damp atmosphere ammonium nitrate crystals tend to deliquesce and cake, cease to flow and form into large agglomerates. To reduce caking, ammonium nitrate crystals are either granulated or "dusted" with finely divided non-hygroscopic additives (lime, phosphate rock, apatite or bone meal, etc.).

On heating, ammonium nitrate loses ammonia. At temperatures above 200°C decomposition of the nitrate speeds up, oxygen is released, the nitrate may explode and is a fire hazard. Decomposition is accelerated in the presence of certain substances (graphite, glass, etc.). Ammonium nitrate may explode in improper storage conditions. It should be remembered that during storage of ammonium nitrate its crystalline structure may undergo transformations from one form to another (recrystallisation); this also promotes caking. When transformation to a new crystalline form occurs there is a considerable increase in the volume of the ammonium nitrate and as a result the bags in which it is stored are split open.

Ammonium nitrate is, as a rule, made at plants which produce ammonia and nitric acid. The manufacture of ammonium nitrate includes the following stages: neutralisation of dilute nitric acid with gaseous ammonia; evaporating the ammonium nitrate solution to a melt containing 98 to 99% NH_4NO_3; granulation.

Neutralisation of dilute nitric acid with gaseous ammonia proceeds by the reaction :

$$HNO_3 + NH_3 \rightarrow NH_4NO_3 + Q$$

Figure 17.3. is a flowsheet of the manufacture of ammonium nitrate. Neutraliser 1 is a vertical cylindrical vessel of stainless steel in which another cylinder, 2, is enclosed. Nitric acid and gaseous ammonia are continuously fed to the bottom part of cylinder 2, in which neutralisation takes place.

The absorption of ammonia by nitric acid is accompanied by an abundant release of heat. The heated solution rises to the top and flows over the brim of cylinder 2 into the space between the first and second cylinders. This space is the evaporating section of the apparatus. Here a considerable fraction of the water is evaporated from the ammonium nitrate solution and as a result the temperature in the neutraliser drops. The so-called juice steam which is released is withdrawn from the apparatus and used as a heating agent for the further evaporation of the solution. After leaving the neutraliser the solution enters the second-stage neutraliser 3 to which the necessary amount of ammonia liquor is added for the final neutralisation of the solution. Then the solution is directed to vacuum evaporator 4, where its concentration is brought up to 97-98%. The melt obtained is fed to granulation tower 5. Granulation is effected by spraying the melt in a hollow reinforced concrete granulation tower, 30 to 35 m high. Cold air is sucked in at the bottom of the tower by fans at the top. The drops of the melt solidify into granules as they fall through the current of cold air, and collect at the bottom of the tower in a hopper, from which they are removed by conveyer *6* for drying, and then go to the storage. Ammonium nitrate is packed in multi-layer paper bags impregnated with bitumen.

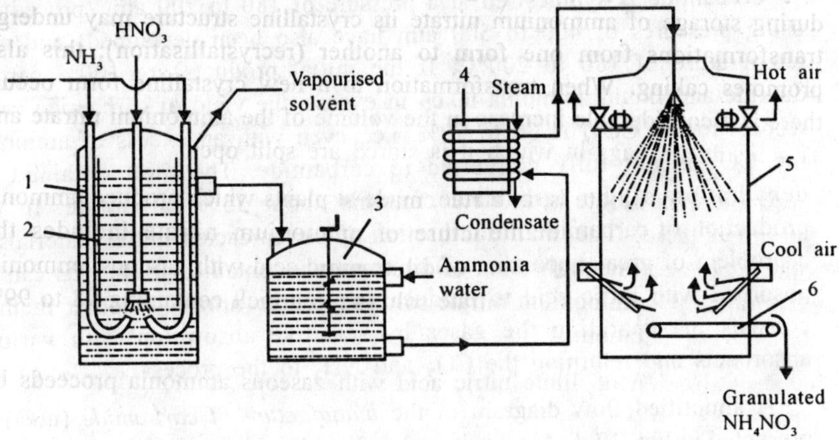

Fig. 17.3. Flow diagram for manufacture of ammonium nitrate manufacturing process.

A single-stage process for manufacturing ammonium nitrate without the use of evaporators has been developed in which ammonia and 60% nitric acid, preheated in heat exchangers, are reacted at a pressure of 4 atm. The vapour-liquid mixture produced in the reactor is separated in an apparatus, from which the melt, which contains 97 to 98% NH_4NO_3, is sent on for granulation.

The manufacture of carbamide (urea)

Carbamide $CO(NH_2)_2$, contains 46% nitrogen, readily available to plants. It is, therefore, a very valuable fertiliser. It is also used as a nitrogenous addition to livestock feed. In industry carbamide is used in making aminoplastics. Carbamide is made from ammonia and carbon dioxide. The reaction proceeds in two stages. In the first stage ammonium carbamate, NH_2COONH_4, is obtained :

$$2NH_3 + CO_2 \rightarrow NH_2COONH_4 + Q$$

In the second stage water is removed from the carbamate (dehydration), and the carbamide is formed :

$$NH_2COONH_4 \rightleftarrows CO(NH_2)_2 + H_2O - Q$$

The carbamide yield increases with the concentration of the initial materials. An excess of NH_3 prevents the formation of undesirable by-products.

Carbamide is synthesised at a pressure of 180 to 200 atm (procedures using pressures of 400 to 500 atm have also been developed) and at a temperature of 180 to 200°C. If the initial products are 100% carbon dioxide and liquid ammonia taken in excess, the yield of carbamide (with respect to the CO_2) is 60 to 70%, i.e., even with an excess of ammonia the CO_2 is not fully converted to carbamide. The melt obtained by dehydration of the carbamate contains about 35% carbamide. In the production of carbamide the utilisation of the unreacted initial materials is a problem of great importance. The ammonia contained in the gas can be absorbed with nitric acid to produce ammonium nitrate. Another method consists in separating the gases by selective absorption with various absorbents and returning the CO_2 and NH_3 to the process.

A simplified flow diagram of the *manufacture of carbamide* (urea) is presented in Fig. 17.4. Ammonia and carbon dioxide enter synthesis column sleeve 2, inside which the reaction takes place. The liquid ammonia first enters the annular space between the outer (1) and inner (2) cylinders. In this way the wall of main cylinder 1, designed to withstand the high pressure, is protected against corrosion by the reaction products. A solution containing about 35% carbamide is withdrawn at the top of the column and after its pressure is reduced to atmospheric it goes to distillation column 3, where the excess ammonia and the products of decomposition of ammonium carbonates are distilled from the carbamide solution. The bottom part of the distillation column is used for heating the carbamide solution, which goes from here to evaporator 4. The melt obtained is collected in reservoir 5, from which it is pumped for crystallisation or to tower 6 for granulation. The granulated carbamide is delivered to the storage premises by conveyer 7.

Manufacture of ammonium sulphate

Ammonium sulphate, $(NH_4)_2SO_4$, contains up to 20-21% nitrogen and is therefore a low-concentration nitrogenous fertiliser. Its repeated use as a fertiliser results in acidification of the soil, and therefore the soil must be neutralised from time to time by introducing lime (liming). Ammonium sulphate is obtained as a by-product when ammonia is removed from coke-oven gas with the aid of suphuric acid :

$$2NH_3+H_2SO_4 \rightarrow (NH_4)_2SO_4+Q$$

Potassium Fertilisers

Potassium fertilisers are obtained from minerals which contain potassium salts. Of greatest importance are the minerals sylvinite, KCl.NaCl, and carnallite, $KCl{\cdot}MgCl_2{\cdot}6H_2O$.

The Verkhnekamsk deposit of potassium salts is the richest in the world. In recent years deposits of potassium salts have been discovered in Byelorussia. Verkhnekamsk sylvinite contains 23 to 30% KCl and 65 to 70% NaCl. Ground sylvinite can be used directly as a fertiliser. However, the high NaCl content of sylvinite restricts its use, since introduction of chlorine compounds into the soil has an adverse effect on some agricultural crops. Large amounts of sylvinite are processed into potassium chloride, which is used as fertiliser either separately or mixed with other fertilisers.

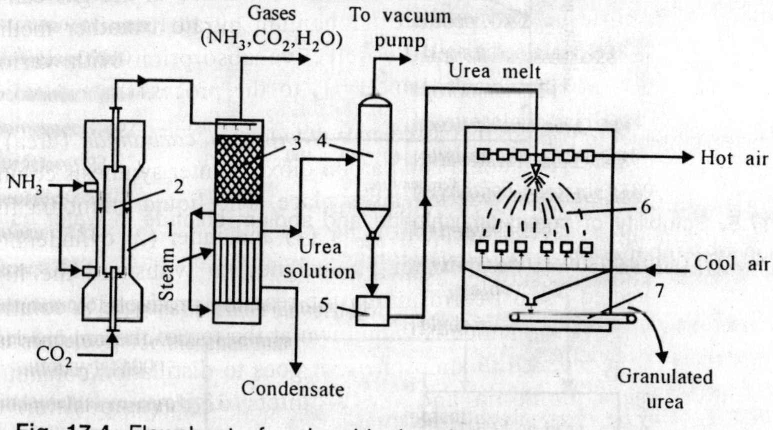

Fig. 17.4. Flowsheet of carbamide (urea) manufacturing process.

The isolation of potassium chloride from sylvinite is based on the different ratio of the solubilities of potassium chloride and sodium chloride in water at different temperatures. This is shown in Fig 17.5. As can be seen, the solubility of NaCl remains practically unchanged from 0 to 100°C, whereas the solubility of KCl increases with the temperature. At about 25°C the solubility curves of KCl and NaCl intersect, i.e., at this temperature the solubilities of the salts become equal. At a temperature of 80 to 100°C there will be more KCl in solution than NaCl. If a solution saturated at 100°C is cooled to 25°C, only KCl will precipitate from it. If the remaining solution is reheated to 100°C, it is no longer saturated with respect to KCl, i.e. it is capable of dissolving more of it. By treatment of new portions of sylvinite with this solution at 100°C KCl only will be dissolved. NaCl remains as a waste product of the production process. Fig. 17.6. is a flowsheet of *separation of sylvinite* into KCl and NaCl.

Potassium chloride can also be obtained from carnallite ($KCl{\cdot}MgCl_2{\cdot}6H_2O$). However, carnallite separation is more complicated as compared with separation of sylvinite. By comprehensive utilisation of carnallite not only potassium chloride can be obtained, but also magnesium salts, metallic magnesium, bromine, and other valuable products. Potassium chloride is also separated from sylvinite and carnallite by flotation procedures.

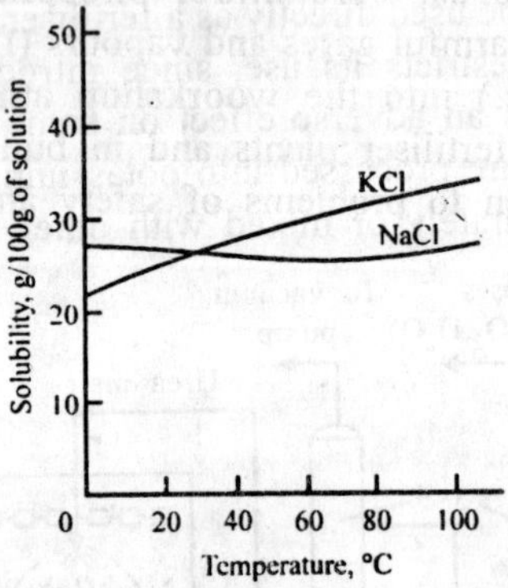

Fig. 17.5. Solubility of potassium chloride and sodium chloride in water as a function of temperature.

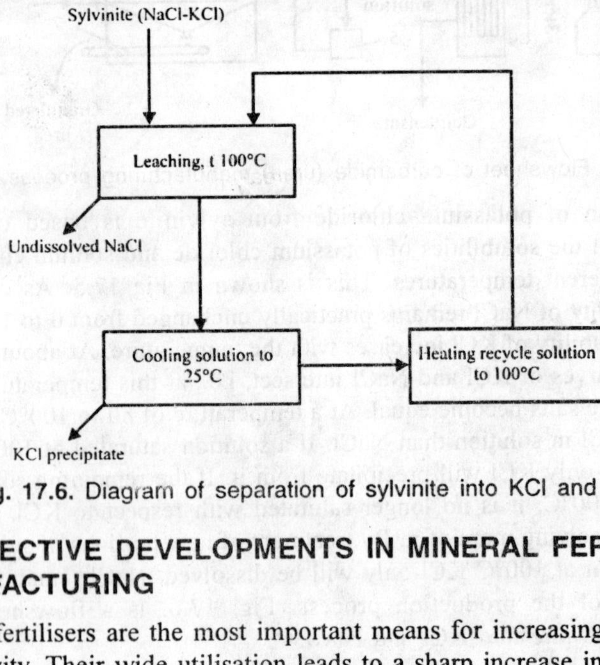

Fig. 17.6. Diagram of separation of sylvinite into KCl and NaCl.

PROSPECTIVE DEVELOPMENTS IN MINERAL FERTILISERS MANUFACTURING

Mineral fertilisers are the most important means for increasing agricultural productivity. Their wide utilisation leads to a sharp increase in farm crops.

The great potentialities of chemistry must be utilised to raise crop yields and the productivity of animal husbandry so as to provide the population of the world with sufficient amounts of foodstuffs and industrial raw materials.

The manufacture of mineral fertilisers involves the consumption of various mineral acids (sulphuric, nitric, phosphoric, etc.), alkalies (soda, ammonia, etc.), the use of high temperatures and pressures (in the production of phosphorus, carbamide, phosphate slags and fused phosphates); release of harmful gases and vapours (fluorine gas, ammonia, nitric acid vapours, etc.) into the woorkshop atmosphere is possible. Therefore, in operating fertiliser plants and in building new ones much attention should be given to problems of safety and improving working conditions.

Chapter 18

Portland Cement, Calcium and Magnesium Compounds

INTRODUCTION

The term cement refers to many different kinds of substances that are used as binders or adhesives. Here it refers to inorganic hydraulic cements (principally portland and related cements) which on hydration form relatively insoluble bonded aggregations of considerable strength and dimensional stability.

Hydraulic cements are manufactured by processing and proportioning suitable raw materials, burning (or clinkering) at a suitable temperature, and grinding the resulting hard nodules called clinker to the fineness required for an adequate rate of hardening by reaction with water. Portland cement consists mainly of tricalcium silicate and dicalcium silicate. Usually two types of raw materials are required: one rich in calcium, such as limestone, chalk, or marl; the other rich in silica, such as clay or shale.

CLINKER CHEMISTRY

Phase Equilibria

During burning in the kiln, about 20–30% of liquid forms in the mix at clinkering temperatures. Reactions occur at surfaces of solids and in the liquid. The crystalline silicate phases formed are separated by the interstitial liquid. The interstitial phases formed from the liquid in normal clinkers during cooling are also completely crystalline to x-rays, although they may be so finally subdivided as to appear glassy under the microscope.

In the relatively small portland cement zone of the equilibrium ternary system, $CaO–Al_2O_3–SiO_2$, almost all cements fall in the high lime portion (about 65% CaO; 21–25% SiO_2; 3–6% Al_2O_3). Cements of lower lime content tend to be slow in hardening and may show trouble from dusting of the clinker by transformation of the β- to γ-C_2S phase, especially if the clinker cooling is very slow. The zone is limited on the high lime side by the need to keep the uncombined CaO at low enough values to prevent excessive expansion due to hydration of the free lime. On the high alumina side, the zone is limited by excessive liquid-phase formation which prevents proper clinker formation in rotary kilns.

Clinker Formation

Portland cements are ordinarily manufactured from raw mixes including components such as calcium carbonate, clay or shale, and sand. As the temperature of the materials increases during their passage through the kiln, the following reactions occur :

(1) evaporation of free water
(2) release of combined water from the clay
(3) decomposition of magnesium carbonate
(4) decomposition of calcium carbonate (calcination)
(5) combination of the lime and clay oxides

Cooling is ordinarily too rapid to maintain the phase equilibria.

Phases Formed in Portland Cements

Most clinker compounds take up small amounts of other components to form solid solutions, e.g., the C_3S solid solution called alite. Under reducing conditions in the kiln, reduced phases such as FeO and calcium sulphide may be formed. Properties of the more common phases are summarised in Table 18.1. The major phases all contain impurities which stabilise the structures formed at high temperatures so that decomposition or transformations do not occur during cooling, as does occur with pure compounds. Portland cement clinker structures vary considerably with composition, particle size of raw materials, and burning conditions, resulting in variations of clinker porosity, crystallite sizes and forms, and aggregations of crystallites.

Raw Material Proportions

The three main considerations in proportioning raw materials for cement clinker are the potential compound composition; the percentage of liquid phase at clinkering temperatures (in the range from 18–25%); and the burnability of the raw mix.

Table 18.1. Phases in portland cement clinker.

Name	*Crystal system*	*Density, g/cm³*	*Moh's hardness*	*Cement chemists' notation*
Alite, Ca_3SiO_5	triclinic			C_3S
	monoclinic	3.14–3.25	ca 4	
	trigonal			C_2S
Belite, Ca_2SiO_4	hexagonal	3.04		
	orthorhombic	3.40		
	monoclinic	3.28	> 4	
	orthorhombic	2.97		
C_3A, $Ca_3Al_2O_6$	cubic	3.04	< 6	C_3A
Ferrite, $Ca_2Al_{<0.7}Fe_{>0.3}O_{3.5}$	orthorhombic	3.74–3.77	ca 5	$C_2A_xF_{1-x}$
Free lime, CaO	cubic	3.08–3.32	3–4	C
Magnesia, MgO	cubic	3.58	5.5–6	M

HYDRATION

Calcium Silicates

Lime and silica are frequently used in various proportions with or without portland cement in the manufacture of calcium silicate hydrate products. Typical compositions in cement chemists' notation include tricalcium silicate hydrate, $C_6S_2H_3$; calciochondrodite, C_5S_2H; α-dicalcium silicate hydrate, C_2SH; afwillite, $C_3S_2H_3$; foshagite, C_4S_3H; xonotlite, C_5S_5H; ranging in density from 2.6–2.98 g/cm^3 Calcium silicate hydrate is not the only variable in composition, but is very poorly crystallised and is generally referred to as calcium silicate hydrate gel (or tobermorite gel) because of the colloidal sizes of the gel particles. Significant changes in the structure of the gel continue over very long periods, changes often having a positive effect on both strength development and reduction of drying shrinkage. Drying and other chemical processes can have significant effects on this structure, such as loss of hydrate water and collapse of the structure to form more stable aggregations of particles.

Tricalcium Aluminate and Ferrite

The hydration of C_3A alone and in the presence of gypsum usually produces well-crystallised reaction products, e.g., C_3AH_6, the cubic calcium aluminium hydrate, and C_4AH_{19} and C_3ACSH_{12} (monosulphate), the hexagonal phases. The formation of highly hydrated trisulphate, ettringite, is responsible for the necessary retardation of hydration of the aluminates in portland cements and the expansion process in expansive cements.

The hydration of the ferrite phase (C_4AF) is of greatest interest in mixtures containing lime and other cement compounds because of the strong tendency to form solid solutions. When the sulphate in solution is very low, solid solutions are formed between the cubic C_3AH_6 and an analogous iron hydrate C_3FH_6. In the presence of water and silica, solid solutions such as $C_3ASH_4.C_3FSH_4$ may be formed.

Other Phases

In cements free lime (CaO) and periclase (MgO) hydrate to the hydroxides. The hydrated silicates formed by portland cements at 100°C are similar to those obtained from lime–silica mixtures or C_3S and C_2S but with a higher C/S ratio of the C–S–N gel.

Hydration at Ordinary Temperatures

Portland cement is generally used at 5–40°C, being temperature extremes avoided. The initial conditions for the hydration reactions are determined by the concentration of the cement particles in the mixing water and the fineness of the cement. The exact course of the early hydration reactions depends mainly on the C_3A, ferrite, and soluble alkali contents of the clinker and the amount of gypsum in the cement. The overall hydration rate increases with the temperature, the fineness of the cement, and to a lesser extent with the water-cement ratio; measurements of the activation energy indicate that the reaction becomes increasingly diffusion controlled.

CEMENT PASTE STRUCTURE AND CONCRETE PROPERTIES

The properties of both fresh and hardened mortars and concretes depend mainly on the cement–water paste properties. The engineering properties of the concrete, e.g., strength, elastic moduli, permeability to water and aggressive solutions, and frost resistance, depend strongly on the water–cement ratio and degree of hydration of the cement. Under sustained loads hardened cements and concrete creep or deform continuously with time, in addition to the initial elastic deformation. Drying of hardened cements results in shrinkage of the paste structure and of concrete members. The slowness of drying and the penetration of the hardened cement by carbon dioxide or chemically aggressive solutions is a result of the small size of the pores.

UNIT OPERATIONS, CHEMICAL CONVERSIONS, ENERGY REQUIREMENTS

Essentially, the unit operations prepare the raw materials in the necessary proportions and in the proper physical state of fineness and intimate contact so that chemical conversions can take place at the calcining temperature in the kiln to form, by double decomposition or neutralisation the compounds like — Dicalcium Silicate, Tricalcium Silicate, Tricalcium aluminate, Tetracalcium aluminoferrite, Magnesium oxide (in free state), CaO, SiO_2, Al_2O_3, Fe_2O_3, MgO, alkali oxides, SO_3, C_3S, C_2S, $CaSO_4$, MgO, K_2O, Na_2O are present in small amounts and form compounds with CaO, Al_2O_3, SiO_2, and SO_3, for example, $KC_{23}S_{12}$, which has properties similar to C_2S. Alkali oxides do not alter the properties of cement, because of the similarity of the properties of the compounds formed in them and those of similar compounds that do not contain them. Other reactions take place, such as dehydration of clay and decarbonisation or calcination of limestone, both being endothermic and having values of 401 and 713 Btu/lb, respectively. The clinker formation is exothermic, with a probable value of 200 Btu/lb of clinker. Probably the net heat requirements are in the neighbourhood of 900 Btu/lb of clinker. The coal consumption, however, indicates an expenditure of 3000 or 4000 Btu/lb of clinker. This heat is evolved in the kiln in carrying out the following reactions, as tabulated under :

Temperature, °C	*Reaction*	*Heat change*
100	Evaporation of free water	Endothermic
500 and above	Evolution of combined water from clay	Endothermic
900 and above	Crystallisation of amorphous dehydration products of clay	Exothermic
900 and above	Evolution of carbon dioxide from calcium carbonate	Endothermic
900–1200	Main reaction between lime and clay	Exothermic
1250–1280	Commencement of liquid formation	Endothermic
1280 and above	Further formation of liquid and completion of formation of cement (compounds)	Probably endothermic on balance

It should be noted that most of the reactions in the kiln proceed in the solid phase but, toward the end, the important fusion occurs. It has been found that no appreciable formation of C_3S occurs below 1250°C, which is given as the temperature at which liquid formation begins. It is this final reaction that yields C_3S, the chief strength-producing constituent of cement, and reduces the free lime to a sufficiently small amount. This final fluid phase amounts to 20 to 30% of the reactions. All these reactions are involved in the "burning of cement."

MANUFACTURING PROCESSES

Wet process and dry process plants produce portland cements. Manufacture consists of quarrying and crushing the rock, including control of the clinker composition by systematic core drillings and selective quarrying; grinding the carefully proportioned materials to high fineness (argillaceous, siliceous, and ferriferous raw-mix components are added to the crushed product; ball mills are used for both wet and dry processes to grind material, although roller mills are used for dry process only); subjecting the raw-mix to pyro-processing in a rotary kiln when, during the burning process, the high temperatures cause vapourisation of the alkalies, sulphur, and halides (kilns are either wet process, dry process, suspension preheaters, or precalciners); and grinding the resulting clinker to a fine powder via open-circuit or closed-circuit grinding.

MANUFACTURING PROCEDURES, WET AND DRY PROCESSES

In both processes closed-circuit grinding is preferred to open-circuit grinding in preparing the raw materials because in the former the fines are passed on and the coarse material returned, whereas in the latter the raw material is ground continuously until its mean fineness has reached the desired value (Fig. 18.2 for some grinding hookups). The *wet process*, though the original one, is being displaced by the *dry process*, especially for new plants, because of the saving in heat, accurate control, and mixing of the raw mixture it affords. During past dry process plants accounted for 58% of the total cement production capacity. These are illustrated in the generalised flowchart in Figs. 18.1. and 18.2. In the wet process the solid material, after dry crushing, is reduced to a fine state of division in wet tube or ball mills and passes as a slurry through bowl classifiers or screens. The slurry is pumped to correcting tanks, where rotating arms make the mixture homogeneous and allow the final adjustment in composition to be made. In some plants, this slurry is filtered in a continuous rotary filter and fed into the kiln. The *dry process* is especially applicable to natural cement rock and to mixtures of limestone and clay, shale, or slate. In this process the materials may be roughly crushed, passed through gyratory and hammer mills, dried, sized, and more finely ground in tube mills, followed by air separators. Proportioning equipment is included in the sequences. Before entering the kiln a thorough mixing and blending by air or otherwise takes place (Figs. 18.1. and 18.2).

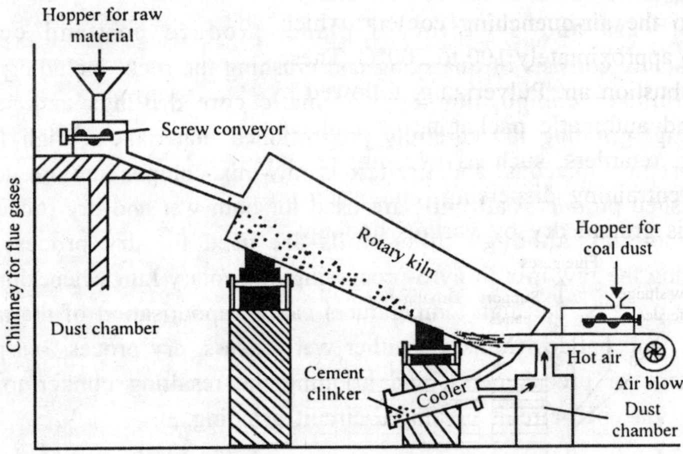

Fig. 18.1. Flow diagram for manufacture of portland cement.

This dry, powdered material is fed directly to rotary kilns, where the previously mentioned chemical reactions take place. Heat is provided by burning oil, gas, or pulverised coal, using preheated air from cooling the clinker. The tendency in recent years has been to lengthen the rotary kiln in order to increase its thermal efficiency. Dry-process kilns may be as short as 150 ft, but in the wet process, 300- to 600-ft kilns are not uncommon. The internal diameter is usually from 8 to 20 ft. The kilns are rotated at from 1/2 to 2 rpm depending on size. The kilns are slightly inclined, so that materials fed in at the upper end travel slowly to the lower firing end, taking from 1 to 3 hour. In order to obtain greater heat economy, part of the water is removed from wet process slurry. Some of the methods used employ slurry filters and Dorr thickeners. Efficient air pollution control equipment such as baghouses or electrostatic precipitators are now required for kilns. Waste-heat boilers are sometimes used to conserve heat and are particularly economical for dry process cement, since the waste gases from the kiln are hotter than those from the wet process and may reach 800°C. Because the lining of the kiln has to withstand severe abrasions and chemical attack at the high temperatures in the clinkering zone, the choice of a refractory lining is difficult. For this reason high-alumina and high-magnesia bricks are widely used. Computers are now used to improve kiln control.

The final product formed consists of hard, granular masses from 1/8 to 3/4 inches in size, called *clinker*. The clinker is discharged from the rotating kiln into the air-quenching coolers, which quickly bring its temperature down to approximately 100 to 200°C. These coolers simultaneously preheat the combustion air. Pulverizing, followed by fine grinding in the tube ball mills and automatic packaging, completes the process. During the fine grinding, retarders, such as gypsum, plaster, or calcium lignosulphonate, and air-entraining dispersing, and water-proofing agents are added. The clinker is ground dry by various hookups.

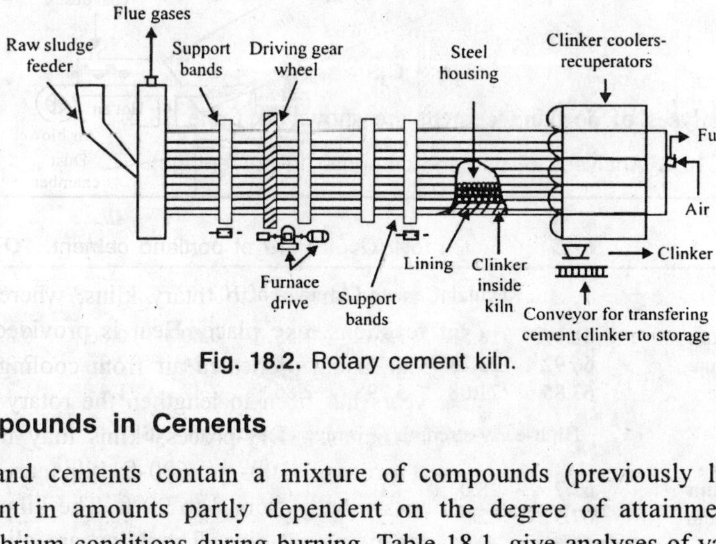

Fig. 18.2. Rotary cement kiln.

Compounds in Cements

Portland cements contain a mixture of compounds (previously listed) present in amounts partly dependent on the degree of attainment of equilibrium conditions during burning. Table 18.1. give analyses of various types of portland cement and the average composition of about 102 regular cements. From these analyses it is seen that portland cement composition approaches a rough approximation to the system $CaO.SiO_2$ and successively closer approximations to the systems $CaO.SiO_2.Al_2O_3$, $CaO.SiO_2.Al_2O_3.$ Fe_2O_3, and $CaO.SiO_2.Al_2O_3.Fe_2O_3.MgO$. A complete understanding of portland cement would require knowledge of the phase equilibrium relations of the high-lime portions of all the two-, three-, four-, and five-component systems involved. Of these, all 12 of the principal two- and three-component systems, and the parts of the four-component systems, CaO-SiO_2-Al_2O_3-Fe_2O_3 and $CaO.SiO_2.Al_2O_3.MgO$, in which portland cement compositions are located, are known. The change in CaO percentage from 61.17 to 66.92 alters the C_3S percentage from 35.3 to 70.6 and that of C_2S from 0 to 33.2.

Setting and Hardening of Cement

Although many theories have been proposed to explain the setting and hardening of cement, it is generally agreed that *hydration* and *hydrolysis* are involved. The hydration products have very low solubility in water. If this were not true, concrete in contact with water would be rapidly attacked. In recent years much attention has been given to the heat evolved during the hydration of cement. The various compounds contribute to the heat of hardening (basis, equal weights, i.e., gram for gram) after 28 days, as follows:

$$C_3A > C_3S > C_4AF > C_2S$$

Analyses of portland cement are shown in Table 18.2.

Table. 18.2. Analyses of portland cements (In percentage).

	CaO	*SiO_2*	*Al_2O_3*	*Fe_2O_3*	*MgO*	*Alkali oxides*	*SO_3*
			Regular Cement (Average of 102)				
Minimum	61.17	18.58	3.86	1.53	0.60	0.66	0.82
Maximum	66.92	23.26	7.44	6.18	5.24	2.9	2.26
Average	63.85	21.08	5.79	2.86	2.47	1.4	1.73
			High-early-strength (Average of 8): High C_3S				
Minimum	62.7	18.0	4.1	1.7	....	....	2.2
Maximum	67.5	22.9	7.5	4.2	....	....	2.7
Average	64.6	19.9	6.0	2.6	....	....	2.3
			Low-heat-of-hardening (Average of 5): Lower C_3S and C_3A, Higher C_2S and C_4AF				
Minimum	59.3	21.9	3.3	1.9	....	....	1.6
Maximum	61.5	26.4	5.4	5.7	....	....	1.9
Average	60.2	23.8	4.9	4.9	....	....	1.7

The chemical specifications of portland cement are given in Table 18.3.

Table 18.3. Chemical specifications for portland cements.

Constituent	*Regular, Type I*	*Moderate-heat-of-hardening, Type II*	*High-early-strength, Type III*	*Low-heat-of-hydration, Type IV*	*Sulphate-resisting, Type V*
Silicon dioxide (SiO_2), min %	....	21.0			
Aluminium oxide (Al_2O_3), max %	....	6.0	....	....	
Ferric oxide (Fe_2O_3) max %	....	6.0	....	6.5	
Magnesium oxide (MgO), max %	5.0	5.0	5.0	5.0	4.0
When $3CaO{\cdot}Al_2O_3$ is 8% or less	2.5	2.5	3.0	2.3	2.3
When $3CaO{\cdot}Al_2O_3$ is more than 8%	3.0	....	4.0		
Loss on ignition, max %	3.0	3.0	3.0	2.3	3.0
Insoluble residue, max %	0.75	0.75	0.75	0.75	0.75
Tricalcium silicate (3CaO. SiO_2), max %	....	....	....	35	
Dicalcium silicate (2CaO. SiO_2), min %	....	....	....	40	
Tricalcium aluminate (3CaO. Al_2O_3), %	....	8	15	7	5

Special Cements

For many corrosive conditions portland cement is unsuitable or not economical. Hence many special cements have been developed, of which many types are industrially important. The types of special cements based on organic plastics now generally known as *adhesives*. *Sulphur cements* have been available commercially as simple mixtures of fillers, and as homogeneous plasticised filled sulphur ingots possessing low coefficients of expansion. Sulphur cements are resistant to non-oxidising acids and salts but should not be used in the presence of alkalies, oils, greases, or solvents. The crystalline change in sulphur structure at 200°F limits their use. Thiokol-plasticised, silica-filled sulphur cements have been accepted as a standard material for joining bricks, tile, and cast-iron pipe.

Silica-filled chemically setting *silicate cements* withstand a higher temperature (1000°F) than the preceding cements. Usually 2 parts by weight of finely divided silica powder is used to 1 part of sodium silicate

(35 to 40°Bé). Silicate cements are resistant to all concentrations of inorganic acids except hydrofluoric. Silicate cements are not suitable at pH values above 7 or in the presence of crystal-forming systems. Two typical applications are the joining of bricks in chromic acid reaction tanks and in alum tanks.

LIME

Lime itself may be used for medicinal purposes, insecticides, plant and animal food, gas absorption, precipitation, dehydration, and causticising. It is employed as a reagent in the sulphite process for paper-making, dehairing hides, the manufacture of high-grade steel and cement, water softening, recovery of by-product ammonia, and the manufacture of soap, rubber, varnish, refractories, and sand-lime brick. Lime is indispensable for use with mortar and plaster and serves as a basic raw material in the production of calcium salts and the improving the quality of certain soils. Either directly or indirectly, limestone and lime are employed in more industries than any other natural substance. Lime is sold as a high-calcium quicklime containing not less than 90% CaO and from 0 to 5% magnesia; small percentages of calcium carbonate, silica, alumina, and ferric oxide are present as impurities. The suitability of lime for any particular use depends on its composition and physical properties, all of which can be controlled by the selection of the limestone and the details of the manufacturing process. Much lime must be finely ground before use.

Depending on composition, there are several distinct types of limes. Hydraulic limes are obtained from the burning of limestone containing clay, and the nature of the product obtained after contact with water varies from putty to set cement. High-calcium content limes harden only with the absorption of carbon dioxide from the air, which is a slow process; hydraulic limes also harden slowly, but they can be used under water. For chemical purposes high-calcium is required, except for the sulphite paper process, where magnesiun lime works better. Although in many sections of the country high-calcium lime is preferred by the building industry for the manufacture of mortar lime plaster, there are places where limestone containing magnesium is burned or where even dolomitic stone is calcined. Typical compositions of such stones are from 35 to 45% CaO and from 10 to 25% MgO. These products, called *magnesian* limes, or dolimes, are favoured by some plasterers, who claim they work better under the travel. In the metallurgical field, "refractory lime", as dead-burned dolomite or as raw dolomite, is employed as a refractory patching material in open-hearth furnaces; it is applied between heats to repair scored and washed spots on the bottom of the furnace. Hydrated lime is finding increased favour in the chemical building trades over the less stable quicklime, despite its increased weight. Quicklime is almost invariably slaked or hydrated before use. Because of the better slaking and the opportunity to remove impurities, factory hydrate is purer and more uniform than slaked lime prepared on the job.

LIME MANUFACTURE

Lime has always been a cheap commodity because limestone deposits are readily available in many states of India. It is produced from limestone near centres of consumption so that freight costs are low.

Raw Materials

The carbonates of calcium and magnesium are obtained from deposits of limestone, marble, chalk, dolomite, or oyster shells. For chemical usage, a rather pure limestone is preferred as a starting material because of the high-calcium lime that results. Quarries are chosen which furnish a rock that contains as impurities low percentages of silica, clay, or iron. Such impurities are important because the lime may react with the silica and alumina to give calcium silicates or calcium alumino-silicates which possess not undesirable hydraulic properties. The lumps sometimes found in "overburned" or dead-burned lime result from changes in the calcium oxide itself, as well as from certain impurities acted upon by excess heat, recognised as masses of relatively inert, semi-vitrified material. On the other hand, it often happens that rather pure limestone is calcined insufficiently, and lumps of calcium carbonate are left in the lime. The lime is called "under-burned" lime.

Energy Changes, Physical Operations and Chemical Conversions

Energy is required for blasting out limestone, for transporting and sizing it, and for the burning or calcining process.

The reactions involved are for

Calcining:

$CaCO_3(c) \rightleftharpoons CaO(c) + CO_2(g) \qquad \Delta H_{1200\text{–}1300°C} = +\ 4.25$ million Btu/ton of lime produced

Hydrating:

$CaO(c) + H_2O(l) \longrightarrow Ca(OH)_2(c) \quad \Delta H = -\ 15.9$ kcal

During calcining the volume contracts, and during hydrating it swells. For calcination the average fuel ratios, using bituminous coal, are 3.23 lb of lime from 1 lb of coal in shaft kilns and 3.37 lb in rotary kilns. As is shown above, the calcination reaction is reversible. Below 650°C the equilibrium decomposition pressure of CO_2 is quite small. Between 650 and 900°C the decomposition pressure increases rapidly and reaches 1 atm at about 900°C. In most operating kilns the partial pressure of CO_2 in the gases in direct contact with the outside of the lumps is less than 1 atm; therefore initial decomposition may take place at temperatures somewhat less than 900°C. The decomposition temperature at the centre of the lump is probably well above 900°C, since there the partial pressure of the CO_2 not only is equal to or near the total pressure, but also must be high enough to cause the gas to move out of the lump, where it can pass into the gas stream. The total *heat* required for *calcining* per ton of lime produced may be divided into two parts: sensible to heat to raise the rock to decomposition temperature, and latent heat of dissociation. Theoretical heat requirements per ton of lime produced, if the rock is heated only to a calcining temperature of 900°C, are approximately 1.3 million Btu for sensible heat and 2.6 million Btu for latent heat. Actual calcining operations, because of practical considerations, e.g., lump size and time, require that the rock be heated to between 1200 and 1300°C, thereby increasing sensible-heat requirements by about 350,000 Btu. Then practical heat requirements are approximately 4.25 million Btu/ton of lime produced in a vertical kiln. About 40% is sensible heat; the rest is latent heat of decomposition. (Fig. 18.3.)

(1) Blasting down of limestone from a quarry face or occasionally from underground veins.

(2) Transportation from the quarry to mills, generally by an industrial railroad.

(3) Crushing and sizing of the stone in jaw and gyratory crushers.

(4) Screening to remove various sizes (e.g., a 4- to 8-inches stone implies that all pieces passing a 4 inches screen or retained on the 8 inches screen have been separated out).

(5) Carting of large stones to top of vertical kilns.

(6) Conveying of small rocks to a rotary kiln.

(7) Conveying of fines to a pulveriser to make powdered limestone for agricultural and other demands.

(8) Burning of limestone according to size, in vertical kilns to give lump lime, or in horizontal rotary kilns to furnish fine lime.

(9) Packaging of the finished lime in barrels or sheet-iron drums or conveying it to a hydrator.

(10) Hydration of the lime.

(11) Packaging of slaked lime in 50-lb paper bags.

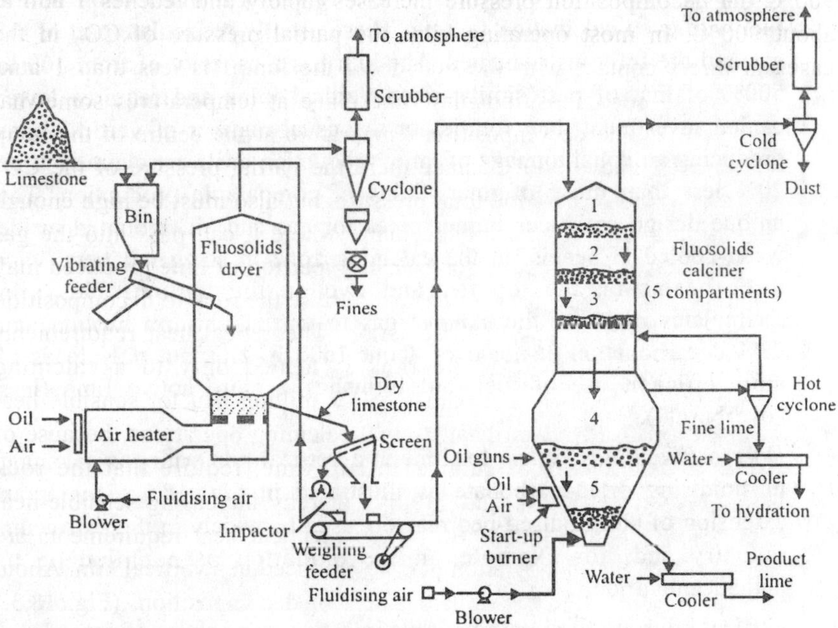

Fig. 18.3. Dorrco FluoSolids system for producing lime from pulverised limestone or calcium carbonate sludge.

Manufacturing Procedures

For the burning of lump limestones, upright kilns are usually employed. However, past difficulties with uneven heating and the blocking of burning gas in the vertical kiln have been largely eliminated by recent chemical engineering changes, as exemplified in Fig. 18.3 with the improvements described. The horizontal revolving kiln, although costing considerably more in capital investment and in fuel, also has been improved to reduce these additional charges per ton of goods produced. Both types of lime kilns have steel shells lined with refractory brick. The following chemical processes have been or are employed in the burning of calcium carbonate to lime:

1. *Vertical shaft kilns with outside heating hearths.* These were the more primitive form and are now obsolete.

2. *Vertical kilns burning a mixed feed* made up of limestone and coke. Here the fuel should be well integrated with the limestone. Such kilns are used in the Solvay soda ash process to provide the high CO_2 gas, as well as lime, needed for this manufacture.
3. *Union Carbide Metals vertical kilns* with gas firing using special multiport water-cooled burners to ensure uniform distribution of the fuel with even calcination to lime. These usually recirculate part of the gas and are fully instrumented. The kiln has a capacity of from 300 to 500% of that of past similar-sized vertical kilns and requires lower capital investment than rotaries or the usual number of vertical kilns producing an equal tonnage of lime; production costs are claimed to be 20% less than those of rotary kilns of comparable production. The unique design embraces burner pipes for gas and air, mounted inside water-cooled H beams, in the calcining zone in staggered tiers, with three beams on the top tier and two on the bottom. This kiln recirculates much of the exhaust gas to ensure uniform heating and 98% decarbonation at not over about 1600°F. This not only leads to more efficient use of fuel, but furnishes a more active lime (less "overburnt").
4. *Dorrco FluoSolids system* for lime in a vertical kiln with gas-calcining of fluidising dried carbonate as illustrated in Fig. 18.3. This is an extension of the fluidised-bed reactors used so widely in the petroleum industry and, for example, in the oxidation of naphthalene to phthalicanhydride.
5. Vertical kilns are fired with multiple injection and control of natural gas with recirculation of air.
6. *Rotary horizontal kilns,* commonly 150 ft long, but occasionally extended to as much as 400 ft (11 ft in diameter) to ensure better overall fuel consumption. The modern horizontal kiln has various devices to properly size and preheat the entering limestone. A rotary counter-current cooler for the product lime is used to preheat air for the burning. These kilns operate on fines or small lumps of limestone or marble or moist precipitated calcium carbonate sludge, any of which would block the efficient and uniform burning of the fuel in a vertical kiln. A kiln is burning lime carbonate slurry from causticising plant as underflow from clarifiers or thickeners. This has 35% solids and is put through two Bird centrifuges, reducing moisture to about 35%. The slurry is introduced into the kiln by screw conveyors at 400°F and the front end temperature is 1760 to 1800°F. Other devices used to lessen the heat requirements of a rotary are the travelling grate-kiln system of

natural gas combination of a short high-temperature rotary kiln discharging into a modified vertical kiln consisting of heating (air-gas), soaking, and cooling zones.

GYPSUM

Gypsum is a mineral that occurs in large deposits throughout the world. It is hydrated calcium sulphate, with the formula $CaSO_4{\cdot}2H_2O$. When heated slightly, the following reaction occurs :

$$CaSO_4{\cdot}2H_2O(c) \longrightarrow CaSO_2{\cdot}\ \frac{1}{2}\ H_2O(c) + 1\frac{1}{2}\ H_2O(g) \qquad \Delta H_{25°C} = +\ 16.5\ \text{kcal}$$

If the heating is at a higher temperature, gypsum loses all its water and becomes anhydrous calcium sulphate, or *anhydrite*. Calcined gypsum can be made into wall plaster by the addition of a filler material such as asbestos, wood pulp, or sand. Without additions, it is plaster of paris and is used for making casts and for plaster; it is hydraulic and hardens under water, but is also slightly soluble in water, hence must not be used in moist exposures.

Calcination of Gypsum

The usual method of calcination of gypsum consists in grinding the mineral and placing it in large calciners holding 10 to 25 tons. The temperature is raised to about 120 to 150°C, with constant agitation to maintain a uniform temperature. The material in the kettle, known to the public as *plaster of paris* and to the manufacturer as *first-settle plaster*, may be withdrawn and sold at this point, or it may be heated further to 190°C to give a material known as *second-settle plaster.* First-settle plaster is approximately the half hydrate $CaSO_4$. 1/2 H_2O, and the second form is anhydrous. Practically all gypsum plaster sold is in the form of first-settle plaster mixed with sand or wood pulp. The second form is used in the manufacture of plasterboard and other gypsum products. Gypsum may also be calcined in rotary kilns similar to those used for limestone.

Hardening of Plaster

This hardening of plaster is essentially a hydration chemical conversion, as represented by the equation.

$$CaSO_4{\cdot}\ \frac{1}{2}\ H_2O(c) + 1\frac{1}{2}\ H_2O(l) \longrightarrow CaSO_4{\cdot}2H_2O(c) \quad \Delta H = -\ 0.7\ \text{kcal}$$

This equation is the reverse of that for the dehydration of gypsum. The plaster sets and hardens because the liquid water reacts to form a solid, crystalline hydrate. With reference to Fig. 18.4, it is apparent that hydration with liquid water takes place at temperatures below about 99°C and that the gypsum must be heated above 99°C for practical dehydration. Commercial plaster usually contains some glue in the water used or a material such as hair or tankage from the stockyards to retard the setting time in order to give the plasterer opportunity to apply the material. Modern equipment collects dust at about 99.8% efficiency.

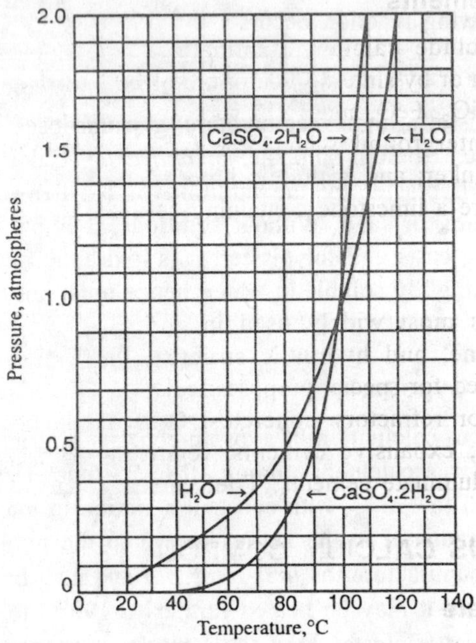

Fig. 18.4. Dehydration pressure of gypsum and vapour pressure of water.

Special Purpose and Blended Cements

These are manufactured essentially by the same processes as ordinary portland cements but have specific compositional and process differences. For example, white cements are made from raw materials of very low iron content; regulated set cements are made with fluorite (CaF_2) additions which also acts as fluxing agents or mineralisers to reduce burning temperatures; expansive cements depend on aluminate and sulphate-phases which result in more ettringite formation during hydration than in normal portland cements; oil-well cements are manufactured for sluggish reactivity and thus levels of C_3A and alkali sulphates are kept low and hydration-retarding additions are employed; blended cements differ in the finish-grinding processes where the cement clinker is interground with pozzolans (natural materials such as diatomaceous earths, opaline cherts, and shales), granulated blast-furnace slag, or a variety of materials; masonry cements are finely interground mixtures, including limestones, hydrated lime, natural cement, pozzolans, clays and air-entraining agents.

Non-Portland Cements

These cements include calcium aluminate cements, manufactured by heating until molten or by sintering a mixture of limestone and bauxite with small amounts of SiO_2, FeO, and TiO_2; supersulphated cements, containing about 80% slag interground with 15% gypsum or anhydrite and 5% portland cement clinker; and hydraulic limes, produced by heating below sintering temperature a limestone containing considerable clay.

Uses

Portland cement is most widely used in concrete for construction of underground, marine, and hydraulic structures. Masonry and oil-well cements are produced for special purposes; calcium aluminate cements are extensively used for refractory concretes. Other cements used are air-entraining cements; expansive cements; regulated set cements; natural cements; calcium aluminate cements; Trief cements; and hydraulic limes.

MISCELLANEOUS CALCIUM COMPOUNDS

Calcium Carbonate

Preparation

(i) Mined from natural surface deposits; (ii) Precipitated (made synthetically) by reaction of calcium chloride and sodium carbonate in water solution, or by passing carbon dioxide through a suspension of hydrated lime ($Ca(OH)_2$) in water.

Properties

White powder or colourless crystals, odourless, tasteless; sp. gr. 2.7–2.95; decomposes at 825°C; non-combustible; non-toxic; very slightly soluble in water (a few parts per million); soluble in acids with evolution of carbon dioxide. Calcium carbonate is one of the most stable, common, and widely dispersed materials. It occurs in nature as aragonite, oyster shells, calcite, chalk, limestone, marble, marl, and travertine; especially in India (structural limestone), Vermont (marble), Italy (travertine), and England (chalk). A nuisance particulate. Tolerance, 5 mg per cubic metre of air.

Uses

Source of lime, neutralising agent; filler and extender in rubber, plastics, paints; opacifying agent in paper; fortification of bread; putty; tooth powders; antacid; whitewash; portland cement; SO_2 removal from stack gases; metallurgical flux; analytical chemistry; CO_2 generation laboratory. (*Note:* Calcium carbonate is a major cause of boiler scale when hard water is used in heating systems.)

Calcium Sulphide

Preparation

It can be prepared by strong heating of pulverised calcium sulphate and charcoal.

Properties

Yellow to light-gray powder with odour of hydrogen sulphide in moist air; unpleasant alkaline taste. Gradually decomposes in moist air or in weak acids. Decomposed by acids; slightly soluble in water high partial decomposition; insoluble in alcohol. Sp. gr. 2.6.

Uses

Luminous paint; depilatory; preparation of arsenic-free hydrogen sulphide; lubricant additive; ore dressing and flotation agent; phosphorus.

Calcium Arsenate

Calcium arsenate is produced by the reaction of calcium chloride, calcium hydroxide, and sodium arsenate or lime and arsenic acid:

$$2CaCl_2(aq) + Ca(OH)_2(c) + 2Na_2HAsO_4\ (aq) \longrightarrow Ca_3(AsO_4)_2(c) + 4NaCl(aq) + 2H_2O(l)$$

Halide Salts

Calcium chloride is obtained commercially as a by-product of chemical manufacture and from natural brines which contain more or less magnesium chloride. For this reason, and since large tonnages are available, it is a very cheap chemical. Its main applications are in solutions used to lay dust on highways (because it is deliquescent and remains moist) and in low-temperature refrigeration. Calcium bromide and iodide have properties similar to those of the chloride. They are prepared by the action of the halogen acids on calcium oxide or calcium carbonate. They are sold as hexahydrates for use in medicine and photography. Calcium fluoride occurs naturally as a fluospar.

MISCELLANEOUS MAGNESIUM COMPOUNDS

Magnesium Oxychloride Cement (Sorel cement). A mixture of magnesium chloride and magnesium oxide that reacts with water to form a solid mass, presumed to be magnesium oxychloride. Fillers such as wood flour, sawdust, sand, powdered stone, talc, etc., are usually present. A variety of proprietary mixtures are available. Strength ranges from 7,000 to 10,000 psi. Copper powder minimises water solubility.

Magnesium Carbonate

The term magnesium carbonate is generally reserved for the synthetic, pure variety. The naturally occurring material is called magnesite.

Preparation

(1) Mined as natural material.

(2) Carbonation of MgO or $Mg(OH)_2$ with CO_2.

(3) Reaction of a soluble Mg salt solution with sodium carbonate or bicarbonate.

Properties

Light, bulky white powder; bulk density about 4 lb per cubic foot; actual sp. gr. about 3.0; decomposes at 350°C. Refractive index about 1.52. Soluble in acids; very slightly soluble in water; insoluble in alcohol. Non-toxic. Non-combustible.

Uses

Magnesium salts; heat insulation and refractory; rubber reinforcing agent; inks; glass; pharmaceuticals, dentrifices and cosmetics; free-running table salts; antacid; making magnesium citrate; filtering medium. Used in foods as drying agent, colour retention agent, anticaking agent, carrier.

Magnesium Chloride

Preparation

Action of hydrochloride acid on magnesium oxide or hydroxide, especially the latter when precipitated from sea water or brines (Great Salt Lake). It can be purified by recrystallisation.

Properties

Colourless or white crystals; deliquescent. Sp. gr. (a) 2.32, (b) 1.56; m.p. (a) 708°C, (b) loses $2H_2O$ at 100°C; if heated rapidly, melts at 116–118°C; b.p. (a) 1412°C, (b) decomposes to oxychloride. Soluble in water and alcohol. Combustible. Moderately toxic by ingestion.

Uses

Source of magnesium metal; disinfectants; fire extinguishers; fire-proofing wood; magnesium oxychloride cement; refrigerating brines; ceramics; cooling drilling tools, textiles (size, dressing and filling of cotton and woollen fabrics, tread lubricant; carbonisation of wool); paper manufacture; road dust-laying compounds; floor sweeping compounds; flocculating agent; catalyst.

Magnesium Oxide (Magnesia)

Two forms are produced, one a light fluffy material prepared by a relatively low-temperature dehydration of the hydroxide, the other a dense material made by high-temperature furnacing of the oxide after it has been formed from the carbonate or hydroxide.

Preparation

(1) By-calcining magnesium carbonate or magnesium hydroxide;

(2) By treating magnesium chloride with lime and heating, or by heating it in air; and

(3) From sea water via the hydroxide.

Properties

White powder, either light or heavy depending upon whether it is prepared by heating magnesium carbonate or the basic magnesium carbonate; sp. gr. about 3.6 (varies); m.p. 2800°C; b.p. 3600°C; slightly soluble in water; soluble in acids and ammonium salt solutions. Non-combustible. Toxic by inhalation of fume. Tolerance (as Mg), 10 mg per cubic metre of air.

Uses

Refractories, especially for steel furnace linings; polycrystalline ceramic for aircraft windshields; electrical insulation; pharmaceuticals and cosmetics; inorganic rubber accelerator; oxychloride and oxysulphate cements; paper manufacture; fertilisers; removal of sulphur dioxide from stack gases; adsorption and catalysis; semi-conductors; pharmaceuticals; food and feed additive.

Magnesium Hydroxide

Preparation

Precipitation from a solution of a magnesium salt by sodium hydroxide; precipitation from sea water with lime. It occurs naturally as brucite.

Properties

White powder; odourless. Soluble in solutions of ammonia salts and dilute acids; almost insoluble in water and alcohol; sp. gr. 2.36; m.p. decomposes at 350°C. Non-toxic, non-combustible.

Uses

Intermediate for obtaining magnesium metal; sugar refining; medicine (antacid, laxative); residual fuel oil additive; sulphite pulp; uranium processing; dentifrices; in foods as drying agent, colour retention agent; frozen deserts.

Magnesium Sulphate

Preparation

(i) Action of sulphuric acid on magnesium oxide, hydroxide or carbonate; and (ii) mined in a high degree of purity.

Properties

Colourless crystals; saline, bitter taste; neutral to litmus; sp. gr. (a) 2.65; (b) 1.678; (a) decomposes 1124°C; (b) loses $6H_2O$ at 150°C; $7H_2O$ at 200°C; soluble in glycerol; very soluble in water; sparingly soluble in alcohol. Low toxicity. Non-combustible.

Uses

Fire-proofing; textiles (warp-sizing and loading cotton goods, weighting silk, dyeing and calico printing); mineral waters; catalyst carrier; ceramics; fertilisers; paper (sizing); cosmetic lotions; dietary supplement.

Magnesium Silicate

Preparation

It is prepared by interaction of a magnesium salt and a soluble silicate.

Properties

Fine, white powder; insoluble in water or alcohol. An absorbent. Non-combustible; may be toxic by inhalation.

Uses

Rubber filler; ceramics; glass; refractories; absorbent for crude oil spills; manufacture of permanently dry resins and resinous compositions; paints, varnishes, and paper (filler); animal and vegetable oils (bleaching agent); odour absorbent; filter medium; catalyst and catalyst carrier; anticaking agent in foods.

Chapter 19

Ceramic and Refractories

INTRODUCTION

Ceramics comprise all engineering materials or products (or portions thereof) that are chemically inorganic, except metals and alloys, and are usually rendered serviceable through high temperature processing. Ceramic materials are normally composed of both cationic and anionic species; their primary difference from other materials is in the nature of their chemical bonding. They are frequently termed ionic solids, i.e., possessing ionic bonding.

Modern ceramics encompass a wide range of materials ranging from single crystals and dense polycrystalline materials through glass-bonded aggregates to insulating foams and wholly vitreous substances. The need for high performance ceramic materials has increased steadily in the last ten to twenty years. As a class of materials, ceramics are better electrical and thermal insulators and more stable in chemical and thermal environments than metals, are appreciably stronger in compression than in tension, and exhibit greater rigidity, hardness, and temperature stability than polymers.

RAW MATERIALS

The principal raw materials of the ceramic industry are clay (including shale and mudstone), silica, and feldspar. Other raw materials include a wide variety of rocks, minerals and synthetic compounds used to manufacture diverse products.

Clays

Clay minerals are of three principal families: kaolinite, $Al_4Si_4O_{10}(OH)_8$, montmorillonite, $X_yAl_2(Al_ySi_{4-y}O_{10})(OH)_2$ where X is usually Na, Mg, or

Al, and illite, $K_y(AlFeMg_4Mg)(Al_ySi_{8-y})O_{20}(OH)_4$. Closely associated minerals are gibbsite, $Al(OH)_3$, diaspore, $HAlO_2$, and bauxite (of indefinite composition but usually given as $Al_2O_3.2H_2O$ which is an intermediate between the first two). All clays have as the major constituents one or more of these minerals.

The kaolinite group includes kaolinite, halloysite, dickite, and nacrite. The montmorillonite group includes montmorillonite, nontronite, beidellite, hectorite, and saponite. The illite group, similar to muscovite but containing less potassium, more silica, and more combined water, includes the illites, the hydromicas, phengite, brammallite, glaucomite, and celadonite. There is a wide range of substitutions that occur in each family. All three groups of minerals are classified as sheet silicas. In addition, most clays will have one or more accessory minerals, e.g., quartz, muscovite, biotite, limonite, hydrous micas, feldspar, vermiculite, and/or carbonaceous matter.

The clays used in the ceramic industry are usually referred to by names that reflect their use. For example: a pottery clay is used to make pottery; a sewer-pipe clay is used for sewer pipe; fire-clays (or refractory clay) are used to manufacture fireclay bricks, crucibles, refractory mortars, etc. china clay is used to manufacture whitewares, slip clays are used for glazing stoneware and poreclain; and brick clays are used to manufacture common brick and face brick. Exceptions to this pattern do occur. For example, kaolins are clays consisting chiefly of kaolinite (Table 19.1). The most refractory of all clays, with softening points of 1700–1785°C, it is used extensively in whitewares as the chief ingredient of most porcelain and china bodies. Ball clays are any sedimentary clay of high plasticity. They are composed primarily of kaolinite with varying amounts of impurities and some organic matter. They are used whenever increased workability is needed.

There are three common types of *feldspar*, potash ($K_2O{\cdot}Al_2O_3{\cdot}6SiO_2$), soda ($Na_2O{\cdot}Al_2O_3{\cdot}6SiO_2$), and lime ($CaO{\cdot}Al_2O_3{\cdot}6SiO_2$), all of which are used in ceramic products to some extent. Feldspar is of great importance as a fluxing constituent in ceramic formulae. It may exist in the clay as mined, or it may be added as needed.

The third main ceramic constituent is sand or *flint.* Its essential properties for the ceramic industries are summarised, along with the similar characteristics of clay and feldspar, in Table 19.2. For light-coloured ceramic products, sand with a low iron content should be chosen.

Table 19.1. Analyses of typical clays.

Typical kaolins, %	SiO_2	Al_2O_3	Fe_2O_3	CaO	MgO	Alkalies	H_2O
Sedimentary (Fla.)[a]	47.0	36.8	0.8	0.2	0.2	0.2	15.0
Crude residual (N.C.)	62.4	26.5	1.4	0.6		1.0	8.5
Washed residual (N.C.)[b]	45.8	36.5	0.3	0.5		0.3	13.4
English china clay	48.3	37.6	0.5	0.1		1.6	12.0
Tennessee ball clay, 1	46.9	33.2	2.0	0.3	0.4	0.7	
English ball clay	49.0	32.1	2.3	0.4	0.2	3.3	
New Jersey Ball clay[c]	45.6	38.9	1.1		0.1	0.2	

[a] Also 0.2% TiO_2.
[b] Also 11% FeO.
[c] Also 1.3% TiO_2.

Table 19.2. Basic raw materials for ceramics.

	Kaolinite	Feldspar	Sand or flint
Formula	$Al_2O_3 \cdot 2SiO_2 \cdot 2H_2O$	$K_2O \cdot Al_2O_3 \cdot 6SiO_2$	SiO_2
Plasticity	Plastic	Non-plastic	Nonplastic
Fusibility	Refractory*	Easily fusible binder	Refractory*
Melting point	3245°F, 1785°C	2100°F, 1150°C	3110°F, 1710°C
Shrinkage on burning	Much shrinkage	Fuses	No shrinkage

* Infusible at highest temperature of coal fire (1400°C).

In addition to the three principal raw materials, a wide variety of other minerals, salts, and oxides is used as fluxing agents and special refractory ingredients. Some of the more *common fluxing agents* which lower temperatures are :

Borax ($Na_2B_4O_7 \cdot 10H_2O$)	Fluorspar (CaF_2)
Boric acid (H_3BO_3)	Cryolite (Na_3AlF_6)
Soda ash (Na_2CO_3)	Iron oxides
Sodium nitrate ($NaNO_3$)	Antimony oxides
Pearl ash (K_2CO_3)	Lead oxides
Nepheline syenite	Lithium minerals
Calcined bones	Barium minerals
Apatite [$Ca_5(F,Cl,OH)(PO_4)_3$]	

Some of more common *special refractory ingredients* are :

Alumina (Al_2O_3)
Olivine [$(FeO,MgO)_2SiO_2$]
Chromite ($FeO.Cr_2O_3$)
Aluminium silicates ($Al_2O_3 \cdot SiO_2$) (kyanite, sillimanite, andalusite)
Dumoritierite ($8Al_2O_3 \cdot B_2O_3 \cdot 6SiO_2 \cdot H_2O$)
Magnesite ($MgCO_3$)
Lime (CaO) and limestone ($CaCO_3$)
Zirconia (ZrO_2)
Titania (TiO_2)
Hydrous magnesium silicates, e.g., talc ($3MgO \cdot 4SiO_2 \cdot H_2O$)
Carborundum (SiC)
Mullite ($3Al_2O_3 \cdot 2SiO_2$)
Dolomite [$CaMg(CO_3)_2$]
Thoria (ThO_2)

Many other raw materials are used in various combinations.

CHEMICAL CONVERSIONS, INCLUDING BASIC CERAMIC CHEMISTRY

All ceramic products are made by combining various amounts of the foregoing raw materials, shaping and heating to firing temperatures. These temperatures may be as low as 700°C for some overglazes or as high as 2000°C for many vitrifications. Such temperatures cause a number of reactions, which are the chemical bases for the *chemical conversions:*

(1) Dehydration, or "chemical water smoking," at 150 to 650°C.

(2) Calcination, e.g., of $CaCO_3$ at 600 to 900°C.

(3) Oxidation of ferrous iron and organic matter at 350 to 900°C.

(4) Silicate formation at 900°C and higher.

Some of the initial chemical changes are relatively simple, such as calcination of $CaCO_3$ and the dehydrations and decompositions of kaolinite. Other reactions, such as silicate formations, are quite complex and change with the temperature and constituent ratios. Phase-rule studies as exemplified by these figures have been of *revolutionary importance* in interpreting empirical observations in the ceramic industries and in making predictions for improvements. For instance, the $Al_2O_3 \cdot SiO_2$ system have led to the important development of processes for mullite refractories. This diagram shows that any percentage of liquefaction can be obtained, dependent on a definite temperature, except at the monovariant points. Thus, if the progressive melting is kept from going too far by controlling the rise in temperature, sufficient solid skeletal material will remain to hold the hot mass together. This $Al_2O_3 \cdot SiO_2$ diagram shows that mullite is the only stable compound of alumina and silica at high temperatures.

Ceramic products are all more or less refractory, i.e., resistant to heat, and the degree of refractoriness of a given product is determined by the relative quantities of refractory oxides and fluxing oxides. The principal refractory oxides are SiO_2, $Al_2O_3 \cdot CaO$, and MgO, with ZrO_2, TiO_2. Cr_2O_3, and BeO used less commonly. The principal fluxing oxides are Na_2O, K_2O, B_2O_3, and SnO_2, with fluorides also used as fluxes in certain compositions.

The common ingredient of all ceramic products is clay (kaolinite, usually), and therefore the chemical reactions which occur on heating clay are quite important. The first effect of the heat is to drive off the water of hydration; this occurs at about 600 to 650°C and absorbs much heat, leaving an amorphous mixture of alumina and silica, as shown by x-ray studies.

$$Al_2O_3 \cdot 2SiO_2 \cdot 2H_2O \longrightarrow Al_2O_3 + 2SiO_2 + 2H_2O$$

In fact, a large proportion of the alumina can be extracted with hydrochloric acid at this stage. As heating is continued, the amorphous alumina changes quite sharply at 940°C to a crystalline form of alumina, γ-alumina, with the evolution of considerable heat. At a slightly higher temperature, beginning at about 1000°C, the alumina and silica combine to form mullite ($3Al_2O_3 \cdot 2SiO_2$). At a still higher temperature, the remaining silica is converted into crystalline cristobalite. Therefore the fundamental overall reaction in the heating of clay is as follows

$$\underset{\text{Kaolinite}}{3(Al_2O_3 \cdot 2SiO_2 \cdot 2H_2O)} \longrightarrow \underset{\text{Mullite}}{3Al_2O_3 \cdot 2SiO_2} + \underset{\text{Cristobalite}}{4SiO_2} + 6H_2O$$

The presence of fluxes tends to lower the temperature of formation of mullite and of attainment of the equilibrium conditions given in the equilibrium diagram.

An actual ceramic body contains many more ingredients than clay itself. Hence the chemical reactions are more involved, and there will be other chemical species besides mullite and cristobalite present in the final product. For example, various silicates and aluminates of calcium, magnesium, and possibly alkali metals may be present. However, the alkali portion of feldspar and most of the fluxing agents become part of the glassy, or vitreous phase of the ceramic body. All ceramic bodies undergo a certain amount of vitrification, or glass formation, during heating, and the degree of vitrification depends upon the relative amounts of refractory and fluxing oxides in the composition, the temperature, and the time of heating. The vitreous phase imparts desirable properties to some ceramic bodies, e.g., by acting as a bond and imparting translucency in chinaware. Even in refractories some vitrification is desirable to act as a bond, but extensive vitrification destroys the refractory property. Thus it is seen any ceramic body is composed of a vitreous matrix plus crystals, of which mullite and cristobalite are two of the most important.

The *degree of vitrification*, or the progressive reduction in porosity, provides the basis for a useful classification of ceramic products as follows:

(1) *Whitewares*. Varying amounts of fluxes, heat at moderately high temperatures, varying vitrification.

(2) *Heavy-clay products*. Abundant fluxes, heat at low temperatures, little vitrification.

(3) *Refractories*. Few fluxes, heat at high temperatures, little vitrification.

(4) *Enamels*. Very abundant fluxes, heat at moderate temperatures, complete vitrification.

(5) *Glass*. Moderate fluxes, heat at high temperatures, complete vitrification.

WHITEWARES

Whiteware is a generic term for ceramic products which are usually white and of fine texture. These are based on selected grades of clay bonded together with varying amounts of fluxes and heated to a moderately high temperature in a kiln (1200–1500°C). Because of the different amounts and kinds of fluxes, there is a corresponding variation in the degree of vitrification among whitewares, from earthenware to vitrified china. These may be broadly defined as follows :

(1) *Earthenware*, sometimes called *semi-vitreous* dinnerware, is porous and nontranslucent with a soft glaze.

(2) *Chinaware* is a vitrified translucent ware with a medium glaze which resists abrasion to degree : it is used for non-technical purposes.

(3) *Porcelain* is a vitrified translucent ware with a hard glaze which resists abrasion to the maximum degree. It includes chemical, inslulating, and dental porcelain.

(4) *Stoneware*, one of the oldest of ceramic wares, was in use long before porcelain was developed: in fact, it may be regarded as a crude porcelain, the raw material being of a poorer grade and not so carefully fabricated.

(4) *Sanitary ware*, formerly made from clay, was usually porous; hence a vitreous composition is presently used. Prefirred and sized vitreous grog is sometimes included with the triaxial composition.

(5) *Whiteware tiles*, available in a number of special types, are generally classified as *floor tiles*, resistant to abrasion and impervious to stain

penetration, glazed or unglazed, and as *wall tiles* (in a variety of colours and textures), which also have a hard, permanent surface.

Manufacture of Porcelain

To represent a typical manufacturing procedure in this group, porcelain as defined above is chosen. There are three lines of production: *wet-process porcelain*, used for production of fine-grained, highly glazed insulators for high-voltage pieces; and *cast porcelain*, necessary for the making of pieces too large or too intricate for the other two methods. These three processes are based on the same raw materials, the differences in manufacturing being largely in the drying and forming (shaping) steps. The wet process is illustrated in Fig 19.1. This flowchart may be broken down into the following sequences :

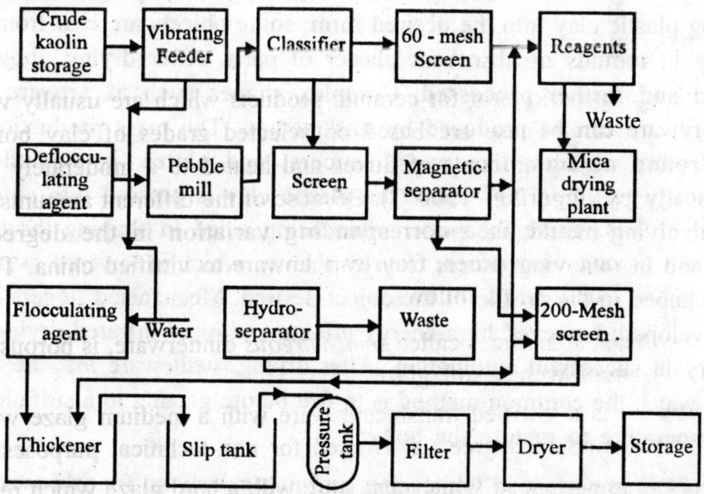

Fig. 19.1. Manufacture of porcelain by wet process.

Raw materials of proper proportions and properties to furnish porcelain of the desired quality are weighed from overhead hoppers into the weighing car.

Feldspar, clays, and flint are mixed with water in the blunder (clay-water mixer) and then passed over a magnetic separator, screened, and stored.

Most of the water is removed (and wasted) in the filter press. All the air is taken out in the pug mill, assisted by a vacuum and by slicing knives. This results is denser and stronger porcelain.

The prepared clay is formed into blanks in a hydraulic press or by hot-pressing in suitable moulds. The blanks are preliminarily dried, trimmed, and finally completely dried, all under carefully controlled conditions. A high surface lustre is secured by glazing with selected materials. The vitrification of the body and the glaze is carried out in tunnel kilns, with exact controls of temperature and movement.

The porcelain articles are protected by being placed in saggers fitted one on top of the other in cars. This represents a one-fire process wherein body and glaze are fired simultaneously. The porcelain pieces are rigidly tested electrically and inspected before storage for sale.

Much *tableware* is manufactured by more complicated procedures than illustrated by the somewhat related porcelain product. Some objects are shaped by being thrown on the potter's wheel where skilled hands work the revolving plastic clay into the desired form; some objects are *cast* from the clay slip in moulds of absorbent plaster of paris. After drying, they are removed and further processed. Complex shapes such as artware and laboratoryware can be produced by this method. The mass production of simple round objects, like cups, saucers, and plates, is carried out economically by *jiggering*, where the plastic clay is pressed into or on a single revolving mould, the potter often being aided in shaping the outer surface and in removing excess clay by a lever, which is lowered over the mould, shaped in the profile of the object desired. Mechanised jiggers have been developed; however, high-energy mixing and rapid infrared drying are necessary in successful automation. After drying, whiteware may be fired in three ways; the common method is to fire before glazing to a sufficiently high temperature to mature the body.

Glazing is important in whitewares and particularly so for tableware. A glaze is a thin coating of glass melted onto the surface of more-or-less porous ceramic ware. It contains ingredients of two distinct types in different proportions: refractory materials such as feldspar, silica, and china clay, and fluxes such as soda, potash, fluorspar, and borax. Different combinations of these materials and the different temperatures at which they are fixed give a wide range of texture and quality. Nepheline syenite permits firing at a lower temperature. The glaze must be bonded to the ware, and its coefficient of expansion must be sufficiently close to that of the ware to avoid defects such as "crazing" and "shivering." The glaze may be put on by dipping, spraying, pouring, or brushing. Decoration of such ware may be "underglaze" or "overglaze." *Glost firing* is the technical term for the firing of the glaze. Earthenware should be glazed between 1050 and 1100°C; stoneware between 1250 and 1300°C.

Non-clay Minerals

Silica

Silica occurs in the form of quartz, tridymite, cristobalite, vitreous silica, cryptocrystalline forms, hydrated silica, and diatomite. Principal sources of silica used in the ceramic industry are the sandstones, quartzites, and sands. Principal uses of silica in ceramic industry are in glass, fused silica, brick, whiteware bodies, whiteware glazes, and enamels. Diatomite is used for insulating refractories up to a temperature of ca 900°C; when bonded with clay, it may be shaped and used for insulation up to a temperature of ca 900°C; when bonded with clay, it may be shaped and used for insulation up to a temperature of 1370°C.

Feldspar

Feldspars are used chiefly as fluxes and sources of Al_2O_3, SiO_2, alkalies (K_2O, Na_2O), and CaO. The plagioclase feldspars vary in composition from albite, $NaAlSi_3O_8$, to anorthite, $CaAl_2Si_2O_8$, in a continuous series to solid solutions. Orthoclase and microcline feldspar ($KAlSi_3O_8$) are referred to as potash feldspar, used extensively in whiteware bodies. Anorthoclase, $(Na,K)AlSi_3O_8$, is a combination of albite and potash feldspar. High soda feldspars are used in glasses and glazes. All feldspars contain some free silica which is not detrimental as long as it is controlled.

Other non-clay minerals

Other non-clay minerals include nepheline syenite, a rock that contains a large percentage of the mineral nephelite, $(Na,K)_2Al_2Si_2O_8$, along with some soda and potash feldspars; lime, derived by calcination of limestone chiefly calcium carbonate, $CaCO_3$, which is used in glazes, enamels, and glasses (e.g., whiting is a high purity calcium carbonate) magnesium silicates such as talc (used as a flux with clays), asbestos (used chiefly for insulating refractories) (used for refractory products; fluxing minerals which lower the vitrification temperature, the melting temperature or the reaction temperature, e.g., lithium minerals spodumene, $Li_2Al_2Si_4O_{12}$, lepidolite, $[K_2(Li_3Al_3)–(Al_2Si_6O_{20})\ (OH,F)_4]$, amblygonite, $Li_2F_2Al_2$-P_2O_8, and petralite, $LiA_1Si_4O_{10}$, as well as barium minerals such as barite, $BaSO_4$, and whiterite, $BaCO_3$; refractory minerals such as zirconium minerals, hydrated alumina minerals, titania, TiO_2, thoria, ThO_2, and graphite, one of the most refractory materials available. More minerals in this group are magnesite, dolomite, gypsum, chromite, the aluminium silicates, and pyrophyllite.

Special Materials

Super refractories

Super refractories include the carbides of elements such as silicon, boron, zirconium, hafnium, tantalum, vanadium, molybdenum, tungsten, and niobium, nitrides, sulphides, single and complex oxides.

Nuclear ceramic materials

Nuclear ceramic materials being researched include uranium oxide, UO_2, uranium carbide, nitrides, sulphides, phosphides, and combinations.

Electronic ceramic materials

Electronic ceramic materials include ferrites with improved magnetic properties, titanates, principally $BaTiO_3$.

In preparing materials for forming, optimal particle size and size distribution may be obtained by crushing and grinding the various materials, separating the various size fractions by screening, and then blending the desired size fractions of each material.

Forming Processes

Forming impacts to ceramic ware permanent shape, and temporary size (altered by shrinkage) and strength (which increases during drying and firing). Forming methods used in the production of ceramics are plastic deformation (e.g., extrusion, dry pressing and all hot forming methods) and casting (e.g., slip and fusion casting).

In preparing materials for forming, optimal particle size and size distribution, may be obtained by crushing and grinding the various materials, separating the various size fractions by screening, and then blending the desired size fractions of each material.

Plastic Forming Processes

In cold forming, the ware is made oversize and shrinks during firing and drying. In hot forming (e.g., hot pressing, hot extrusion, hot rolling, forging and swaging, hot isostatic pressing, and chemical vapour deposition), the ceramic ware is made close to its final size.

Extrusion processes

Extrusion processes are used to form ware of any shape that has an axis normal to a fixed cross-section. It is widely used in the structural clay products industry and to a lesser extent in the whiteware and refractories industries. Completely non-plastic materials can be extruded with the addition of suitable plasticisers such as gums, starches, poly(vinyl alcohol), waxes, and wax emulsions. Extrusion usually comprises three stages: pugging, deairing, and extrusion through a die.

Soft plastic forming

Soft plastic forming is usually used to produce shapes with symmetrical circular cross-sections from materials with 20–30% water and significant percentages of plastic clays. Common processes include jiggering, used principally in the whiteware industries, and hand moulding.

Dry pressing

Dry pressing is used when many parts of relatively simple shapes are to be made, e.g., in the whiteware, refractories, and abrasive industries and in the production of cermets (ceramic–metal composites). The parts must have no undercuts, be fairly uniform in thickness, and have a length not normally exceeding twicc their diameter. The material to be dry pressed is usually in the form of granules, either dry or containing up to 12% water. Isostatic presses and vibratory forming are used to overcome the problems of non-uniform density which arise

Casting processes

For slip casting, a suspension of the raw materials in a liquid such as water is prepared by blunging or ball milling. This slip is poured into a porous plaster of Paris mould, with the interior surfaces of the mould conforming to the exterior surface of the desired ware. As the plaster absorbs water from the slip, solid particles are deposited on the interior surface of the mould. Slip casting permits the formation of complex shapes, and is widely used in the whiteware industries to make art objects and dinnerware that is not adaptable to jiggering, and in the refractories industry to make crucibles and other special shapes. Slip-cast moulds are relatively inexpensive and reusable. Special techniques can be used to form materials without plasticity; thin sheets of ceramics can be made by the doctor-blade process, and complex shapes can be made by injection moulding.

THERMAL TREATMENT

Thermal treatment is an essential step in the manufacturing of ceramic products. Materials that are stable at room temperature have to be raised to relatively high temperatures (ranging from 700°C for enamels to ca 1650°C for alumina ceramics) for reactions to take place.

Methods of Thermal Treatment

Thermal treatment is a smoothly varying time–temperature profile, rather than a particular temperature held for some time period. To determine firing temperature and quality control, thermocouples, radiation pyrometers, and optical pyrometers are used; expendable pyrometric cones are used for a large class of materials, e.g., structural clay products, porcelains, sanitary ware. Microstructures and properties of many speciality ceramics vary according to time–temperature profile and often require precise monitoring of thermal gradients and temperature.

Extreme temperatures are achieved in protective atmospheres using refractory metals; the upper limit is the temperature at which vapourisation occurs, e.g., for molybdenum, 1800–2200°C; for tungsten, 3000°C. Above 3000°C, graphite is used; plasma devices are used up to 15,000°C, e.g., for dynamic thermal processing of ceramic particulates. Atmospheric control of the furnace is required for oxide ceramics to control the volatile components, and for non-oxide ceramics, nitrogen and inert gases are used to exclude oxygen.

Kilns

A tunnel kiln is the most prevalent equipment for thermal treatment; it is used for structural clay products, refractories, tile, porcelain, dinnerware, and electronic components. A periodic kiln is ordinarily used for small-scale production. A rotary kiln is used for manufacturing products such as cement and calcined lime or dolomite. Arc melting is used for processing materials such as alumina, magnesia, zirconia, and mullite.

Hot-pressing processes

Hot-pressing processes, which incorporate forming and thermal treatment in one step, are used to obtain a dense ceramic in chemical systems that cannot be fully densified using conventional firing methods; it is limited to very simple shapes and slow rates of production.

Physical and Chemical Changes During Thermal Treatment

Drying and binder removal occurs between 0–400°C. The ware loses physically held water and organic binders which are essential for the forming operation.

Firing occurs at higher temperatures, and involves these changes: dissociation, the loss of carbon dioxide by carbonate constituents of ceramic compositions; compound formation, the continued heating of clays which results in the formation of mullite, $3Al_2O_3.2SiO_2$; polymorphic transformation of materials capable of existing in more than one crystallographic arrangement, e.g., silica and zirconia; sintering, the fabrication of a product usually involving densification by pore removal through diffusional mechanisms; and vitrification.

PROPERTIES AND APPLICATIONS

Composition and Microstructure

Composition is the chemical and mineralogical makeup of the total substance of the material (including its deliberate and accidental impurities); microstructure is the crystallographic structure of each component or phase and, in particular, the size, shape, and distribution of each of the phases present (including the unoccupied space or porosity). These are determinative parameters fixed by the combined response of the starting materials to all the processing steps.

A ceramic materials is a crystalline, vitreous, or mixed type solid which may be monophase (composed of a single substance) or polyphase (containing two or more discrete substances); monophase crystalline ceramics may be monocrystals. Most ceramics are polycrystalline masses with abrupt changes in orientation or composition occurring across each grain or phase boundary. Many properties of ceramics are surface-sensitive, both with respect to mechanical condition and/or chemical and atmospheric environment.

Chemical Properties of Ceramic Materials

Ceramic materials based on oxides, carbides, nitrides, and borides exhibit a high degree of chemical stability, e.g., oxides are extremely stable over wide ranges of temperature and environmental conditions. The stability of non-oxide ceramic compounds results from the compactness of the crystal structure, the directed chemical bonding (generally ionic or covalent, or of mixed ionic-covalent character), and of the high field strengths associated with the relatively small, highly charged cations encountered in refractory ceramics. Thermodynamically, the most stable ceramic will be that having the greatest negative free energy of formation from the elements, e.g., yttria, thoria, alumina, beryllia, magnesia, and stabilised zirconia. Most refractory ceramics have very low vapour pressures in the most stable valence state, but the suboxides tend to be volatile (SiO, Al_2O, etc). Interactions between dissimilar ceramics, or between ceramics and metals, become increasingly likely as operating temperature is increased, or at low pressure. Even a system of materials with high melting points can form fluid liquids at temperatures much below either melting point; e.g., Al_2O_3 melts at 2050°C and CaO at 2500°C, but when in contact they form a reactive, eutectic liquid at 1450°C. The presence of a few per cent impurities can significantly lower the temperature of liquid formation.

Optical Properties

Many ceramic substances are optically transparent in single crystal or vitreous forms. Optical properties are closely linked to composition and structure of the ceramic; e.g., the degree of polarisation is a function of ion size, binding energy, and crystallographic direction, and is measured by the index of refraction. Also, the refractive index of SiO_2 is 1.55; of MgO 1.74; of $ZrSiO_4$ 1.95; of TiO_2 2.71; and for LiF is 1.392. Principal and very practical sources of unfilled energy bands are impurities, or deliberate colorant oxide additives such as cobalt blue and chrome green. The microstructure of a polycrystalline ceramic tends to dominate the useful optical properties of the base material. Polyphase polycrystalline materials may develop additional scattering owing to differences in refraction across boundaries, dispersed phases in particular acting as opacifiers; dispersions of SnO_2, $ZrSiO_4$, TiO_2 and/or Sb_2O_3 are commercially employed as opacifiers and whiteners in enamels, glazes, and glasses. Highly translucent, almost transparent, pore-free polycrystalline ceramics are produced by sintering or the hot-pressing methods, and require careful control of impurities. Control of impurities can minimise internal absorption.

Thermal Properties

The thermal properties of greatest interest include specific heat, thermal conductivity, and thermal expansion. The principal properties improving thermal shock resistance are low thermal expansion, high thermal conductivity, and high strength. Ceramic materials in general do not have particularly good thermal shock resistance because of their brittle nature. They are used in many applications where ductile materials could not be used because of superior high temperature properties. Materials such as fused silica, cordierite, and lithium aluminosilicates have very low thermal expansion and good thermal shock resistance despite being weak when compared to other ceramic materials.

Elastic Properties

Elastic properties of solids arise from the interaction of mechanical distortion, called strain, with the periodicities of the atomic or ionic structure. Elastic moduli measure the force involved in achieving recoverable, small, unit displacements of atoms or ions from their equilibrium positions, and are dependent upon the type of bonding and the electron density configuration of the crystals. For isotropic substances such as glass, and for randomised polycrystalline materials having pseudoisotropic behaviour, elastic properties can be adequately expressed in terms of :

$$E = \text{Young's modulus of elasticity} = \frac{\sigma}{\varepsilon}$$

$$G = \text{Shear modulus} = \frac{\tau}{\gamma}$$

$$\mu = \text{Poisson's ratio} = \frac{E^{\gamma}}{2G} - 1$$

where σ = normal stress, ⊒ = shear stress, ε = normal strain, γ = shear strain.

Strength

Ceramics are brittle materials, strong in compression but much weaker in transverse bending or in tension. As in glass at room temperature deformation is entirely elastic up to the point of failure and brittle fractures attributed to stress-induced catastrophic extension of pre-existing defects, the well-known Griffith cracks. The theoretical strengths of ceramic materials, derived from atomic bonding consideration, are far larger than the observed tensile strengths which are explained by the presence of flaws. The crack-tip areas, or microflaws from which fractures originate, are produced during the fabrication process and can be identified by careful microscopy; the largest flaw is often the one that affects strength.

At high temperatures, ceramics can deform plastically, a mechanical behaviour similar to metals, and in creep. Ceramic material selected for resistance to creep at high temperatures may be rather weak and prone to brittle fracture at low temperatures, and subject to excessive deformation at higher temperatures

Engineering with ceramics

A minimum service-life guarantee can be determined from knowledge of crack propagation behaviour. The value obtained for strength in a test is highly dependent on how the tests are made; the type of flaw present at the surface of the material can determinate if these flaws are larger than those inherent in the bulk materials.

Electrical and Magnetic Properties

Ceramics are increasingly used as electronic and magnetic components. Most oxide ceramics in their normal valence states are insulators while most monoxide ceramics are semiconductors; e.g., SiC is much used as a high temperature heating element and as a voltage-dependent resistor to ground. Ceramics containing transition metal ions may display pronounced magnetic effects; in contrast to magnetic metals, they do not act as electrical conductors.

Magnetic materials are classified as hard (permanent magnets) or soft (materials that may be magnetised and demagnetised). The main types are the spinel ferrites, rare earth garnets, and hexagonal ferrites.

Composites and Cermets

An important segment of the ceramic industry is concerned with complex materials or components of which the ceramic phase is only a part; another part is metallic or plastic, and the useful properties of the combination are different from those of either phase alone. Composites contain at least one phase which is macroscopic in at least one dimension (fibrous shape). Cermet contains ceramic and metallic phases randomly shaped and initimately dispersed one within the other on a microstructural scale.

Uses

All useful properties of a ceramic must be provided in advance; failure in service will necessitate replacing the piece *in toto*. Many ceramics are produced in volume as standard items, e.g., refractory bricks and shapes, crucibles, furnace tubes, and hermetic seals. Ceramic-metal composites and glass-fibre-reinforced plastics are employed for aerospace hardware. Ceramic fibre-reinforced metals, metal-impregnated ceramics, and metal-ceramic laminates are being researched. The largest group of cermets are carbides or cemented carbides.

CERAMIC AS ELECTRICAL MATERIALS

The electrical characteristics of ceramic materials vary greatly; oxides, borides, nitrides, and carbides have metallic conductivity in some cases but behave like semi-conductors in others. Many ceramics are good insulators even at high temperatures. The atomic processes are very different for the various conduction modes and the transport of current may be caused by the motion of electrons or electron holes, or by ions.

Semiconductor materials are used in many specialised applications such as resistance heating elements; semiconductor devices such as rectifiers, photocells, transistors, thermistors, detectors, and modulators. Ceramics are equally important as electrical insulators; porcelains are used for both low- and high-voltage insulation. Consequently, the entire range of electrical conduction properties should be considered.

Electrical Conduction Phenomena

Materials are usually classed according to the specific conductivity mode, such as insulators which have low concentrations and low mobility of carriers. Metallic conductors such as borides, carbides, some oxides, etc., have a high conductivity value which is not a strong (exponential) function of temperature. Insulators have a low conductivity which increases exponentially with temperature; for example, the conductivity (Ω.m) at room temperature of SiO_2 glass is $< 10^{-12}$; of low voltage porcelain 10^{-10}–10^{-12}; of fire-brick clay 10^{-6}. Semiconductors are intermediate and have an exponential temperature dependence. The conductivity $(\Omega.m)^{-1}$ of dense silicon carbide is 10; of boron carbide 200; of pure germanium 10^4. Note that conductivity is the inverse of resistivity ($\sigma_i = 1/\rho_i$) and the two are used interchangeably. The resistivity in SI units is Ω.m but much of the literature remains in units of Ω.cm.

The total conductivity is that owing to the net motion of electrons, holes, and each of the ions :

$$\sigma = \sigma_1 + \sigma_2 + \sigma_3 \; ... + \sigma_i$$

The fraction of the total conductivity (at a specific temperature, composition, etc) owing to the particle is called the transference number of i: $t_i = \sigma_i/\sigma$.

Ionic Conduction in Ceramics

Ions in compounds or glasses such as oxides and halides constitute an ever-present charge carrier that can contribute to electrical conductivity.

Electrical conductivity resulting from ion migration is important in many ceramic materials. Its analysis requires a determination of the concentration and mobility of charge carriers. For an ion to move through the lattice under the driving force of an electric field, it must have sufficient thermal energy to pass over a free-energy barrier, ΔG^+, the intermediate position between lattice sites. The result is a mobility that depends on the ion jumping from site to site. The ion mobility (for the rocksalt crystal structure) is expressed as :

$$\mu_i = \frac{4ez_i a^2 \nu}{kT} e^{-\Delta G^+/kT}$$

where a is the lattice parameter (ca 3×10^{-8} cm), ez_i is the charge on the ion, and ν is the vibrational frequency (ca 10^{13}/s).

When aliovalent (different valence) impurities are added to an ionic solid, the crystal lattice compensates by forming defects that maintain both electrical neutraliy and the anion-to-cation ratio of the host lattice. For example, addition of a mol of CaO to ZrO_2 requires the formation of a mol of oxygen vacancies. If this concentration is larger than the oxygen vacancies created by the thermal effects, then the conductivity from the motion of doubly charged oxygen ions is directly proportional to the concentration of CaO.

In polycrystalline materials, ion transport within the grain boundary must also be considered. For oxides with close-packet oxygens, the O-ion almost always diffuses much faster in the grain boundary than in the bulk. Other examples are less clear except that second phases at boundaries often promote higher ion diffusivities.

Several types of compounds show exceptionally high ionic conductivity and have recently become of technological interest. Such phases fall into three broad groups: (i) halides and chalcogenides of silver and copper in which the metal atom is disordered over several alternative sites; (ii) oxides with the β-alumina structure in which a monovalent cation is mobile; and (iii) oxides of the fluorite structure in which large concentrations of defects are caused by either a variable-valence cation or solid solution with a second cation of lower valence, e.g., CaO–ZrO_2 or Y_2O_3–ZrO_2.

Electron Conduction in Ceramics

The relatively high mobility of conducting electrons or electron holes when present contributes appreciably to electrical conductivity. In some cases, metallic levels of conductivity result; in others, the electronic contribution is extremely small. In all cases, the electrical conductivity can be interpreted in terms of carrier concentrations and carrier mobilities. Metallic conduction occurs in a few cases of transition-metal oxides such as ReO_3, CrO_2, VO, TiO, and ReO_2; the doped perovskite structures $LaTiO_3$, $CaVO_3$, $CaMoO_3$, $BaTiO_3$, $SrCrO_3$, $SrFeO_3$, $LaNiO_3$, $LaFeO_3$, $LaCoO_3$, and $LaCrO_3$; many of the bronzes, Na_xWO_3, La_xWO_3, and $Na_xNb_2O_5$; and some of the spinels, $Li_{0.5}\ In_{0.5}\ Cr_2S_4$. It occurs at specific dopant concentrations.

Two types of scattering affect the motion of electrons and holes. Lattice scattering results from thermal vibrations of the lattice and increases with the increasing amplitude of vibrations at higher temperatures, $T^{-3/2}$. A second source of scattering is the presence of impurities which distort the periodicity of the lattice, $T^{+3/2}$. The total mobility is proportional to the sum of these two terms. The temperature dependence of the mobility term for quasi-free electrons and holes is much smaller than that for their concentrations (exponential). As a result, the conductivity has a temperature dependence chiefly determined by the concentration term.

Non-stoichiometric and solute-controlled electronic ceramics

Most oxide semiconductors are either doped with impurities to create extrinsic defects or annealed under conditions in which they become non-stoichiometric. Although these effects have been carefully studied in oxides, the precise nature of the low mobility value is often difficult to measure. Because the impurity atoms introduce new localised energy levels for electrons, intermediate between the valence and conduction bands, impurities strongly influence the properties of semiconductors. If the new energy levels are unoccupied and lie close to the energy of the top of the valence band, electrons are easily excited out of the filled band into these new acceptor levels. This leaves an electron hole in the valence band which can contribute to electrical conductivity. Positive-carried (*p*-type) oxide conductors commonly arise as a result of non-stoichiometric composition with a decreased metal content (e.g., $Cu_{2-x}O$) and are sometimes called deficit semiconductors. If the impurity additions have filled electron levels close to the energy level of the conduction band, electrons may be excited from impurity atoms into the conduction band; these are called donor levels. The electron excited into the conduction band is able to contribute to conductivity. Negative-carried (*n*-type) oxide conductors commonly result from a non-stoichiometric composition with an excess metal content (e.g., $Zn_{1+x}O$) and are sometimes called excess semiconductors.

Silicon carbide can be doped with boron to provide acceptor levels within the band gap (ca .03 eV above the valence band), thus making it a *p*-type conductor, or nitrogen can be added to provide donor levels and *n*-type conduction (ca 0.07 eV below the conduction band).

Another method of obtaining semiconductor with controlled resistivity which avoids solutions of two or more compounds with widely different conductivity. Magnetite, Fe_3O_4, is an excellent conductor with a specific resistance of about 10^{-4} Ω·m, compared to values of about 10^8 Ω·m for most stoichiometric transition-element oxides. The good electrical conductivity of magnetite is a function of the random location of Fe^{2+} and Fe^{3+} ions on octahedral sites which allows electron transfer from cation to cation.

In general, a condition for appreciable conductivity in the spinel structure is the presence of ions having multiple valence states at equivalent crystallographic sites. Examples of solid-solution semiconductors include Fe_3O_4–$MgCr_2O_4$, Fe_3O_4–$FeAl_2O_4$, and Fe_3O_4–$FeCr_2O_4$. The temperature coefficient or activation energy increases with increasing resistivity. Semiconductor materials of this type with materials like $MgAl_2O_4$, $MgCr_2O_4$, and Zn_2TiO_4 as the non-conducting component can be prepared with a controlled temperature coefficient of resistivity. Semiconductors made in this way are used as thermistors.

Ceramics with high electronic conductivity or with non-linear behaviour

As previously indicated, many oxides are metallic conductors. The perovskite- and rutile-structure oxides are the most frequently studied. The *d*- or *f*-electron states of the transition-metal ions overlap, causing a partially filled band of ca 10^{28} electrons/m^3 for conduction. The mobility of the electrons is usually less than 1 cm^2/(V.s). These materials are technologically significant as electrodes and in microelectronics. The ReO_2 and RuO_2 compounds are used in organic and inorganic pastes as electrical components and are applied as thick films followed by a firing step. The material most often studied in doped barium titanate, $BaTiO_3$, which is used as a barrier layer capacitor, integranular capacitor, and non-linear positive-temperature-coefficient (PTC) resistor.

Mixed Conduction in Ceramics

The conductivity of many ceramic materials is significantly influenced by both ionic and electronic defects. Highly resistive ceramics such as MgO and Al_2O_3 fall into this category; however, good electrical conductors may be of a mixed type. The CeO–ZrO_2 system is an example of mixed conduction. The ionic transference number varies from 0.03 to 1.0 depending on temperature and composition.

COLOURANTS FOR CERAMICS

Ceramics colourants can be divided into two broad groups: colourants used in conjunction with clay-based products such as dinnerware, tile, sanitaryware, etc., and colorants used in the manufacture of glass (unusually containers). In the manufacture of glass, wherein silica is the major ingredient, the presence of iron as an impurity imparts a decided coloured tint under normal conditions. Since this discolouration is generally considered to be undesirable, decolourisers are used as components of the glass batch. In clear glasses, colour is produced primarily by transition elements, alone or in combination. In effect, the glass, most often a silicate composition, acts as a solvent for solute colouring ions that produce a colour characteristic of the element(s) involved.

Colour can also be produced by developing a colloidal suspension in a glass matrix. Metals such as copper (or possibly Cu_2O), gold, and silver, and pigmenting materials such as cadmium sulphoselenides are known to produce strong colours and are used commercially to this end. Translucent or opaque glasses, most often white, can be produced by the precipitation of a crystalline phase during the cooling of a glass, by separation of an immiscible vitreous phase, or by precipitation of a crystalline phase from the glass during a secondary heating.

Both temperature and firing atmosphere can have marked effects on colour. Also, chemical composition of the matrix in which the colour is developed can have a significant effect on the colour produced. The average particle diameter and particle size distribution of ceramic products as well. Generally, if a colourant is reduced to too small a size, the resultant colour will be weakened; the colourant will be more readily attacked by the matrix with a probable total loss of colour.

Because of the inherent relatedness of ceramic colourants, it is possible to utilise similar manufacturing methods in their preparation. Close control is exercised in the selection of raw materials, which may be metallic oxides or some salts of appropriate metals, plus normally inert substances (from the point of view of not developing colour by themselves) such as silica, alumina, zirconia, calcium carbonate, clays, borax, etc. Certain mineralisers, such as some of the alkali or alkaline-earth halides, may be added to promote the formation of desired crystalline species in a subsequent thermal treatment. Depending on the colourant being prepared, both very narrow and relatively broad impurity tolerances are encountered.

Following a weighing operation, the raw materials are thoroughly blended. The end-use colouring capabilities of the raw-material mixtures are developed through a high temperature calcining operation. Following calcination the colorant is ground or milled to the optimum size desired. Finally, colourants may be blended either with other colourants to produce an intermediate hue (e.g., the ZrO_2–SiO_2 family) or with inert materials, such as silica, alumina, kaolin, etc., to produce different tints or tones of the same hue. The ultimate control utilised in the manufacture of ceramic colourants is the final appearance in an application. There are no serious problems with health or safety in the use and manufacture of ceramic colourants.

Colourant Systems

Black

Most black ceramic colourants are formed by a calcination of oxides (usually) of selected elements that develop a spinel structure when heated.

Blue

Cobalt is the major element used in producing blue colours for the ceramic industry; vanadium, ZrO_2, and SiO_2 are also used. Blue colours in glass, including some glazes, also can be developed by the addition of copper oxide. Characteristically, the resultant colour is bluish-green, somewhat between those colours produced by cobalt and chromium oxides when they are used separately.

Brown

A wide range of brown colours can be produced from combinations of chromium, iron, and zinc, most usually from equal parts of the oxide forms, although metal salts of these elements can be used as well.

Gray

Gray colours in ceramics can be obtained by use of one of the black colourants with a suitable inert diluent or a combination of tin and antimony oxides.

Green

Chromium oxide is the main ingredient in most green ceramic colourants and is used to prepare most green-coloured glasses.

Pink

Several basic colourant families are available for use in the manufacture of products requiring a pink coloration. These can be enriched, darkened, or lightened through the use of additives. Probably the oldest composition series is a combination of chromium, tin, and calcium to form the so-called chrome-tin pinks. Combinations of chromium and aluminium oxides will also yield a pink-to-ruby colour. Another series involves the use of iron in combination with zirconium and silicon. Very delicate pinks can be produced in vitreous systems by colloidal suspensions of gold or selenium.

Red

Iron oxide is one of the oldest red colourants and also one of the most powerful. By far the most prevalent red colourant series for lower temperature applications, i.e., over glaze colours, glass enamels, porcelain enamels, and coloured glazes, is a family of pigments based on cadmium, selenium, and sulphur.

White

In ceramic bodies white achromatic colours are developed as a result of the use of materials relatively free from impuritics. However, most so-called white bodies have a slight cream tint resulting often from minor quantities of iron and titanium dioxide that are often associated with kaolins, ball clays, feldspar, and silica, the usual ingredients in most whiteware products. Iron is also a contaminant in many other materials as well.

Yellow

A number of different compounds are in commercial use to produce yellow colours in ceramics; one of the most widely used is cadmium sulphide. A combination of lead and antimony oxides or metallic salts that is stable to ca 1100°C is useful in most ceramic applications in that very strong yellow colours can be achieved. Tin-vanadium yellows are usually prepared from tin oxide and a compound, such as ammonium vanadate, by calcining at 1100–1200°C. The last system comprises a combination of zirconium, silicon, and praseodymium. The main colourant for producing a very pure yellow in glass is CdS which is characterised by a sharp absorption edge.

REFRACTORIES

Refractories are materials that resist the action of hot environments by containing heat energy and hot or molten materials. The type of refractories that are used in any particular application depends upon the critical requirements of the process. There is no well-established line of demarcation between those materials that are and those that are not refractory, although the ability to withstand temperatures above 1100°C without softening has been cited as a practical requirement of industrial refractory materials.

Physical Forms

Refractories may be preformed (shaped) or formed and installed on site. Common forms include brick, setter tile and kiln furniture, fusion-cast shapes, cast and hand-moulded refractories, insulating refractories, castables and gunning mixes, plastic refractories and ramming mixes, mortars, composite refractories, and refractory coatings.

Raw Materials

Today, high purity natural raw materials are increasingly in demand as well as synthetically prepared refractory grain made from combinations of high purity and beneficiated raw materials. The material produced upon firing raw as-mined minerals or synthetic blends is called grain, clinker, co-clinker, or grog. Raw materials include silica, fireclay, high alumina diaspore clays, calcium aluminate cements, zirconia, and other basic raw materials, e.g., magnesite, dolomite, forsterite, chrome ore, silicon carbide, beryllia and thoria, and carbon and graphite.

General Properties

General properties of pure refractory materials are listed in Table 19.3.

Phase diagrams can be used to predict the reactants between refractories and various solid, liquid, and gaseous reactions. However, phase diagrams are derived from phase equilibria of relatively simple pure compounds and real systems are highly complex and may contain a large number of minor impurities that significantly affect equilibria. Furthermore, equilibrium between the reacting phases may not be reached under actual service conditions, and the physical environment of a product may be more influential to its life than the chemical environment.

Physical properties

The important physical properties of some refractories are listed in Table 19.4.

Mechanical properties

The physical properties of a particular refractory product depend upon its constituents and the manner in which they were assembled. The physical properties may be varied to suit specific applications; for example, for thermal insulation, highly porous products are employed, whereas dense products are used for slagging or abrasive conditions.

Table 19.3. Properties of pure refractory materials.

Material	Formula	Mp, °C	True specific gravity g/cm³	Mean specific heat J/(kg.K)[a]	Temp range, °C	Thermal conductivity, W/(m.K) at 500°C	Thermal conductivity, W/(m.K) at 1000°C	Linear thermal expansion coefficient per °C × 10^6, from 20–1000°C
Aluminium oxide	Al_2O_3	2015	3.97	795.5	25–1800	10.9	6.2	8.6
Beryllium oxide	BeO	2550	3.01	1004.8	25–1200	65.4	20.3–	9.1
Calcium oxide	CaO	2600	3.32	753.6	25–1800	8.0	7.8	13.0
Magnesium oxide	MgO	2800	3.58	921.1	25–2100	13.9	7.0	14.2
Silicon dioxide[b]	SiO_2		2.20	753.6	25–2000	1.6	2.1	0.5
Thorium oxide	ThO_2	3300	10.01	251.2	25–1800	5.1	3.0	9.4
Titanium oxide	TiO_2	1840	4.24	711.8	25–1800	3.8	3.3	8.0
Uranium oxide	UO_2	2878	10.90	251.2	25–1500	5.1	3.4	
Zirconium oxide[c]	ZrO_2	2677	5.90	460.6	25–1100	2.1	2.3	
Mullite	$3Al_2O_3.2SiO_2$	1850[d]	3.16	628.0	25–1500	4.4	4.0	4.5
Spinel	$MgO.Al_2O_3$	2135	3.58	795.5		9.1	5.8	6.7
Forsterite	$2MgO.SiO_2$	1885	3.22	837.4		3.1	2.4	9.5
Zircon	$ZrO_2.SiO_2$	2340–2550[e]	4.60	544.3		4.3	4.1	4.0
Carbon	C		2.10	1046.7	25–1300	13.4	9.9	4.0
Silicon carbide	SiC	3990[f]	3.21	795.5	25–1300	22.5	23.7	5.2

[a] To convert J to cal, divide by 4.184.
[b] Silica glass.
[c] Cubic, stabilised with CaO.
[d] Congruent.
[e] Incongruent.
[f] Dissociates above 2450°C in reducing atmosphere and is readily oxidised above 1650°C.

Thermal properties

Refractories, like most other solids, expand upon heating, but much less than most metals. The degree of expansion depends upon the chemical composition.

Manufacture

Processing

Initial processing may include an extensive survey of the raw material deposit, selective mining, stockpiling by grade, and beneficiation techniques such as weathering, grinding, washing, heavy-media separation, froth flotation, etc. Further steps include crushing and grinding, screening, mixing, forming, drying, and burning.

Speciality refractories

Bulk refractory products include gunning, ramming, or plastic mixes, granular materials, hydraulic-setting castables, and mortars. These products are generally made from the same raw materials as their brick counterparts.

Table. 19.4. Typical ranges of physical properties of alumina, silica and zirconia refractory brick.

Type	*PCE*[a]	*Modulus of rupture MPa*[b]	*Deformation under 172 kPa*[b] *load, % linear change after 1 1/2 hours*		*Linear reheat change*[c]		*Bulk density*
			Temperature, °C	*Change, %*	*at °C*	*%*	*g/cm*^{3}
Silica		2.8–11.2	1650	0			1.60–1.80
Fireclay							
semisilica	27–31	2.1–4.2	1450	0.1–2			1.80–2.10
medium duty	29–31	7.0–11.2	1450	1–6			2.11–2.20
high duty	31–33	2.8–21.0	1450	0.5–15	1600	0–1 S	2.13–2.30
super duty	33–34	2.8–24.0	1450	0–9	1600	0–1 S	2.28–2.48
High alumina							
50% Al_2O_3	34–35	7.0–11.2	1450	2–6	1600	0–1 E	2.27–2.43
60%	36–37	7.0–11.2	1450	1–4	1600	0–5 E	2.10–2.49
70%	37–38	7.0–11.2	1450	1–3	1600	1–7 E	2.20–2.66
80%	38–39	7.0–12.6					2.50–2.90
85%[d]	38–39	21.0–35.0	1450	0.3–3	1600	0–2 E	2.70–2.92
90%	40–41	14.0–35.0	1700	0–1	1700	0–1 E	2.67–3.10

(Contd...)

Type	*PCE*[a]	*Modulus of rupture MPa*[b]	*Deformation under 172 kPa*[b] *load, % linear change after 1 1/2 hours Temperature, °C*	*Change, %*	*Linear reheat change*[c] *at °C*	*%*	*Bulk density g/cm*^{3}
100%	41–42	12.6–21.0	1650	1–2	1700	0–0.5 S	2.84–3.10
Zircon			1600	2–5	1540	0	3.77
Molten cast							
Al_2O_3–ZrO_2–SiO_2							3.70–3.74
Al_2O_3–SiO_2							3.00–3.20
Al_2O_3 high soda							2.89
Al_2O_3 low soda							3.50

[a] Pyrometer cone equivalent, as determined by ASTM C 24.
[b] To convert MPa to psi, multiply by 145; for kPa, multiply by 0.145.
[c] S = Shrinkage, E = expansion.
[d] Phosphate bonded.

Health and Safety Factors

Because industrial refractories are by their very nature stable materials, they usually do not constitute a physiological hazard. This statement does not apply, however, to unusual refractories that might contain heavy metals or radioactive oxides such as thoria, urania, and plutonia, or binders or additives that may be toxic. Inhalation of certain fine dusts may constitute a health hazard.

Selection and Uses

Any manufacturing process requiring refractories depends upon proper selection and installation. When selecting refractories, the environmental conditions are evaluated first, then the functions to be served, and finally the expected length of service. All factors pertaining to the operation service, design, and construction of equipment must be related to the physical and chemical properties of the various classes of refractories.

By far the most common industrial refractories are those composed at single or mixed oxides of Al, Ca, Cr, Mg, Si and Zr. These oxides exhibit relatively high degrees of stability under both reducing and oxidising conditions. Carbon, graphite, and silicon carbide have been used both alone and in combination with these oxides. Refractories made from the above materials are used in ton-lot quantities in industrial applications. Other refractory oxides, nitrides, borides, and silicides are used in relatively small quantities for speciality applications in the nuclear, electronic, and aerospace industries.

Silica refractories

Uses include open-hearth roof linings, refractories for coke ovens, coreless-induction foundry furnaces, and fused-silica technical ceramic products.

Semisilica refractories

Uses include shapes for open-hearth stoves and checkers.

Fireclay refractories

Fireclay refractories are used in kilns, ladles, and heat regenerators, acid-slag-resistant applications, boilers, blast furnances, and rotary kilns.

High alumina refractories

These refractories are used in kilns, ladles, and furnaces that operate at temperatures or under conditions for which fireclay refractories are not suited.

Chrome refractories

These refractories are used in non-ferrous metallurgical furnaces, rotary-kiln linings, secondary refining vessels, such as argon-oxygen decarborisers (AODs) and glass-tank regenerators.

Magnesite refractories

These refractories are used in lining and maintenance of steelmaking and refining vessels and checkers.

Dolomite refractories

Dolomite refractories are primarily used in linings of BOF vessels and refining vessels, and in ladles and cement kilns.

Spinel refractories

Spinel refractories are used in cement kilns and steel-ladle linings.

Forsterite refractories

Uses include non-ferrous-metal-furnace roofs and glass-tank refractories not in contact with the melt, i.e., checkers, ports, and uptakes.

Silicon carbide refractories

Silicon carbide has a wide range of refractory uses including chemical tanks and drains, kiln furniture, abrasion-resistance linings, blast-furnace linings, and non-ferrous metallurgical crucibles and furnace linings.

Zirconia refractories

The most common zirconia-containing refractories are made from zircon sand and are used mostly for glass-tank paver brick.

REFRACTORY COATINGS

Refractory coatings denote those metallic, refractory-compound (i.e., oxides, carbides, nitrides) and metal-ceramic coatings associated with high temperature service as contrasted to coatings used for decorative or corrosion-resistant applications. They also denote coatings of high melting materials that are used in other than high temperature applications. A coating may be defined as a near-surface region with properties that differ significantly from the bulk of the substrate.

The highest melting refractory metals are tungsten, tantalum, molybdenum, and niobium, although titanium, hafnium, zirconium, chromium, vanadium, platinum, rhodium, ruthenium, iridium, osmium, and rhenium may be included. Many of these metals do not resist air oxidation; hence, very few if any, are used in their elemental form for high temperature protection. However, bulk alloys based on nickel, iron, and cobalt with alloying elements such as chromium, titanium, aluminium, vanadium, tantalum, molybdenum, silicon, and tungsten are used extensively in high temperature service. Some modern high temperature oxidation- and corrosion-resistant coatings have compositions similar to the high temperature bulk alloys and are applied by thermal spraying, evaporation, or sputtering. The protection mechanism for these high temperature alloy coatings is based on adherent impervious surface films of the Al_2O_3, SiO_2, CrO_2, or a spinel-type that grows on high temperature exposure to air.

Refractory coatings also include materials with high melting points, e.g., silicides, borides, carbides, nitrides, or oxides, and combinations such as oxy-carbides, etc., In addition, mixtures of metals and refractory compounds (sometimes called metallides) of various microstructural configurations (i.e., laminates, dispersed phases, etc.) can also be classified as refractory coatings.

In some cases, a coating is a new material that is deposited onto the substrate by a variety of methods. It is then called a deposited or overlay coating. In other cases, the coating may be produced by altering the surface material to produce a surfaces layer composed of both the added and substrate materials. This is called a conversion coating, cementation coating, diffusion coating, or chemical-conversion coating when chemical changes in the surface are involved. Coatings may also be formed by altering the properties of the surfaces by melting and quenching, mechanical deformation, or other processes that changes the properties without changing the composition.

All coating methods consist of three basic steps: synthesis or generation of the coating species or precursor at the source; transport from the source to the substrate; and nucleation and growth of the coating on the substrate. These steps can be completely independent of each other or may be superimposed on each other, depending on the coating process. A process in which the steps can be varied independently and controlled offers great flexibility, and a larger variety of materials can be deposited.

Numerous schemes can be devised to classify deposition species, i.e., atoms and molecules, liquid droplets, bulk quantities, or the use of a surface-modification process (Table 19.5).

The coating has to adhere to the substrate. The bonding may be mechanical as a result of the interlocking between the asperities on the surface and the coating. In diffusion or chemical bonding, the substrate and the coating material interdiffuse at the interface.

Table 19.5. Coating methods.

Atomistic deposition	*Particulate deposition*	*Bulk coatings*	*Surface modification*
Electrolytic environment	thermal spraying	wetting processes	chemical conversion
electroplating	plasma spraying	painting	electrolytic
electroless plating	D-gun	dip-coating	anodisation (oxides)
fused-salt electrolysis	flame spraying	electrostatic-spraying	fused salts
chemical displacement	fusion coatings	printing	chemical-liquid
Vacuum environment	thick-film ink	spin coating	chemical-vapour
vacuum evaporation	enameling	cladding	thermal
ion-beam deposition	electrophoretic	explosive	plasma
molecular-beam epitaxy	impact plating	roll bonding	leaching
Plasma environment		overlaying	mechanical
sputter deposition		weld-coating	shot peening
activated reactive			

(Contd ...)

Atomistic deposition	*Particulate deposition*	*Bulk coatings*	*Surface modification*
evaporation		liquid-phase epitaxy	thermal
plasma polymerisation			surface enrichment
ion plating			diffusion from bulk
Chemical-vapour Environment			sputtering
Chemical-vapour Deposition			ion implantation
reduction			
decomposition			
plasma enhanced			
spray pyrolysis			

Residual stresses, which are always present in coatings, arise from thermal-expansion mismatch between coating and substrate, or growth stresses caused by imperfections in the coating that are built in during the process of film growth.

Coatings may be permeable either to the atmosphere or the substrate material. Diffusion of oxygen through a coating can result in gaseous products that may rupture the coating. The design of a coating that might equilibrate with its substrate during use is based on the phase diagram of the system.

Microstructure of Coatings

The microstructure of bulk coatings resembles the normal microstructure of metals and alloys produced by melt solidification. The micro structure of particulate-deposited materials resembles a cross between rapidly solidified bulk materials with severe deformation and powder compacts produced by pressing and sintering. A special feature of particulate coatings is a significant degree of porosity (ca. 2–20 vol%) which strongly affects the properties of the deposit.

The microstructure and imperfection content of coatings produced by atomistic deposition processes can be varied over a very wide range to produce structures and properties similar to or totally different from bulk-processed materials.

Applications

Coatings can be classified into six categories: chemically functional, mechanically functional, optically functional, electrically functional, biomedical, and decorative. In addition, there are some unique applications in the aerospace programmes, such as the ablative coatings of pyrolytic carbon and graphite- and silica-based materials for protection of nose cones and the space shuttle during re-entry. Another unique energy-related application is the coating of low-Z elements such as TiC for the first wall of thermonuclear reactors to minimise contamination of the plasma.

REFRACTORY FIBRES

The refractory fibre defines a wide range of amorphous and polycrystalline synthetic fibres used at temperatures generally above 1093°C. Chemically, these fibres can be separated into oxide and non-oxide fibres. The former include alumina-silica fibres and chemical modifications of the alumina-silica system, high silica fibres (> 99% SiO_2), polycrystalline zirconia, and alumina fibres. The diameters of these fibres are 0.5–10 μm (av ca 3 μm). Their length, as manufactured, ranges from 1 cm to continuous filaments, depending upon the chemical composition and manufacturing technique. Such fibres may contain up to ca 50 wt% unfiberised particles. Commonly referred to as shot, these particles are the result of melt fibrisation usually associated with the manufacture of alumina-silica fibres. The presence of shot reduces the thermal efficiency of fibrous systems. Shot particles are not generated by the manufacturing techniques used for high silica and polycrystalline fibres, and consequently, these fibres usually contain < 5 wt% unfibrised material. Refractory fibres are manufactured in the form of loose wool. From this state, they can be needled into flexible blanket form, combined with organic binders and pressed into flexible or rigid felts, fabricated into rope, textile, and paper forms, and vacuum-formed into a variety of intricate, rigid shapes.

The non-oxide forms, silicon carbide, silicon nitride, boron nitride, carbon, or graphite have diameters of ca 0.5–50 μm. Generally, non-oxide fibres are much shorter than oxide fibres except for carbon, graphite, and boron fibres, which are manufactured as continuous filaments. Carbon, graphite, and boron fibres are used for reinforcement in plastics in discontinuous from and for filament winding in continuous form. In addition to moderate temperature resistance, these fibres have extremely high elastic modulus and tensile strength. Carbon and graphite fibres cannot be accurately classified as refractory fibres because they are oxidised above ca 400°C. This is also true of boron fibres which form liquid boron oxide at approximately 560°C. Most silicon carbide, silicon nitride, and boron nitride fibres are relatively short, ranging in length from single-crystal whiskers of less than 1 mm to fibres as long as 5 cm. Diameters are ca 0.1–10 μm. These fibres are used mostly to reinforce composites of plastics, glass, metals, and ceramics. They also have limited applications as insulation in and around rocket nozzles and in nuclear-fusion technology requiring resistance to short-term temperatures above 2000°C. For inert atmosphere or vacuum applications, carbon-bonded carbon-fibre composites provide effective thermal insulation up to 2500°C.

Health and Safety Factors

Synthetic or man-made refractory fibres do not appear to pose the health hazards of naturally occurring mineral fibres like asbestos. Nonetheless, proper respirators should be worn in environments of excessive exposure to refractory fibres.

Chapter 20

Glass Industries

INTRODUCTION

Glass was formed naturally from common elements in the earth's crust long before anyone ever thought of experimenting with its composition, molding its shape, or putting it to the myriad of uses that it enjoys in the world today.

Glass technology has evolved for six thousand years, and some of today's principles date back to early times. This includes what is today known about the structure of glass, its composition, properties, method of manufacture, and uses.

Common usage of the term glass follows the definition of Morey: "Glass is an inorganic substance in a condition that is continuous with, and analogous to, the liquid state of that substance, but which, as the result of a reversible change in viscosity during cooling, has attained so high a degree of viscosity as to be, for all practical purposes, rigid." Both organic and inorganic materials may form glasses if their structure is non-crystalline, i.e., if they lack long-range order. This includes some plastics, metals, and organic liquids. In principle, rapid cooling could prevent crystallisation of any substance if the final temperature is sufficiently low to prevent structural rearrangement.

Glass is not merely a supercooled liquid. This distinction is illustrated by the volume-temperature diagram shown in Fig. 20.1 When a liquid that normally does not form a glass is cooled, it crystallises at a slightly below the melting point (path A). If there are insufficient crystal nuclei or if the viscosity is too high to allow sufficient crystallisation rates, undercooling of the liquid can occur. However, the viscosity of the liquid rapidly increases with decreasing temperatures, and atomic rearrangement slows down more than would be typical for the supercooled liquid. This results in the deviation from the metastable equilibrium curve that is depicted by paths B and C in Fig. 20.1. This change in slope with temperature is characteristic of glass. Fig. 20.1. shows that the point of intersection of the two slopes defines a transformation point (glass-transition temperature) *Tg*, for a given cooling rate. Practical limitations on cooling rate define the transformation range *Tg* → T'_g as the temperature range in which the cooling rate can affect the structure-sensitive properties such as density, refractive index, and volume resistivity. The structure, which is frozen in during the glass transformation, persists at all lower temperature. Thus, a glass has a configurational or a "fictive" temperature that may differ from its actual temperature. The fictive temperature is the temperature at which the glass structure would have been the equilibrium structure. It describes the structure of a glass as it relates to the cooling rate. A fast-quenched glass would have a higher fictive temperature than a slowly cooled glass. Glasses can be prepared by methods other than cooling from a liquid state, including solution evaporation, reactive sputtering, vapour deposition, neutron bombardment, and shock-wave vitrification.

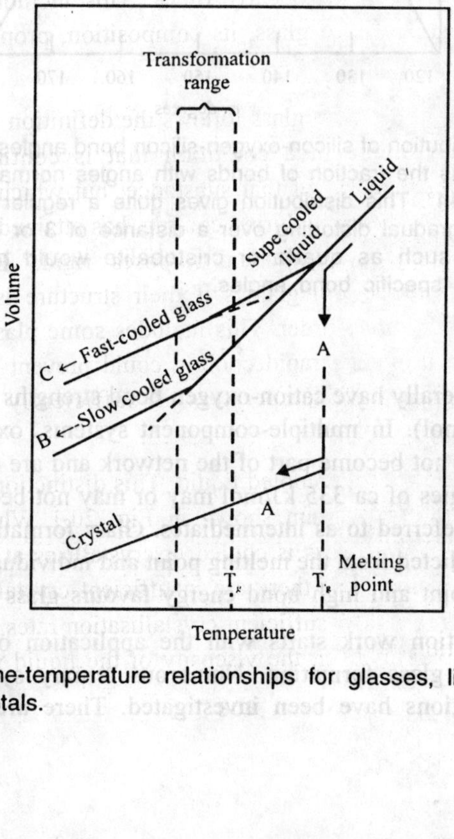

Fig. 20.1. Volume-temperature relationships for glasses, liquids, supercooled liquids, and crystals.

STRUCTURE

The basic structural unit of silicate glasses is the silicon-oxygen tetrahedron in which a silicon atom is tetrahedrally coordinated to four surrounding oxygen atoms. Oxygens shared between two tetrahedra are called bridging oxygens. The similarity between crystalline and vitreous structure on the basis of silicon-oxygen-silicon bond angles and, hence, the relative orientations of the silicon tetrahedra is pointed out in Fig. 20.2.

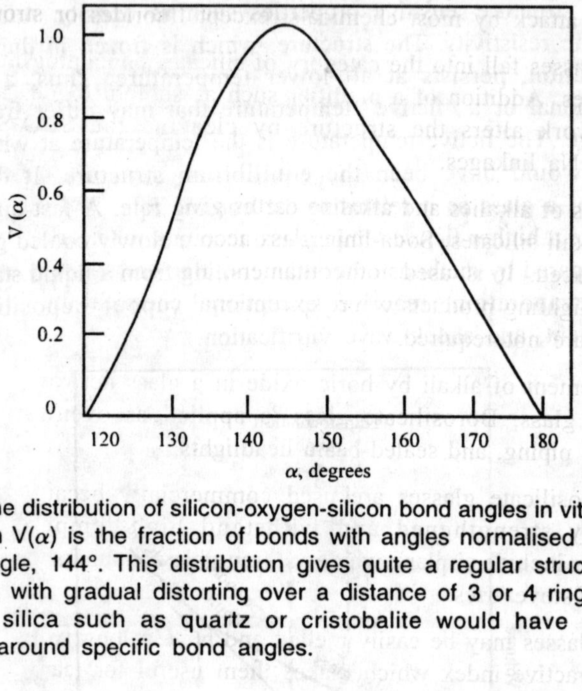

Fig. 20.2. The distribution of silicon-oxygen-silicon bond angles in vitreous silica. The function V(α) is the fraction of bonds with angles normalised to the most probable angle, 144°. This distribution gives quite a regular structure on the short range, with gradual distorting over a distance of 3 or 4 rings (2-3 nm). Crystalline silica such as quartz or cristobalite would have a narrower distribution around specific bond angles.

Composition

Glass formers generally have cation-oxygen bond strengths greater than 335 kJ/mol (80 kcal/mol). In multiple-component systems, oxides with lower bond strengths do not become part of the network and are called modifiers. Oxides with energies of ca 335 kJ/mol may or may not become part of the network and are referred to as intermediates. Glass formation of individual oxides can be predicted from the melting point and individual bond energies. A low melting point and high bond energy favours glass formation.

Glass-composition work starts with the application of structural and bonding rules of glass formation. Numerous ternary systems and their glass-forming regions have been investigated. There are three types of ternaries: single-former and two modifiers; two formers and one modifier; and three glass formers.

Single-Phase Glasses

Vitreous silica is the most important single-component glass. Highly cross-linked vitreous silica is viscous and has a thermal expansion coefficient within the 0-300°C range of about $5.5 \times 10^{-7}/°C$. It is an excellent dielectric and resists attack by most chemicals, except fluorides or strong alkali.

Most glasses fall into the category of silicates containing modifiers and intermediates. Addition of a modifier such as sodium oxide, Na_2O, to the silica network alters the structure by cleaving the Si-O-Si bonds to form Si-O.Na linkages.

Mixtures of alkalies and alkaline earths give glasses of higher durability than the alkali silicates. Soda-lime glass accounts for nearly 90% of all the glass produced. It is used for containers, flat glass, pressed and blown ware, and lighting products where exceptional chemical durability and heat resistance are not required.

Replacement of alkali by boric oxide in a glass network give a lower-expansion glass. Borosilicate glass is applied as ovenware, laboratory equipment, piping, and sealed-beam headlights.

Aluminosilicate glasses are used commercially because they can be chemically strengthened and withstand high temperatures. Thus, applications include airplane windows, frangible containers, lamp envelopes, and top-of-stove uses.

Lead glasses may be easily melted and have a long working range and a high refractive index which makes them useful for lead crystal, optical glass, and hand-formed art ware.

Borates, including vitreous B_2O_3, have been studied more than any other glass-forming system with the exception of silicates. The very low durability of borate glasses precludes their use in all except the most special applications. Low molecular weight Lindemann glasses ($Li_2O.BeO.B_2O_3$) were developed as x-ray-transmitting glasses.

Phosphate glasses, like borates, tend to have low durability. Important commercial applications of phosphate glasses do exist, however. Almost-clear, heat-absorbing glasses with several per cent iron oxide are possible. Phosphate-based glasses also are more resistant to fluoride than silicate glasses. Some of the optical glasses produced by Schott, Hoya, Owens-Illinois, and Corning-Sovirel use phosphate as the primary glass former.

Germanium, arsenic, and antimony oxides all form stable glasses and their structures have been predicted.

Glasses based upon sulphur, selenium, or tellurium rather than oxygen are well known. These glasses, although often opaque to visible light transmit ir radiation of a much longer wavelength than oxide systems, and many are also semiconductors.

Although zinc chloride glasses are known, BeF_2-containing glasses are the more common of the halide glasses.

Under highly specialised conditions, the crystalline structure of some metals and alloys can be suppressed and they form glasses. These amorphous metals can be made from transition-metal alloys, e.g., nickel-zirconium, or transition or noble metals in combination with metalloid elements, e.g., alloys of palladium and silicon or alloys of iron, phosphorus, and carbon.

Glass-Ceramics and Phase-Separated Glasses

Glass-ceramics are melted and formed by conventional glass manufacturing techniques and then given a subsequent heat treatment (ceram) to transform then into fine-grained crystalline materials. By definition, glass-ceramics are more than 50% crystalline after heat treatment; frequently, the product is more than 95% crystalline. The operation is generally accompanied by an increase in viscosity which increases the product's use temperature.

Aluminosilicate glass-ceramics are the most useful commercial products. The presence of alkali or alkaline earth oxides affects the type of crystal structure. The morphology is regulated by a nucleating agent, usually TiO_2 or TiO_2 + ZrO_2, and a heat treatment. Many other components are added that further optimise the crystalline phases and the glass-ceramic properties.

The formation of a glass-ceramic is extremely complex, but generally follows a four-step sequence: (i) a dispersed amorphous phase, structurally incompatible with the host glass, usually highly unstable, and enriched in one or two key oxides (e.g., TiO_2 or ZrO_2), forms on either cooling or reheating the glass; (ii) primary crystalline nuclei, which are often titanates or zirconates, form either heterogeneously at the phase boundaries or homogeneously within the second phase; (iii) a metastable crystalline phase heterogeneously nucleats on the primary crystallites and grows, generally at the expense of the glassy second phase. This produces a typically fine-grained, metastable, solid-solution material; and (iv) the metastable material breaks down to the final, stable, fine-grained crystalline structure by means of isochemical phase transformations, reactions between metastable phases, or exsolution. Devitrification is the uncontrolled formation of crystals in glass during melting, forming, or secondary processing. It can adversely affect the optical properties, mechanical strength, and sometimes the chemical durability of the glass. Devitrification is most likely to occur in glasses in which the optimum temperatures for maximum nucleation rate and for maximum growth rate nearly coincide. If these glasses are held too long in this critical temperature range or are cooled too slowly through it, the glass starts to crystallise.

Glasses that derive their colour, optical transparency, or chemical durability from a small amount of a finely dispersed second phase are termed phase-separated glasses. They are distinguished from glass-ceramics by virtue of their predominantly glassy character. Therefore, the control of the nucleation and morphology of the second phase is not so critical.

Photosensitive glasses employ structural transformations, i.e., precipitation and crystallisation, which are sensitised by exposure to electromagnetic radiation, usually ultraviolet, but sometimes visible or x-ray radiation. The transformation is generally initiated by the photoreduction of a metal ion. Subsequent or simultaneous heat treatment morphologically transforms the glass in the regions exposed to the radiation leading, in general, to a change in optical or physical properties.

Properties

The viscosity of a glass determines its melting, forming, and annealing procedures as well as the limitations of its use at high temperature. Viscosity is ordinarily measured between 10^{13} and 10 Pa·s (10^{14} and 100 P); at room temperature, it is greater than 10^{19} Pa·s (10^{20} P). The rapid but smooth change of viscosity with temperature is shown for several glasses in Fig. 20.3. Reference points on the viscosity-temperature curve in Fig. 20.3 have been chosen to characterise properties of an individual glass and to facilitate comparisons of similar glasses. Those used most frequently are the working, softening, annealing, and strain points. The working point of a glass is the temperature at which its viscosity is exactly 1 kPa·s (10^4 P). At this viscosity, glass is sufficiently fluid for forming by most common methods. Glass generally deforms under its own weight at the softening point. Residual stress can be annealed out of the glass in a matter of minutes at the annealing point and in a matter of hours at the strain point.

The thermal expansion of a glass determines the range of materials to which it can be sealed safely. It also affects the glass's ability to survive thermal shock or cycling. A glass's usefulness as a heat exchanger or as a thermal barrier and its ease of melting and forming depend upon its heat-transfer properties and emissivity. The upper use temperature is a function of all of these properties.

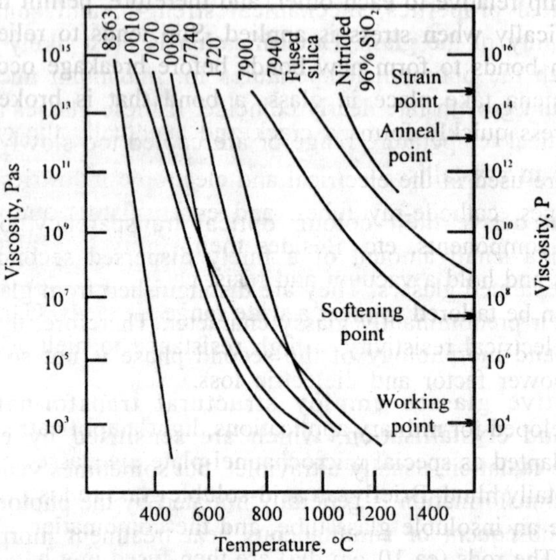

Fig. 20.3. Viscosity vs temperature for some commercial glasses.

Stresses caused by steady-state thermal gradients may or may not cause failure, depending on the degree of constraint imposed by some parts of the item upon others or by the external mounting. Thus, under minimum constraint and maximum uniformity of gradient through the thickness, very large temperature differences can be tolerated.

At room temperature, the thermal conductivity of glasses is 0.67-1.21 W/(m·K), the most common composition being near the upper end of the range. Thermal conductivity of glass-ceramics is 1.7-3.8 W/(m·K). At a mean temperatures of 200°C, the values are greater by approximately 25%.

The liquid's temperature determines the susceptibility of a glass to devitrification and, therefore, influences its forming limitations and often its heat-treating requirements.

Gas permeation though glass is of crucial importance to any high-vacuum system. To permit the advent of high intensity tungsten-halogen lamps, the lamp envelopes must contain gas under pressure for a long time at high temperatures. The permeation rate varies with temperature, glass type and composition, and gas.

Because of its amorphous structure, glass is brittle, reasonably abrasion-resistant, and almost perfectly elastic as long as its temperature is low enough to prevent viscous flow. Glasses do not contain crystallographic planes that slip relative to each other, and therefore, permit the material to deform plastically when stress is applied. Slip tends to relieve stress and allow broken bonds to form new bonds before breakage occurs. Since no such phenomena take place in glass, a bond that is broken because of excessive stress quickly forms a crack and eventually the glass fractures.

Glasses are used in the electrical and electronic industries as insulators, lamp envelopes, cathode-ray tubes, and encapsulators and protectors for microcircuit components, etc. Besides their ability to seal to metals and other glasses and hold a vacuum and resist chemical attack, their electrical properties can be tailored to meet a wide range of needs. Generally, a glass has a high electrical resistivity, a high resistance to dielectric breakdown, and a low power factor and dielectric loss.

First developed for military applications, light-amplification devices have now been adapted as special microchannel-plate eyeglasses for people who are almost totally blind. Briefly, an acid-soluble glass rod is first selected and placed inside an insoluble glass tube, and the combination is drawn into a coaxial rod. The rods (ca 10 μm dia) are then fused in a bundle, sliced into thin plates, polished, and leached in acids so that the core glass is removed, and thus formed into a microchannel glass plate. The cladding glass is then rendered electronically conducting on the surface, and electrodes are deposited and assembled into the device. In combination with a photocathode located in the front end of the tube and a phosphor screen located at the exit end, a highly efficient light-amplification device can be made.

The properties of some glasses and glass-ceramics make them more suitable for microwave application than others. The relative suitability of a glass or glass-ceramic material is expressed by a factor of merit that is proportional to the material's strength and inversely proportional to its Young's modulus, expansion coefficient, and rate of microwave-energy absorption.

The resistance of glass to chemical corrosion is frequently the reason for its use. However, the durability of a glass varies from highly soluble to highly durable, depending upon its composition and the solvent considered. Glasses low in silica are usually not durable to acid. Hydro-fluoric acid, however, attacks even high silica glasses. Comparisons are usually based upon measurements of weight loss, changes in surface quality, or the analysis of solutions that were in contact with a glass.

Optical glasses are usually described in terms of their refractive index at the sodium D line (589.3 nm) and their *v* value (or Abbé number) which is a measure of the dispersion or variation of index with wave-length. These data are incorporated by Schott into a six-digit numbering system used to identify optical glasses. Glasses with index $n_D < 1.60$ and a *v* value of ≥ 55 are defined as crown glasses; those with a *v* value below ca 50 are defined as flint glasses. Glasses with $n_D > 1.60$ are defined as crown glasses if *v* is ≥ 50. Crowns are usually alkali silicate glasses and flints are lead alkali silicates.

The spectral transmission of glass is determined by reflection at the glass surfaces and the optical absorption within the glass. Overall transmission of a flat sample at a particular wavelength is equal to $(1 - R)^2 e^{-\beta t}$, where β is the absorption coefficient, *t* the thickness of glass, and *R* the air–glass reflection coefficient. Transmission is a function of wavelength. It may be controlled by the type of glass used or by the control or addition of colouring additives, melting atmosphere, melting temperatures, and cooling schedules.

When strained, glass becomes doubly refracting, or birefringent. Stress-optical effects are presumed to be elastic; therefore, strain is proportional to stress. The stress-optical coefficient, *B*, or Brewster's constant, is a measure of this proportionality, and varies greatly with glass composition.

Interaction of glass with low-energy visible and ultraviolet radiation may result in the alteration of the electronic states. These changes may cause colouration or luminescence effects, e.g., photochromic or photosensitive glass. When changes in colour are produced by sunlight, the effect is frequently referred to as solarisation. Effects produced by ions, gamma rays, and x-rays are relevant to glasses used as electronic tube envelopes, radiation-shielding windows, and dosimeters.

Manufacture and Processing

Most glass articles are manufactured by a process in which raw materials are converted at high temperatures to a homogeneous melt that is then formed into the articles. The flow diagram in Fig. 20.4 summarises the details of conventional glass manufacturing. The vapour deposition of SiO_2 from a flame fed with $SiCl_4$ and oxygen is the basis for manufacturing high-purity glass used for blanks that are redrawn into optical-waveguide fibres. Fused silica items that cannot be formed from viscous melts of SiO_2 or quartz are prepared by vapour deposition. Raw materials are selected according to purity, supply, pollution potential, ease of melting, and cost. Sand is the most common ingredient. Limestone is the source of calcium and magnesium. Powdered anthracite coal is a common reducing agent, and common colourants for glass include iron, chromium, cerium, cobalt, and nickel.

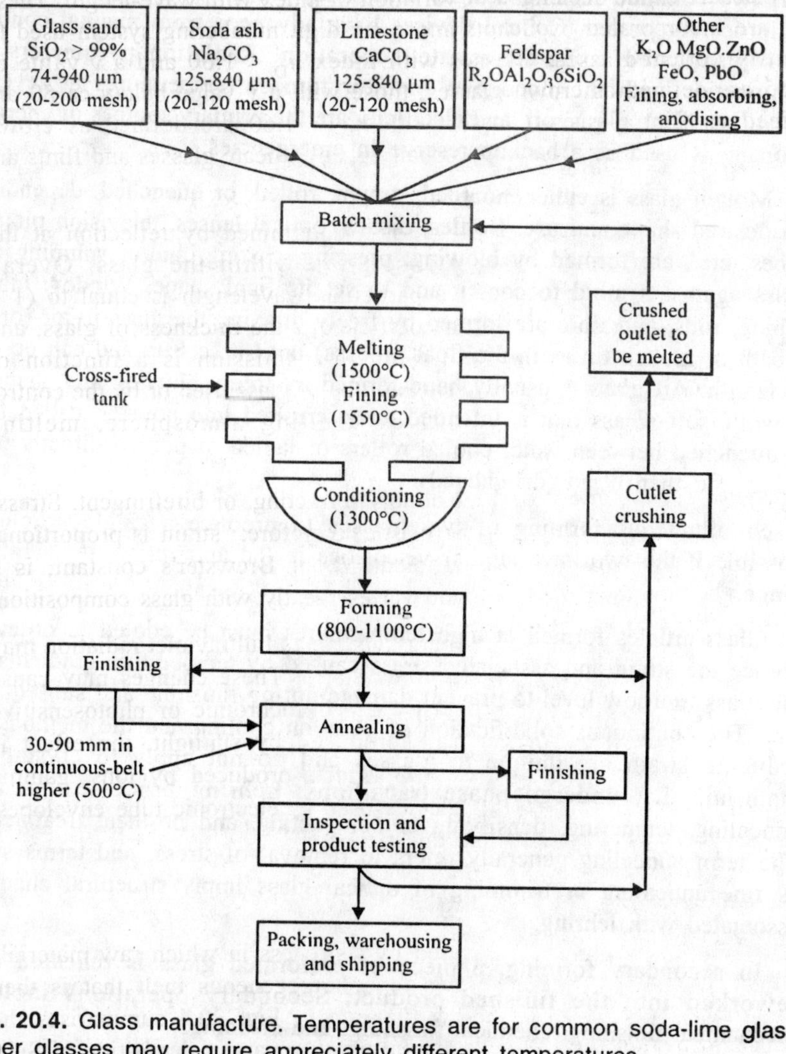

Fig. 20.4. Glass manufacture. Temperatures are for common soda-lime glass. Other glasses may require appreciately different temperatures.

Melting and fining depend on the batch materials interacting with each other at the proper time and in the proper order. Thus, extreme care must be taken to obtain materials of optimum grain size, to weigh them carefully, and mix them intimately. The efficiency of the melting operation and the uniformity and quality of the glass product are very often determined in the mix house. Batch handling systems vary widely through the industry, from manual to fully automatic. Melting units range from small pot furnaces for manual production to large, continuous tanks for rapid machine forming. The two common feeding designs used today are the screw feeder and the reciprocating pusher. Control devices have advanced from manual control to sophisticated computer-assisted operation. Radiation pyrometers in conjunction with thermocouples monitor furnace temperatures at several locations. Natural gas, oil, and electricity are the primary sources of energy; propane is used as a backup reserve in emergencies.

Molten glass is either moulded, drawn, rolled, or quenched, depending on desired shape and use. Bottles, dishes, optical lenses, television picture tubes, etc., are formed by blowing, pressing, casting, and/or spinning the glass against a mold to cool it and to set its final shape. Window glass, tubing, rods, and fibre are formed by freely drawing the glass in air (or in a bath of molten tin as in the float process) until it sets up and can be cut to length. Art glass is usually hand-formed by freely blowing and shaping it while soft. Glass that is intended to be crushed into powder, called frit, is quenched between water-cooled rollers or ladled or poured directly into water and then dried (dri-gauged).

Simultaneous forming of two glasses to produce a single article is possible if the two have similar viscosities in the appropriate temperature range.

Glass articles formed at high temperatures must be cooled in order to reduce the strain and associated stress caused by temperature gradients in the glass to a low level to prevent damage during finishing and subsequent use. The continuous solidification of glass on cooling and the methods of reducing strain are unique to glasses and do not apply to crystalline materials that undergo phase transitions. Lehring processes include annealing, tempering, densifying or compacting, and postheat treatments. The term annealing generally refers to removal of stress, and terms such as fine annealing or trimming of optical glass imply structural changes associated with lehring.

In secondary forming, a piece of preformed glass is reheated and reworked into the finished product. Secondary operations include mechanical finishing, chemical finishing, and cleaning.

GLASS-CERAMICS

Glass-ceramics are polycrystalline solids produced by the controlled crystallisation of glasses. They are primarily silicate-based materials that can be formed by highly automated glass-forming processes and converted to a ceramic-like product by the proper heat treatment.

Glass-Ceramic Process

Fig. 20.5 is a flow chart of a glass-ceramic manufacturing process. By definition, a glass-ceramic initially must be a glass. Thus, the first stages of the manufacturing process are similar to the processing of most commercial glasses; however, there are significant differences in the postforming stages. The environmentally related effects resemble those encountered in conventional glass processing. Pollution problems are generally minimal except for emissions during the melting of some compositions.

Postforming

The shaped body may be annealed (cooled slowly to avoid stresses) and inspected prior to heat treatment (Fig. 20.5), or the shaped object may be put directly into a heat-treatment furnace. The heat-treatment process may be continuous or intermittent.

Nucleation

One of the main factors that permits the formation of a conventional glass without crystallisation is the very slow nucleation rate. Although some glass compositions are self-nucleating, most of the commercial glass-ceramic compositions rely on the presence of a nucleating agent, usually a noble metal or a transition-metal oxide, to initiate internal crystallisation.

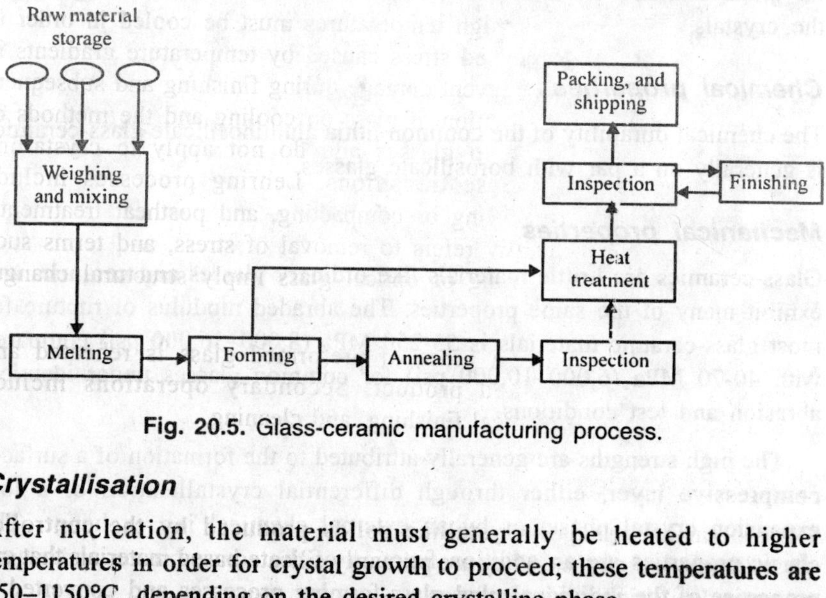

Fig. 20.5. Glass-ceramic manufacturing process.

Crystallisation

After nucleation, the material must generally be heated to higher temperatures in order for crystal growth to proceed; these temperatures are 750–1150°C, depending on the desired crystalline phase.

After heat treatment, the material may undergo additional finishing steps such as cutting, grinding, polishing or decorating, or it may be packaged directly for shipment.

Properties

The properties of a glass-ceramic are dependent on such microstructural parameters as: the amount and properties of the crystalline phase(s); the crystal sizes, shapes, and orientations; and the amount, composition, and distribution of the glassy phase.

Thermal properties

Glass-ceramic materials have been reported with thermal-expansion coefficients ranging from ca (–60 to 200) × 10^{-7}/°C. The thermal-expansion properties, particularly the low expansions, are the primary properties that have been exploited commercially. The thermal-conductivity range is ca 1.7–5.4W/(m·K).

Optical properties

Transparent and opaque glass-ceramics can be made. The degree of transparency is a function of the optical properties of the crystalline species, the grain size, and the difference of the refractive indexes of the glass and the crystals.

Chemical properties

The chemical durability of the common lithia aluminosilicate glass-ceramics is generally on a par with borosilicate glasses.

Mechanical properties

Glass-ceramics are brittle materials like ordinary glasses and ceramics and exhibit many of the same properties. The abraded modulus of rupture for most glass-ceramic materials is 55-250 MPa (8,000–36,000 psi) compared with 40-70 MPa (6,000–10,000 psi) for common glasses under identical abrasion and test conditions.

The high strengths are generally attributed to the formation of a surface-compressive layer, either through differential crystallisation of a low expansion crystal phase, or by an external chemical ion exchange. The elastic properties are an addition function of the concentration and elastic properties of the individual phases.

Electrical properties

Most glass-ceramics are insulators.

Magnetic properties

Magnetic phases can be crystallised from a glass, but magnetic glass-ceramics are not commercially available.

Commercial Applications

Lithium silicate systems have been used for the manufacture of a photosensitive glass that can be chemically machined. The commercial lithia aluminosilicate glass-ceramics are among the best glass-ceramic materials in terms of formability. The combination of the magnesia aluminosilicates properties and formability had led to their use as radomes. In addition to machinability, mica glass-ceramics are strong, thermally shock-resistant and have good dielectric properties.

Glass-ceramic materials called slagcerams have been produced by the crystallisation of glasses made from metallurgical slags. Products from these slags such as floor and wall coverings, are attractive because of the low cost and the availability of the raw materials, and have been manufactured in several countries for many years.

GLASSY METALS

The most familiar glasses are oxide glasses such as silicate or borate glasses. However, they are not the only members of the family of glasses. A recent addition is the group of glassy metals, or metallic glasses, also called amorphous alloys. They offer excellent possibilities in the electric, electronic, and structural industries. Several compositions of glassy metals, particularly those containing about 80 atom% transition metals and about 20 atom% metalloids, such as B, Si, P, C, are now available in commercial quantities.

Unlike crystalline solids, glasses do not have a long-range ordered atomic structure. They are in essence frozen liquids. Most glassy metals cannot be obtained by simply cooling liquid metals at rates commonly employed in metals industries because the samples would crystallise. However, when liquid metals are quenched at a very high rate, often exceeding 10^5 K/s, it is possible to avoid crystallisation and thus obtain glassy metals that largely preserve the atomic structure of liquid metals. The properties of glassy metals are sometimes similar to those of crystalline metals. It is, therefore, not possible to judge unambiguously whether a solid is crystalline or amorphous by observing its macroscopic properties alone. Glassiness can be determined either directly by studying the atomic structure with diffraction methods, or indirectly by carefully studying the thermodynamic properties.

Properties

The mechanical properties of a glassy metal are characterised by a tensile strength that exceeds 1% of Young's modulus, the formation of shear bands, ductile flow on the fracture surface, and in many cases, embrittlement after an annealing that does not cause crystallisation. Most glassy metals show an appreciable amount of mechanical creep at temperatures as low as 200°C. Because of their ductility, glassy metals can be processed mechanically by cutting or punching. Many glassy metals containing transition metals or rare earths, or both, are ferromagnetic or ferrimagnetic (Table 20.1.).

Table 20.1. Representative alloy composition of glassy metals.

Composition	*Remarks*
$Fe_{80}B_{20}$	Metglas 2605
$Fe_{40}Ni_{40}P_{14}B_6$	Metglas 2826
$Fe_{82}B_{10}Si_8$	Metglas 2605S
$Pd_{80}Si_{20}$	
$Pd_{77.5}Si_{16.5}Cu_6$	very stable alloy, easy to vitrify
$Pt_{60}Ni_{15}P_{25}$	very stable alloy, easy to vitrify
$Cu_{60}Zr_{40}$	
$Mg_{70}Zn_{30}$	
$Gd_{17}Co_{83}$	sputtered sample can support magnetic bubbles
$Mo_{64}Re_{16}P_{10}B_{10}$	superconducting below 8.7 K
$W_{60}Ir_{20}B_{20}$	crystallisation temperature above 1200 K

Glassy metals usually have high electrical resistivities. The temperature dependence of the resistivity is often small, and can be either positive or negative or show a weak minimum. Various metal-metal glasses such as $La_{30}Au_{20}$ and metal-metalloid glasses such as $(Mo_xRu_{1-x})_{80}P_{20}$, show superconductivity.

Glassy metals containing Cr, such as chromium-iron alloys, are extremely corrosion resistant because of the formation of a dense chromium oxo-hydroxide layer on the surface of the metal.

Thermal Behaviour

Glassy metals are thermodynamically unstable and crystallise when they are heated. Crystallisation is usually accompanied by drastic property changes. Furthermore, the glassy metals as produced are not in a stable state within the amorphous phase because of rapid quenching. An annealing treatment near the glass-transition temperature (T_g) can also result in substantial property changes even if it does not lead to crystallisation. This stabilisation phenomena is called structural relaxation.

Formation

Any liquid can be made into glass if the quenching rate is high enough to avoid the minimum time of the crystalllisation curve shown in Fig. 20.6. In practice, the quenching rate that may be achieved is restricted by the limited heat conductivity and heat capacity of the sample. All the glassy metals stable at RT are alloys with two, three, or more constituent elements.

Preparation

A number of methods are used to prepare glassy metals from the corresponding crystalline alloys but they generally belong to one of three categories. The first is the liquid-quenching method mentioned earlier. The second is vapour deposition including the ion-sputtering technique. The third is the wet chemical technique of electrodeposition or electroless chemical deposition.

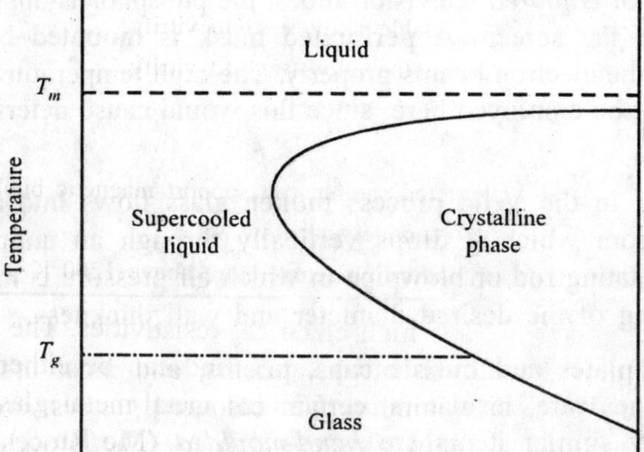

Fig. 20.6. A schematic time-temperature-transformation (TTT) diagram for solidification of liquid; T_m and T_g denote the melting and the glass-transition temperatures, respectively.

The liquid-quenching methods are, in general, simple and suited to large-scale mass production but limited as to the composition ranges.

The vapour-deposition technique affords a wider range, particularly the ion-sputtering method. The equipment for vapour deposition is complex and expensive and the deposition rate is significantly lower (by a factor of 10^6 to 10^{10}) than the production rate in liquid quenching.

Liquid-quenching techniques include splat quenching, the roller quenching technique, the Pond-Maddin technique, and the melt-spinning technique. Vapour-quenching techniques include the thermal vapour-deposition method and the ion-sputtering technique.

Applications

Applications of glassy metals include those as a permalloy replacement, in power devices, bubble memory devices, and delay lines. Mechanical applications include their use as strengthening materials, cutting tools, and brazing materials.

Televisions tubes are now made as large 27 inches across and consist of three principal parts, the face phosphorescent screen, on which the picture is produced, the envelope, and the electron gun. The phosphor is applied to the face screen of the envelope either by settling or dusting. Manufacture of the glass envelope was difficult until centrifugal casting was invented, which uses a revolving mold to produce a much more uniform wall thickness. The glass parts are sealed together, using a gas flame or gas and electricity. For *coloured* television tubes, the phosphor is applied to the inner surface of the screen. A perforated mask is mounted behind the screen to direct the electron beams properly. The high temperature involved in sealing cannot be employed here, since this would cause deterioration of the phosphor.

Glass tubing. In the Vello process molten glass flows into a drawing compartment, from which it drops vertically through an annular space surrounding a rotating rod or blowpipe in which air pressure is maintained, to produce tubing of the desired diameter and wall thickness.

Glass tower plates and bubble caps, prisms, and most other optical glass, most kitchenware, insulators, certain coloured glasses, architectural glass, and many similar items are *hand-moulded*. The process consists essentially in drawing a quantity of glass, known as the *gather*, from the pot or tank and carrying it to the mould. Here the exact quantity of glass required is cut off with a pair of shears, and the ram of the mold driven home by hand or by hydraulic pressure. Certain glass forming is carried out by semi-automatic methods which involve a combination of the machine and hand-moulding processes previously described. Volumetric flasks and cylindrical Pyrex sections for towers are fabricated in this manner.

Annealing

To reduce strain, it is necessary to *anneal all glass* objects, whether they are formed by machine- or hand-moulding methods. Briefly, annealing involves two operations: (i) holding a mass of glass above a certain critical temperature long enough to reduce internal strain by plastic flow to less than a predetermined maximum; and (ii) cooling the mass to room temperature slowly enough to hold the strain below this maximum. The lehr, or annealing oven, is nothing more than a carefully designed heated chamber in which the rate of cooling can be controlled so as to meet the foregoing requirements. The establishment of a quantitative relationship between stress and birefringence caused by the stress has enabled glass technologists to design glass to meet certain conditions of mechanical and thermal stress. With the foregoing data as a basis, engineers have produced continuous-annealing equipment, with automatic temperature regulation and controlled circulation, which permits better annealing at a lower fuel cost and with less loss of product.

Finishing

All types of annealed glass must undergo certain finishing operations which, though relatively simple, are very important. These include cleaning, grinding, polishing, cutting, sandblasting, enamelling, grading, and gaging. Although all these are not required for every glass object, one or more is almost always necessary.

MANUFACTURE OF SPECIAL GLASSES

Research and developments are at the heart of the new and improved types and properties of glass. This section illustrates some of the new glass products that have resulted.

Fused Silica Glass

This glass is also known as quartz glass or vitreous silica glass. It is made by high temperature pyrolysis of silicon tetrachloride. It is very resistant thermally as well as chemically. The most important properties of fused silica glass are: (i) lowest specific gravity (2.07); (ii) high softening point (2600°F); (iii) low coefficient of cubical thermal expansion; (iv) highest annealing point (1965°F); (v) unusual chemical stability; and (vi) transparency to ultra violet radiation. Because of its low thermal expansion, it is used for telescope mirrors.

High Silica Glass

The high silica, called Vycor approaches fused silica in some of its properties. It contains about 96% SiO_2 and 3% B_2O_3 and the rest alumina and alkali. It can be prepared from borosilicate glass which on heating separates silica as a separate layer. The latter, following a HCl treatment at 98°C and a second heat treatment furnished a glass containing about 96% glass, known as high silica glass. The most important properties of high silica glass are: (i) highest softening point (2732°F); (ii) high annealing point which is second after fused silica glass; (iii) high chemical durability and extremely stable to all acids, except hydrofluoric acid (HF); (iv) lower specific gravity than any common glass, but higher than that of fused silica glass; (v) low thermal conductivity; (vi) transparency to nearly all ranges of wave length; and (vii) great capacity to resist structural changes during sudden temperature fluctuations.

Due to high durability and high heat resistance power, it is widely used in the manufacture of laboratory wares such as condensers, crucibles etc.

Optical Glass

Optical glass includes only those glasses with high homogeneity and special composition which have definite optical characteristics of such accuracy as to permit their use in scientific instruments. Most important requirements of an optical glass are: (i) its composition should be according to the required optical properties; (ii) the batch should produce glass of sufficiently low viscosity; (iii) the glass should not devitrify even upon long annealing; (iv) it should produce nearly a colourless product without the use of colouring material; (v) it should be free from bubbles; (vi) it should have advantageous grinding and polishing properties; and (vii) it should be capable of withstanding the action of atmosphere.

Optical or Crookes glass generally contain phosphorus and lead silicates with little cerium oxide. Cerium oxide is capable of absorbing ultra-violet light. In general, optical glasses are relatively soft and have low melting points. The important properties of optical glass are: (i) low viscosity; (ii) absence of bubbles; (iii) no devitrification on long annealing; and (iv) capacity to sustain optical clarity.

Optical glass is used in making optical instruments as lenses of telescopes, microscopes etc.

Borosilicate Glass

This glass is an example of optical glass and contains about 13-28% of B_2O_3 and 80-87% silica. It has low expansion coefficients, superior resistance to shock, high electrical resistance and high chemical stability. The composition of typical borosilicate glass is: SiO_2 = 80.5%, B_2O_3 = 13%, Al_2O_3 = 3%, K_2O = 3% and Na_2O = 0.5%. A glass of low thermal coefficient of expansion and high chemical resistance is obtained by substitution of alkali (Na_2O) and basic alkaline-earth oxides (CaO) of the sodalime glass by boron and aluminium oxides. These glasses are used in the manufacture of baking dishes, laboratory glass ware, pipelines, insulators and washers.

Lead Glasses

These glasses are obtained by mixing lead oxide, silica and alkali, with lead contents as high as 92%. The important properties of lead glass are: (i) high refractive index (ii) low hardness; (iii) high dispersion power; and (iv) high density. The density of high lead glasses can be as high as 8.0. They are widely used in optical work, because of their high index of refraction and high dispersion. They are also used in the manufacture of electric bulbs, neon tubing and radiators, because of high electric resistance of the glass.

Soda Lime Glass

The soda lime glass has the following composition: SiO_2 = 70 to 74%: CaO = 10 to 13% and Na_2O = 13 to 16%. It is used in the manufacture of containers of all kinds, flat glass, tumblers, table ware and automobiles such as window, plate, float, wire etc. They are sufficiently viscous and they melt with more difficulty but are more chemically resistant.

Alkali Silicates

These are water soluble glasses and used only as solutions. They are prepared by fusing together the sand or silica and soda ash (Na_2CO_3). The product known as sodium silicate, solidifies to a glassy mass, known as water glass. A kind of soluble glass can be manufactured by fusing pure sand (SiO_2) with Na_2CO_3 or Na_2SO_4 and some charçoal in a reverberatory furnace and heating the product with water in an autoclave to form glass like mass known as water glass.

$$Na_2CO_3 + SiO_2 \rightarrow Na_2SiO_3 + CO_2$$

$$Na_2SO_4 + 4C \rightarrow Na_2S + 4CO$$

$$Na_2S + 3Na_2SO_4 + 4SiO_2 \rightarrow 4Na_2SiO_3 + 4SO_2$$

Water glass is used as an adhesive in paper, in the manufacture of corrugated-paper boxes, in fire proofing and in egg preservation.

Coloured Glass

Coloured glasses are used for decoration, technical and scientific purposes. Coloured glass can be prepared in pot furnace from batch material containing inorganic colouring agents (usually metallic oxides).

Colour is produced by the absorption of certain light frequencies by colouring agents in solution. Colouring agents used for this purpose are the oxides of transition elements, such as those of Ti, V, Cr, Mn, Fe, Co, Na and Cu. Such type of glasses may be those in which the colour is either due to chemical structural environment or due to differences in state of oxidation. For example, NiO is soluble in sodium lead glass yielding a brown colour, but in potash glass it dissolves to give a heliotrope. This NiO is an example of the agent which produces colour due to chemical structural environment. Chromium oxides produce colour due to differences in state of oxidation, e.g., they produce colours ranging from green to orange, depending upon the properties of basic oxide, Cr_2O_3 to the acidic oxide and the composition to the glass (whether it is acidic or basic): (i) colour is due to coloidal particles precipitated within the colourless glass as a result of heat treatment. For example, precipitation of colloidal gold, producing gold ruby glass; and (ii) colour is due to microscopic or large particles which may be coloured themselves. For example, SeO_2 (selenium reds) used in traffic lights, lantern glodes etc.

Opal Glass

Opal glass or translucent glass is clear when melted, but becomes opalescent as the glass is shaped. This is probably due to the separation and suspension of minute particles of NaF and CaF_2 in the medium, which disperse the light passing through them. These glasses are used for the transmission of specified wavelengths and for table wares. The batch material for the manufacture of opal glass consists of Na_2CO_3, CaO, MgO, ZnO, CaF_2, SiO_2 and $KAlSi_3O_8$. The proportion of different materials differ in different samples. The composition of one such opal glass is as under.

SiO_2 = 68–70%,	Al_2O_3 = 3–4%,	CaO = 4–6%
ZnO = 2–3%,	Na_2O = 15–16%,	CaF_2 = 3–4%

Safety Glass

There are two main classes of safety glass. Laminated and tempered or strengthened glass. The laminated glass may be prepared by placing a non brittle plastic sheet between two thin plate glass sheets. The three sheets are pressed together under moderate heat to seal the edges. They are then heated under pressure in an autoclave to bring the plastic sheets in absolute contact with the glass sheet. Ultimately the three sheets are made into one single sheet. The edges of the sandwich may be sealed with a water resistant material. The plastics used for this purpose are proxylin or cellulose acetate, polyvinyl acetate and polyvinyl butyracetal. The letter has widely been used and has a capacity to elongate 300 per cent.

The tempered glass is very strong and tough, and possesses high internal stresses. The already formed glass vessel or sheet to be strengthened by tempering or annealing is heated to 800°F, just below its softening point, and then quenched in air or molten salt, or oil. The quenching produces a hard surface, which resists bending and shattering forces. The interior pulls on the outside surface, putting it into compression, and the interior develops compensating tension and provides a very good strength to the glass. Such type of glass is used for doors and windows of automobiles and for pipes. A third type of safety glass is produced as sheet glass by incorporating a steel wire net in the molten glass prior to formation of the sheet.

Fibre Glass

Such type of glass is used in textiles and reinforcing and can be spun into yarn, gathered into a mat, and made into insulation and a great variety of other products may be made with it (such as pipe with plastic bond). Fibre glass is nothing but molten glass processed mechanically to a flexible thread of filament. A hot platinum nozzle filled with molten glass forces out the fluid in the form of a thin continuous thread which when cought by a rapidly moving disc gets converted into fibre through elongation and twist given by the disc. Glass fibre, known as staple fibre has an average length of 19–12 inches and diameter of about 0.0028 inches. Fibre glass is not brittle, but possesses high tensile strength. Fabrics made of glass fibre are bad conductors of heat and electricity and are non-inflammable. Hence articles made of fibre glass are fire-proof.

Glass Laminates

For the manufacture of glass laminates, sheets of glass fabric are soaked in a solution of phenol formaldehyde and placed one above the other. The laminated sheets are then allowed to cure under heat and pressure, as a result of which hard, light and strong laminated sheet are obtained. Besides sheets, tubes and other shapes are also available commercially. It is used as a material of construction, for making light weapons, and insulating materials, because it is highly non-inflammable and insulating. The most important properties of laminated glass are: (i) it is shatter-proof; and (ii) it is shock-proof and can withstand sudden changes of temperature and some pressure without undergoing cracking.

A bullet resistant laminated glass is obtained by pressing together several layers of glass with vinyl resins in alternate layers. The thickness of bullet resistant laminated glass generally varies from 12.7 mm to 76.5 mm and these are prepared for specific purposes.

Glass Wool

Glass wool is obtained by the action of steam jets on dripping streams of molten glass down and small holes. The glass wool (which consists of innumerable tiny fine fibres) formed in this manner is carried away by a conveyor belt on which it is hurled. Glass wool is a fibrous wool-like material composed of intermingled fine threads of filaments of glass and has a number of important properties. For example: (i) it is non-combustible; (ii) it is fire proof; (iii) it has low electrical conductivity; (iv) it is heat proof; (v) it does not absorb moisture or water; (vi) It is chemically resistant to a number of chemicals; (vii) it has a low density of 65 kg/m^3; (viii) it has low thermal conductivity; and (ix) it has a high tensile strength (about 8 times that of steel).

Glass wool is used for heat insulation purposes in ovens, metal pipelines, motors, vacuum cleaners, walls and roofs of houses and in domestic appliances. It is also used in the filtration of corrosive liquids such as acids and alkalies, because it is chemical resistant. It is also used for sound insulation and for the manufacture of fibre glass, by blending with plastic resins.

Flint Glass

It is clear transparent potash lead glass used in making electric bulbs and optical instruments.

Bottle Glass

It is the cheapest type of soda lime glass, used in making bottles.

Jena Glass

It is a mixture of zinc and barium borosilicates and is used in making laboratory glass ware.

Pyrex Glass

This glass is a mixture of sodium aluminium borosilicates, representing a type known as hard glass. It is used in the manufacture of costly and high temperature laboratory equipment, thermometers etc.

Rare Earth Glass

It is found that unusual colours and characteristics are developed when some rare earth oxides are mixed with glass. For example, Alexandrite, heliolyte and royalite glasses, known as moser glass contain 4–5% Nd_2O_3, small amounts of Nd_2O_3 and Pr_2O_3 and small amount of Nd_2O_3 and Se respectively and have unique capacity of displaying two different colours at two different thickness (in the same wavelength region) or at two different wave length regions (in the same thickness). For example, heliolyte gas (5% Nd_2O_3 + 5% Pr_2O_3) is olive green in thin layers but brown red in thick layers. Alexandrite (4–5% Nd_2O_3) is bluish purple in day light and brilliant red in lamp light. The neodymium glasses has recently been used in the study of lasers.

Photosensitive Glass

This is high alumina soda lime glass which is sensitive in the u.v. region. It can reproduce the object in live natural colours from the ordinary black and white negative, recorded in the visible region. The positive on glass is developed by thermal treatment only at 540°–550°C. The desired photo activity in u.v. region can be obtained by admixture of high alumina soda lime glass with small amounts of Cu_2O, NaCN, SnO_2 abeitic acid in appropriate amounts. A blue colour is promoted by NaCN in absence of tin oxide. In presence of tin oxide an impression in red is observed. By manupulating the ingredients in glass, brown and yellow images can also be possible. A potash-alumina glass, mixed with $LiSiO_3$, cerium and silver salts in appropriate proportions have also been used as photosensitive glass.

Photochromic Glass

Such glasses have a large number of microscopic particles of silver halides trapped in the three dimensional silicate net work in fixed concentration. On exposure to light, temporary colour centres consisting of silver particles only are produced and these add up quickly producing total darkness. The intensity of darkness depends upon the concentration of Ag. Because reversible darkening is controlled by the radiations in the u.v. region, quite abundant in day light, the photoblackening does not occur markedly in the lamp light in the night.

Froast Glass

Such glasses are made by sand blasting a sheet of common glass.

Ground Glass

Soda lime glass, the surface of which has been rendered translucent either by rubbing with emery powder and water or with emery powder and turpentine is commonly known as ground glass.

Insulating Glass

It is transparent glass that can be prepared by two more plates of glass separated by 6–13 mm thick gap, filled with dehydrated air and then hermetically sealing around the edges. Insulating glass provides high insulation against heat and so used as a thermal insulating material.

Vitreosil (99.5% Silica Glass)

It can be prepared by heating pure sand (SiO_2) to its melting point (above 1750°C). It can not be shaped easily, because of high viscosity of the glass and it also contains some bubbles because of the absence of any flux in its preparation. The final product is translucent (softening temperature about 1650°C). Its thermal expansion is lowest, being only 0.55×10^{-6} cm.cm^{-1} $^0C^{-1}$. If vitreosil glass is heated for a long time above its melting point, finally a transparent glass, known as clear silica glass is obtained. This glass is used mainly for wares for chemical plants, for chemical laboratory wares and for electrical insulating materials in electrical heaters and furnaces etc.

96% silica glass can also be prepared by first producing a borosilicate glass of typical composition, which is then shaped into the desired forms, somewhat bigger than the final dimensions needed. The articles are then heat treated to separate glass into two layers one consisting of mainly of silica and the other of alkali oxides and borates. The alkali oxides and boron oxide are dissolved by dipping the article in ‚t acid. As a result, fine porous structures consisting of 96% SiO_2, 3% B_2O_3 and traces of other metals is obtained. The glass shapes are washed carefully and annealed to 1200°C. As a result, some 14% shrinkage takes place and hard as well as firm shapes are obtained. These are almost gas tight and translucent. 96% silica glass thus obtained may be heated to higher temperatures to get transparent

The coefficient of expansion of 96% silica glass is very low and hence it is highly resistant to thermal shock. Its softening temperature is high (about 1500°C). These glasses are resistant to chemicals like acids and dilute alkalies, but corroded by hydrofluoric acid, hot phosphoric acid and concentrated alkalies. 96% silica glass are widely used in those circumstances where high resistance to temperature is required. They are used for the construction of chemical plants. crucibles, induction furnace linings, electrical insulators etc. 99.5 silica glass or vitreosil is especially used for the construction of pipe lines for hot concentrated acids.

Chapter 21

Nuclear Industries

INTRODUCTION

Modern civilisation demands energy in many forms. Coal and petroleum are the conventional sources of energy. These are fossil fuels and non-renewable sources of energy. At present we mainly depend upon them for our energy requirement. However, the deposits of fossil fuel may not last more than a few decades to supply the demand of power. Therefore, attempts are being made to explore the non-conventional sources of energy. Though the hydroelectric power is the renewable source, the power generation depends upon the monsoon and cannot meet the ever increasing demand. Solar energy, nuclear energy, wind energy and the tidal energy are the alternate sources of energy. In the years to come the use of solar energy for power generation will be inevitable. But the present technology is inadequate to make use of the solar energy for power generation economically, on large scale. Similarly, the use of wind and tidal energies have also some limitations. Nuclear fuels are potential sources of energy.

Nuclear fuel is a material that undergoes nuclear fission on bombarding with neutrons, releasing enormous amount of energy. ^{233}U, ^{235}U and ^{239}Pu are the nuclear fuels. These isotopes are called *fissile*. ^{232}Th and ^{238}U do not undergo nuclear fission but on bombarding with neutrons undergo a series of nuclear disintegration reactions and produce fissile materials.

Examples :

$${}^{232}_{90}Th + {}^{1}_{0}n \longrightarrow {}^{233}_{90}Th + \gamma$$

$${}^{233}_{91}Th \longrightarrow {}^{233}_{91}Pa + {}^{0}_{-1}e \quad \text{and} \quad {}^{233}_{91}Pa \longrightarrow {}^{233}_{92}u + {}^{0}e_{-1}$$

${}^{233}_{92}U$ undergoes nuclear fission. Hence it is a nuclear fuel.

Similarly, $^{238}_{92}$U produce $^{239}_{94}$Pu which undergoes nuclear fission.

The material which do not undergo nuclear fission, but can produce fissile materials are called fertile materials. Thus, ^{232}Th and $^{238}_{92}$U are fertile materials.

1 kilogram of uranium on complete fission can produce energy equivalent to 20 × 10^6 KWH, while 1 kilogram of coal on complete combustion can produce only 2 KWH of power. The process of nuclear fusion also produces a very large amount of energy. However, at present there are many problems in controlling the nuclear fusion process. The major problem associated with the nuclear fission process is the disposal of the hazardous radio active wastes.

URANIUM AND URANIUM COMPOUNDS

Uranium, at no. 92, at wt 238.03, is a member of the actinide series of transition elements.

Occurrence

Uranium is present in the earth's crust at ca 2 ppm. Acidic rocks with a high silicate content, such as granite, have a uranium content above average, whereas the contents of sedimentary and basic rocks, such as basalts, are below average. However, 90% of the world's known uranium sources are contained in conglomerates and sandstone. Primary uranium minerals, crystallised from low-melting rocks, are generally black and contain uranium in a valence lower than six. Secondary minerals are yellow, greenish-yellow, bright green, or orange, and contain uranium in the valence +6.

Uraninite and pitchblende differ in physical form; their composition ranges from $UO_{2.0}$ to $UO_{2.67}$. Uraninite is well crystallised; pitchblende is either amorphous or consists of very fine crystals. Gummite, $UO_3.nH_2O$, and becquerelite, $2UO_3.3H_2O$, are typical hydrated oxides. They are fairly common, but commercially unimportant. Carnotite, $K_2O.2UO_3.V_2O_5.3H_2O$, and tyuyamunite, $CaO.2UO_2.V_2O_5.8H_2O$, are the commercially important minerals. Phosphates are represented by autonite, $Ca(UO_2PO_4)_2.8H_2O$, and torbernite, $Cu(UO_2PO_4)_2.12H_2O$. Uranium arsenate minerals are found in European deposits. Numerous organic complexes are found in sedimentary deposits, e.g., thucholite and carburan.

Resources

Reasonable assumed resources (RAR) contain deposits of such size, grade, and configuration that recovery is within the given production cost ranges with the current mining and processing technology. This type of resource is estimated at 1.8×10^6 metric tons. Estimated additional resources (EAR) occur as extensions of well-known deposits, little-explored deposits, or undiscovered deposits; they are estimated at 2.6×10^6 tons.

Properties

Uranium is a dense, lustrous metal that resembles iron; it is ductile, malleable, and weakly magnetic. In air, it tarnishes rapidly, and in a very short time, even a polished surface becomes coated with a dark-coloured layer of oxide. In the solid state, uranium exists in three allotropic modifications. The thermodynamic properties have been determined with great accuracy (Table 21.1.). Thermal conductivity ranges from 0.251 W/(cm·K) at 309 K to 0.326 W/(cm·K) at 673 K. Spectroscopic properties have been studied in great detail and more than 30,000 lines of the arc-and-sparc-emission spectra have been catalogued.

Table 21.1. Thermodynamic properties of uranium metal.

Function or parameter	*Value or equation*
Entropy of α-U, J/K[a]	50.21 ± 0.12
Entropy of U(g), J/K[a]	199.6
Enthalpy of fusion, kJ/mol[a]	8.326 ± 0.54
Enthalpy of sublimation, kJ/mol[a]	1062.73
Free energy of vapourisation, G, kJ/mol[a]	
solid to gas	525.3 – 0.137
liquid to gas	487.6 ± 0.11
Normal (extrapolated) boiling point. K	3818
Vapour pressure of liquid[b]	$\log p\ (\text{kPa}) = -\dfrac{(25{,}230 \pm 370)}{T} + (7.72 \pm 0.17)$

[a] To convert J to cal divide by 4.184.

[b] To convert kPa to mm Hg, multiply by 7.5.

Of the four oxidation states, +3, +4, and +6 are stable enough to be of practical importance; +5 is of minor importance. The alternation between +4 and +6 states is utilised in the extraction of uranium from ores and in purification.

Isotopes

There are fifteen known isotopes of uranium, not counting the isomeric states; ^{234}U, ^{235}U, and ^{238}U exist in nature. The natural isotope composition usually varies. In addition to ^{235}U, an artificial isotope, ^{233}U, was found to be fissionable. Today, it is an industrial product produced in kilogram quantities by the thermal breeding reaction.

$$^{232}\text{Th} + n \xrightarrow{\gamma} {}^{233}\text{Th} \xrightarrow{\beta^-} {}^{233}\text{Pa} \xrightarrow{\beta^-} {}^{233}\text{U}$$

The product is isolated by a radio-chemical process.

Extraction from Ore

Conventional ore-dressing techniques have not been successful in the preconcentration of uranium minerals. Gravity separation is sometimes possible. Electrostatic methods generally give low recoveries in low concentrations. Flotation usually gives satisfactory concentrations.

A high temperature roasting or calcining operation before leaching is frequently desirable, and the characteristics of many ores are thus improved. Treatment with acids or alkalies converts the uranium contained in the ore to water-soluble species. Most mills use acid leaching. Alkaline or carbonate leaching is used for ores with high lime content, which would require excessive consumption of acid.

The crude uranium isolated in the leach liquors requires additional purification. Direct or selective precipitation has not been commercially successful in acid leach systems. However, precipitation may be used to enrich the uranium in side streams. Anionic sulphato or carbonato complexes may be absorbed from leach liquors on anion-exchange resins. The uranium is eluted with a saline or acid solution.

Some uranium ores exhibit extremely poor filtering and settling characteristics after leaching. To avoid large liquid-solid separation equipment; the ion-exchange process has been modified to extract uranium directly from the leach pulp (resin-in-pulp process).

Solvent extraction is widely used in the recovery of uranium from ores. Contrary to ion exchange, solvent extraction can be operated in a continuous countercurrent flow. The preferred extractants are di(2-ethylhexyl) phosphate (D2EHPA) and dodecyl phosphate (DDPA). The product of the extraction process is a purified uranium solution. Uranium is usually precipitated directly from solutions resulting from carbonate leaching. From acid solutions, uranium is precipitated by neutralisation with ammonia or magnesia. With the former, a precipitate of composition

$(NH_4)_2(UO_2)_2SO_4(OH)_4 \cdot nH_2O$ is obtained. So-called yellow cake with a higher uranium content is obtained with magnesia.

Conversion and Purification

The crude product from the refinery is purified for nuclear applications. Yellow cake is dissolved in nitric acid and extracted with tributyl phosphate. The uranium is back-extracted with recycled concentrated acid containing < 1% HNO_3 or with deionised water. Evaporation gives uranyl nitrate hexahydrate, which is pyrolysed to UO_3. Reduction with hydrogen gives the dioxide, which is converted to UF_4 by treatment with HF. Direct fluorination of the tetrafluoride, also called green salt, gives the hexafluoride.

For reactor applications, enriched UF_6 from an isotope-separation plant is converted to other compounds or the metal.

Manufacture of the Metal

Uranium metal is produced by reduction of the tetrafluoride in a bomb at ca 700°C (Ames process).

$$UF_4(s) + \begin{cases} 2Ca \\ 2Mg \end{cases}$$

$$\rightarrow U + \begin{cases} 2CaF_2 & \Delta H^\circ = -560.5 \text{kJ/mol} \ (-134 \text{ kcal/mol}) \\ 2MgF_2 & \Delta H^\circ = -349.4 \text{kJ/mol} \ (-83.51 \text{ kcal/mol}) \end{cases}$$

Most nuclear reactors built for the generation of electric power are based on uranium fuel enriched in ^{235}U. Reactors using natural uranium, such as the Hanford production reactors or some power reactors, do not require enriched uranium. Purified natural uranium fabricated in rods (slugs) may be used.

Only the gaseous-diffusion process is used for the separation of ^{235}U and ^{238}U on an industrial scale. Highly purified gaseous uranium hexafluoride is pumped through a series of diffusion aggregates arrayed in cells in a cascade pattern. Enormous amounts of electric power are required. The high capital cost and power consumption of gaseous-diffusion plants have led to the investigation of centrifugal separation of ^{235}U and ^{238}U.

Electromagnetic separation, one of the most powerful methods, so-called calutron separators. The calutron techniques has been used to separate pure samples of ^{234}U, ^{236}U, and stable isotopes of many other elements.

Other methods for isotope separation were developed, but none advanced beyond the pilot-plant stage. Laser excitation appears very promising. Uranium hexafluoride vapour is ionised by means of a tunable rhodamine-B laser in such a manner that ^{235}U atoms are selectively excited to form UF_5 molecules which are collected on a sonic impactor (MLIS-process).

Isolation of specific isotopes

The uranium isotopes ^{232}U, ^{233}U, and ^{234}U are the daughters of the actinide isotopes ^{232}Pa, ^{233}Pa, and ^{238}Pu, respectively, which may be obtained pure. For the production of ^{232}U, proactinium-231 is bombarded with neutrons. The ^{232}U can be separated by an ion-exchange process. Uranium-233 is obtained by bombardment of thorium-232 with slow neutrons. Uranium-234 is the daughter product of plutonium-238 which, in turn, is produced by neutron bombardment of neptunium-237.

Uranium Compounds

Uranium metal, heated to 150–200°C in a hydrogen atmosphere, gives uranium trihydride, UH_3, a black powder. Rare isotopes of hydrogen, i.e., tritium, can be stored in the form of hydrides. Deuterium or tritium are absorbed on uranium turnings heated to 200°C and then may safely be stored as solids. Upon heating to 500°C in vacuum, the isotope is released as a highly pure gas.

Fluorides

Uranium forms seven binary fluorides: UF_3, UF_4, U_4F_{17}, U_2F_9, α-UF_5, β-UF_5, and UF_6. The trifluoride is prepared from stoichiometric amounts of UF_4 and uranium metal. The tetrafluoride is prepared by hydrofluorination of UO_2 with excess gaseous HF at ca 550°C. The uranium (IV, V) fluorides, U_4F_{17} and U_2F_9, are obtained from the tetrafluoride and the hexafluorides at 200°C and 2.36 kPa (17.7 mm Hg). These two compounds are observed in gas centrifuges as decomposition products. The grayish-white α-pentafluoride is obtained from the tetra- and hexafluorides by reduction with HBr at 80–100°C; at 150–200°C, the yellowish-white β-pentafluoride is obtained. The hexafluoride is prepared by direct fluorination of the tetrafluoride; its properties are given in Table 21.2. It is produced on a large scale as feed for the gaseous-diffusion, separation-nozzle, and gas-centrifuge processes. A large number of ternary uranium fluorides are composed of a binary fluoride and an alkali or alkaline-earth fluoride.

Table 21.2. Properties of uranium hexafluoride.

Property	*Value*
Triple point at 151 kPa[a], °C	64.052
Sublimation point, °C	56.4
Density	
solid, g/cm³	5.09
liquid, g/mL	6.63
Heat of formation, solid at 25°C, kJ/mol[b]	−2158.9
Heat of vapourisation at 64.01°C, kJ/mol[b]	28.899
Heat of fusion at 64.01°C, kJ/mol[b]	19.196
Heat of sublimation at 64.01°C, kJ/mol[b]	48.095

[a] To convert kPa to mm Hg, multiply by 7.5.
[b] To convert J to cal, divide by 4.184.

Chlorides

The trichloride, an olive-green solid, is prepared from the trihydride and HCl at 250–300°C. The tetrachloride, a dark green solid, is best prepared from UO_3 and hexachloropropene.

$$UH_3 + 3HCl \rightarrow UCl_3 + 3H_2$$

$$UO_3 + 3\ CCl_3CCl = CCl_2 \rightarrow UCl_4 + 3CCl_2 = CClCOCl + Cl_2$$

The pentachloride is prepared by treatment of the trioxide with carbon tetrachloride. The hexachloride, a dark green solid, is obtained by disproportionation of the pentachloride in vacuum.

Bromides and iodides

The tribromide, a reddish-brown crystalline material, is prepared from the trihydride and HBr or from the elements. The dark brown tetrabromide is prepared by heating uranium turnings in a stream of nitrogen with bromine vapour. The pentabromide is unstable. The tri- and tetraiodides are made from the elements; the product depends on the temperature and the pressure of the iodine.

Oxides

The uranium-oxygen system is extremely complicated. Alterations of the oxidation state are significant in refining ore concentrates. The dioxide, mp ca 2800°C, is very stable and occurs in nature as pitchblende. The tetrauranium enneaoxide, U_4O_9, forms black crystals and occurs in three modifications. The triuranium octoxide, U_3O_8, is a component of a complicated phase system.

The trioxide, UO_3, exists in six polymorphic modifications that differ in crystallographic properties and colour. It is readily obtained from the thermal decomposition of various uranyl compounds. The monoxide, UO, forms very slowly above 2000°C in mixtures of UO_2 and uranium. The hydrated peroxide, $UO_4{\cdot}xH_2O$, is utilised in purification technology. It may be precipitated from uranyl nitrate solution with H_2O_2. Several hydrated oxides have been investigated, e.g., $UO_3.2H_2O$, $UO_3.H_2O$, and $UO_3.0.5H_2O$.

NUCLEAR FISSION

We have seen that very heavy nuclei have lower binding energies than the nuclei with intermediate masses and hence tend to undergo nuclear fission reactions.

The first nuclear fission reaction was carried out in 1939 by O. Hahn and Strassmann in Germany by bombarding ^{235}U isotope with slow neutrons. ^{235}U isotope on bombarding with slow neutrons is converted into an unstable intermediate ^{236}U which splits up into two or more fragments and also emitting two or more neutrons with the evolution of enormous amount of energy.

$$ {}^{235}_{92}U + {}^{1}_{0}n \longrightarrow [{}^{236}_{92}U] \longrightarrow {}^{140}_{56}Ba + {}^{94}_{36}Kr + 2\,{}^{1}_{0}n + \text{Energy} $$

The process of splitting up of heavier nuclei such as ^{235}U by bombarding with slow neutrons into two or more smaller nuclei of approximately equal masses is known as *nuclear fission.*

A fission process is much more complicated. A given nucleus undergoing fission may split in number of different ways depending upon the energy of the neutrons. For example,

$$ {}^{235}_{92}U + {}^{1}_{0}n \longrightarrow {}^{236}_{92}U \longrightarrow {}^{90}_{37}Rb + {}^{144}_{55}Cs + 2\,{}^{1}_{0}n $$

$$ {}^{235}_{92}U + {}^{1}_{0}n \longrightarrow {}^{236}_{92}U \longrightarrow {}^{87}_{35}Br + {}^{146}_{57}La + 3\,{}^{1}_{0}n $$

$$ {}^{235}_{92}U + {}^{1}_{0}n \longrightarrow {}^{236}_{92}U \longrightarrow {}^{96}_{36}Kr + {}^{137}_{56}Ba + 3\,{}^{1}_{0}n $$

Important Feature of Nuclear Fission

Chain reaction

During the fission process more number of neutrons are released than the number of neutrons required to initiate the fission. Every neutron so released, in principle, can cause the fission of more atoms and this can continue as long as fissionable material is available. Such a reaction is called as *chain reaction.* Thus nuclear fission reaction is a self sustained reaction. Since, the fission process is also accompanied by the release of large amount of energy, by properly controlling the process, it can be used as a source of energy (Fig. 21.1).

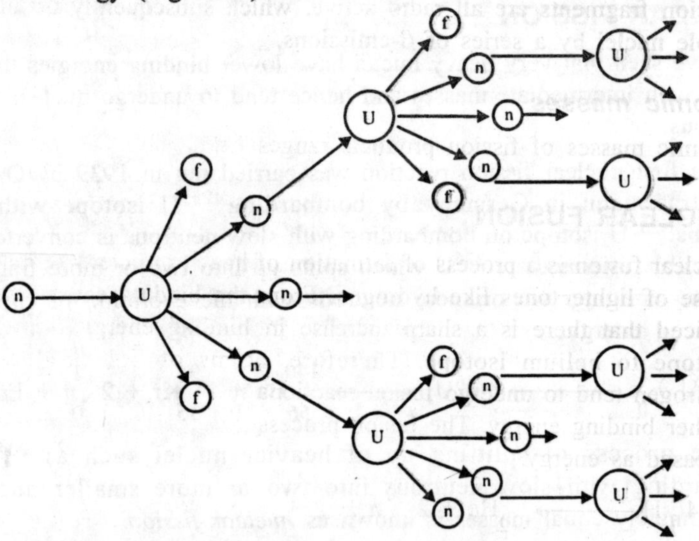

Fig. 21.1. Schematic representation of chain reaction. (n = neutron; u = uranium; f = fission fragments).

Energy released during fission process

The sum of the masses of the fission fragments and neutrons set free during the fission process is found to be less than the sum of the masses of the reactant nuclei. The mass lost is converted into energy and released in the form of heat. This amounts to release of enormous amount of energy during fission process.

$$^{235}_{92}U + ^{1}_{0}n \longrightarrow ^{95}_{42}Mo + ^{139}_{57}La + 2\,^{1}_{0}n + 7\,^{0}_{-1}e + \text{Energy}$$

$235.0439 + 1.0087 \qquad\qquad 94.9057 + 138.9061 + 2.0174$

$= 236.0439$ amu $\qquad\qquad = 235.8592$ amu

$\therefore$ Mass loss (ΔM) = 0.2234 amu

$\therefore$ Fission energy = $0.2234 \times 931 \simeq 208$ MeV

or Fission energy = $0.2234 \times 14.93 \times 10^{-11}$ J

$= 33.35 \times 10^{-10}$ J

Fission fragments

Fission fragments are all radio active, which subsequently disintegrate to stable nuclei by a series of β-emissions.

Atomic masses

Atomic masses of fission products ranges from about 70 to 160.

NUCLEAR FUSION

Nuclear fusion is a process of formation of heavier nuclei like helium from those of lighter ones like hydrogen. From the binding energy values it is noticed that there is a sharp increase in binding energy from hydrogen isotope to helium isotope. Therefore, atoms of lighter elements like hydrogen tend to undergo fusion reactions to form heavier nucleus having higher binding energy. The fusion process involves loss of mass which is released as energy.

$$4\,{}_{1}^{1}H \longrightarrow {}_{2}^{4}He + 2\,{}_{+1}^{0}e + \gamma$$

$\Delta M = 0.02871$ amu

Energy released = 26.7 MeV

However, to initiate the fusion reaction, extremely high temperature of the order of 10^8k is required to overcome the coulombic barrier. These reactions are called *thermonuclear reactions*.

Main Features of Fusion Process

(1) To initiate the fusion process extremely high temperature is required.

(2) During the fusion process enormous amount of energy is released as heat which makes it possible to maintain the fusion process, i.e., a chain reaction is set.

(3) Fusion products are not radio active.

Drawbacks of fusion process

(1) It is very difficult to provide the very high temperature to initiate the process.

(2) At this temperature the atom is fully ionised, producing a plasma of +ve and –ve ions. It is difficult to confine the plasma state till fusion takes place. One method is to confine a fusion reaction in so called magnetic bottles. The plasma is kept within magnetic field so that it does not come in contact with any material.

(3) It is not possible to control the fusion process. Comparison of Nuclear Fission and Fusion process is shown in Table 21.3.

Table 21.3. Comparison of nuclear fission and fusion process.

Fission	*Fusion*
Splitting up of the heavy nuclei like ^{235}U by bombarding with slow neutrons, into two or more light fragments.	Formation of one nucleus like He by the combination of two light nuclei like hydrogen.
Fission process is brought about by neutrons.	Fusion is brought about by extremely high temperature.
More number of neutrons are emitted which causes chain reaction.	Fusion is accompanied by the increase in temperature which maintains the fusion process, positrons are emitted.
Enormous amount of energy is released.	Energy released is much higher than the fission process.
The fission fragments formed are radio active.	The fusion products are not radio active.
Fission process can be controlled. The energy released can be used to produce electricity.	Fusion process cannot be controlled. At present the process is used in producing nuclear weapons like hydrogen bomb.

MASS DEFECT AND BINDING ENERGY

A nucleus of an atom is mainly composed of Z protons, where Z is the atomic number of the element and (A–Z) neutrons where A is the mass number. Protons and neutrons present in a nucleus are collectively known as nucleons. Binding energy is an important characteristic of a nucleus which determine the stability of a nucleus.

Binding energy is the energy which holds the nucleons together in the nucleus. When the nucleons come within the range of nuclear forces a part of their mass is lost resulting in the formation of the nucleus. Thus the actual mass of every nucleus is less than the sum of the masses of the nucleons present in the nucleus. *This difference in mass is known as mass defect* (ΔM). If M is actual atomic mass of the isotope and M' is the actual sum of the masses of constituent protons and neutrons, the mass defect is given by

$$\Delta M = M' - M \quad \text{...(21.1)}$$

The mass of an electron being negligibly small compared to the mass of a proton, its mass may be ignored.

If Z is the atomic number and A is the mass number, the number of protons is equal to Z and number of neutrons is equal to (A–Z). Now the equation (21.1) can be written as

$$\Delta M = Zm_p + (A - Z)m_n - M \quad \text{...(21.2)}$$

where m_p and m_n are the masses of protons and neutrons respectively.

Binding Energy

When the nucleons are bound in the nucleus, the loss in mass (ΔM) is converted into an equivalent amount of energy, which is called the *binding energy* (binding energy is the energy which holds the nucleons together in nucleus) and given by the Einstein's mass–energy relationship.

$$\text{Binding energy} = \Delta M.c^2$$

where 'c' is the velocity of light, the value being $c = 2.998 \times 10^8\ ms^{-1}$.

Binding energy is expressed in electron volts, or million electron volts (MeV). Generally the atomic masses are expressed in atomic mass units (amu). 1 amu is equal to 1/12th the mass of an atom of C–12 isotope. According to Avogadro's number 1g atom (i.e., 12 g) of C–12 isotope contain 6.02×10^{23} carbon atoms. According to definition, the mass of one atom of C–12 isotope

$$= \frac{12}{6.02 \times 10^{23}} g = \frac{12}{6.02} \times 10^{-26} kg$$

$$\therefore 1 \text{ amu} = \frac{1}{12} \times \frac{12}{6.02} \times 10^{-26} \text{ kg} = 1.66 \times 10^{-26} \text{ kg} = 1.66 \times 10^{-26} \text{ kg}$$

Energy corresponding to 1 amu is therefore

$$E = 1.66 \times 10^{-27} \text{ kg} \times [2.998 \times 10^8 \text{ ms}^{-1}]^2$$

$$= 14.93 \times 10^{-11} \text{ J } [\because J = kg\ m^2s^{-2}]$$

$$\approx 931.0 \text{ MeV} \qquad [\because 1.6 \times 10^{-13} J = 1 \text{ MeV}]$$

Thus $$1 \text{ amu} \equiv 14.93 \times 10^{-11} \text{ J} \equiv 931 \text{ MeV} \quad ...(21.3)$$

∴ when the mass defect is 1 amu the energy released is 931 MeV

or $$B.E = \Delta M \times 931 \text{ MeV} \quad ...(21.4)$$

The average binding energy per nucleon is given by the expression

$$B.E / \text{Nucleon} = \frac{B.E}{\text{Total no. of nucleons}} \quad ...(21.5)$$

BINDING ENERGY AND NUCLEAR STABILITY

The stabilities of nuclei depend upon their binding energies per nucleon. The nuclei with higher binding energies per nucleon are relatively more stable than those with smaller values of binding energy per nucleon. The tendency of an atom to undergo fussion or fission can be predicted with the help of the plot of binding energy per nucleon versus the mass number. From the plot three important conclusions can be drawn :

(1) The binding energy per nucleon rises sharply from the isotopes of hydrogen to those of next heavier elements like helium or lithium. Therefore, if two atoms of hydrogen are fused to form a heavy nuclei like helium, a large amount of energy is released (Fig. 21.2). Such a process is called nuclear fusion.

(2) The binding energy per nucleon reaches maximum at mass number approximately 60 and the elements with intermediate mass numbers 50 to 150 have higher binding energies than those of heavier elements are relatively stable.

(3) With the further increase in mass number beyond 60, there is a decrease in the binding energies. Therefore, the heavier elements like uranium tend to split up into nuclei of intermediate masses with the release of large amounts of energy. Such processes are called nuclear fission.

Problems on binding energy :

[Mass of proton = 1.0073 amu and

Mass of neutron = 1.0087 amu]

(1) The atomic mass of ${}^{4}_{2}He$ is 4.0017 amu. Calculate the binding energy per nucleon.

Z of He = 2 and (A – Z) of He = (4 – 2) = 2

$$\Delta M = Zm_p + (A - Z)m_n - M$$

$= 2 \times 1.0073 + 2 \times 1.0087 — 4.0017$ amu $= 0.0303$ amu

$BE = \Delta M \times 931$ MeV

$= 0.0303 \times 931 = 28.20$ MeV

$$\text{Binding energy per nucleon} = \frac{\text{BE}}{\text{Total number of nucleons}}$$

$$= \frac{28.20}{4} = 7.05 \text{ MeV.}$$

(2) Calculate the binding energy per nucleon in MeV of $^{56}_{26}Fe$. Atomic mass of $^{56}_{26}Fe$ = 55.934 amu. Z = 26, A - Z = 56–26 = 30 and M = 55.934 amu.

$\therefore \Delta M = Zm_p + (A - Z)\, m_n - M$

$= (26 \times 1.0073 + 30 \times 1.0087 – 55.934)$ amu $= 0.5168$ amu

Total BE $= \Delta M \times 931$ MeV $= 0.5168 \times 931$ MeV

$= 481.14$ MeV.

$$\text{B.E / Nucleon} = \frac{481.14}{56} = 8.59 \text{ MeV } [\because \text{Protons + neutrons} = 56]$$

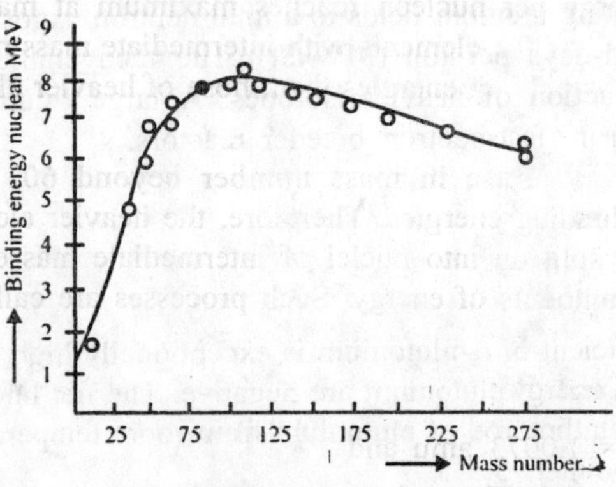

Fig. 21.2. Binding energy curve

PLUTONIUM AND ITS COMPOUNDS

Plutonium

Plutonium, Pu, is element number 94 in the periodic table. It is a member of the actinide series and is metallic. Isotopes of mass number 232 through

246 have been identified and all are radioactive. The most important isotope is plutonium-239; also of importance is plutonium-238.

The large energy release associated with the fission reaction is the most significant property of plutonium. The energy can be applied in electric-power generating reactors, industrial explosives, or military explosives. Large quantities of plutonium are produced in uranium-fueled power reactors. Approximately 200 kg Pu are produced per 1000 MW_e of electric power.

The isotope ^{238}Pu also is of technical importance because of the high heat of its constant radioactive decay. Such radiation is used as fuel in small terrestrial and space nuclear-power sources. Plutonium is unique in that it is the first element to be synthesised.

Sources

From a practical viewpoint, plutonium does not occur in natural ores. All of the fifteen plutonium isotopes have been synthesized and all are radioactive. Technologically, plutonium-239 is the most important isotope. It is characterised by a high fission reaction cross-section and is abundant in irradiated natural uranium.

Commercial electric-power generating reactors generally produce plutonium by irradiating uranium fields to a total neutron exposure of more than 5000 megawatt-days per ton (MW·d/t). The recoverable plutonium contains a larger fraction of heavier isotopes. A large future source of plutonium will be from fast-neutron breeder reactors.

Physical Properties

Thermal

The expansion coefficient of α-plutonium is exceptionally high for a metal, whereas those of δ- and δ'-plutonium are negative. The net linear increase in heating a polycrystalline rod of plutonium from room temperature to just below the melting point is 5.5%.

Thermodynamic

The thermal conductivity of plutonium-24? is 0.084 and 0.155 W/(m.K) (0.020 and 0.037 cal/(s.cm °C)) for the α- and β-phase, respectively.

Electrical and magnetic

The electrical resistivity of plutonium is high in all modifications as a result of the band structure in metallic plutonium.

Radioactive self-heating

Because of their radioactivity, all plutonium isotopes generate heat. The self-heating of a ^{239}Pu metal sphere has been determined: (1.923 + 0.019) × 10^{-3} W/g (1.824 + 0.18 Btu/(s·g)).

Spectroscopic

Isotope shifts of the 238, 239, and 240 isotopes were obtained in the spark spectrum, reexamined with hollow-cathode spectral data, and supplemented by ^{241}Pu data. Based on the shifts of 20 lines of Pu I, and assuming the 238–240 interval as unity, the shift values obtained are given below :

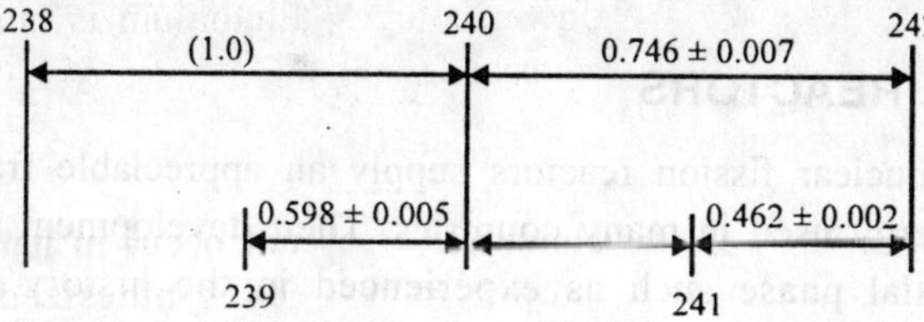

Chemical Properties

Plutonium is the fifth member of the actinide series. Its electronic structure is $1s^2 2s^2 2p^6 3s^2 3p^6 3d^{10} 4s^2 4p^6 4d^{10} 4f^{14} 5s^2 5p^6 5d^{10} 5f^{5-6} 6s^2 6p^6 6d^{0-1} 7s^2$.

Plutonium is an active metal and, as such, it can be made by the usual methods for active metals such as electrolysis in fused salts, reduction of its halides with an active metal such as Li, Ca, Ba, or reduction of the oxide with Ca in a $CaCl_2$ melt. Plutonium forms compounds with all the non-metallic elements except the rare gases. The halogens and halogen acids form Pu halides, other chalcogens form chalcogenides, and CO forms a carbide. Nitrides are formed with NH_3 and N_2 and and hydrides with H_2.

The metal dissolves readily in concentrated HCl, H_3PO_4, HI, $HClO_4$, or HSO_3NH_2.

The corrosion of plutonium has been summarised. Plutonium oxidises very slowly in dry air, i.e., $< 10^{-2}$ mm/yr; the rate is accelerated by water vapour.

Extractive metallurgy and conversion chemistry

The production methods used for the separation of Pu from U and fission products are liquid-liquid extraction and ion exchange.

Plutonium Compounds

Plutonium forms compounds with many of the metallic elements and all of the non-metallic elements except the helium-group gases. The refractory compounds are of interest as potential fast-breeder reactor fuels. Commercially important compounds include the oxides, halides, and oxalates.

Health and Safety Factors

The principal hazards of plutonium are those posed by its radioactivity, its nuclear critical potential, and its chemical reactivity in the metallic state.

NUCLEAR REACTORS

Engineered nuclear fission reactors supply an appreciable fraction of the electrical energy used in many countries. Their development and use is in a controversial phase such as experienced in the history of industrial development with the use of coal, steam power for railways and ships, the internal-combustion engine, and even electricity itself.

One remarkable characteristic (established early in the history of nuclear reactors operating at high power density) is that, after a shutdown, it may not be possible to restart the chain reaction for 15–40 hour. This is owing to transient poisoning of the fission-chain reaction by one of the fission products, ^{135}Xe.

Safety Aspects

Much attention has been paid to safety in establishing the technology of nuclear reactors. The training of operators derived from accumulated experience with these systems must not be underestimated. Although a trained mind is essential in a complex situation, certainly in action requires accurate and adequate information from the necessary monitoring instruments. Containment of the fission products is the basic requirement for minimising exposure to radiation.

Heat Removal

The rate of the chain reaction can increase to levels at which the safe removal of the heat from fission and, later, from residual fission products becomes the principal engineering concern. Thus, many coolants or caloporteurs have been used, including gases, especially air, CO_2, He, and dry steam; liquids, especially water, single-phase pressurised water, two-phase boiling water or fog, heavy water, organic liquids, especially terphenyl and hydrogenated terphenyl, which is liquid at room temperature; the alkali metals, sodium, Na-K alloy, potassium, and lithium; and mercury, molten bismuth, and molten salts, especially fluorides.

Fuels

Depending on the purpose of the reactor, many fuels have been used, although UO_2, ThO_2, PuO_2, and mixtures of these as sintered, high density pellets have been used extensively.

Special Materials

Tritium is produced in reactors not only as a direct fission product, but also by the action of fast neutrons on carbon, nitrogen, and other light elements. Helium is troublesome because it promotes the nucleation of bubbles, especially in nickel and nickel alloys. Zirconium hydride is used in TRIGA (Training Reactor, Isotopes General Atomic) as a solid moderator.

CLEAR FUEL RESERVES

Light-water reactors are the chief type of nuclear power reactors in reaction. They are fueled with uranium that has been enriched to ca 3% ^{235}U from the naturally occurring concentration of 0.71%. Some experimental thorium-fueled reactors are in operation, but they are not expected to be in significant use until well after 2000.

Uranium

The earth's crust contains 2–3 ppm uranium, alkalic igneous rocks. Uranium combines readily with oxygen. Its solubility and distribution in rocks and ore deposits depend largely on its valence state: Uranium is highly soluble in the six-valent state and relatively insoluble in the four-valent state. Uraninite, the most common mineral in uranium deposits, contains the four-valent ion.

Resources

WOCA (world outside centrally planned economies) reasonably assumed uranium resources are ca 2×10^6 metric tons.

Availability

A principal factor in considering fuel-resources availabilities the time and effort needed to discover and develop the resources attainable production levels.

Demand

The greater demand would be predominantly for light-water reaction without fuel recycle; the lower-demand cases would involve substantial employment of liquid-metal fast-breeder reactors.

Low grade uranium resources

Low grade resources include shales, water, mill tailings, and enrichment-plant tails.

Thorium

Resources

The crustal concentration of thorium is 10–20 ppm. The element is dispersed widely in variable amounts in a number of different types. It is commonly associated with rare earths, niobium, titanium, zirconium.

Demand and supply

The demand for thorium in recent years is limited to a small amount used for non-nuclear purposes, e.g., lamp mantles, catalysts, refractories, and in high temperature alloys, significant market for thorium as a nuclear fuel has developed.

WATER CHEMISTRY OF LIGHT-WATER REACTORS

In a (PWR) pressurised-water reactors a closed circuit of high pressure, high temperature water transfers heat from the reactors core to once-through (OTSG) or recirculating U-tube steam generators, where steam is produced. Current secondary-cycle-chemistry control practices parallel those of fossil plants; however, impurity levels are maintained much lower. Initially, secondary water chemistry specified for U-tube generators was based on sodium phosphates. Subsequent to observations of corrosion attack, all volatile treatment (AVT) was adopted at most units. Several new types of corrosion have been observed subsequently. Corrosion of

OTSGs has not been as extensive or as serious a problem as with U-tube steam generators. A summary of initial AVT specifications is given in Tables 21.4 and 21.5.

Table 21.4. Initial PWR feedwater specifications for All-Volatile-Treatment (AVT) operation[a].

Property	*Babcock & Wilcox*	*Combustion engineering*	*Westinghouse*
pH	9.3–9.5[b]	8.8–9.2[c]	8.8–9.2[c]
	8.5–9.3[d]	9.2–9.5[e]	up to 9.6[e]
O_2, ppb	< 7	< 10	< 5
shutdown		100	
N_2H_4, ppb	20–100	10–50	$[O_2] + 5$
Acid conductivity, μS/cm	< 0.5	< 0.5	
Abnormal[f]	< 0.5 to ≤ 1 (24 h)	> 1.5 (4 h)	
	< 1 to ≤ 2 (12 h)		
shutdown	> 2		
Iron, ppb	< 10	< 10	< 10
Copper, ppb	< 2	< 10	< 5
Total solids, ppb	< 50		
Total silica (as SiO_2),			
ppb	< 20	< 10	
Ammonia, ppm		< 1	< 0.5

[a] Tabulated values are for normal operation unless noted otherwise.

[b] Carbon-steel feedwater heater tubes or combinations of carbon steel and stainless-steel feedwater and/or reheater tubes.

[c] With copper alloys in feedwater heaters, reheaters, or condenser.

[d] With stainless-steel feedwater heater tubes and stainless-steel or copper-nickel reheater tubes.

[e] With no copper alloys in feedwater heaters, reheaters, or condenser.

[f] Corrective action or shutdown recommended within indicated time.

Boric acid is added to the PWR primary system to compensate for fuel consumption and to control reactor power. A hydrogen overpressure is maintained, and small amounts of base are added to reduce corrosion rates not only to address component-integrity concerns but also to minimise shutdown radiation levels.

In a BWR (boiling-water reactor), steam is generated on the surface of the nuclear fuel and is used to drive the turbine generator. Since chemical additions are not made, materials corrosion is dependent primarily on coolant oxygen concentrations, which are governed by radiolytic decomposition of water and the steam-liquid oxygen equilibrium.

BWR chemistry specifications are given in Table 21.6. Because of the importance of maintaining fuel integrity in the direct cycle and of minimising radiation levels on out-of-core surfaces, transport of corrosion products to the reactor by the feedwater [illegible] of considerable importance.

Table 21.5. Initial PWR recirculating steam generator blowdown specifications for AVT operation.

		Westinghouse	
	Combustion engineering	*Freshwater*	*Seawater or brackish water*
pH	8.2–9.2	8.5–9.0	8.5–9.0
abnormal[a]	< 7.5 or 79.5 (4 h)	8.5–9.2 (2 wk)	8.0–9.2 (2 wk)
shutdown	10.5	< 8.5 or > 9.4	< 8.0 or > 9.4
Specific conductivity, μS/cm	< 7		
abnormal[a]	> 15 (4 h)		
Suspended solids, ppm	< 1		
abnormal[a]	> 10		
Free hydroxide, ppm		< 0.05	< 0.05
abnormal[a]		> 0.05 to ≤ 0.34 (24 h)	≤ 0.05 to ≤ 0.34 (24 h)
shutdown	5	> 0.34	> 0.34
Silica, ppm	< 1		
abnormal[a]	> 10 (4 h)		
Cation conductivity, μS/cm		< 2.0	< 2.0
abnormal[a]		> 2 but ≤ 7 (2 wk)	> 2 but ≤ 120 (2 wk)
shutdown		> 7	> 120

[a] Corrective action or shutdown recommended within indicated time.

Recently, incidences of intergranular stress-corrosion cracking (IGSCC) in stainless-steel piping have reduced the availability of BWR systems significantly. Numerous remedies for the problem are being evaluated and applied.

Table 21.6. BWR water-quality specifications for normal and abnormal conditions.

	General Electric
Reactor water	
Conductivity, μS/cm	≤ 1
abnormal (2 wk/yr allowed)	> 1 but ≤ 10
shutdown	> 10
Chloride, ppm	≤ 0.2
abnormal (2 wk/yr allowed)	> 0.2 but ≤ 0.5
shutdown	> 0.5
pH	5.6 to 8.6[a]
Feedwater	
conductivity, μS/cm	≤ 0.1
pH	6.5 to 7.5
metals, ppb	< 15 total
	< 2 Cu
oxygen, ppb	> 20 but < 200

[a] Inferred from conductivity limits.

ISOTOPE SEPARATION

The difficulty and high cost of isotope separation has limited the use of separated isotopes in nuclear reactors to specific cases where no non isotopic substitutes are available. The most important example is ^{235}U, the most abundant, naturally occurring, fissionable material. Other isotopes that are separated for nuclear use are ^{2}H or deuterium, D, which isused as D_2O as a neutron moderator in nuclear reactors and is a probable reactant in thermonuclear reactors; ^{3}H or tritium, T, which is produced in nuclear reactors and is a reactant in thermonuclear reactors under development; boron, ^{10}B, whose high neutron cross-section is used in control rods and safety devices for reactors; and lithium ^{7}Li, which is used in reactor cooling water systems because of its low thermal neutron cross-section.

FUEL-ELEMENT FABRICATION

The fuel assembly forms the basic unit of nuclear-reactor core. Fuel assemblies are fabricated by processing fertile and fissile materials into suitable chemical and physical forms, encapsulating the materials in a protective metallic sheath to produce a fuel rod, and assembling the fuel rods into the required configuration. Different reactor types, e.g., water-cooled, gas-cooled, and sodium-cooled breeder reactors, require different fuel-assembly designs.

CHEMICAL REPROCESSING

Fuel elements that are discharged from light-water nuclear-power reactors (LWRs) typically contain uranium, 0.79 wt% ^{235}U and 0.8 wt% ^{239}Pu, which can be recovered by chemical reprocessing. However, the economic incentive for reprocessing fuel for recycle into light-water reactors is marginal. Recycle of breeder-reactor fuel is, of course, essential to the breeder concept. Chemical reprocessing must provide for the high recovery of uranium and plutonium, and their separation from each other and from hazardous radioactive contaminants. Reprocessing implies the efficient separation of fertile materials, e.g., ^{238}U or ^{232}Th; from one another, e.g., ^{239}Pu from ^{235}U and/or ^{233}U; from the highly radioactive fission products; and from materials that are undesirable neutron poisons, including certain fission-product isotopes.

Purex Process

The process used is almost exclusively some form of the Purex aqueous solvent extraction process involving tributyl phosphate. Although other solvents have been employed and other processes, i.e., precipitation, ion exchange, and volatility processes, have been used, the Purex process and a variant for thorium-based materials, i.e., the Thorex process, are almost universally accepted.

The fuel is initially dissolved in nitric acid after an economically optimised period to allow radioactive decay of most of the short-lived isotopes that account for most of the heat and the high intensity gamma radiation associated with spent fuel elements. Preprocessing decay time, which largely eliminates troublesome fission products, e.g., ^{131}I, generally is one to two years for thermal-reactor fuels and three to nine months for breeder fuels.

Dissolution of the solid fuel is accompanied by complete release of the rare-gas fission products, krypton and xenon, into the off-gas system. Since the dissolver overheads also ordinarily contain fractions of some fission products, e.g., iodine, tritium, ruthenium, carbon-14 as CO_2, and tellurium, the off-gas usually is treated to remove or minimise the discharge of these gaseous fractions, as well as entrained particles or aerosols.

The multicycle Purex process depends primarily on countercurrent liquid-liquid extraction techniques that involve the transfers of a large number and variety of solutes between aqueous nitrate dissolver solution and the immiscible, less dense, organic solution. The extraction is repeated several times to achieve the desired degree of purification.

Typically, recoveries of > 99.8% of plutonium and uranium and decontamination factors of > 10^6 are achieved.

FAST BREEDER REACTORS

A breeder reactor is a nuclear reactor that produces more nuclear fuel or fissile material than it consumes. Although the emphasis for the past 30 years has been on the liquid-metal fast breeder reactor (LMFBR) and the development of the uranium-plutonium fuel cycle, use of the thorium-uranium fuel cycle for breeder reactors also has been studied, primarily for thermal reactor systems. Consideration has been given to the use of this cycle in fast breeder reactors.

Design

A fast breeder reactor can be designed in core and blanket configurations to produce more fissionable material than it consumes. This is achieved by converting fertile material, e.g., ^{238}U or ^{232}Th, in the blanket into ^{239}Pu and ^{233}U, respectively. Thus, in a typical LMFBR, the central region of the reactor (Fig. 21.3)., i.e., the reactor core, contains the plutonium-bearing assemblies which are comprised of a mixture of plutonium fuel and fertile uranium oxides. The blanket region surrounds the core and holds the uranium-bearing assemblies. Most of the fission occurs in the core, and most of the breeding occurs in the blanket. At intervals, the reactor is shut down to remove selected core and blanket assemblies and to replace them with fresh ones. The spent assemblies then can be reprocessed chemically to reclaim the fuel and the fertile materials for eventual recycling into new and existing reactors.

Breeder reactor types

The ability of a reactor to sustain a sufficient conversion or breeding ratio depends on a number of engineering and physical factors that influence the reactor's neutron environment. Average neutron energy is influenced by reactor core geometry, the composition of the fuel, and the presence of other materials, e.g., the coolant and construction materials. Design decisions can reflect the different options available; the various factors often are interrelated, and a number of permutations and combinations are possible. The liquid-metal fast breeder reactor (LMFBR), the gas-cooled fast reactor (GCFR), and the light-water breeder reactor (LWBR), have received the most attention. Primary emphasis has been on the LMFBR, which is the most highly developed of these reactor types, and on the uranium/plutonium fuel cycle; eleven experimental and demonstration LMFBRs are now in operation worldwide.

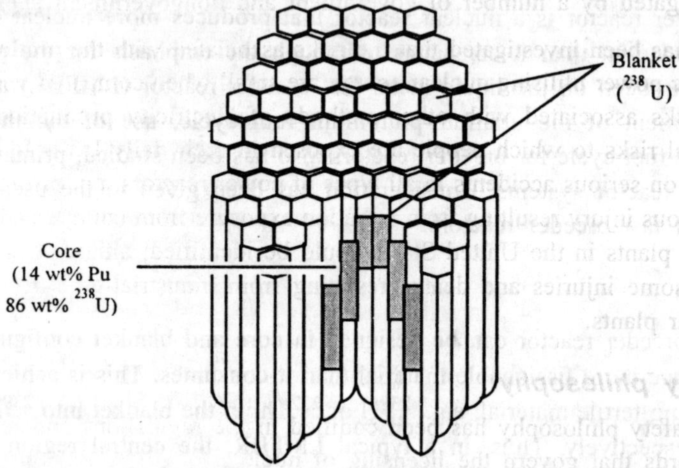

Fig. 21.3. Core and blanket arrangement for a typical LMFBR.

SAFETY IN NUCLEAR FACILITIES

In many countries, concern for the safety of nuclear power plants is curtailing their growth rate. The Nuclear Regulatory Commission licences and regulates nuclear facilities such that there is no undue restraint to the health and safety of the public. One of many requirements is that each nuclear plant must limit occupational exposures to 5 rem/yr and exposures in unrestricted areas to 0.5 rem/yr. In practice, the public exposure at nuclear plant sites generally is some orders of magnitude less. However, it generally is assumed that, for the purpose of evaluating the risks of radiation, the probability of including cancer is proportional to dose, no matter how small the dose. If the preceding hypothesis is true (and it may not be at very low dose), the benefits of the activity giving the exposure must be weighed against the risks and the costs of the low level exposure. Furthermore, the average public exposure from nuclear power plants is 0.1 mrem/yr, which is very small in comparison to the routine exposure of ca 200 mrem/yr for the average citizen.

Nuclear Fuel Cycle

The nuclear power plant that produces electricity is only one element among many in the nuclear fuel cycle. The other elements include mining, milling, transportation, conversion, enrichment, fuel fabrication, reprocessing, and waste disposal, and each has associated risks that must be considered when assessing the total risk associated with generating electricity by nuclear power. Such risks have been reported in numerous studies and are being investigated by a number of government and nongovernment agencies.

It has been investigated that the risks associated with the production of electric power utilising nuclear energy are small when compared with either the risks associated with other methods of electricity production or the societal risks to which people are exposed in their daily lives. In a recent report on serious accidents in all types of power reactors, not a single death or serious injury resulting from radiation exposure from commercial nuclear power plants in the United States could be identified; although there have been some injuries and deaths resulting from industrial-type accidents in nuclear plants.

Safety philosophy

The safety philosophy has been codified in the regulations and regulatory standards that govern the licensing of nuclear facilities. Implicit in these regulations is the concept of defense in depth, with which three levels of safety are identified. Furthermore, the risk from nuclear plants is shown to be much less than that from most other accepted risks such as fires, plane crashes, dam failures, etc.

The first level of safety is achieved by designing a plant for maximum safety in normal operation and maximum tolerance for system malfunctions. The second level of safety is based on the assumption that incidents occur despite care in design, construction, and operation. Safety systems should be provided to protect operators and the public and to prevent or minimise damage when such incidents occur. The third level of safety is the provision of additional safety systems as appropriate, based on the evaluation of the effects of hypothetical accidents, where some protective systems are assumed to fail simultaneously during the accident that they are intended to control.

Other safety considerations in design include quality assurance, surveillance testing, redundancy and coincidence, and indenpendence and diversity.

Safety features

A nuclear power plant abunds with safety features that are installed either to fulfill operational requirements (including safety) of the licensee, or to meet the preceding safety, philosophy as interpreted by the NRC. Some of the more significant safety features are listed in Table 21.7.

Table 21.7. PWR safety features, functions and purposes.

System	*Function*	*Purpose*
Reactivity control		
safety rods	shut down reactor	stops fission-heat generation
boron injection	maintain reactor shut-down	prevents fission-heat generation from recurring
Pressure control		
pressuriser	maintain system pressure	maintains cooling in liquid state
relief valves	prevent system over-pressure	prevents equipment leaks and/or rupture
Emergency core cooling		
high-pressure injection	maintain coolant at pressure	maintains cooling at pressure
accumulators	provide cooling during large pipe failure	maintains cooling during transient blowdown event
low pressure injection	maintain cooling of core after blodwdown	prevents core from overheating
Emergency heat removal		
auxiliary feedwater system	provide heat sink for reactor coolant	ensures core coolability
decay-heat removal system	maintain coolness of core	prevents core from overheating
Containment		
isolation system	enclosure of radiation	minimises release of radioactivity to the environment
spray system	control containment temperature and pressure	minimises containment leakage
recirculation system	cool and cleanup containment atmosphere	minimises release of radioactivity to the environment
hydrogen recombination	recombine hydrogen with oxygen	minimises explosion potential
Vital auxiliaries		
emergency power system	supply power for emergency needs	permits plant safety systems to function in the event of loss of normal power

(Contd...)

System	Function	Purpose
component cooling water	supply cooling water to equipment needed in an emergency	permits plant safety systems to function in the event of an emergency
instrument air	supply instrument air for emergency needs	permits plant safety systems to function in the event of loss of normal instrument air

Accidents

The occurrence and consequences of accidents in nuclear power plants is the ultimate measure of the effective application of the previously described safety concepts to reactor technology.

Other Nuclear Fuel Cycle Facilities

Because the nuclear power plant constitutes over 99% of the radiation risk resulting from nuclear fuel cycle, the radiation risk from the balance of the fuel cycle is small and is distributed among many facilities. There are three potential risk mechanisms: inadvertent critically, inhalation exposure, and direct radiation. The last mechanism is accommodated readily by adequate radiation shielding, which is well documented.

SPECIAL ENGINEERING FOR RADIOCHEMICAL PLANTS

Concerns that are unique to the nuclear industry result from the handling of radioactive and potentially fissionable materials. The special engineering considerations are categorised in terms of the design for radiation resistance, the nuclear safety design, the mitigation of occupational radiation exposure, and the minimisation of environmental impact.

Radiation Resistance

Although a spectrum of radiation types, e.g., alpha particles, beta particles X-rays, gamma rays, and neutrons, is emitted from spent nuclear fuel, the primary equipment-design consideration is the effect of gamma radiation, especially on organic materials.

Specially developed organic materials for use in gamma radiation fields have lifetimes that are improved by a factor of 10–100 over the standard materials intended to function for ca 100–10,000 hour. They generally are used when no design replacement is available. Local shielding can be provided around the organic material as a means of extending its useful life. Inorganic lubricants, e.g., molybdenum disulphide, can also be substituted for organic lubricants.

Safety Design

Fissionable material, i.e., the reusable fuels for nuclear reactors, including ^{233}U from the thorium fuel cycle or ^{235}Pu from the uranium fuel cycle, are capable of spontaneous criticality if accumulated in adequate concentration, volume, and geometry. A safe configuration is an infintely long cylinder, e.g., 11.5 cm dia of solution or 4.4 cm dia of solid uranium or plutonium. Other means of preventing nuclear criticality are by limiting the mass of fissile material present; limiting the concentration of fissile material in solution; eliminating or controlling materials that moderate the neutrons and thus reduce the number required; or using liquid or solid poisons or neutrons absorbers that greatly increase the number of neutrons required to initiate or maintain the chain reaction or nuclear criticality.

Occupational Radiation Exposure

The massive shielding surrounding the process equipment usually is made of concrete and is 1–2 m thick, depending on the intensity of the radiation being shielded and the density of the concrete. Since the shielding thickeness required for alpha, beta, and gamma radiation is a density function, lead and steel are used where the wall thickness must be limited. Visual observation is provided with windows of leaded glass, which are available in densities that are nearly that of steel, with periscopes that penetrate the shielding, or with television cameras. The shielding is provided to protect the plant workers.

Since radioactive gases and fine airborne radioactive particles are generated during reprocessing, the ventilation system for the plant is an important design factor. Conventional ventilation systems maintain a flow of air from the worker-occupied areas to the areas of high contamination to the clean-up systems. In the clean-up systems, particulates are removed and the radioactive gas content is reduced to a safe level before being removed from the plant through the stack.

Environmental Impact

Plant design, as related to environmental impact, refers to the special design requirements that are imposed to ensure that the general population and environment are properly protected. Two key considerations are control and reduction of contaminants from planned or routine releases, and prevention or containment and, mitigation of unplanned releases from accident situations.

Remote Operation and Maintenance

The equipment and facilities for radio-chemical use are designed to have high efficiencies, high reliability, and the capabilities, for remote operation and maintenance. The fundamental keys in the analysis of a remote design are a concept of equipment that, for its stated use, produces the required product from a defined input; selection of construction materials that are compatible with the internal environment (initial form, intermediate products, and final product) and external environment (radiation, atmosphere, and temperature) in which the machine operates; the capability to monitor critical functions so that indication of operational status is available constantly; a modular machine design to provide for adjustment and replacement of key machine segments; and space and structural strength to accommodate remote manipulation.

HEAVY WATER FOR NUCLEAR POWER INDUSTRY

Hydrogen has three isotopes, protium or light hydrogen with a mass number of 1, deuterium or heavy hydrogen with a mass number of 2, and tritium with a mass number of 3. The name, "protium" for the lightest isotope of hydrogen has not become popular. So, we will call this isotope hydrogen in the rest of this article though this word actually denotes the element constituting the mixture of all the three isotopes. Hydrogen is quite abundant in nature in the combined form, the most familiar of its compound being water. Deuterium also occurs in nature in very small quantities. Natural hydrogen contains deuterium to the extent of about 1 in 6000. Tritium is a radioactive isotope and exists only in extremely minute amounts in nature, of the order of 1 in 10^{17}, i.e., 1 in a hundred million billion. Just as hydrogen combines with oxygen to form its oxide, H_2O, called water, deuterium combines with oxygen to form its oxide, D_2O which is called heavy water. Heavy water is an isotopic form of water and occurs in natural water to the same extent as does deuterium in natural hydrogen.

Evidence of the presence of more than one isotope of hydrogen came to light as a result of the difference in its atomic weights determined chemically and physically with the mass spectrograph.

Properties of Heavy Water

Isotopes differ in their nuclear and physical propeties, but may be regarded as identical in their chemical properties. The relative mass difference between hydrogen and deuterium is the greatest of all the known isotopes and is about 100 per cent. Consequently, hydrogen and deuterium show a larger difference in their physical properties than do two different isotopes of other element. Light water, H_2O and heavy water, D_2O exhibit similar difference in different isotopes of any other element. Light water, H_2O and heavy water, D_2O exhibit similar differences in their physical propeties Table 21.8. gives some of the important physical properties of heavy water and light water.

In chemical properties, the differences between heavy water and the light water exist mainly in the kinetic aspects of reaction, i.e., in the rates of reaction. Reactions with heavy water proceed, in general, slower than the corresponding reactions with light water. In biological systems, that is plants and animals, the substitutions of deuterium for hydrogen can alter the delicately balanced life processes substantially. Hence, it could have a toxic effect on some of the living organisms. Studies have shown that mammals can tolerate replacement of deuterium by hydrogen upto 15%. Severe toxic effects result when the replacement exceeds 30%.

Table 21.8. Properties of deuterium and protium oxides.

Property	*Heavy water* D_2O	*Light water* H_2O
Boiling point (°C)	101.4	100.0
Melting point (°C)	3.82	0.00
Specific gravity at 25°C	1.107748	0.997044
Temperature of maximum density (°C)	11.22	4.08
Heat of vapourisation (calories per mole)	9966	9700
Dielectric constant	82.0	80.5
Viscosity at 20°C (centipoise)	1.103	1.002
Refractive index	1.32844	1.33300
Surface tension (dynes/cm)	67.8	72.8

Analysis of Heavy Water

Ordinary water and heavy water have substantially similar chemical properties but differ slightly in physical propeties. A mixture of two or more substances which are chemically different can be easily analysed by means of their varying chemical properties, and the concentration of each component readily and accurately determined. Mixtures of heavy water and light water cannot be so analysed. Hence analysis of heavy water constitutes a challenging task to the analyst. He has to take recourse to very carefully controlled and tedious methods of physical measurements or to the use of expensive, sophisticated instruments such as infrared spectrophotometers or mass spectrometers for accurate analysis. We have seen from Table 21.8 that the density or specific gravity of heavy water is about 10% higher than that of light water. Measurement of density with a pyknometer, under very carefully controlled conditions of temperature, is the absolute method for determination of heavy water in water. One normally employs this technique to standardise a sample and use it as a primary standard for the much easier, relative methods briefly described below.

One of the less familiar physical characteristics in which deuterium oxide differs considerably from hydrogen oxide is in the absorption of infrared light. Infrared radiation forms part of the electromagnetic spectrum of radiation which in the decreasing order of frequency (or increasing order of wavelength) consists of: (i) gamma rays; (ii) X-rays; (iii) ultraviolet rays; (iv) visible light; (v) infrared radiation, and so on. Light with wavelengths less than about 1 μm and greater than about 0.8 μm (1 μm = 10^{-6}m) fall in the infrared region of the spectrum. Heavy water and light water show distinct differences in their infrared absorption spectra. Making use of these differences one can quantitatively measure the concentration of heavy water with a high degree of accuracy. The technique is simple, but employs infrared spectrophotometers which are relatively expensive, sophisticated, optical instruments. A good deal of training is required before an operator can practise this technique to get reliable results. This is a relative method, and analyses are carried out with the help of calibration curves made with sample of known concentrations of heavy water (standards). The other methods for the isotopic analysis include measurements of mass spectra, refractive index, thermal conductivity and viscosity. They vary considerably in the case of operation, accuracy, and sensitivity.

Production of Heavy Water

One way of enriching natural water in heavy water is by electrolysis. Just as the isotopic analysis of water depends on the small differences in the physical properties of the isotopic waters, the methods of separating heavy water (D_2O) from light water also depend on such differences. In addition, the small differences expected in the chemical properties of deuterium and hydrogen lend themselves for use in separating deuterium by means of exchange reactions between deuterium and hydrogen in their compounds. The techniques of enriching heavy water of deuterium fall into two major classes: (i) those based on differences in physical properties, mainly boiling point and vapour pressure; and (ii) those based on exchange reactions between deuterium and hydrogen compounds. The methods that can be employed for commercial production of heavy water include: (i) fractional distillation of water; (ii) fractional distillation of liquid hydrogen; and (iii) exchange reactions between compounds of hydrogen and deuterium. In all these methods, a parameter of interest is the separation factor which tells us how well the separation of the isotopes occurs during the process. The separation factor is mathematically expressed as follows :

$$S = \frac{(H)_g\ (H)_l}{(D)_g\ (D)_l}, \text{ where}$$

S = separation factor, $(H)_g$ = concentration of hydrogen in gaseous phase, $(D)_g$ = concentration of deuterium in gaseous phase, $(H)_l$ = concentration of hydrogen in liquid phase, and $(D)_l$ = concentration of deutetrium in liquid phase. Table 21.9. gives the separation factor decreases with temperature.

Table. 21.9. Separation factor for various methods of enriching deuterium.

Method	*Separation factor at 30°C*
Distillation of water	1.1
Distillation of liquid hydrogen	1.5 at 260°C
Exchange between hydrogen sulphide and water	2.3
Exchange between hydrogen and ammonia	3.5
Exchange between hydrogen and water	3.7
Electrolysis of water	9.5

All the above methods have relative advantages and disadvantages. As they have to be used to concentrate deuterium or heavy water D_2O from its natural concentration of about 0.015% (atom%) to nearly 100%, i.e., an enrichment of about 6700 times, all these processes consume tremendous amounts of energy. They are, therefore, expensive and so is the product, heavy water.

As India has chosen nuclear reactors with natural uranium as fuel and moderated and cooled with heavy water as the basic type for its nuclear power programme, we have planned and put up heavy water production plants to cater to the needs of our nuclear power industry.

Table 21.9 shows that electrolysis has the highest separation factor. But the consumption of electrical energy is very high if this process is used for enrichment from the natural concentration to 100% heavy water. The chemical exchange between hydrogen and water has the next highest separation factor. This process has the disadvantage that the rate of reaction is too low to enable it to be of practical value; no suitable catalyst to accelerate this reaction has yet been found. Exchange between hydrogen and water is of industrial value as the rate of reaction can be speeded up with potassium amide as a homogeneous catalyst. Exchange between hydrogen sulphide and water proceeds with acceptable speed even without a catalyst and lends itself to ready industrial exploitation. Distillation of liquid hydrogen and water are relatively simpler and proven methods but their, separation factor is low.

We, in India, at present have five heavy water production plants in various stages of construction, commissioning and production. They are at Nangal in Punjab, Anushakti near Kota in Rajasthan, Baroda in Gujarat, Talcher in Orissa, and Tuticorin in Tamil Nadu. The Nangal plant uses electrolysis followed by hydrogen distillation. The one near Kota employs chemical exchange reaction between hydrogen sulphide and water. Baroda, Talcher and Tuticorin plants make use of hydrogen-ammonia exchange reactions.

Uses

Among the purely scientific uses of heavy water, the most important is its application as a tracer in the study of reaction mechanisms in the course of physical, chemical and biological processes. Heavy water, D_2O, permits the preparation of many deuterated compounds by direct or exchange reactions with it. Thus, alkali metal oxides such as sodium oxide (Na_2O) and calcium oxide (CaO), when dissolved in heavy water, yield corresponding deuterohydroxides (deuteroxides or deuterated hydroxides), NaOD and $Ca(OD)_2$ "moderator" in nuclear reactors constitutes its most dramatic technological application. A thermal nuclear reactor works on the fission chain reaction in uranium. Uranium has three natural isotopes: U–238 (99.3%), U–235 (0.7%) and U–234 (negligible). Uranium–238 and U–235 undergo fission reaction when bombarded by neutrons. The nuclei of these isotopes absorb the neutron and become excited compound nuclei. Then they split into the two heavy fragments along with the emission of two or three neutrons. Fig. 21.4. pictorially represents this process. The phenomenon of fission in uranium can be made a convenient source of energy if a chain reaction of fission can be sustained in a controllable fashion. This is precisely what happens in the core of a nuclear reactor.

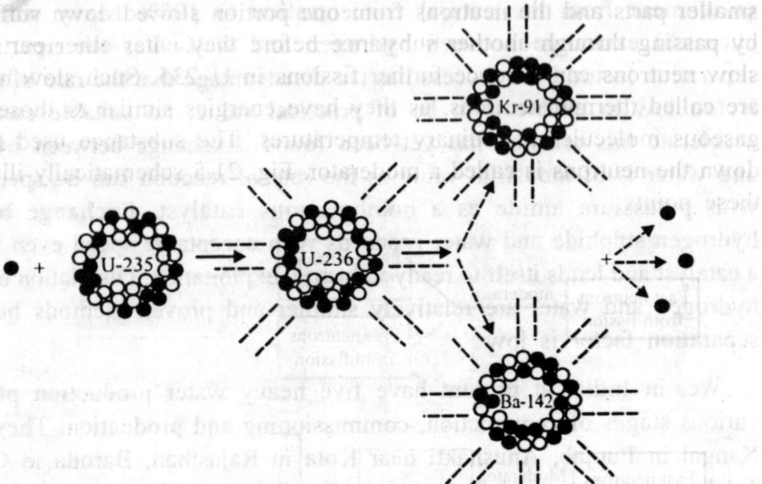

Fig. 21.4. Fissioning of a uranium-235 nucleus by a bombarding neutron.

When a uranium nucleus fissions, it emits two to three fast neutrons. If at least one of these neutrons can be made to cause fission of another uranium nucleus and one neutron from this second fission to cause fission in a third nucleus and so on, a chain reaction with a constant number of fissions per unit time will result. In a lump of natural uranium, neutrons emitted in the fissioning of a uranium nucleus can have three alternative paths: (i) the neutrons can escape from the body of the metal; (ii) they can be captured by other uranium nuclei without undergoing fission (non-fission capture); and (iii) they can fission other uranium nuclei. The escape can be reduced by increasing the size of the lump so that its surface-to-volume ratio decreases. The probability that a neutron emitted in a fission reacts with another uranium nucleus depends on its speed and hence its energy. This probability increases with decreasing neutron energy. The probablility for a nuclear reaction to take place is called the cross-section for that reaction. As the neutron energy decreases, the cross-sections for both fission and non-fission capture increase. However, U–238 cannot be fissioned by neutrons of low energy. So, after having been slowed down below a particular neutron energy, neutrons cannot fission U–238, though they can continue to fission U–235. For a controlled and sustained chain reaction in natural uranium, we have to depend upon its small U–235 content. As the neutron energy decreases, the non-fission capture cross-section for U–238 becomes so high that most of the neutrons from the spontaneous fissions end up in this reaction. That is why a lump of natural uranium cannot sustain a fission chain reaction. If this lump is divided into smaller parts and the neutrons from one portion slowed down sufficiently by passing through another substance before they enter other parts, these slow neutrons can produce further fissions in U–235. Such slow neutrons are called thermal neutrons, as they have energies similar to those of the gaseous molecules at ordinary temperatures. The substance used to slow down the neutrons is called a moderator. Fig. 21.5 schematically illustrates these points.

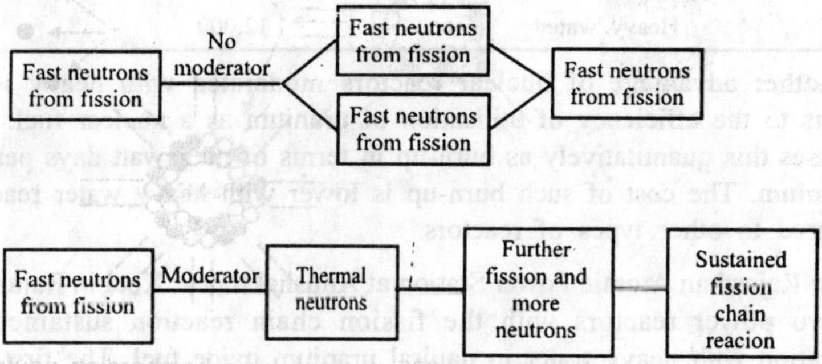

Fig. 21.5. The effect of moderator in bringing about a fission chain reaction in uranium.

A moderator should slow down the neutron sufficiently before it passes to the next part of uranium without interacting with it in other ways. The neutrons lose their energy by colliding with the atomic nuclei in the moderator by elastic collisions in which the kinetic energy is distributed between the colliding nuclei, and the total kinetic energy is conserved. The amount of energy lost by a neutron in colliding with a nucleus is maximum when the target nucleus has the same mass as the neutron. Thus, light water in which the hydrogen nuclei have almost the same mass as neutrons should be the best moderator. But, unfortunately, hydrogen also absorbs neutrons. So, though its moderating power is high, its moderating ratio which is equal to the ratio of the moderating power (a function of the loss of enegy per collision) to the macroscopic neutron capture cross-section (a function of the probability a neutron capture), is not so high. Of the possible moderators for a nuclear reactor, heavy water has the highest moderating ratios as shown in Table 21.10.

Because of its high moderating ratio, heavy water, when used as a moderator, permits a sustained fission chain reaction in natural uranium. Moderators with less moderating ratios require enriched uranium, i.e., uranium in which the concentration of U–235 has been increased by an isotope enrichment process. Such enrichment of uranium is a capital intensive and difficult technology.

Table 21.10. Moderating ratio of some important moderators.

Substance	*Moderating ratio*
Light water	72
Beryllium	159
Carbon	170
Heavy water	12,000

Another advantage of nuclear reactors moderated with heavy water pertains to the efficiency of utilisation of uranium as a nuclear fuel. One expresses this quantitatively as burn-up in terms of megawatt days per ton of uranium. The cost of such burn-up is lower with heavy water reactors compared to other types of reactors.

The Rajasthan Atomic Power Station at Anushakti near Kota in Rajasthan has two power reactors with the fission chain reaction sustained by moderation with heavy water in natural uranium oxide fuel. The first unit of this station, also the first nuclear power plant of this type in India, has been in operation since 1972. Its second unit went into operation in 1980. A similar reactor of another twin-unit nuclear power station has been commissioned at Kalpakkam near Chennai in Tamil Nadu. A third such nuclear power station is under construction at Narora in Uttar Pradesh, and the fourth station is at Kakarapar near Surat in Gujarat.

Just as we can split heavy nuclei to obtain energy so can we make light nuclei to combine to form heavier nuclei and get energy in that process too. Such processes are called fusion reactions. Fusion as a commercial source of power poses many technological problems which are still to be solved. However, this could be a future source of energy to mankind, and deuterium will again play a major role in it.

Chapter 22

Iron and Steel

INTRODUCTION

Iron is the most abundant metal in the earth crust. It occurs widely distributed in nature and constitutes about 4.5% of the earth's crust. For the manufacture of pig iron, in addition to a suitable iron ore, we need flux grade of limestone and metallurgical coke. The pig iron is then prepared by charging a proper iron ore, limestone and coke in necessary proportion into the blast furnace. Steel from pig iron is produced in an open hearth furnace or in a Bassemer converter.

Thus for the production of iron and steel three main raw materials are required, the iron ore, coal and limestone. The important iron ores are :

(1) *Haematite* (Fe_2O_3): Haematite contains about 70% iron. This is also called as red ore because the colour varies from bluish grey to dark red. In India only haematite ore is used for the production of pig iron. Haematite is smooth and easily reducible.

(2) *Magnetite* (Fe_3O_4): Magnetite contains about 72% iron. This is called black ore because the colour varies from steel grey to black.

(3) *Limonite* ($Fe_2O_3.3H_2O$): Limonite contains about 60% iron. This is called as brown ore because the colour varies from yellowish brown to brown.

IMPURITIES IN IRON ORE AND THEIR EFFECTS

The principal impurities found in the iron ore are silica and alumina. These impurities not only reduce the iron content in the ore but also increase the cost of production of iron and steel. The presence of $CaCO_3$ or $MgCO_3$ in the gangue would decrease the amount of flux and hence reduce the cost of production of steel. Sometimes the concentration of silica, alumina and limestone are such that ore is self fusing.

Sulphur and phosphorus, which exert a very injurious influence on the properties of iron and steel, are present in varying proportions in iron ore. The former is present in the form of FeS_2 and the latter in the form of $Fe_3(PO_4)_2$. Sulphur may vary from 0.1 to 0.4 per cent and phosphorus from 0.01 to 1.5 per cent. Some sulphur and phosphorus is also contributed by fuel and flux. Good steel must not normally contain more than 0.05 per cent sulphur or phosphorus. Sulphur up to about 1 percent in iron ore can be got rid of in the blast furnace in the form of slag. Phosphorus on the other hand cannot be got rid of in the form of slag in the blast furnace and a small amount of it present in the charge whether derived from the ore or flux will be found in pig iron. This phosphorus beyond about 0.05 per cent has a marked effect in lowering the value of an iron ore unless the ore is to be used for manufacturing pig iron for foundry purposes only, where high phosphorus gives fluidity to the metal.

A small amount of manganese from a trace to about 1.5 per cent is always present as an impurity in most of the iron ores. Its presence is desirable because it forms manganese sulphide (MnS) which is less harmful than FeS.

Fuel

The fuel in the steel production serves two purposes. Firstly, it provides energy to meet the charge ana secondly, reduces iron ore to metallic iron by combining with the oxygen of the ore. The fuel used is metallurgical coke which is produced from coal by heating it form 70–80 hours in closed chamber out of contact with air at a temperature of 1100°C. The resulting coke is obtained in hard porous lumbs and consisting of carbon of about 70–90% and ash 10–25%.

Fluxes

Usually limestone or dolomite is used as flux. Its functions are :

(1) To form a fluid slag with the coke ash, ore gangue and many other changed impurities.

(2) To form a slag of such chemical composition that it will provide a degree of control of sulphur content of the iron.

The objectionable impurities in limestone and dolomite are SiO_2, sulphur and phosphorus. The lower the silica, sulphur and phosphorus the more valuable is the flux.

MANUFACTURE OF PIG IRON

Pig iron is produced by smelting iron ore with flux and fuel in a blast furnace. The blast furnace is a tall steel shell or shaft of circular cross section lined with hard fire bricks. The refractory for the lining is so chosen as to cause minimum failure due to carbon disposition, alkali attack and abrasive action of the charge. Blast furnace has a peculiar shape to enable the charge to slide down readily. Blast furnace consists of a steel shaft about 25 metre high, the base of furnace is called hearth and the broadest portion is called bosh. A schematic diagram of the blast furnace is shown in Fig. 22.1.

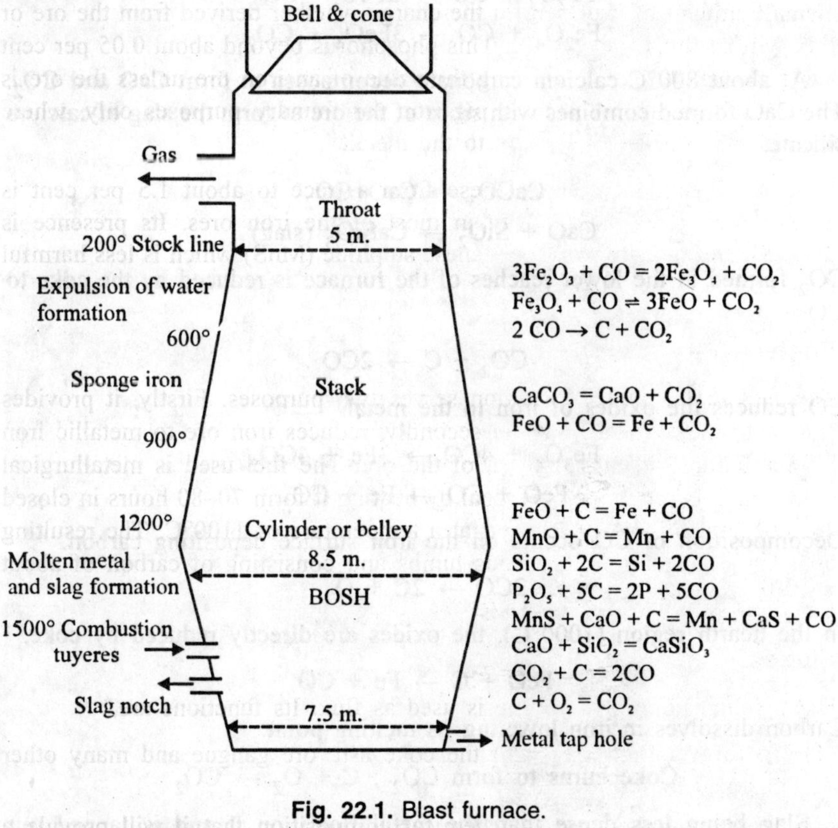

Fig. 22.1. Blast furnace.

As already mentioned, the charge consists of ore, limestone and coke. Limestone decomposes to form CaO which combines with the slilica present in the ore to form the slag of calcium silicate. Limestone is therefore the flux. The reactions occuring in the blast furnace depend on the temperature prevailing in the various zones. Temperature increases as we go down the furnace. It is around 400°C at the throat (top) and around 1600°C at hearth.

Near the throat region dehydration and decomposition (in the case of carbonate ore) occur.

In the shaft region (500–1000°C) the main reactions are the reduction of Fe_2O_3 to lower oxides. The reducing agent is CO which is formed in the lower part of the furnace by the burning of the coke.

$$3Fe_2O_3 + CO \rightarrow 2Fe_3O_4 + CO_2$$
$$Fe_2O_3 + CO \rightarrow 2FeO + CO_2$$
$$Fe_2O_4 + CO \rightarrow 3FeO_4 + CO_2$$

At about 800°C calcium carbonate decomposes to form CaO and CO_2. The CaO formed combines with silica of the ore to form the slag of calcium silicate.

$$CaCO_3 \rightarrow CaO + CO_2$$
$$CaO + SiO_2 \rightarrow CaSiO_3 \text{ (slag)}$$

CO_2 formed in the lower reaches of the furnace is reduced by the coke to CO.

$$CO_2 + C \rightarrow 2CO$$

CO reduces the oxides of iron to the metal.

$$Fe_3O_4 + 4CO \rightarrow 3Fe + 4CO_2$$
$$FeO + CO \rightarrow Fe + CO_2$$

Decomposition of CO occurs on the iron surface depositing carbon.

$$2CO \rightarrow 2C + O_2$$

In the hearth region (1000°C), the oxides are directly reduced by coke.

$$FeO + C \rightarrow Fe + CO$$

Carbon dissolves in iron lowering its melting point.

Coke burns to form CO_2 : $C + O_2 \rightarrow CO_2$

Slag being less dense than the metal floats on it and is tapped out through the slag-notch. Slag is used in cement industry. The molten iron used to be poured into vessels which looked like a litter of pigs. Hence the molten iron is called pig iron. When pig iron is remelted in a cupola and cast into ingots, it is named cast iron.

The hot exhust gases from the blast furnace contain about 15 to 25% CO_2 and the rest N_2. It is used to preheat the air blast. The eventual exhaust gas is used in boilers for raising steam.

Disposal of Pig Iron and Slag

The iron coming out from the blast furnace is disposed off in one of the ways: (i) Cast in sand beds; (ii) Cast in cast iron moulds in a pig casting machine; and (iii) Taken to the steel works in the molten condition.

In the first method sand beds are prepared in front of the furnace and moulds are made in sand to receive the molten iron. A certain amount of sand adheres to the pig and is not desirable in steel production as silica is to be fluxed with extra fuel and flux.

Casting in cast iron moulds is usually carried out in pig casting machines where a series of moulds is carried on a chain. The metal is poured into the mould and by the time it reaches the end, it gets solidified and drops into the wagon to be carried away. The empty mould is given a spray of lime and water, the latter dries away leaving a thin coating of lime to prevent sticking of molten iron with the mould.

In the third method, the iron is transported to the steel plants where it is poured into the big mixer converter for steel production.

The slag is either mixed with flowing water to quench it or is sent to a settling tank where water is mixed, slag gets settled and clean water is reused for the purpose. The slag is used for cement manufacture, railway ballast or floor tiles, etc.

Cast Iron

Cast iron is produced by pouring the liquid metal into moulds of desired size and shape. The metal gets cooled and solidified and takes the desired shape. The pig iron and scrap iron are melted in a cupola or air furnace. Cupola is a cylindrical shell of boiler plates lines with fire brick and resting on a square bed plate. The opening of the bed plates is closed by two semi-circular cast iron hinged doors opening downwards. These support the charge and help in cleaning etc. From the bottom of the cupola, air for combustion is blown in. At the bottom are iron tap hole and slag tap holes. Cupolas are cleaned, repaired and then lighted with the help of wood, waste oil and then charge of coke is poured in from the top of wood etc. When the bed charge is well ignited iron is put in. The charge of coke with every charge of iron is about 15 cm deep. About 6 to 7 kg iron is used per kg of coke. The blast is turned on, iron melted and poured into moulds. Sometimes for longer operation of cupola, flux is added.

Properties of cast iron

Cast iron is an impure variety of iron and contains the highest percentage of carbon and traces of impurities like sulphur, phosphorus, manganese, and silica. On account of impurities, the cast iron or pig iron is very weak and can not be shaped into different articles by making use of processes like forging or hammering. The condition of carbon in cast iron may be free carbon (or graphite) or combined carbon (as Fe_3C, cementite). The characteristics of cast iron mainly depend upon the conditions of carbon in which it exists. In fact, pig iron is the direct metallic product from the blast furnace without any change and cast iron is prepared from pig iron with or without refining or alloying treatment. The pig iron is melted and then cast into moulds of the desired shapes. It is then called cast iron. The chemical composition of cast iron as well as pig iron is same. Cast iron can also be prepared from mixtures of pig iron with steel or scrap or alloying elements. The average composition of cast iron is : Iron 93–95%, carbon 2.5–4.5%, silica 0.6–2.8%, phosphorus 0.4–1%.

Varieties of Cast Iron

The various varieties of cast iron are given and discussed below:

(1) *White cast iron:* This variety of cast iron contains carbon in the combined state as iron carbide (Fe_3C) and a very small amount of free carbon as graphite. It is obtained by cooling by molten pig iron quickly, because if rate of cooling is more, cementite (or iron carbide, Fe_3C) precipitates and imparts hardness to cast iron. White cast iron is produced when silicon content is low. It is also called chilled cast iron, because it is obtained when molten metal is cooled rapidly. It is called white cast iron as it contains little or no graphite. White cast iron can be prepared by two methods :

(a) By casting grey iron in such a manner that it is chilled to cool so rapidly that it produces a white surface layer.

(b) By adjusting the proportion of carbon and silicon in the composition as low as practicable.

White cast iron is manufactured by melting low phosphorus pig iron and steel scrap together in an air furnace or in a cupola furnace. Best results, are however, obtained by duplexing or triplexing processes which consist in adopting the combination of cupola, air furnace, Bessemer converter and electric furnace.

Because of hardness, brittleness and poor machinability, white cast iron has not been used widely. It is, however, used where good abrasion resistance is needed, e.g., in railway carriage brakes. White cast iron is generally used in making malleable cast iron.

(2) *Grey cast iron:* This cast iron variety contains carbon as free carbon in the form of graphite and very small amount of Fe_3C (cementite). It is obtained by cooling the molten pig iron slowly, because when rate of cooling is slow formation of cementite is less and the hardness is not enough. Silica when present in large quantity (2–3%) favours decomposition of cementite to graphite and ferrite resulting cast iron of better quality, known as grey cast iron.

Grey cast iron is thus produced when silicon content is high and molten metal is cooled slowly. It possesses a greyish colour because of evenly distributed graphite. A cast iron in which all the carbon is in graphite form, is a soft, easily machinable metal insensitive to notches, having high damping capacity and high compressive strength. It can be cast into moulds of intricate shapes and possesses good fusibility and brittleness. Grey cast iron possess low tensile strength, ductility and impact strength (low toughness), since carbon in it is present in the form of weak flakes or thin plate like crystals. Consequently they act as gaps within strong metal, iron. Because of the presence of graphite flakes, the vibrations of machines are also transformed easily and consequently, grey cast iron is used for making bases for heavy machines. Grey cast irons have no well defined yield limit and modulus of elasticity. The approximate composition of grey cast iron is:

Iron = 92–93%, Carbon (graphite) = 3.5–4%, carbon (combined) = 0.5%, silicon = 2–3%, phosphorus = 0.15–1.0%, sulphur = 0.02–0.15% and manganese = 0.4 to 1.0%.

Grey cast iron is produced by melting together low quality foundry pig, scrapped castings and coke in a cupola. When this type of cast iron is fractured, it gives a grey appearance and hence it is called grey cast iron. Such type of cast iron can be manufactured with all its carbon in the form of free graphite flakes, but it is not always desirable. Generally, in castings 0.8% of the carbon is in the Fe_3C form and the rest 2–4% is in the graphite flakes form. Even a series of cast irons can be made ranging from cast iron containing all carbon in the graphite form to cast iron containing a good percentage of carbon in the combined form such as Fe_3C. The variation in properties is appreciable because carbides harden and give strength to the iron.

Manganese removes sulphur and provides increased tensile strength and hardness. Pure iron melts at 1530°C, whereas cast iron melts at 1150°–1225°C. Phosphorus increases the fluidity of cast iron and makes it brittle when cold, whereas sulphur makes it brittle when red hot.

Cast iron is the cheapest and is not of much use except for making weights etc. Most of the cast iron used in industry is grey cast iron due to the presence of graphite. It is used for making base plates, lathe bodies and in foundries for casting a variety of items. More than 75% of the cast iron is converted into steel.

(3) *Malleable cast iron:* Cast iron is hard and brittle and can not be welded. Malleable cast iron is usually prepared by slow cooling of white cast iron of special composition so that the hardness is less. The average composition of malleable cast iron is C = 2–2.5%, Si = 0.8–1.1%%, S = less than 0.10%, P = 0.1–0.2% and Mn = 3%. To produce malleable cast iron, the white cast iron is subjected to the process of annealing, which consists in heating the white cast iron in an oven slowly to 1600°F or 875°C and keeping this temperature for 25–30 hours, depending upon size, and then slowly cooling it at a rate of 10°F per hour to room temperature. During heating, Fe_3C decomposes to give graphite in the form of small round particles, called nodules. Annealing is generally carried out by two methods.

(a) The castings are packed in an oxidising material. As a result, some carbon is removed. Such malleabilised castings are known as white heart.

(b) Annealing where castings are packed in an inert material, such as a ferrous sulphate scale or slag. Such malleabilised castings are known as black heart and consist almost entirely of graphite and ferrite.

Because of the absence of hard Fe_3C, the malleable cast iron is neither hard nor brittle. Its weakening effects are reduced because of the presence of graphite in nodular form (not as large flakes). This cast iron is known as malleable, but it cannot be drawn into wires or made into sheets by hammering. Malleable cast iron possesses good toughness, shock resistance, ductility and machinability (The properties possessed by cast iron and mild steels) and finds use in making automobile parts (e.g. pedals, hubs, shearing gear housings, rear axle housings etc), agriculture machineries, railway equipments, carpentry tools etc.

In fact, in malleable cast iron, carbon content is low but not less than 2% as the freezing temperature would be high below this percentage. Sulphur should be quite low otherwise it would retard the conversion of cementite to graphite and ferrite. Phosphorus has no appreciable effect.

(4) *Ductile cast iron:* Ductile cast iron can be produced by adding a small amount of magnesium (0.1–0.5%) or cerium to the molten cast iron at the ladle. As a result, graphite content is rendered nodular or spheroidal in form and gets dispersed throughout the material. The approximate composition of ductile cast iron is: Mg = 0.1–0.5%, Mn = 0.5 - 0.10%, Ni = 3.0 - 3.5%, P = 0.1%, C = 3.5% - 4%, Si = 1.0 - 4.0%. Magnesium controls the graphite form and nickel as well as manganese increase the strength but decrease the ductility. Sulphur content in ductile cast iron is low and so the manganese content is also kept low. Silicon is used as alloying element.

Ductile cast iron has high strength, fluidity, wear resistance, weldability, castability and machinability and generally employed for making cylinder heads, cylinder valves, compressors etc.

(5) *Alloyed cast iron:* In such cast irons one or more alloying elements are incorporated inorder to increase the utility of the metal. Alloyed cast irons are obtained either in cupolas by melting together the alloying compounds or heating the alloying compounds in an air furnace or electric furnace or by adding the alloying metal to the pouring ladle after drawing the molten iron from the surface. Best results are obtained by heating alloying compounds in an air furnace or electric furnace. The usual alloying elements are Ni and Cr, although Mo, V etc. are also used. Nickel is added upto 5% for increasing hardness, resistance to wear, corrosion resistance and machinability. Chromium is added upto 3% to make the cast iron extremely hard, tough, strong and corrosion resistant. Chromium checks the formation of graphite and promotes the formation of carbides. Higher percentage of chromium harden the iron by increasing the percentage of combined carbon.

When nickel and chromium are added in the ration of 3:1, they make their graphite and carbide forming tendencies to neutralise and the cast iron so obtained has improved grain refinement, hardness and strength.

Molybdenum, when added upto 1.5%, improves the strength and wear resistance of cast iron, but reduces machinability. Molybdenum decreases the rate of graphitisation, retards critical transformation and thus improves uniformity of structure. Vanadium to the extent of 0.5% improves carbide formation and hence increases the hardness and strength of cast iron to an appreciable extent.

WROUGHT IRON

Wrought iron is highly refined metallic iron containing a little iron silicate slag, which is distributed throughout the iron in a fine fibrous and film form. It's carbon content varies from 0.12 to 0.15 per cent. Wrought iron is manufactured by two methods :

Puddling Process

The furnace is rectangular in shape and has a chimney on one side. The fuel (soft coal or coke) is burned on a grating situated on the side opposite to that of the chimney. The flames issuing from the grating are diverted by the arch-shaped root and concentrated on the hearth to heat the charge resting in the hearth. The hearth which is shaped like a saucer is lined with iron ore. The iron ore lining acts as the oxidising material.

Pig iron having the following composition : 1 to 1.5 per cent silicon, 1 per cent manganese, 3.50 per cent carbon, 0.30 per cent phosphorus and 0.06 per cent sulphur, is charged into the hearth through the charging door and melted by the flames. To facilitate oxidising reactions, almost pure iron or mill scale is added. First the silicon and manganese get oxidised as per the following reactions:

$$2\,FeO + Si \rightarrow 2Fe + SiO_2$$

$$FeO + Mn \rightarrow Fe + MnO$$

$$FeO + SiO_2 \rightarrow FeO.SiO_2$$

$$MnO + SiO_2 \rightarrow MnO.SiO_2$$

After the removal of silicon and manganese, the temperature is kept moderately low to facilitate the removal of as much phosphorus as possible:

$$2P + 5FeO \rightarrow P_2O_5 + 5Fe$$

$$P_2O_5 + 3FeO \rightarrow (FeO)_3.P_2O_5$$

Immediately after a substantial part of the phosphorus goes to slag, the temperature of the bath is raised as quickly as possible. Then at this stage carbon is oxidised to carbon monoxide.

$$2C + O_2 \rightarrow 2CO$$

$$2CO + O_2 \rightarrow 2CO_2$$

Some of the sulphur also gets oxidised and escapes as a gas.

The melting point of iron rises progressively (as the impurities are oxidised) and solidification starts with the formation of a pasty material. Considerable amount of slag gas admixed with the pasty wrought iron. The pasty wrought iron is removed from the furnace in the shape of round balls and then squeezed and shaped into bars under a power hammer or a squeezing machine. During squeezing operation most of the slag is ejected out. The bars so produced are then taken to a rolling mill where they are rolled into the desired sections.

Astron or Byers Process

In this process, the pig iron, iron oxide and silica are first melted in a cupola furnace and refining operations are carried out in a Bessemer converter. At the same time, a quantity of iron silicate slag is prepared in an open hearth furnace. The refined iron made in the Bessemer converter is poured at a predetermined rate into a ladle containing the molten slag (temperature of the slag is lower than the freezing point of iron) already prepared. Since the slag is maintained at a temperature considerably lower than the freezing point of iron, the iron is continuously and rapidly solidified producing small globules of the metal (the escaping gases during solidification shatter the metal into small fragments).

The excess slag is poured off and the remaining mass of iron and slag is taken to a press. The press squeezes out the surplus slag and a rectangular block known as bloom as formed. The hot bloom is immediately passed through rolling mills to produce wrought iron products of different shapes and sizes.

Uses of Wrought Iron

Wrought iron is mostly used for making small size water pipes and fittings, corrugated sheets and ornamental sheets and also used for making chains, bolts, nails, anchors etc.

STEEL

Steel is an alloy of iron and carbon, and strictly speaking it is a solid solution of carbon in iron. The carbon content of steel ranges from 2.025 to 1.5%. Below this range it is wrought iron. Since steel contains carbon in a range intermediate between wrought iron and cast iron, its preparation from cast iron involves removal of all carbon followed by the addition of the required amount of carbon. Any method of manufacture of steel is based on this principle only. Carbon and other impurities are oxidised away and then additions made. Processes are conducted in the liquid state.

Production of Steel

For converting the pig iron into steel, following methods can be used: (i) Cementation; (ii) Crucible process; (iii) Bessemer converter process; (iv) Open-hearth process; (v) L.D. process; (vi) Duplex process; (vii) KALDO process; (viii) Spray steel making process ; and (ix) Electric process.

Cementation process

In this process, bars of wrought iron are arranged in a brick compartment of the furnace with charcoal between and around them. The furnace is heated (950°C) to redness and maintained for 7 to 10 days at that temperature. The bars are then allowed to cool down. During heating the iron bars absorb carbon from the charcoal. The amount of carbon absorbed depends on the time and temperature. Outer layers of bars absorb more carbon than interior ones. The bars so obtained are very hetrogeneous in nature and their chemical composition is not certain. However, the homogeneity can be increased to a certain extent by hammering and forging process. During hammering and forging process the bars are reheated and forged into smaller sections. Several such bars are bundled together and beaten down under a hammer at bright red heat. This will make the bars to get welded together and become one piece. This method of treatment is to make the steel as uniform in composition as possible. This process is an obsolete now.

Crucible process

This process is also an obsolete one now but in the past it was regarded as one producing the finest quality steels. It is actually a refining process for making steel obtained from cementation process, more homogeneous and uniform in structure. Steel pieces from cementation process are charged in the crucible kept in a furnace which is heated up previously. The pieces melt down in about an hour time. The slag present in the metal comes to the surface and is skimmed off. The molten metal is then poured into ladies or moulds.

Bessemer process

The Bessemer process may be either of acid or basic type. The Bessemer converter is shown in Fig. 22.2.

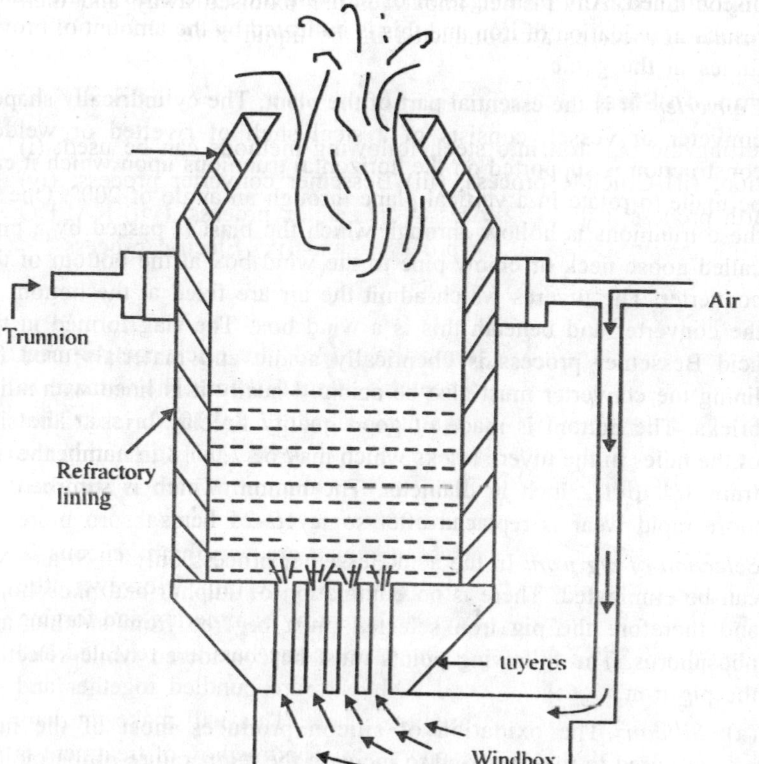

Fig. 22.2. Bessemer converter.

Acid bessemer process

(1) *General principle:* When a current of air is blown through molten pig iron which contains carbon, silicon and manganese, these elements together with some iron get oxidised. The carbon escapes as a gas while the silicon and manganese along with any iron that may be oxidised from mixed ferrous manganese silicate slag. The oxidation of these elements, especially silicon, produces a large amounts of heat which not only maintains the temperature and fluidity of the molten metal but actually increase its temperature. All the reactions taking place in the converter are reflected in the flame which issues from the mouth of the converter. During the early part of the blow when the manganese and silicon are being oxidised, the flame is short and of low luminosity as the blow proceeds and the carbon is eliminated. When decarborisation is completed, the flame drops and blowing is

discontinued. Any prolongation of blowing after the drop of the flame results in oxidation of iron and this is indicated by the amount of brown fumes in the game.

(2) *Converter:* It is the essential part of the plant. The cylindrically shaped conveter or vessel, consists of a steel shell of rivetted or welded construction is supported on the horizontal trunnious upon which it can be made to rotate in a vertical plane through an angle of 200°. One of these trunnions is hollow, through which the blast is passed by a pipe called goose neck or elbow pipe to the wind box at the bottom of the converter. The tuyeres which admit the air are fixed at the bottom of the converter and beneath this is a wind box. The slag formed in the acid Bessemer process is chemically acidic and materials used for lining the converter must also be acidic. Usually it is lined with silica bricks. The bottom is made of good quality fireclay bricks. The size of the holes in the tuyere bricks which may be 7 to 18 in number varies from 1/4 to 1/2 inch in diameter. The bottom which is subjected to more rapid wear is replaced after some 15–25 heats.

(3) *Selection of pig iron:* In the acid Bessemer process only C, Si and Mn can be eliminated. There is no elimination of sulphur and phosphorus and therefore the pig iron selected must be free from sulphur and phosphorus. The following points must be considered while selecting the pig iron.

(a) *Silicon:* The oxidation of silicon produces most of the heat required to maintain and to increase the temperature of the molten metal. The silicon content could be about 2.5 per cent. If the silicon content is less than 1 per cent then the metal is liable to blow cold owing to insufficient heat being produced by the oxidation of silicon. When the silicon exceeds above 3 per cent the heat produced by the oxidation may be excessive and overheats the metal. Silicon goes to the slag as silicates of iron and manganese and therefore higher the silicon content the greater the loss of iron in the slag.

(b) *Carbon:* Carbon is usually maintained at about 1 per cent higher than the silicon content and 3 to 3.5 per cent is considered ideal. The higher the carbon, the more prolonged will be the blow.

(c) *Manganese:* This to some extent controls the type of slag produced and is thus of considerable importance. To produce a dry slag the Mn should not exceeds half the percentage of silicon. The Manganese should be in the range of 0.7 to 1.0 per cent. If the manganese exceeds this figure. It delays the process and tends to produce a wet slag. When working with a very high manganese pig iron there is a tendency for the MnO formed during the early stage of the blow to attack the lining causing excessive wear.

(d) *Sulphur and Phosphorus:* Sulphur and phosphorus are not eliminated during acid Bessemer process and they should not exceed 0.05 per cent.

After completion of the previous blows the converter is turned to a horizontal position. The molten pig iron is then poured into the converter from the ladle. The air blast is turned on while the converter remains in this horizontal position. In this position the tuyeres are exposed above the bath level to prevent the passage of metal into the tuyeres and into wind box. As the converter is rotated into a vertical position the air passes through the molten metal and iron is first oxidised to Fe_3O_4.

$$3Fe + 2O_2 \rightarrow Fe_3O_4$$

This Fe_3O_4 then reacts with the impurities

$$2Si + Fe_3O_4 \rightarrow 3Fe + 2SiO_2$$

$$4Mn + Fe_3O_4 \rightarrow 3Fe + 4MnO$$

$$4C + Fe_3O_4 \rightarrow 3Fe + 4CO$$

Some of the higher oxides of iron are only partially reduced

$$2Fe_3O_4 + Si \rightarrow 6FeO + SiO_2$$

The ferrous and manganese oxides then react with SiO_2 to form slag. This first period or the silicon blow begins at the blast is turned on and the converter is turned to vertical position. During this period sharp transparent flame extends from the mouth of the vessel. As the blowing continues the flame starts to lengthen after about 4 minutes. This indicates the removal of silicon and manganese. Then the second period or carbon blow begins. It is during the second period that the flame attains its full brilliance and length extending as much as 30 feet beyond the mouth of the converter. This flame results from the evolved CO-burning into CO_2 as it comes in contact with the air at the mouth of the converter. When most of the carbon has been removed the flame diminishes and becomes less luminous until the flame drops, which indicates the complete removal of carbon. The determination of this period causes reddish fumes indicating the oxidation of iron.

Immediately after the flame drops, the converter should be turned down and the metal is poured into a teeming ladle. This time interval from end point to turn down is referred to as the after blow which is of extreme importance because it determines the physical properties of the products.

As the metal is being poured into the teeming ladle the necessary finishing and alloy additions are made to meet the steel specifications.

Basic Bessemer process or Thomas process

An improvement to the Bessemer process was made by removing phosphorus with the addition of lime and by lining the converter with basic refractory bricks such as magnesia or dolomite. When selecting pig iron for the basic Bessemer process the following factors must be considered:

(1) *Carbon:* It is readily removable and the amount present is of little importance.

(2) *Silicon:* It is readily removable but its presence prolongs the blow. A high silicon content means an increased quantity of lime to reutralise the SiO_2 formed. A good basic iron should not exceed 1 per cent silicon and ideally is about 0.5 per cent.

(3) *Sulphur:* This passes into the slag as a mixed sulphide of calcium and manganese. This should be below 0.1 per cent.

(4) *Manganese:* The removal of this element is lower than in acid process due to the presence of less silicon. The ideal percentage of manganese is 0.70–1.5 per cent.

(v) *Phosphorus:* It is the heat producer in the basic process and it may vary from 1.8–2.5 per cent is considered to be ideal.

The oxygen of the air combines with iron to form Fe_3O_4 which brings about the oxidation of the other impurities. The silicon is more rapidly and completely removed because of the lower initial silicon and also the lime readily combines with the silica so formed. Reactions in the converter are given as under:

(1) *Removal of manganese:* The removal of Mn is slower than in the acid process because of two factors.

(a) Mn passes into the slag as $MnO.SiO_2$ and there is a small concentration of SiO_2 available.

(b) Lime tends to replace MnO from its combination with SiO_2.

$$2MnO.SiO_2 + 2CaO \rightarrow 2CaO.SiO_2 + 2MnO$$

(2) *Removal of sulphur:* This is removed throughout the process and passes into the slag as a mixed sulphide of Ca and Mn.

$$FeS + Mn \rightarrow MnS + Fe$$

$$FeS + CaO \rightarrow CaS + FeO$$

The removal of sulphur is accelerated by a high Mn content in the original pig iron. A falling temperature also favours the forward reaction. If the pig iron is high in sulphur then it should be partially desulphurised before charging to the converter by the addition of soda ash.

(3) *Removal of phosphorus:* This is not removed until all the other impurities have been eliminated. When the carbon has been eliminated, the flame drops and after blow commences during which the phosphorus is rapidly removed. This may be represented by the following equations.

$$8Fe_3P + 5Fe_3O_4 \rightarrow 39Fe + 4P_2O_5$$

The P_2O_5 formed immediately combines with CaO and forms tetra basic phosphate of lime.

$$P_2O_5 + 4CaO \rightarrow 4CaO.P_2O_5$$

With the converter in horizontal position the necessary lime of about 300 pounds per ton of iron is added from an over head hopper and then molten iron is poured into the converter. The blast is then commenced and the converter is rotated into a vertical position (Blast pressure is 30 Psi). At first there is very little flame. After the Si and some of the Mn have been removed the flame lengthens and glows in luminosity due to carbon elimination. Then the flame drops indicating the end of decarburisation. The blow is continued, during which period the phosphorus is removed. This period is characterised by dense brown fumes. The total duration of the blow is 15 to 16 minutes including an after blow of 3 to 5 minutes. Following the after blow the vessel is finally turned down and slag is run off into a slag pot before the alloying additions are made. This is essential to prevent the reversion of phosphorus from the slag to the metal.

In basic Bessemer process the finishing additions are usually made as the blown metal is running into the ladle. The Si and Mn are added in the form of Fe-Mn, Fe-Si-Mn.

Open hearth process

Usually basic open-hearth process is used. The converter is a wrought iron reverberatory furnace flanked on either side by checkerwork firebrick chambers. The schematic diagram is shown in Fig. 22.3. Initially natural gas and air are preheated in a chamber on one side. The hot gases coming out. Difference among wrought iron, pig iron and steel is given in Table 22.1.

Table 22.1. Difference among wrought iron, steel iron and pig iron.

Properties	*Cast iron*	*Steel*	*Wrought iron*
% of carbon	1.7–4.3	0.25–1.5	005–0.25
Melting point (°C)	1200	1300 to 1400	1500
Brittleness or malleability	Brittle	Malleable & brittle	malleable
Hardness	Hard	Hard & soft	Soft
Tempering	Cannot be tempered	Can be tempered	Cannot be tempered
Welding	Cannot be welded	can be welded	Can be welded
Magnetisation	Cannot be permanently magnetised	Can be permanently magnetised	Cannot be magnetised

Uses

Wrought iron is used for making electromagnetic cores, railway couplings, chains and anchors. Mild steel is used for making springs, railway tracks, carnes, joists and surgical instruments. Cast iron is used in the manufacture of wrought iron and steel and in foundries. Comparison between open-hearth and Bessemer process is given in Table 22.2.

Table 22.2. Comparison between open-hearth and Bessemer process.

Bessemer process	*Open-hearth process*
It is a quick process, takes only 20 mins.	It is a slow process, takes 12 hrs. for a heat
The charge is pig iron	The charge is pig iron, ore and scrap
Output per heat is low because the converter is small	Output per heat is high because the furnace is big.
No fuel is needed	Producer gas is used
Composition cannot be controlled because temp. cannot be adjusted	Composition can be controlled because temperature can be adjusted during the long heat.
Steel is inferior quality	Steel of excellent quality is obtained

From the furnace pass through the battery of chambers on the other side heating them as they pass along. This technique of regenerative heat recovery is a unique feature of this steelmaking process.

The charge consists of pig iron, iron ore and scrap. The oxidation of the impurities is carried out by iron ore and not by air. It could used a large amount of scrap (upto 50% of the charge can be scrap). In contrast to Bessemer process the O.H. method is slow (takes around 12 hours for a heat). Initial installation costs are heavy; but the amount of pig iron that can be processed is huge (500 tons per charge). It slowness allows time to do analysis of the material being cooked and to adjusts the composition of the bath. Thus O.H. steel is superior in quality to Bessemer steel. Since the bath has a larger surface area than that in the Bessemer, closer contact of the metal and slag occurs with the consequent improvement in the removal of impurities. First the Si and Mn are removed; then C is oxidised; CO causes effervescence; when the boiling has subsided, deoxidiser (spiegeleisen) is added to bring the carbon to the required level. In the basic O.H. process all the phosphorus can be removed before all the carbon is removed. This is not possible in basic Bessemer. This enables steels of wider range of carbon content to be made.

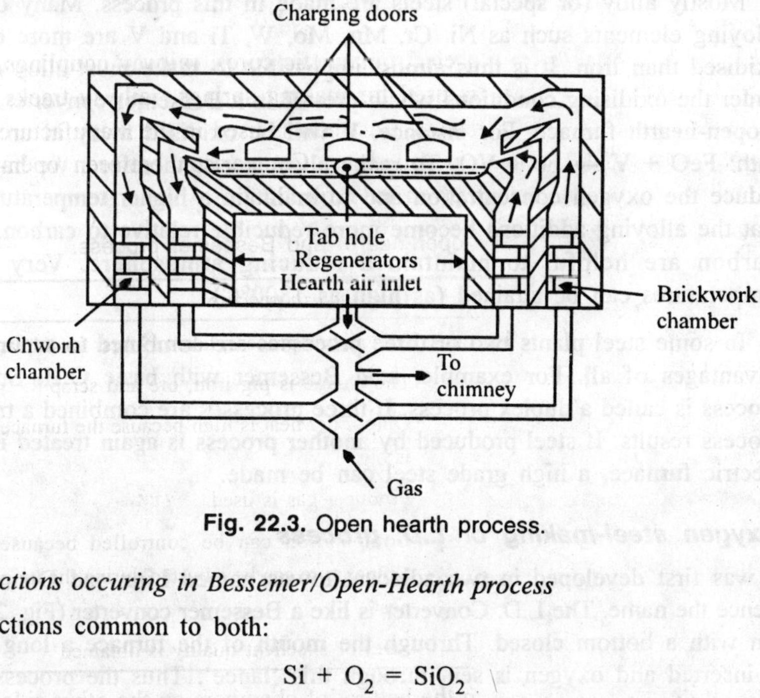

Fig. 22.3. Open hearth process.

Reactions occuring in Bessemer/Open-Hearth process

Reactions common to both:

$$Si + O_2 = SiO_2$$

$$4P + 5O_2 \rightarrow 2P_2O_5$$

$$S + O_2 \rightarrow SO_2$$

$$MnO + SiO_2 \rightarrow MnSiO_3$$

Reaction occurring only in Bessemer:

$$C + O_2 \rightarrow CO_2$$

Reaction occurring only in O-H process :

$$Fe_2O_3 + 3C \rightarrow 2Fe + 3CO \text{ (reduction of iron ore)}$$

$$Fe_3O_4 + 4C \rightarrow 3Fe + 4CO \text{ (reduction of scrap)}$$

A comparison of the two processes is given in Table 22.2.

Electric process

This is basically a remelting process and not a refining process. Molten pig iron is charged or scrap is melted by electric arc between graphite electrodes. No air is sent. No chemical fuel is used. Slag floats on the surface and the slag is poured out by tilting the furnace. Then lime and anthracite coke are thrown in to form a new reducing slag. Now alloying additions are made to the bath.

Mostly alloy (or special) steels are made in this process. Many of the alloying elements such as Ni, Cr, Mn, Mo, W, Ti and V are more easily oxidised than iron. It is thus almost impossible to make most alloy steels under the oxidising conditions whcih prevail in a Bessemer converter or in a open-hearth furnace. For instance, V will be oxidised away in an O.H. bath: $FeO + V \rightarrow Fe + VO$. To make alloy steels, therefore, we have to reduce the oxygen concentration and to maintain a higher temperature so that the alloying additions become more reducible relative to carbon. The Carbon are helpful to maintain a reducing atmosphere. Very high temperatures can be attained (as high as 1800°C).

In some steel plants two or three processes are combined to secure the advantages of all. For example, acid Bessemer with basic O.H. Such a process is called a duplex process. If three processes are combined a triplex process results. If steel produced by another process is again treated in the electric furnace, a high grade steel can be made.

Oxygen steel-making or L.D. process

It was first developed in two adjacent towns named Linz and Donowitz. Hence the name. The L.D. Converter is like a Bessemer converter (Fig. 22.4), but with a bottom closed. Through the mouth of the furnace a long tube is inserted and oxygen is sent through this 'lance'. Thus the process can be considered as an oxygen-rich Bessemer process. Reactions are quicker than in Bessemer and the temperatures attained are higher. Charging of the pig iron, removal of slag and addition of deoxidiser (spiegeleisen) are the same as in Bessemer. But the quality of steel is better, especially in that it is free of nitrogen. For treating high-phosphorus in that it is free of nitrogen. For treating high-phosphorus pig irons the L.D. process was modified by injecting lime with the oxygen jet. L.D. process had made the Bessemer and O.H. processes obsolescent.

The most advanced technique in steelmaking is 'spray refining' developed by British Iron and Steel Research Association. But it has not gained wide commercial acceptance.

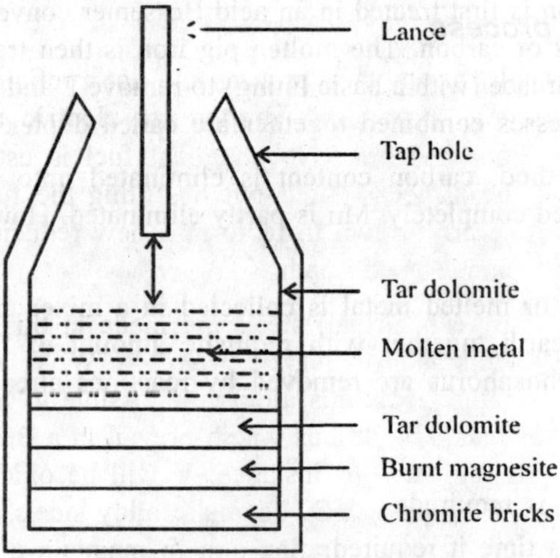

Fig. 22.4. L.D. process.

Types of Steel

Steel are broadly classified as plain carbon steels and alloy or special steels. Plain carbon steel does not contain any alloying element in more than 2 or 3%. It is further classified on the basis of its carbon content. Mild steels contain between 0.25 to 0.4% carbon and hard steels contain from 0.5 to 1.5% carbon. Steels are classified into three types on the basis of their carbon contents.

(1) Very mild or dead-mild steels containing upto 0.25% carbon, used for making boiler plates, rivets, bolts, wire, nails, etc.

(2) Low or medium carbon steels containing upto 0.6% carbon, used for making railway axles, rails, springs, machine parts, etc.

(3) High carbon steels containing between 0.6 and 1.5% carbon, used for making tools, saws, knives, files, ball bearings, etc.

Duplex process

The combination of Bessemer process and open-hearth process is known as duplex process. The high phosphorus pig iron can not be treated by basic

Bessemer process. Similarly, pig iron containing less than 1.5% P can not be treated by the same process, because 1.5% P is the minimum requirement for the basic Bessemer process. Thus basic-open hearth process was applied to these irons, which is quite lengthy process.

To convert such high phosphorus pig iron to steel in half the time, the molten pig iron is first treated in an acid Bessemer converter to remove Si, Mn and a part of carbon. The molten pig iron is then transferred to basic-open hearth furnace (with a basic lining) to remove P and remaining carbon. The two processes combined together are called duplex (double) process.

In this method, carbon content is eliminated upto 0.03%. Silicon is almost removed completely. Mn is partly eliminated. However, P and S are not eliminated.

The liquid or melted metal is collected in a mixer and charged in the basic open hearth together with requisite amount of flux and ore. The sulphur and phosphorus are removed by reactions already discussed.

Advantages

(1) Less flux is required.
(2) Much less time it required.
(3) No steel scrap is utilised.
(4) It is an all hot metal process.
(5) It is advantageous when steel scrap is in short supply.

EFFECT OF ELEMENTS ON STEELS

(1) *Carbon:* Without carbon there is no steel. Carbon hardens and strengthens steel. The tensile strength increases with increasing carbon content upto 0.8%. Upton this limit the strength increase is 70 kg. cm^{-2} for every 0.01% carbon. Ductility and malleability decrease with increasing carbon content. Carbon makes iron less permeable to magnetic field.

(2) *Silicon:* Plain carbon steel contains around 0.2% silicon which increases strength without compromising on ductility. In high carbon steels it converts cementite (Fe_3C) to iron and graphite which reduces hardness and brittleness.

(3) *Sulphur:* Sulphur renders steel hot-short, i.e., brittle at high temperatures. It combines with iron to form FeS the crystals of which are incompatible with those of iron. Manganese is a good scavanger for sulphur and hence presence of manganese of offsets the harmful

effects of sulphur. If sulphur is present upton 0.3%, the steel is machinable and is suitable for making screwstock such as nuts, bolts and threaded joints.

(4) *Phosphorus:* Phosphorus renders steel cold-short, i.e., brittle even at room temperature. Instead of dissolving uniformly it segregates. Hence the effect. However, within 0.1% range it increases strength, hardness and corrosion-resistance.

Nitrogen is present in steel as a hard nitride and is undesirable if the steel is to be used for making boiler plates because nitrided steel suffers from caustic embrittlement. Titanium is a good denitrider for steel. Oxygen present as FeO causes fracture under static stresses. Aluminium or titanium is prescribed as deoxidiser.

Broadly speaking the effects of these elements on cast iron are the same as their effects on steel.

Alloying Elements in Steel

Alloy steel or special steel is a steel which contains Si, Cr, Ni, Co, Mn, V, Ti, W, etc. in appreciable amounts, generally in more than 3% modifying the properties of the steel by their presence. Alloy steels have specific uses. Alloying additions are made to steel with the following purposes in view: (i) increasing hardness; (ii) improving miscibility in iron; (iii) stabilisation on one of the solid solutions of carbon in iron, such as austenite or ferrite; (iv) imparting corrosion resistance; and (v) inducing formation of carbides.

Silicon increases hardness and magnetic suscepitibility. Chromium imparts corrosion resistance through passivation. Nickel improves shock resistance and corrosion resistance. It also increases the strength. The temperature range for heat treatment is also widened by the pressure of nickel in steel. Manganese is a deoxidiser and desulphuriser in steel. In large percentages manganese makes steel tough and impact resistant, Vanadium is a deoxidiser next only to aluminium. Titanium is a denitrider as well as deoxidiser. Tungsten increases the hardness of steel very much. Further, in tungsten steel hardness is maintained at high temperatures. Molybdenum is similar to chromium and tungsten in effect.

Silicon steels are used in chromium and tungsten in effect. Steel containing 18% Cr and 8% Ni is used for making corrosion resistant vessels. Nickel steel is used for making turbine blades. Invar, a nickel steel, has very low coefficient of thermal expansion and hence used for making pendulums. Manganese steels are used for making jaw-crushers. Tungsten steels are the choice materials for high speed tools.

Heat Treatment of Steel

This is an operation or a series of operations involving heating and cooling of steel in the solid state for the purpose of obtaining desired end properties of a material. The various processes can be classified as: (i) annealing; (ii) normalising; (iii) hardening; (iv) tempering; and (v) case-hardening.

In all these processes the steel is heated fairly slowly to a predetermined temperature, and then cooled, and it is the rate of cooling which determines the resultant structure of steel.

Annealing

Steel is heated to a specified temperature, held at this temperature for a period of time and then cooled at a controlled rate. A number of treatments are classified under this term. These may be employed to reduce hardness, remove residual stresses, improve toughness, restore ductility or to alter the electrical or magnetic properties.

In full annealing steel is heated to 30 to 60°C above the austenite-ferrite transformation temperature, held at this temperature for a fixed time and then cooled slowly to below the eutectic temperature. For each inch of thickness the steel is kept for one hour. Cooling rate is usually 10 to 30°C per hour. A stress-relief anneal is done for removing residual stresses in castings. A process anneal is done to restore the ductility. Spheroidisation is another type of annealing for preparing screw stock.

Normalising

It is different from annealing in the method of cooling. Here cooling is done in still air. This is more rapid than full annealing. The surface finish of a normalised steel is better than that of a full annealed sample. Thin sections show a finer refinement than thicker sections.

Hardening

When a piece of steel is cooled rapidly from above its critical temperature it becomes considerably hard. The degree of hardness depends on initial quenching, temperature, size of the sample, nature of the quenching medium (oil or water) and the agitation of the medium. Water-quenching gives a hard surface called martensite. During this process iron changes from face-centred cubic to body-centred cubic.

Tempering

A fully hardened steel may be reheated to restore its toughness and ductility and cooled slowly. During this process the initial stresses are relieved and brittleness reduced. As the steel is heated, the oxide film which begins to form on a bright, clean surface first assumes a pale-yellow colour, then passes through a number of interference colours till it becomes a dark blue hue. The temperature ranges from 22°C to 300°C. For each temperature region of tempering the steel processed has a specific use. Strong and hard steel is made at 400°C— tempering; tough and ductile steel is made at 600°C— tempering. Steel tempered at 200–300°C is used for making blades and tools. Tempering at 300°C yields a steel suitable for making springs, saws, etc.

Case hardening

In this surface modification process steel is kept in a drum filled with carbon powder and a little of barium carbonate. The box is sealed and heated to about 900 to 950°C when α-iron is converted into γ-iron which dissolves the carbon to form austenite. the duration of heating depends on the depth of hardness required. The article is then quenched in oil or water. The case or surface becomes hard, while the core or the interior is still tough. In this process the article is heated in a stream of gaseous hydrocarbons, in which case the method is known as gas-carburising.

Nitriding

In this process the steel is heated in a stream of ammonia. Nitrogen formed by decomposition of NH_3 enters the surface of the steel and forms iron nitride. This makes the surface hard. A little of alloying elements such as Cr, Mo, V, etc. should be present for a successful nitriding. Since the hardening does not depend on the type of iron (α-iron or γ-iron), the hardness of the article does not varnish at lower temperatures unlike in carburising.

In cyaniding the steel article is heated in a both of NaCN containing a little NaCl. At a temperature of about 900°C both C and N become available. Thus both carburising and nitriding occur simultaneously.

Chapter 23

Energy Conservation in Chemical Process Industries

INTRODUCTION

Energy is one of the most vital inputs for economic development of any nation. Per capita consumption of energy is an index which refers to the economic development of any country and India ranks poorly in this index with its energy consumption which is about 1/10th of the average per capita consumption of the world. The industrial sector is the major energy-consuming sector in India and uses about 50 per cent of the total commercial energy in the country.

In the changing scenario of globalisation the chemical industry in India is facing increasing levels of competition from international manufacturers. While India is lowly placed in terms of per capita energy consumption, the specific energy consumption level for most of the products in India is much higher than the global average. This is mainly because industry in India is not using the cutting edge of the technology and the installed capacities are smaller, resulting in energy consumption, which is higher than that in the international arena. High unit cost of energy in India is an additional adverse factor for the Indian Chemical Industry. In order to be a global player, it is essential to remain competitive in the world market. This requires incorporation of innovative technologies that are more energy efficient together with renovation of the existing equipment and properly monitoring and managing the energy consumption. A focused effort on energy management and conservation will go a long way in getting the best out of the energy actually consumed and the topic of energy management and conservation in chemical industries is of immense importance as far as India is concerned.

Chemical industry is the fountainhead for the manufacture of a variety of chemical products beginning with products like soda ash, paper, plastics, petrochemicals, petroleum products, organic and fine chemicals, fertilisers, pharmaceuticals and some speciality chemicals like isotopes (heavy water for example). All these products are manufactured using very large number of processes and operations.

The level of processing and the energy consumption depends on the complexity of the process involved. The energy consumption when expressed as equivalent steam consumption varies for a simple process from a tonne of steam per tonne of the product to 10–15 tonnes of steam per tonne of product for a more difficult process (like in case of separation of propylene from propane or ethyl benzene from benzene or in electrolytic production of chlorine) to 17000 tonnes of steam per tonne of product in case of production of heavy water (which is like any other chemical industry). A very large amount of energy consumption in case of heavy water is basically due to low abundance of deuterium in nature and large amount of enrichment required.

In all chemical processing systems, there has been tremendous innovations world over for making the process more and more energy efficient and it is reflected in the Table 23.1.

Table 23.1. Specific energy consumptions of typical chemicals in Mkcals/MT.

Product	*Early '90s*	*End '90s*
Ammonia	10	7.0
Urea	7	5.8
Methanol	18	8.8
Petrochemicals	2.6	2.1
Viscose fibre yarn	21.7	15.5
Heavy water	12200	10000

The chemical industry all over the world felt adverse impact of rising energy prices in 1999. The global hydrocarbon and energy cost index was 30 per cent higher in the third quarter of 1999 compared to the same period in 1998. The scenario in India also was not better and the continuously spiraling petroleum costs paint a bad future for the industry. The only solace is in the fact that there is a great amount of awareness of the linking of energy and economics and energy conservation is now getting a major boost in almost all the chemical industries.

There is considerable scope for reducing the wasteful consumption and improvement in energy efficiency. The use of older processes and equipment and poor overall integration of the process systems of a typical unit often results in large energy consumption. Thus there is an urgent need to have a focused study of the existing consumption level with efficiency of the systems and management of the energy consumption.

PROCESS DESIGN AND ENERGY CONSUMPTION

The concept of maximisation of production with lower energy consumption is catching on. The basic process in chemical industry is the separation process and separation based on difference in volatility accounts for more than 80 per cent. The distillation is used for separating the easiest of hydrocarbon components to the most difficult ones like deuterium oxide (heavy water) from natural water. The energy required for these processes widely vary from say 1–16 tonne steam per tonne of product to 17000 tonne steam per tonne of product in case of heavy water. A mere saving of even 10 per cent in the energy consumed will bring down the cost of product substantially.

The energy consumption of any process is required to be looked at from the overall thermodynamics. For example, a typical distillation process shows the thermodynamic efficiency to be less than 10 per cent and this means that a large amount of energy is merely wasted. Innovative concepts like cascading or staging, reactive distillation etc. have already been used in the complex process for heavy water production. Innovative system integration using pinch analysis for the heat exchange network has also helped in reducing the wastage by maximising the thermal efficiencies.

The heavy water process being a highly demanding one, use a lot of innovations that were not seen in a regular process flow sheet and the specific consumption has fallen from 540 GJ/kg (distillation of water) to 30–40 GJ/kg for the reactive distillation process. In the conventional chemical industry ammonia plant design merits as an example as a lot of innovations have taken place—there by reducing the energy consumption level from 22 Mkcal/kg to 7 Mkcal/kg presently. For example, in the field of heavy water and ammonia where such energy conservation measures have been taken up are given here.

Heavy Water Process

Deuterium is an isotope of hydrogen and heavy water is deuterium oxide. Due to its nuclear properties heavy water is used as moderator and coolant in the pressurised heavy water reactors (PHWRs). A typical 220 MW reactor requires about 275 tonnes of heavy water as the initial charge which remains in the system. Heavy water is present in natural water in extremely small concentration (about 144 ppm) and due to its isotopic nature it is very difficult to separate it from normal water.

The low abundance of heavy water coupled with a very poor separation factor (separation factor is defined as concentration of deuterium in liquid phase to that in vapour phase) make the heavy water production an extremely difficult task. The requirement of large tonnage of the product with very high purity not only make the plants highly complex in their configuration, but highly energy intensive also. The average energy consumption figures (design figures) for the two main technologies followed in India are given below.

Process	*Sp. energy consumption*
$H_2S–H_2O$	45 GJ/kg
$NH_3–H_2$	33 GJ/kg

The GJ (Geiga Joules) represent the total energy both electrical and thermal where the electrical energy is converted into thermal using appropriate heat rate figures. In terms of equivalent steam consumption, 45 GJ/kg refers to approximately 17 tonnes of HP steam (30 kg/cm^2) per kg of heavy water. In terms of cost, the $H_2S–H_2O$ bithermal process works out cheaper especially since 65 per cent of the total energy is in the form of steam which is supplied either through a captive power plant or nuclear power station while 80 per cent of energy in case of monothermal $NH_3–H_2$ is in the form of electrical power. Thus a Geiga Joule in case of bithermal costs about Rs. 120, while it costs more than Rs. 250 in case of monothermal process.

Ammonia Production

The ammonia technology is one of the very old processes using *Haber-Bosch* process producing ammonia from hydrogen and nitrogen with iron catalyst. Energy is the largest single component in ammonia production cost and energy efficiency has been the principal goals of technological developments for years. From the earliest coke based plants where capacity was very low, which necessitated a number of parallel trains, the technology got revolutionised due to the introduction of steam reforming process. Integrated single train ammonia process was introduced in the early sixties and the concept of steam turbine drive for the synthesis gas compressor was the innovative design which helped in sharp fall in the specific energy consumption. The specific energy was around at 22 Mkcal/

tonne for the early plants and the innovations in process technology has improved the specific energy consumption to about 7 Mkcal/tonne which is near to the theoretical value of 5 Mkcal/tonne.

The basic ammonia process is built around the highly energy intensive primary reformation reaction section in a fired reformer and recovery of the process waste heat for producing steam, to drive the synthesis gas compressor which increases the pressure of the ammonia synthesis gas to the ammonia loop pressure. The concept of adiabatic reforming exchanger for the directly fired primary reformer, use of PSA for separation of CO_2 direct gas turbine coupled compressors, innovative ammonia converters, hydrogen gas recovery from the purge gas and co-generation mode captive power plant are the latest innovations tried in ammonia technologies. The introduction of smaller size catalyst and the use of radial and axial-radial converters have helped in reduction of the ammonia synthesis loop pressure from 600 bars to as low as 120 bars—an astonishing 80 per cent reduction. Success has already been achieved in using new catalysts based on rare earth metals which may further bring down the synthesis loop pressures to lower than 100 bars.

The above illustrations indicate that proper review of the technology and integration of energy efficient processes and equipment is essential in the design stage for achieving the minimum energy consumption. This calls for special efforts to reduce the energy consumptions especially in energy intensive process industries. In case of heavy water even 1 per cent reduction in specific energy consumption will result in an annual saving of Rs. 2.77 crore—taking all the operating heavy water plants into account. Heavy water contributes anything between 15 to 30 per cent towards cost of generation of nuclear power depending upon vintage of the plant. Considering the fact that energy constitutes 70 per cent of the works cost of heavy water production, the importance of energy conservation in HWPs can be well understood.

ENERGY MANAGEMENT PRINCIPLES

The energy conservation campaign carried out by Heavy Water Board (HWB) both at the corporate level as well as the individual plant level has resulted in spectacular results and today HWB is in a position to share this experience with the industry in general and chemical process plants in particular. It may be mentioned that HWB has bagged the ICMA award for energy reduction for the year 1999–2000. The energy management principles followed in HWB are applicable to other chemical industries as well and are illustrated below. The energy management must begin from the very top and should have the following six essential elements:

(1) A well defined energy policy.

(2) Organisational structure to implement the policy.

(3) Methodology for energy management.

(4) A motivated staff.

(5) Comprehensive information systems.

(6) Investment.

Energy Policy

A well defined energy policy is extremely important for a successful energy conservation activity. The corporate management should take a decision to formulate an energy policy and a matching action plan for implementation. HWB management took such a decision and the matching action plan. The energy policy needs to be translated into specific objectives and these objectives should also be made known to all the people concerned. The energy policy stated is to pay focused attention to every aspect of plant operation and maintenance activities with a view to reduce energy consumption and also identify and implement plant modifications which can result in energy saving with specified limit on payback period for the capital invested. In HWB 1999–2000 was declared as 'energy conservation year' with the target of reduction in overall specific energy consumption by 10 per cent over the best achieved earlier period. The best achieved specific energy consumption was taken as the weighted average value considering all the heavy water plants for the previous five year (1994–95 to 1998–99) at the corporate level. However, the target value for each of the separate plants was taken as 90 per cent of its own best achieved earlier. In the case of other chemical industries the bench marked figures of specific energy consumption available may be used at the corporate level with the individual targets based on the specific values for the individual plants. The achieved specific energy consumption at the heavy water plants was reviewed on a fixed frequency basis and corrective actions taken for meeting the specified target value.

It is needless to mention that a strong energy policy coupled with a very strong commitment to reduce the energy must come from the very top in any organisation. Multipronged efforts are needed to effect any significant results in energy conservation. In the case of HWB the task was more challenging because of different technologies involved and multi location of the plants. Even the plants based on the same process and set up at different points of time have variations from one another due to improvements effected in technology with the experience of the previous plants and naturally have different energy consumption norms. The strategy was to maximise production from the more energy efficient plants and to re-optimise the operating parameters of others to achieve the optimum level of production resulting in the minimum energy consumption level. For example, HWP Manuguru based on H_2S–H_2O technology and having its own captive power plant which was the latest installed plant was operated to maximise the production levels while the plant at Kota based on the same technology but of somewhat older design was operated with re-optimised operating parameters. This was done by developing a mathematical model to arrive at the operating parameters which will result in minimum specific energy consumption.

Organisation Structures for Implementing the Energy Conservation (Encon) Policy

Along with the energy policy, an organisational structure with well defined responsibilities is also necessary for implementing the energy conservation measures. In HWB a well defined organisation for energy conservation was put in place with an apex body comprising of senior officers and the chief executive in the central office and a co-ordinator at each plant who will co-ordinate between central office and plant. At every plant a monitoring body was also identified along with the local co-ordinator. Interaction between the apex body and the site level body was held on a continuous basis and periodic reviews were carried out in order to monitor the specific energy consumption and take corrective action. Awareness programmes for all the employees were organised in the form of workshops where experts from various fields of energy conservation were invited to discuss with the employees.

At the plant level the implementation of the energy conservation was primarily vested with the General Manager who in turn through his production team kept a complete control on energy consumption figures. The technical services section of the plant assisted in carrying out periodic energy audits and implementing Encon measures. Certain capital schemes aimed at conserving energy were also taken up for implementation in consultation with the design group in central office as well as technical services section of the individual plants.

Methodology Adopted for Energy Management

Having declared 1999–2000 as energy conservation year in all the heavy water plants and adequate awareness among the plant personnel created, the following methodology was adopted for energy management. As a first step, it was decided to identify the areas where energy is getting wasted and where the scope existed for energy conservation. A comprehensive energy audit was carried out and this exercise brought out the following as the factors contributing to avoidable energy consumption:

(1) Plant trips and downtime due to equipment failure and instrument malfunctioning.

(2) Fouling of heat-transfer equipment.

(3) Inadequate operation attention to parameters having bearing on production and energy consumption.

(4) Energy losses through leakages in utility systems, poor condition of thermal insulation, high temperature of the gases, excessive blow-down, etc.

(5) Operation of rotating equipment at off-best efficiency point.

(6) Running additional pumps and compressors to take care of the trips of one of the running equipment.

(7) Over consumption of utilities like instrument air etc.

(8) Erroneous measurements of consumptions and other parameters.

Based on the energy audit, corrective actions were identified for implementation. The continuous monitoring on a daily and a monthly basis was set up with targets and abnormalities found in energy consumption pattern were analysed and corrective actions taken.

Motivation

Even with a good energy policy and organisational structure to support this policy, if the motivation is not existing at all the levels, it would be difficult to meet the set targets. Formal and informal channels of communication need to be established and the operating staff right up to the operator level need to be informed and motivated in their efforts and involvement in the energy conservation activity. Incentives in the form of weightage for energy conservation in production linked incentive scheme as well as best plant performance awards also provided motivation towards energy conservation programmes. A very high degree of motivation could be achieved by bringing awareness among the employees of the importance of energy conservation activity at all levels.

Information System

A very good information system which will provide adequate information to the plant personnel on a day to day basis on the energy consumption

patterns is essential. The HWB could develop an effective information system for analysing the various results. At every plant the weekly energy consumption figures and the target values were displayed at strategic points. As mentioned earlier, continuous monitoring of the energy consumption figures and also the specific energy values were carried out at the corporate office. Various banners were displayed bringing out the importance of energy conservation.

Investment

The entire energy management will have limited impact if the organisation is not able to implement various energy conservation schemes identified after a thorough review of the plant systems. This is important especially in older plants which were designed around two and a half decades earlier. HWB has undertaken a number of capital schemes for improving energy efficiency. Some of the important schemes implemented are indicated below:

(1) Recirculation of effluent water.

(2) Waste heat recovery from the effluent water.

(3) Hydraulic system optimisation and trimming of the pump impellers wherever excess margin exists.

(4) Implementation of variable speed drive.

(5) Integration of utility systems including refrigeration system.

(6) Review of shut-down and start-up procedures.

It is important to mention that a cost benefit analysis needs to be done in order to arrive at viability of each of the schemes. While pay-back period is one such criteria, one may have to look at other rigorous financial analyses like internal rate of returns (IRR) etc. while evaluating the various alternatives available.

Thus energy conservation is not a fancy any more. Failure to conserve energy results in inefficient utilisation of resources and leads to environmental problems necessitating additional cost for effluent treatment. All the chemical plants put together consume approximately 500 million tonnes of coal equivalent of energy and even a one percentage saving can do wonders in reducing the environmental impact.

References

Adams J.A., and **Rogers D.F.**, *Computer-Aided Heat Transfer Analysis*, McGraw-Hill, New York.

Backhurst J.R. and **Parker T.H.**, *Process Plant Design*, Elsevier, New York.

Baum B. and **Parker C.H.**, *Solid Waste Disposal Recycle and Pyrolysis*, Reinhold, New York.

Bauman H.C., *Fundamentals of Cost Engineering in the Chemical Industry*, Reinhold, USA.

Berenson C., *Administration of the Chemical Enterprise*, Wiley-Interscience, New York.

Boynton R.S., *Chemistry and Technology of Lime and Limestone*, John-Willey, USA.

Budnikov P.O., *Technology of Ceramics and Refractories*, Prentice-Hall, UK.

Burke J.E., *Progress in Ceramic Science*, MacMillan-Pergamon, New York.

Camp, T.R., *Water and Its Impurities*, Reinhold, New York.

Chambers R.D., *Fluorine in Organic Chemistry*, Wiley-Interscience, New York.

Chanlett E.T., *Environmental Protection*, McGraw-Hill, New York.

Clauser H.R., *Encyclopaedia of Engineering Materials and Processes*, Reinhold, USA.

Conover J.A., *Automatic Control for Chemical Processes*, Wiley-Interscience, New York.

Considine D. M., *Encyclopaedia of Instrumentation and Control*, McGraw-Hill, New York.

Cooper H.B.H. and **Rossano A.T.**, *Source Testing for Air Pollution Control*, McGraw-Hill, New York.

Coughnour D.R. and **Koppel L.B.**, *Process Systems Analysis and Control*, McGraw-Hill, New York.

Cryotech Proceedings, *The Production and Use of Industrial Gases*, IPC Science and Technology Press, London, UK.

DeGarmo E.P., *Materials and Processes in Manufacturing*, John Willey-Interscience, New York.

Dickinson D., *Practical Waste Treatment and Disposal*, Wiley-Interscience, New York.

Doremus R.H., *Glass Science*, Wiley-Interscience, New York.

Eckenfelder W.W. and **O'Connor D.J.**, *Biological Waste Treatment*, Pergamon, UK.

Evans L., *Selecting Engineering Materials for Chemical Plants*, Wiley-Interscience, New York.

Ewing G.W., *Instrumental Methods of Chemical Analysis*, McGraw-Hill, New York.

Fawcett H.H. and **Wood W.S.**. *Safety and Accident Prevention: Chemical Operation*, Wiley-Interscience, New York.

Francis A.W., *Liquid-Liquid Extraction*, Wiley-Interscience, New York.

Hamilton A. and **Hardy H.L.**, *Industrial Toxicology*, McGraw-Hill, New York.

Hampel, *Electrochemical Encyclopaedia*, Reinhold, UK.

Harvey B.G., *Nuclear Physics and Chemistry*, Prentice-Hall, UK.

Hove J. and **Riley W.C.**, *Ceramics for Advanced Technologies*, Wiley-Interscience, New York.

Jackson F.R., *Recycling and Reclaiming of Municipal Solid Wastes*, Noyes, UK.

Jacobs M.B., *Chemical Analysis of Air Pollutants*, Wiley-Interscience, New York.

James A.N., *Industrial Chemistry*, Noyes, UK.

Jolles Z.E., *Bromine and its Compounds*, Noyes, UK.

Jurah J.M., *Quality Control Handbook*, McGraw-Hill, New York.

Kent J.A., *Riegel's Industrial Chemistry*, Reinhold, USA.

Kingzett, *Chemical Encyclopaedia*, Van Nostrand, USA.

Kirk-Ohmer, *Encyclopaedia of Chemical Technology*, Wiley-Interscience, New York.

Kit B. and **Evered D.S.**, *Rocket Propellant Handbook*, MacMillan, New York.

Kriegel W.W. and **Palmour H.**, *Mechanical Properties of Engineering Ceramics*, Wiley-Interscience, New York.

Laitinen H.A. and **W.E. Harris**, *Chemical Analysis*, McGraw-Hill, New York.

Landau R., *Chemical Plant*, Reinhold, USA.

Levenspiel O., *Chemical Reaction Engineering*, Wiley-Intersicence, New York.

Luyben W.L., *Process Modelling, Simulation and Control for Chemical Engineers*, McGraw-Hill, New York.

Mantell C.L., *Carbon and Graphite Handbook*, Wiley-Interscience, New York.

McCoy J.W., *The Chemical Treatment of Cooling Water*, McGraw-Hill, New York.

Miller S.A., *Acetylene: Its Properties, Manufacture and Uses*, Academic, UK.

Morton M., *Rubber Technology*, Van Nostrand, Reinhold, USA.

Nagiev M.F., *The Theory of Recycle Processes in Chemical Engineering*, MacMillan-Pergamon, UK.

Perry R.H., *Chemical Engineers' Handbook*, McGraw-Hill, New York.

R. Norris Shreve and **Joseph A. Brink**, *Chemical Process Industries*, McGraw-Hill, New York.

Ray W.H., *Process Optimisation*, Wiley-Interscience, New York.

Rhodes T. and **Carroll G.**, *Industrial Instruments for Measurements and Control*, McGraw-Hill, New York.

Rudd D., *Process Synthesis,* Prentice-Hall, New York.

Rudge A.J., *Manufacture and Use of Fluorine and its Compounds*, Oxford, USA.

Sauchelli V., *Chemistry and Technology of Fertilisers*, Reinhold, UK.

Sconce J.S., *Chlorine: Its Manufacture, Properties and Uses*, Reinhold, UK.

Sherwood T.K., *Process Design*, Wiley-Interscience, New York.

Singer F.S., *Industrial Ceramics*, Chemical Publishing, UK.

Siting M., *Catalyst Manufacture: Recovery and Use*, Noyes, UK.

Siting M., *Sulphuric Acid Manufacture and Effluent Control*, Noyes, UK.

Siting M., *Organic Chemical Process Encyclopaedia*, Noyes, UK.

Stephenson R.N., *Introduction to Chemical Process Industries*, Reinhold, USA.

Urbanski T., *Chemistry and Technology of Explosives*, Pergamon, UK.

Van Vlack L.H., *Physical Ceramics for Engineers*, Addison-Wesley, UK.

Van Wazer J.R., *Phosphorus and its Compounds*, Wiley-Interscience, New York.

Weiss S., *Hydrofluoric Acid Manufacture*, Noyes, UK.

Yaffe L., *Nuclear Chemistry*, McGraw-Hill, New York.

Index

A

Absorption, 35

Acetylene, 274

Acidity, 133

Activated aluminas, 363

Activated carbon, 292

Adsorption, 34

Alkali and chlorine products, 469

Alkali silicates, 627

Alkali treatment of boiler water, 115

Alkali-metals and ammonium borates, 378

Alkalinity, 132

Alloying elements in steel, 693

Alumina (aluminium oxide), 354

Aluminium carboxylates, 361

Aluminium chloride hexahydrate, 362

Aluminium chloride, 360

Aluminium hydroxides, 363

Aluminium nitrate, 362

Aluminium sulphate, 357

Alums, 359

Ammonia, 402

Ammonia production, 699

Ammonium chloride, 408

Ammonium compounds, 408

Ammonium nitrate, 409

Ammonium sulphate, 410

Analysis of heavy water, 664

Ancillary equipment, 37

Anhydrous hydrochloric acid, 347

Annealing, 694

Argon, 283

B

Bacteriological tests, 136

Barium and its compounds, 513

Barium carbonate, 521

Barium chloride, 524

Barium chromate, 534

Barium compounds, 521

Barium ferrite, 530

Barium hydroxide, 532

Barium nitrate, 528

Barium stearate, 533

Basic principles of raw materials benefication, 89

Basic types of industrial chemical processes and flow schemes, 84

Bassemer process, 682

B-coli index, 140

B-coli test, 139

Benite, 246

Binding energy, 645

Blanc fixe or precipitated barium sulphate, 531

Bleaching powder, 501

Boiler scale and sludges, 110

Boric acid esters, 381

Boric acid, 376

Boron, 374

Boron compounds, 375
Boron halides, 382
Boron oxides, 375
Boron phosphate, 381
Borosilicate glass, 627
Bottle glass, 630
Bromides and iodides, 640

C
Calcined alumina, 365
Calcium arsenate, 570
Calcium carbonate, 569
Calcium silicates, 554
Calcium sulphide, 570
Carbon and artificial graphite, 290
Carbon black, 299
Carbon dioxide, 264, 287
Case hardening, 695
Cast iron, 675
Catalysts and catalysis, 182
Catalytic processes, 81
Caustic embrittlement, 116
Caustic soda, 497
Ceramic and refractories, 575
Ceramic as electrical material, 591
Chamber process, 332
Characteristics of water, 107
Chemical engineer in the coming years, 66
Chemical kinetics, 16
Chemical precipitation, 174
Chemical reaction equilibrium, 15
Chemical reactors, 29
Chlorides, 640
Chlorination, 148
Chlorine, 492
Clarigester, 176
Classification of unit operations, 2
Clathrates, 416
Claude cycle, 273
Clays, 575
Clinker chemistry, 552
Coagulation of water, 143
Colourants for ceramic, 595
Coloured glass, 628
Common salt, 451,
Communications, 69
Compounds in cements, 559
Comprehensive utilisation of raw materials, 96
Compressors, 41
Conservation of energy, 13
Conservation of mass, 10
Contact of immiscible phases, 23
Copper sulphate, 366
Corrosion of boiler, 120
Crystallisation, 22
Cyanamide, 395

D
Desalination of water, 163
Dewatering, 179
Diamond (natural), 304
Diamond (synthetic), 306
Diazodinitro phenol, 216
Dicyandiamide, 398
Digestion, 178
Disinfection of water, 147
Distillation, 22, 34
Drying, 22
Ducts, 38
Duplex process, 691

E
Effect of boiler scales and sludges, 111
Effect of water on rocks and minerals, 109
Energy management principles, 700
Environmental control, 188
Enzymatic oxidation reactions, 203
Enzyme catalysed selective oxidation

reactions, 198
Exploratory research, 56
Explosive detonation, 214
Explosives and propellants, 213
Extraction, 36

F

Factors promoting corrosion, 118
Fast breeder reactors, 657
Feldspar, 583
Fertiliser, 535
Fibre glass, 629
Filtration of water, 145
Flint glass, 630
Flocculation, 144
Flow sheets, 42
Fluid flow, 21
Fluorides, 639
Fluorine, 351,
Fluxes, 672
Formation of deposits in boiler units, 112
Fuel and power for the chemical industry, 100
Fundamental research, 55
Fused silica glass, 625

G

Gas-gas, 23
Gas-liquid, 23
Gas-solid, 24
General aspects of chemical engineering, 69
Glass industries, 608
Glass laminates, 629
Glass wool, 630
Glass–ceramic, 618
Glassy metals, 621
Graphic flow diagrams, 44
Ground glass, 632
Gypsum, 567

H

Halide salts, 570
Hard and soft water, 150
Hardening, 694
Heat exchangers, 32, 51
Heat transfer, 21
Heat treatment of steel, 694
Heavy water for nuclear power industry, 663
Heavy water process, 698
Helium, 282
High silica glass, 628
Hydration, 554
Hydrochloric acid, 344
Hydrofluoric acid, 348
Hydrogen peroxide, 389
Hydrogen, 255
Hydrogen-ion concentration (pH value), 129
Hydroxides, 371

I

Ideal gas law, 18
Imhoff tanks, 176
Impurities in water, 124
Indian chemical industry today, 7
Industrial carbon, 289
Inorganic lithium compounds, 369
Instrumentation and control, 52
Insulating glass, 632
Iron and steel, 671
Isotopes, 637
Isotope separation, 655

J

Jena glass, 630

K

Kel-chlor process, 495

Knoopscale, 305
Krypton, 284

L

L.D. process, 690
Land disposal, 171
Lead glasses, 527
Lime, 562
Liquid hydrogen, 263
Liquid-liquid extraction, 22
Lithium acetate, 369
Lithium amide, 369
Lithium and lithium compounds, 367
Lithium borate, 370
Lithium carbonate, 370
Lithium metal, 368

M

Magnesium carbonate, 571
Magnesium chloride, 571
Magnesium hydroxide, 573
Magnesium oxide (magnesia), 572
Magnesium silicate, 573
Magnesium sulphate, 573
Manufacture of oleum, 333
Manufacture of pig iron, 673
Manufacture of special glasses, 625
Market research, 64
Mass and energy balances, 10
Mass transfer equipment, 34
Materials for chemical processes, 88
Mechanical physical separation, 22
Mechanism of boiler scale formation, 110
Melamine, 389
Mercury fulminate, 215
Methodology adopted for energy management, 702
Methods of preventing deposit formation, 113
Miscellaneous magnesium compounds, 571
Miscellaneous methods of water treatment, 166
Miscible phases separated by a membrane, 26
Mixing the reactants, 79
Molybdates, 371

N

Natural graphite, 308
Neon, 283
Nitramines, 219
Nitric acid, 416
Nitriding, 695
Nitroaromatics, 221
Nitrocellulose, 218
Nitrogen fertilisers, 545
Nitrogen industries, 394
Nitroguanidine, 221
Nitrous oxide, 287
Non-clay minerals, 583
Normalising, 694
Nuclear fission, 641
Nuclear fuel cycle, 659
Nuclear fusion, 643
Nuclear industries, 634
Nuclear reactors, 650

O

Open hearth process, 687
Optical glass, 628
Organomolybdenum compounds, 372
Oxides and hydroxy compounds, 371
Oxygen, 268
Oxygen steel marking or L.D. process, 690
Oxygenases, 207

P

Permanent hardness, 152
Peroxidases, 209

Phase equilibrium, 19
Phase-transfer catalysts, 188
Phosphate treatment of boiler water, 113
Phosphite and its derivatives, 432
Phosphoric acid, 434
Phosphoric acids and phosphates, 432
Phosphorus acids, 431
Phosphorus and phosphides, 424
Phosphorus compounds, 427
Phosphorus fertilisers, 537
Phosphorus halides, 429
Phosphorus industries, 423
Phosphorus oxides, 431
Phosphorus sulphides, 428
Photochromic glass, 631
Photosensitive glass, 631
Physical tests for water, 124
Picric acid and ammonium picrate, 222
Pipes, 37
Plant design, 61
Plant equipment, 29
Plant technical service, 64
Plutonium and its compounds, 647
Polyphosphates, 441
Polyphosphoric acids, 441
Potassium acid tartrate, 511
Potassium carbonate, 510
Potassium chloride, 504
Potassium fertilisers, 549
Potassium hydroxide, 508
Potassium industries, 504
Potassium nitrate, 511
Potassium sulphate, 511
Pre-scientific chemical industry, 6
Primary explosives, 215
Priming and foaming, 116
Process design and energy consumption, 698
Process design and evolution, 60
Process development, 58
Process diagrams, 42
Process flow diagrams, 44
Process piping and instrumentation flow diagrams, 44
Process research, 57
Product development, 65
Product sales, 64
Production supervision, 62
Propellants, 225
Pumps, 40, 51
Pyrex glass, 631
Pyrophosphates, 442

Q

Quality of water, 121

R

Radon, 285
Rain water, 106
Rare earth elements, 279, 383
Rare earth glass, 631
Rate of chemical processes, 77
RDX and HMX, 220
Reactor types, 30
Reasons for the analysis of water, 123
Refractories, 597
Refractory coatings, 603
Refractory fibres, 606
Removal of permanent hardness, 153
Removal of temporary hardness, 152
Replacement of edible raw materials, 97
River water, 106

S

Safety glass, 584
Safety in nuclear facilities, 658
Scientific chemical industry, 6
Sea water, 107
Secondary explosives, 217

Setting and hardening of cement, 560
Sewage and sewage treatment, 169
Silica, 583
Skimming tank, 176
Sludge treatment and disposal, 177
Soda ash (sodium carbonate), 470
Soda lime glass, 584
Sodium amide, 467
Sodium and potassium nitrates, 420
Sodium bicarbonate, 490
Sodium bisulphate, 464
Sodium borates, 377
Sodium carbonate, 470
Sodium chloride (common salt), 451
Sodium chlorite, 503
Sodium chromate, 387
Sodium cyanide, 467
Sodium ferrocyanide, 468
Sodium hydrosulphide, 465
Sodium hydroxide (caustic soda), 497
Sodium hypochlorite, 502
Sodium meta phosphates, 449
Sodium nitrite, 465
Sodium perborate, 466
Sodium phosphates, 444
Sodium silicate, 460
Sodium sulphate (salt cake), 455
Sodium sulphate decahydrate, 463
Sodium sulphide, 465
Sodium sulphite, 464
Sodium thiosulphate, 458
Sodium tripolyphosphate, 449
Special cement, 561
Spring and well water, 106
Stacks, 41
Steel, 681
Strontium and strontium compounds, 373
Strontium carbonate, 374
Strontium chlorate, 374
Strontium chloride, 374
Sulphur, 314
Sulphur and sulphuric acid, 313
Sulphur dioxide, 286, 317
Sulphur trioxide, 318
Surface water and ground water, 107
Synthetic ammonia, 389
Synthetic diamond industry, 307
Synthetic fuel, 167

T

Technical reading, 70
Technical sales and customers service, 65
Tempering, 695
Temporary hardness, 152
Tetryl, 221
The systematic analysis of chemical industries, 10
Thermochemistry, 15
Total solids, 135
Transalkylations, 192
Transportation equipment, 37
Treatment of boiler water with complexing agents, 114
Tricalcium aluminate and ferrite, 555
Trinitrotoluene, 222

U

Unit operations, 21
Uranium and uranium compounds, 635
Uranium compounds, 639
Urea, 441

V

Valves, 39
Vanadium, 596
Vessels, 51
Vitreosil, 632

W

Waste-water treatment, 414
Water and the chemical industry, 97
Water chemistry of light-water reactors, 652
Water conditioning in chemical process industry, 105
Water for boilers, 106
Water for industrial purposes, 106
Water softening, 149
Water-borne diseases, 142
Whitewares, 580
Wrought iron, 680

X

Xenon, 285
X-ray transmitting glasses, 611

Y

Yarn, 21
Young's modulus, 589, 615

Z

Zeolite process, 159
Zinc borates, 390
Zirconium, 584, 601
Zirconium refractories, 603